Organocatalysts in Polymer Chemistry

Organocatalysts in Polymer Chemistry

Synthesis and Applications

Edited by Zhibo Li

WILEY-VCH

Editor

Prof. Zhibo Li
Qingdao University of Science and Technology
Qingdao, 266042
China

Cover Image: Courtesy of Xiaoyu Guo

Library of Congress Card No.: applied for

British Library Cataloguing-in-Publication Data
A catalogue record for this book is available from the British Library.

Bibliographic information published by the Deutsche Nationalbibliothek
The Deutsche Nationalbibliothek lists this publication in the Deutsche Nationalbibliografie; detailed bibliographic data are available on the Internet at <http://dnb.d-nb.de>.

The manufacturer's authorized representative according to the EU General Product Safety Regulation is WILEY-VCH GmbH, Boschstr. 12, 69469 Weinheim, Germany, e-mail: Product_Safety@wiley.com.

Print ISBN: 978-3-527-35241-8
ePDF ISBN: 978-3-527-84305-3
ePub ISBN: 978-3-527-84306-0
oBook ISBN: 978-3-527-84307-7

Typesetting Straive, Chennai, India
Druck und Bindung: CPI books GmbH Leck, Germany

Contents

Preface

Catalysts not only play a crucial role in increasing the reaction rate, but also have a great impact on the selectivity of desirable reaction over side reactions. In particular, catalysts have been a powerful tools in polymer synthesis, including the ability to control the molecular weight, dispersity, terminal groups, the architecture, stereochemistry and topology of the macromolecule (linear, branched, cyclic, bottle-brush, cross-linking, *etc*) as well as the composition and monomer sequence of the copolymers. Organometallic catalysts consisting of a metal center and organic ligands are the most extensively investigated catalysts for polymer synthesis considering the ability to finely adjust the catalytic activity and selectivity by almost limitless combination of metals and ligands. Organocatalysts have emerged as an alternative to organometallic complexes, especially for the biomedical and electronic applications where metal residues may be detrimental to the material performance. Compared to organometallic catalysts, the advantages of organocatalysts include simple handle, low cost, better stability to oxygen and water, good solubility and easy removal from the polymers. In addition, many organocatalysts are commercially available, thus promoting their widespread applications for polymer synthesis.

Organocatalysis can be traced back to 1894 when Knoevenagel promoted the condensation reaction of formaldehyde and malonate using ethylenediamine as the catalyst [1]. Organocatalysts have been greatly developed, especially in the field of enantioselective organocatalysis. The earliest attempt that can be verified is the enantioselective synthesis of cyanohydrin with quinine alkaloids reported by Bredig and Fiske in 1911 [2]. In 2021, the Novel Prize in Chemistry was awarded to Benjamin List and David W.C. MacMillan for their pioneering contributions in the development of asymmetric organocatalysis. Compared to the organocatalysis on the synthesis of small molecules, the organocatalysis for polymer synthesis is still in its infancy. In 2001, James L. Hedrick *et al.* reported the first organocatalytic "living" ring-opening polymerization of lactide using ethanol as an initiator and 4-(dimethylamino)pyridine (DMAP) or 4-pyrrolidinopyridine (PPY) as the catalyst [3]. Since then, the field of organocatalytic polymerization has undergone a renaissance. In the past two decades, organocatalysts have been greatly developed and achieved extensive applications in various polymer synthesis, including ring-opening polymerization (ROP), ring-opening copolymerization

(ROCP), controlled radical polymerization as well as the depolymerization of commodity polymers.

Some excellent reviews have been published to summarize the recent advancements on organocatalytic polymerization, especially for the ROP of cyclic monomers [4–13]. A book with the title of "Organic Catalysis for Polymerisation" edited by Andrew Dove, Haritz Sardon and Stefan Naumann summarized the achievements of organocatalytic polymerization until 2018 [14]. Given the rapid development of this field, expecially the binary organocatalyst in polymer synthesis, organocatalyst catalyzed photo polymerization, as well as organocatalyst catalyzed depolymerization, we believe it is a suitable time to edit a book to review the state-of-the-art examples for organocatalytic polymerization.

All chapters in this book are written by the outstanding scholars in their field. This book include following topics: ROP of cyclic esters toward degradable polymers, ROCP of epoxide and anhydrides, ROCP of CO_2/COS/CS_2 with epoxides, ROP of cyclicsiloxane, ROP of *N*-carboxyanhydride (NCA) to prepare polypeptides, stereoselective ring-opening (co)polymerization and controlled ATRP and RAFT. We will also cover the applications of organocatalysts in the depolymerization of commodity polyesters such as PET and polycarbonate. Recent progresses of binary organocatalyst and organic Lewis pairs for polymer synthesis will also be discussed. This book will cover the most recent research progresses regarding the applications of organocatalysts in polymer synthesis, particularly in degradable polymers. It will summarize the advances of organocatalysis in polymer chemistry in the past two decades. We hope this book will offer a great guideline for postgraduate students, professors, technicians as well as industrial experts.

December 2024 *Zhibo Li*

References

1 Knoevenagel, E. (1894). Ueber eine Darstellungsweise der Glutarsäure. *Ber. Dtsch. Chem. Ges.* 27: 2345–2346.

2 Bredig, G. and Fiske, P.S. (1912). Durch Katalysatoren bewirkte asymmetrische synthese. *J. Plankton Res.* 46: 7.

3 Nederberg, F., Connor, E.F., Möller, M. et al. (2001). New paradigms for organic catalysts: the first organocatalytic living polymerization. *Angew. Chem. Int. Ed.* 40: 2712–2715.

4 Fukushima, K. and Nozaki, K. (2020). Organocatalysis: a paradigm shift in the synthesis of aliphatic polyesters and polycarbonates. *Macromolecules* 53: 5018–5022.

5 Kiesewetter, M.K., Shin, E.J., Hedrick, J.L., and Waymouth, R.M. (2010). Organocatalysis: opportunities and challenges for polymer synthesis. *Macromolecules* 43: 2093–2107.

6 Kamber, N.E., Jeong, W., Waymouth, R.M. et al. (2007). Organocatalytic ring-opening polymerization. *Chem. Rev.* 107: 5813–5840.

7 Liu, S., Ren, C., Zhao, N. et al. (2018). Phosphazene bases as organocatalysts for ring-opening polymerization of cyclic esters. *Macromol. Rapid Commun.* 39: 1800485.

8 Hu, S., Zhao, J., Zhang, G., and Schlaad, H. (2017). Macromolecular architectures through organocatalysis. *Prog. Polym. Sci.* 74: 34–77.

9 Xu, J., Wang, X., Liu, J. et al. (2022). Ionic H-bonding organocatalysts for the ring-opening polymerization of cyclic esters and cyclic carbonates. *Prog. Polym. Sci.* 125: 101484.

10 Hong, M., Chen, J., and Chen, E.Y.X. (2018). Polymerization of polar monomers mediated by main-group Lewis acid–base pairs. *Chem. Rev.* 118: 10551–10616.

11 Dove, A.P. (2012). Organic catalysis for ring-opening polymerization. *ACS Macro Lett.* 1: 1409–1412.

12 Jehanno, C., Pérez-Madrigal, M.M., Demarteau, J. et al. (2019). Organocatalysis for depolymerisation. *Polym. Chem.* 10: 172–186.

13 Naumann, S. and Dove, A.P. (2015). *N*-heterocyclic carbenes as organocatalysts for polymerizations: trends and frontiers. *Polym. Chem.* 6: 3185–3200.

14 Dove, A., Sardon, H., and Naumann, S. (ed.) (2018). *Organic Catalysis for Polymerisation*. The Royal Society of Chemistry https://doi.org/10.1039/9781788015738.

1

Organocatalyzed Ring-Opening Polymerization of Cyclic Esters Toward Degradable Polymers

Feng Li[1], *Takuya Isono*[1], *and Toshifumi Satoh*[1,2]

[1]*Division of Applied Chemistry, Faculty of Engineering, Hokkaido University, Sapporo 060-8628, Japan*
[2]*List Sustainable Digital Transformation Catalyst Collaboration Research Platform (List-PF), Institute for Chemical Reaction Design and Discovery (ICReDD), Hokkaido University, Sapporo 001-0021, Japan*

1.1 General Introduction

The discovery of ring-opening polymerization (ROP) of cyclic esters to afford polyesters dates back to the 1930s. The hydrolyzable nature of the ester functional group in the polymer chain endows the chain with degradability (e.g. thermal, chemical, bio), rendering polyesters as promising candidates for biomedical applications and as environmentally benign polymer materials. In addition, cyclic esters exhibit polymerizability with an extremely broad scope of catalysts. The ROP of cyclic esters can occur via anionic, cationic, and coordination mechanisms, using different types of catalysts such as transition-metal catalysts, enzymes, and organocatalysts (Figure 1.1). Thus, the ROP of cyclic esters is the first and the most investigated organocatalyst-based polymerization reaction to date.

After extensive investigations over the past two decades, various organocatalysts have been reported to exhibit catalytic activity in the ROP of cyclic esters (Figure 1.2). The typical reaction mechanism for various catalysts is introduced in Section 1.2.

After the numerous initial investigations of novel organocatalysts, driven by the scientific interest in transition-metal-free catalysts, several factors such as catalytic efficiency, selectivity, thermal stability, and safety have been considered in recent works toward meeting the requirement for industrial application. However, metal complexes, such as tin(II) 2-ethylhexanoate ($Sn(Oct)_2$), are still used in industries to produce polyesters. An increasing number of recent studies have indicated the promising future of organocatalysts, even under industrially relevant conditions. In Section 1.3, the paradigm shifts in the organocatalyzed ROP of cyclic esters are illustrated.

β-Butyrolactone (β-BL), lactide (LA), δ-valerolactone (VL), and ε-caprolactone (CL) are among the most studied cyclic ester monomers because of their relatively high ring strain, good polymerizability, biodegradability, and biocompatibility of their corresponding polyesters, poly(β-butyrolactone) (P(β-BL)), poly(lactic acid) (PLA), poly(δ-valerolactone) (PVL), and poly(ε-caprolactone) (PCL) (Figure 1.3a).

Organocatalysts in Polymer Chemistry: Synthesis and Applications, First Edition. Edited by Zhibo Li.

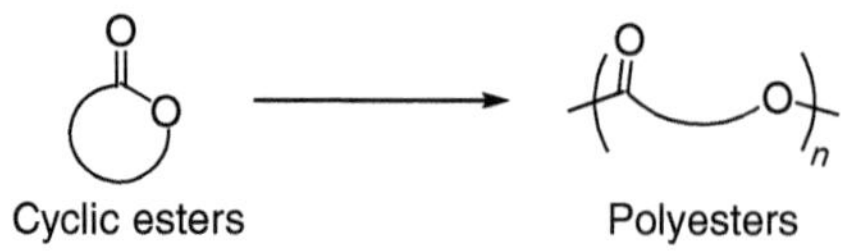

Figure 1.1 ROP of cyclic esters into polyesters.

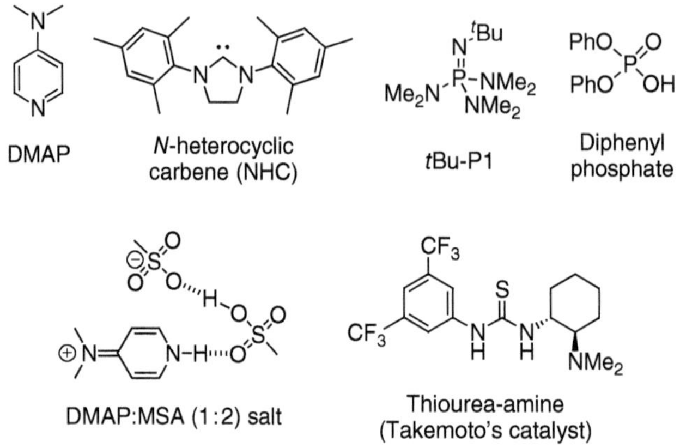

Figure 1.2 Representative organocatalysts for the ROP of cyclic esters.

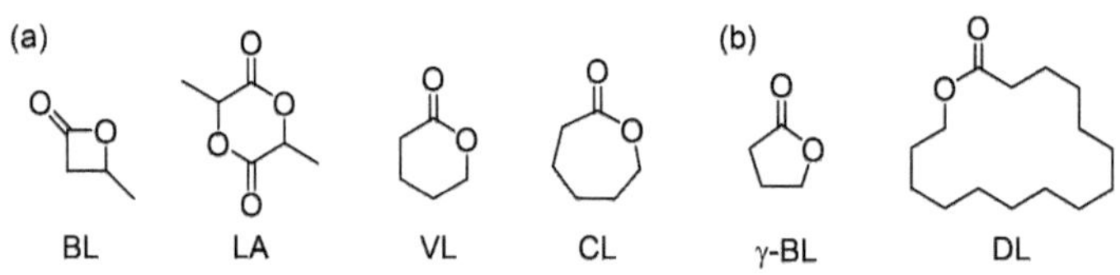

Figure 1.3 Representative cyclic esters with relatively high (a) and low (b) ring strains.

However, five-member and large-ring lactones such as γ-butyrolactone (γ-BL) [1, 2] and ω-pentadecalactone (DL) [3, 4] cannot be polymerized easily owing to their low ring strain (Figure 1.3b). The successful demonstration of controlled polymerization of low ring strain γ-BL under low temperatures by highly active catalysts, including metallic catalysts and organocatalysts, and depolymerization of P(γ-BL) back to γ-BL monomer at an elevated temperature signified the impact of the closed-loop polymerization methodology [1, 2]. Since this seminal work of Hong and Chen, chemically recyclable polyesters with novel monomer designs, especially the ones that can be easily derived from renewable biomass resources, have become an emerging research topic, and the number of research reports has increased rapidly in recent years. Although their research was largely focused on the monomer design, polymer properties, and recyclability, organocatalysts have been used extensively. Polyesters that can degrade easily under environmental conditions are also important, aside from the chemically recyclable polyesters. Therefore, the introduction of other facile functional groups to the main chain of polyesters for enhancing their degradability has also been an important topic in recent years. In Section 1.4, breakthroughs in achieving improved degradability and recyclability are discussed.

This chapter focuses on the features of organocatalysts, cyclic ester monomers, and the corresponding polymers. Utilizing the organocatalyzed ROP of cyclic esters for the synthesis of block copolymers and tailoring a highly complicated polymer architecture design for synthesizing advanced degradable materials are also important research directions; however, they are not the topics of this chapter.

Alkali-metal salts, such as the salts of carboxylic acids, vitamin C, and (thio)ureas, are not completely organic compounds. The reaction mechanisms of these catalysts are similar to those of common organocatalysts, rather than transition-metal catalysts. In addition, sodium and potassium ions are safe and essential for the human body; therefore, they have been categorized as organocatalysts herein.

1.2 Polymerization Mechanism

1.2.1 Nucleophilic Catalysts

Nucleophilic catalysts, e.g. 4-dimethylaminopyridine (DMAP) and *N*-heterocyclic carbenes (NHCs), are widely used in organic chemistry. The ROP of lactide by DMAP, as reported by Hedrick and coworkers in 2001, is recognized as the landmark that initiated the era of organocatalyzed polymerization [5]. Since this seminal work, many other nucleophilic catalysts, such as phosphines [6], amidines [7], and NHCs [8, 9], have been investigated.

The reaction conditions and polymerizable monomers largely depend on the catalysts employed. A quantitative comparison of the nucleophilicities of these catalysts can provide a better understanding. The Mayr reactivity parameters provide a scale for quantitatively evaluating and comparing the nucleophilicities of various nucleophilic catalysts. Four representative nucleophilic organocatalysts are shown in Figure 1.4, whose Mayr nucleophilic parameter *N* increases in the order of triphenyl phosphines, DMAP, 1,8-diazabicyclo[5.4.0]-7-undecene (DBU), and NHCs [10].

Regarding the reaction mechanism of nucleophilic catalysts for the ROP of cyclic esters, the catalytic cycle typically commences with a nucleophilic attack from the catalysts on the carbonyl group of the cyclic esters to open the ring and generate a zwitterionic intermediate. If the reaction is conducted in the presence of an alcohol chain-transfer agent, i.e. ROH, the hydroxyl group can be activated by the anionic site of the zwitterionic intermediate, thus inducing an intramolecular nucleophilic

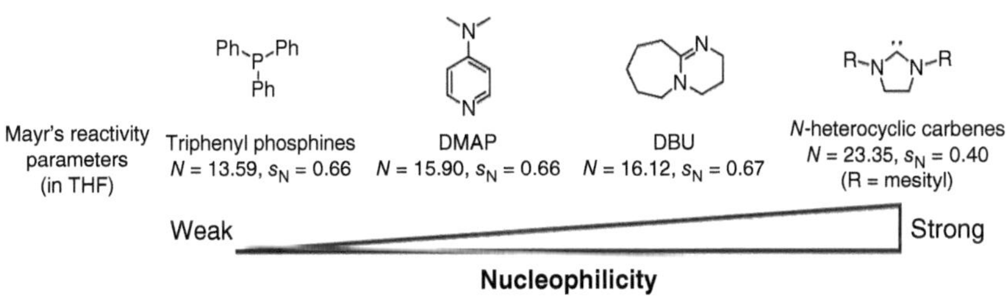

Figure 1.4 Representative nucleophilic catalysts and their nucleophilicities evaluated using the Mayr reactivity parameters.

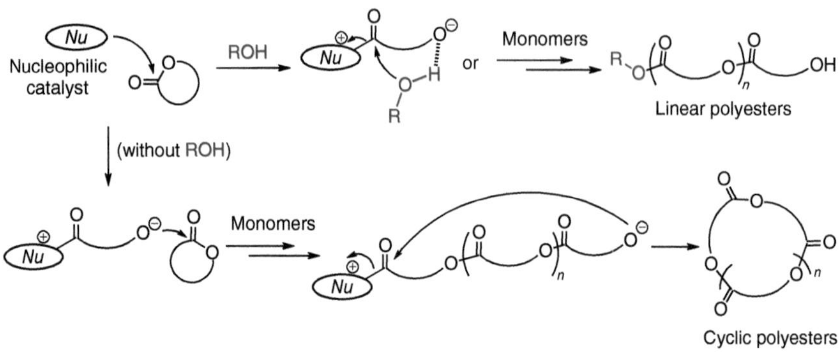

Figure 1.5 General reaction mechanism of the ROP of cyclic esters using nucleophilic catalysts in the presence and absence of alcohol initiators.

attack and releasing the nucleophilic catalyst (Figure 1.5, upper reaction route). The iteration of this process affords linear polyesters. In the absence of ROH, the anionic site, i.e. alkoxide, of the generated zwitterionic intermediate continues to attack other cyclic ester monomers, and at a certain stage, the anionic chain end can nucleophilically attack the cationic activated carbonyl group and release the nucleophilic catalysts, affording cyclic polyesters (Figure 1.5, lower reaction route).

Nucleophilic catalysts typically exhibit a strong basicity; however, their reaction mechanism may not be identified easily. For example, DBU and 1,5,7-triazabicyclo [4.4.0]dec-5-ene (TBD) are moderate/strong Brønsted bases, but they exhibit relatively high nucleophilicity. Therefore, they could follow the mechanisms of either nucleophilic catalysts [11] or Brønsted base catalysts [12].

1.2.2 Base Catalysts

Organobase catalysts constitute an important type of organocatalysts for the ROP of cyclic esters. Base catalysts can be divided into Lewis bases and Brønsted bases. This section focuses on Brønsted bases, which function as proton acceptors. Lewis bases with a high nucleophilicity have been categorized as nucleophilic catalysts in Section 1.2.1. In the ROP of cyclic esters, commonly used Brønsted bases catalysts include amines, amidines, guanidines, and phosphazenes [13–15]. Pyridines and other *N*-containing heterocycles are Brønsted bases as well; however, considering their weak basicity and medium-to-high nucleophilicity, they have been introduced in Section 1.2.1 as nucleophilic catalysts (e.g. DMAP).

The basicity of the abovementioned Brønsted bases spans a wide range across 24 orders of magnitude – from the relatively weak triethylamine ($pK_{BH}{}^{+}$ 18.8 in acetonitrile) to the super strong base *t*-BuP_4 ($pK_{BH}{}^{+}$ 42.7 in acetonitrile) (Figure 1.6) [16–18]. Hence, the basicity of the catalyst can be tuned according to the requirements of the corresponding polymerization reaction. Catalysts with medium-to-strong basicity, e.g. DBU, TBD, and phosphazenes, are most commonly used.

Regarding the reaction mechanism, the chain end activation mechanism is typically considered in the ROP of cyclic esters with alcohols as initiators

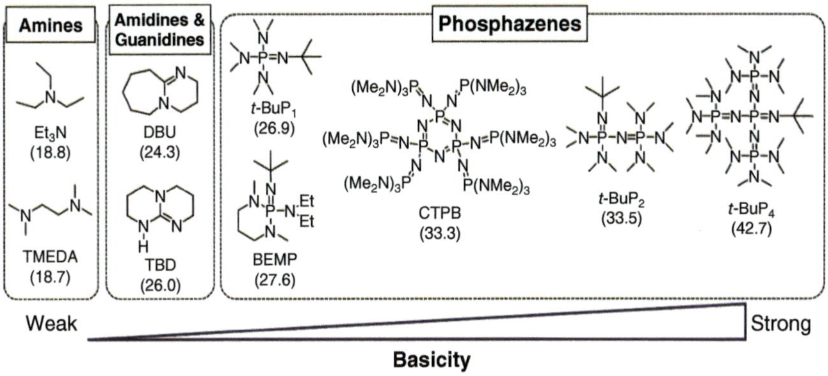

Figure 1.6 Representative Brønsted base catalysts and their basicity (pK_a of the conjugated acid in MeCN).

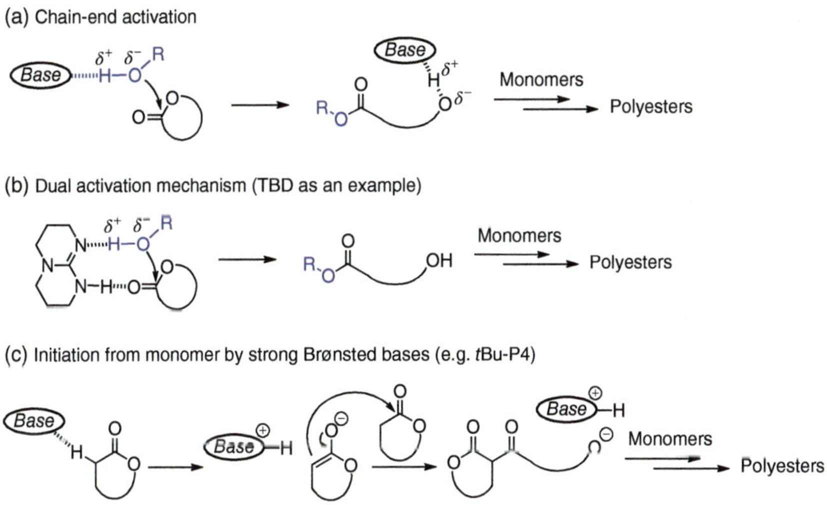

Figure 1.7 General reaction mechanism of the ROP of cyclic esters using Brønsted base catalysts.

(Figure 1.7a) [19]. Through hydrogen bonding, the base activates the –OH group of the alcohol initiators or the propagating chain ends to enhance their nucleophilicity and attack the carbonyl group of the cyclic esters. The higher the basicity of the catalyst, the more prone it is to complete deprotonation toward generating naked alkoxide anions; thus, its reactivity is higher.

When a bifunctional organobase, such as TBD, is used, the dual-activation mechanism via double hydrogen bonding interactions is usually considered (Figure 1.7b) [20]. The Brønsted basic site of the guanidine activates the –OH group, whereas the Brønsted acidic site (N–H) interacts with the carbonyl group of the cyclic esters. When cyclic esters are polymerized without using alcohol initiators, the polymerization could initiate from the monomer via the deprotonation of the α-proton of the esters, thus forming nucleophilic enolate species (Figure 1.7c). In this case, highly basic catalysts, such as phosphazenes, are commonly used [2].

1.2.3 Acid Catalysts

Acid catalysts are among the most important organocatalysts for the ROP of cyclic esters. Based on their characteristics, acid catalysts can be divided into Brønsted acids and Lewis acids. In terms of the organocatalyzed ROP of cyclic esters, Brønsted acid catalysts have been widely reported. Herein, the representative mechanisms and examples are introduced. Simple strong inorganic acids, such as HOTf and HCl, can catalyze the ROP of cyclic esters; however, these reactions were typically not considered organocatalyzed polymerizations at the time of the study [21–25].

Carboxylic, phosphoric, and sulfonic acids are commonly used in the organocatalyzed ROP of cyclic esters based on Brønsted acid catalysts (Figure 1.8). On the basis of the structure and acidity of the catalyst and the polymerization conditions, the reaction mechanisms can be divided into three categories (Figure 1.9). The activated monomer (AM) mechanism is a typical polymerization mechanism for the ROP of cyclic esters (Figure 1.9a). The Brønsted acid activates the carbonyl group of the cyclic ester, rendering it more prone to a nucleophilic attack from the hydroxyl group of either the initiator or the propagating chain end. The nucleophilic attack could also occur at the sp^3 carbon next to the ester group, thus forming an ion-pair intermediate with the counteranion of the Brønsted acid catalysts (Figure 1.9b) [28, 29]. This chain end structure is highly reactive and can accept a nucleophilic attack from another molecule of the cyclic ester monomer. This polymerization mechanism is usually referred to as the activated chain end (ACE). The activated chain ends are usually terminated by quenching the polymerization using alcohols or other nucleophiles. The AM and ACE polymerization mechanisms were proposed a long time ago; however, the dual-activation mechanism of the ROP of cyclic esters by

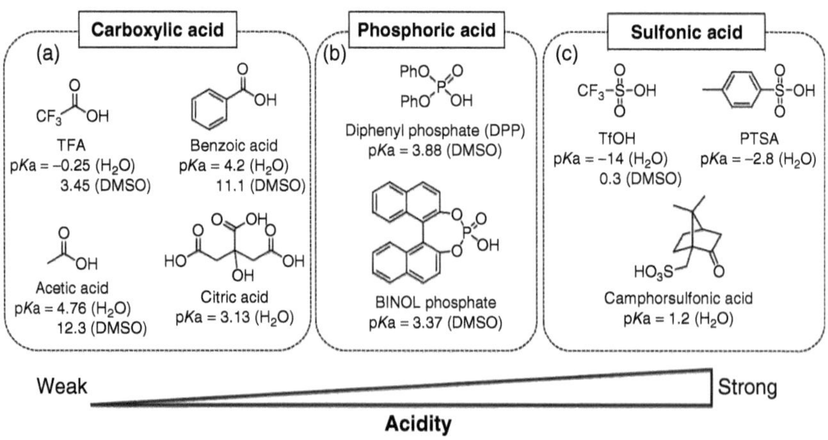

Figure 1.8 Representative Brønsted acid catalysts and their acidities. Source: Bordwell [26] and Christ [27].

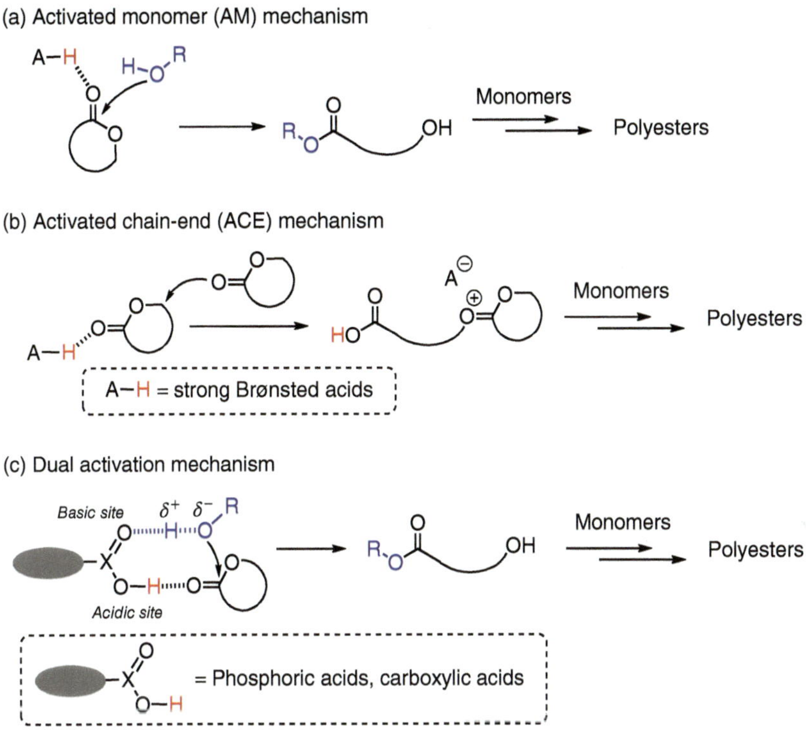

Figure 1.9 General reaction mechanism of the ROP of cyclic esters using Brønsted acid catalysts.

acid catalysts was not recognized until the recent decades (Figure 1.9c) [30, 31]. In the dual-activation mechanism, the Brønsted acidic site of the catalyst activates the carbonyl group of the cyclic ester monomer and renders it more electrophilic; meanwhile, the Brønsted basic site of the catalyst activates the alcohol initiator or chain end, rendering it more nucleophilic. This dual-activation mechanism ensures proximity of the monomer and propagating chain end, thus allowing for smooth progress of the polymerization. Considering the structural features and acidities of the catalysts, the catalysis by sulfonic acids is more likely to proceed via the AM or ACE mechanism because of their high acidities. In the case of carboxylic and phosphoric acids, the polymerization usually proceeds via the dual-activation mechanism because of the weak-to-medium acidities of these two types of catalysts.

1.2.4 Ionic Catalysts

During the initial research on organocatalyzed polymerization, catalysts without charges were used in most studies. In the past decade, salt or ionic organocatalysts have been developed for the ROP of cyclic esters [32]. They exhibit better catalytic activities and thermal stabilities than conventional organocatalysts. These aspects are introduced in Sections 1.3.1 and 1.3.3.

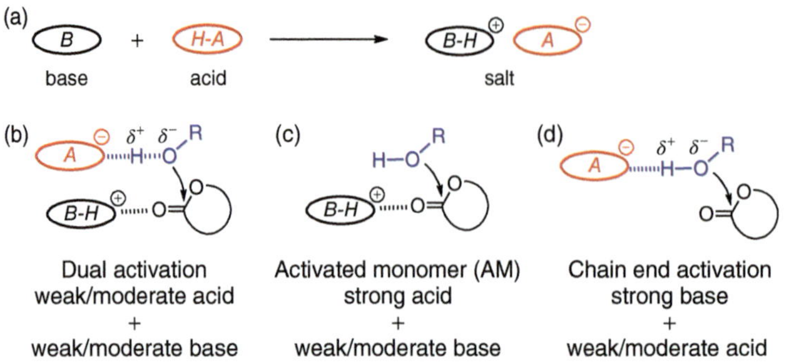

Figure 1.10 Salt catalysts and general reaction mechanism of the ROP of cyclic esters for different types of salt catalysts.

When a Brønsted acid and base are mixed in a ratio of 1 : 1, a salt is formed (Figure 1.10a). However, the produced salt exhibits a different acidity or basicity, depending on the relative strengths of the acidic and basic parts, which in turn determines the reaction mechanism governing the ROP of cyclic esters. A dual-activation mechanism has been typically proposed – The anion activates the alcohol initiator or propagating chain end, whereas the cation activates the cyclic ester monomer (Figure 1.10b). This mechanism is usually observed for a mixture of a weak/moderate acid and weak/moderate base, such as 1 : 1 mixtures of DBU and benzoic acid, DMAP and diphenyl phosphate (DPP), and DMAP and saccharin (Figure 1.11a–c) [33–35]. The conjugated bases and acids of strong acids and bases are weak, respectively; hence, they exhibit weak or negligible interactions with the monomer or propagating chain end. Therefore, the AM mechanism is typically proposed when a salt catalyst constituted by a strong acid and a weak/moderate base is used, e.g. a 1 : 2 mixture of bipyridine and camphorsulfonic acid (Figures 1.10c and 1.11d) [36]. In theory, the chain end activation mechanism can also be considered for combinations of strong bases and weak/moderate acids (Figure 1.10d). However, to the best of our knowledge, this activation has not been reported

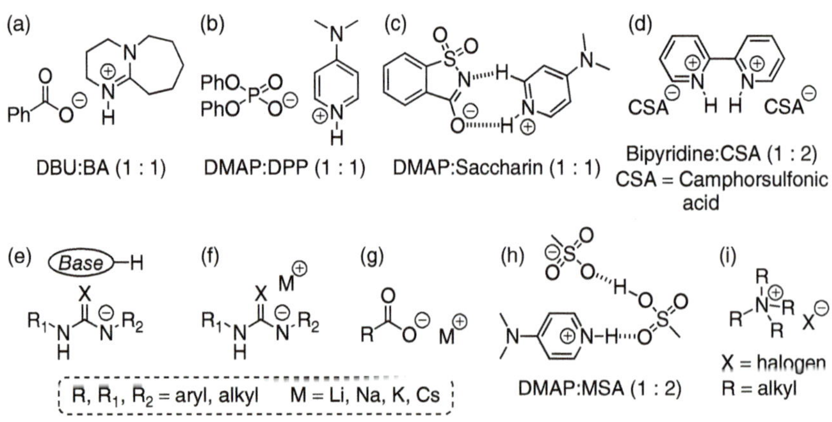

Figure 1.11 Examples of salt and ionic catalysts.

[32]. Ureas, thioureas, and carboxylic acids have relatively weak acidities. Their alkali-metal salts and salts with strong organic bases, such as phosphazenes, have been reported as effective catalysts for the ROP of cyclic esters (Figure 1.11e–g) [37–39]. The dual-activation mechanism has been proposed in these cases because of the bifunctional nature of the anions thereof.

In addition to the salts formed using a 1 : 1 ratio mixture of an acid and a base, mixtures with an unequal ratio, i.e. an excess amount of either an acid or a base, have been used for fine-tuning the reactivity and catalytic performance. For example, the ionic catalyst DMAP: methanesulfonic acid (MSA) (1 : 2) has been used for catalyzing the ROP of L-lactide (LLA) in bulk at an acceptable reaction rate without epimerization (Figure 1.11h) [40]. Tetraalkylammonium halides, which are representative of onium salts, can also function as effective catalysts for the ROP of cyclic esters (Figure 1.11i) [41, 42].

Zwitterions are also referred to as inner salts, in which the cations and the anions are covalently linked together in the same molecule. To the best of our knowledge, zwitterionic organocatalysts have not been used for the ROP of cyclic esters thus far. However, for the ROP of trimethylene carbonate (TMC), trimethyl glycine, which is a natural betaine sourced from sugar beets, has been demonstrated to be an effective and environmentally benign organocatalyst [43].

1.2.5 Bifunctional and Multifunctional Catalysts

Bifunctional and multifunctional organocatalysts are molecules in which two or more catalytically active moieties are linked covalently. Dual-activation is the commonly accepted reaction mechanism by which the ROP of cyclic esters is promoted efficiently. Although the dual-activation model can be realized using monofunctionalized catalysts such as DPP and TBD, the use of bifunctional and multifunctional catalysts could be beneficial for fine-tuning the interactions with both the propagating chain ends and cyclic ester monomers. Thus, both the active propagating chain end and the activated cyclic ester monomer can adopt suitable conformations in the transition state, thereby decreasing the activation energy. Bifunctional and multifunctional catalysts are viewed as the mimicry of enzymes, which catalyze reactions using multiple amino acid residues cooperatively.

Regarding the catalyst design, Takemoto's catalyst, which comprises a thiourea moiety and trialkyl amine group, is one of the most representative bifunctional organocatalysts (Figure 1.12a) [44]. It has been successfully used in the ROP of LA [45]. Because Takemoto's catalyst is chiral, it has also been employed for the stereoselective polymerization of racemic lactides (*rac*-LA) [46].

In the case of Takemoto's catalyst, the weakly acidic thiourea functions as a hydrogen-bond donor that interacts with the carbonyl group of the cyclic esters, thus activating the electrophile; the basic amine group partially deprotonates the –OH group at the propagating chain end or initiators, thus activating the nucleophile (Figure 1.12b). Owing to its efficiency, Takemoto's catalyst has been used as a prototype for developing bifunctional and multifunctional catalysts. Various bifunctional catalysts have been developed by varying the components of either the hydrogen-bond donor or the Brønsted base. For example, upon changing the

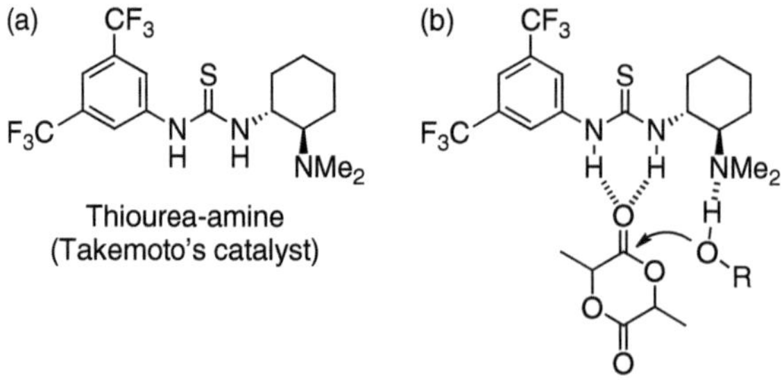

Figure 1.12 (a) Takemoto's catalyst; (b) dual-activation mechanism in the ROP of LA.

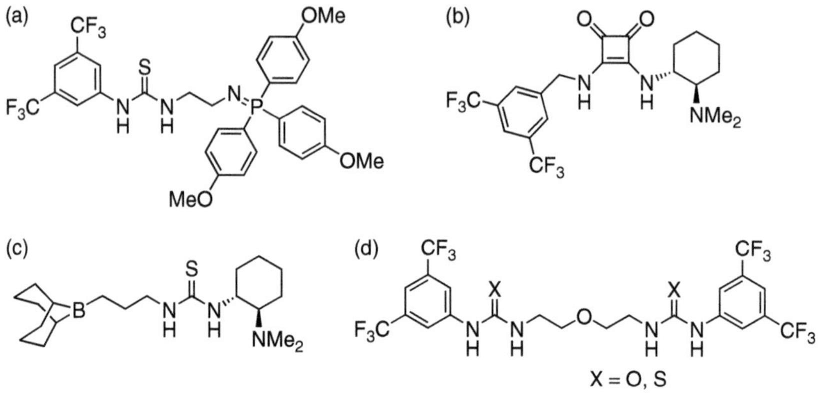

Figure 1.13 Some bifunctional and multifunctional catalysts.

amine moiety in the thiourea–amine catalyst to a more basic iminophosphorane moiety, its catalytic activities in the ROP of LA, VL, and CL are enhanced, while maintaining good control over the dispersity (*Đ*) (Figure 1.13a) [47]. The change in the hydrogen-bond donor moiety from thiourea to squaramide also affords an effective bifunctional catalyst for the ROP of LA (Figure 1.13b) [48]. The borane–thiourea–amine trifunctional organocatalyst can catalyze the synthesis of block copolymers, with PLA as one block, from a monomer mixture in a one-pot, one-step manner (Figure 1.13c), as per a recent report [49]. Mimicry of enzymes by introducing multiple functional groups into a single catalyst molecule could be a promising strategy. However, this approach significantly increases the synthetic complexity of the catalysts and the number of possible structures to be explored to obtain an optimized catalyst candidate. Furthermore, a novel catalyst design concept of using more than one hydrogen-bond donor unit, such as urea or thiourea, has been reported for the ROP of cyclic esters (Figure 1.13d) [50, 51].

1.3 Recent Trends in Organocatalyst Development

Regarding the recent trends in organocatalyst development for the ROP of cyclic esters, the following four aspects are considered to be significant: enhancements in the catalytic efficiency, selectivity, heat tolerance, and safety. These four aspects are introduced in this section.

1.3.1 Higher Catalytic Efficiency

When the ROP of cyclic esters approaches a high percentage of monomer conversion, transesterification can easily occur, leading to a broad dispersity. Before the period of extensive research on organocatalyzed polymerization, strong inorganic bases were used for catalyzing or initiating the ROP of cyclic esters via anionic polymerization. For instance, when the ROP of lactide is initiated by lithium diisopropylamide (LDA) at room temperature in a dioxane solvent, the monomer conversion can reach over 95% within a few minutes, affording PLA with a broad dispersity ($Đ = 1.9$) [52]. In the organocatalyzed ROP of cyclic esters, high selectivity has been considered as a common research objective instead of high reactivity, for a relatively long time.

In recent years, organocatalysts have been successfully developed to achieve fast, selective, and well-controlled ROP of cyclic esters. (Thio)urea anions are good examples. The (thio)urea anions with alkaline-metal cations, such as Na^+ or K^+, were first reported to catalyze the ROP of various cyclic esters in a fast and controlled manner [37, 53]. By fine-tuning the molecular structure of (thio)ureas and the countercation, the ROP of LA, δ-VL, ε-CL, etc., could be completed within seconds, with the monomer conversion reaching >90% and narrow dispersity ($Đ < 1.1$). The reaction proceeds via the dual-activation mechanism, with the (thio)urea anion (N^-) activating the propagating chain end (–OH), whereas the N–H moiety of the (thio)urea anion interacts with the carbonyl group of the cyclic esters (Figure 1.14a). In addition to alkaline-metal bases, various strong organobases, including DBU, MTBD, cyclopropenimine [54], and phosphazene bases, have been investigated in conjunction with (thio)urea for the ROP of cyclic esters (Figure 1.14b). Consequently, when the acidity of (thio)ureas and the basicity of bases attain equilibrium – pK_a of (thio)urea ≈ pK_a of base-H^+ – the best catalytic activity is achieved [55]. Regarding the reaction mechanism, the ROP is proposed to proceed via an anionic mechanism when pK_a of (thio)urea < pK_a of base-H^+ (Figure 1.15a). When pK_a of (thio)urea > pK_a of base-H^+, a cooperative mechanism is proposed (Figure 1.15b).

After these pioneering works, investigations have been expanded by further tuning the catalyst structures, synthesizing statistical and block copolymers, and using them in flow chemistry. For instance, the urea salt with the tetra-*n*-butyl ammonium cation is an effective catalyst for the rapid, selective, and versatile ROP of lactides [56]. Further expansion of the investigation scope of the phosphazene bases in forming urea salts enabled the efficient synthesis of poly(γ-butyrolactone) [57], random poly(lactic-*co*-glycolic acid) [58], and poly(lactic acid)-*b*-poly(alkyl-δ-lactone)-*b*-poly(lactic acid) triblock copolymer as thermoplastic elastomers [38] and pressure-sensitive adhesives [59]. Aside from (thio)urates, the 2,2′-bisindole anion can function as an excellent catalyst to promote the rapid ROP of cyclic esters (Figure 1.15c) [60].

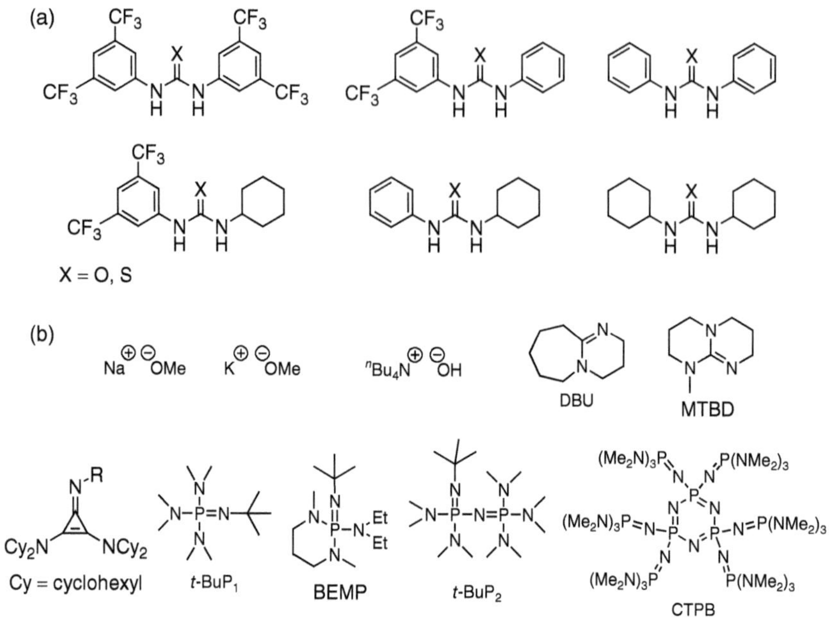

Figure 1.14 Representative structures of (a) ureas, thioureas and (b) bases.

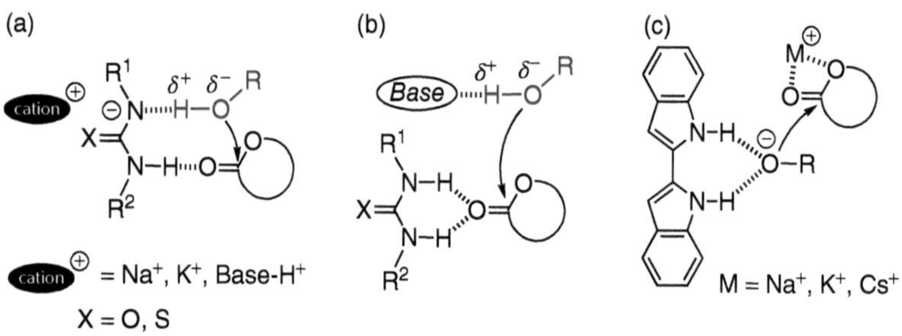

Figure 1.15 (a) Anionic mechanism; (b) cooperative mechanism; (c) reaction mechanism using the 2,2′-bisindole anion.

The progress in rapid and controlled ROP for cyclic esters has opened up new avenues for developing high-throughput synthetic platforms of polymer libraries using flow chemistry techniques [61]. Moreover, by employing continuous-flow reactors, ultrafast ROP of cyclic esters can be achieved using conventional strong inorganic base catalysts, such as KO*t*Bu, affording the polyester products in a well-controlled manner, which cannot be achieved under batch polymerization conditions [62].

1.3.2 Higher Selectivity

High selectivity has been a common, important research goal in the development of novel catalytic reactions and polymerizations. In the ROP of cyclic esters, selectivity

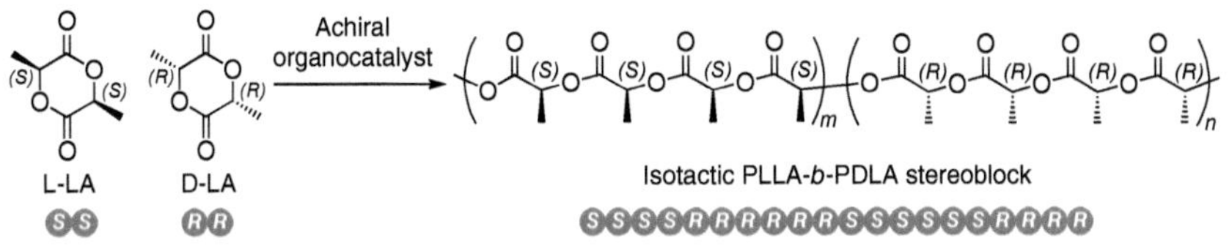

Figure 1.16 ROP of *rac*-LA using achiral organocatalysts.

can include various aspects. This section focuses on enantiomer-selective polymerization and briefly introduces some examples of other aspects.

Among all reports on enantiomer-selective ROP of cyclic esters, the ROP of racemic lactide (*rac*-LA) and racemic *β*-butyrolactone (*rac*-*β*-BL) have been chosen as two representative examples [63–66]. Chiral metal-complex catalysts have been extensively investigated for these two ROPs, and highly selective catalysts have been developed. In the past decade, reports on organocatalysts, especially chiral ones, have emerged.

In the early stages, the ROP of *rac*-LA by achiral organocatalysts was mainly reported, wherein isotactic PLA or isotactic-rich PLA was obtained. The selectivity relies on the chirality of the chain end structure that is combined with the achiral catalyst. The L-LA and D-LA molecules were consumed at the same rate, and the obtained isotactic-rich PLA was typically a multiblock copolymer consisting of PLLA and PDLA stereoblocks (Figure 1.16). Therefore, the formation of stereocomplexes was observed in some cases. NHCs and phosphazenes are representative achiral organocatalysts used for this polymerization [67–69]. However, to achieve a high level of isotacticity, in which the meso dyads (P_m) ≥ 0.90, the polymerization was usually performed at a low temperature (≤ −70 °C).

Compared to the use of achiral organocatalysts in the synthesis of isotactic PLA, the enantiomer-selective polymerization of *rac*-LA based on chiral organocatalysts is more challenging (Figure 1.17). Cinchona alkaloids were the first organocatalysts reported for this polymerization [70]. However, their stereoselectivity factor is low ($s \leq 4.4$) (Figure 1.18a). 1,1′-Bi-2-naphthol (BINOL)-derived chiral phosphoric acids

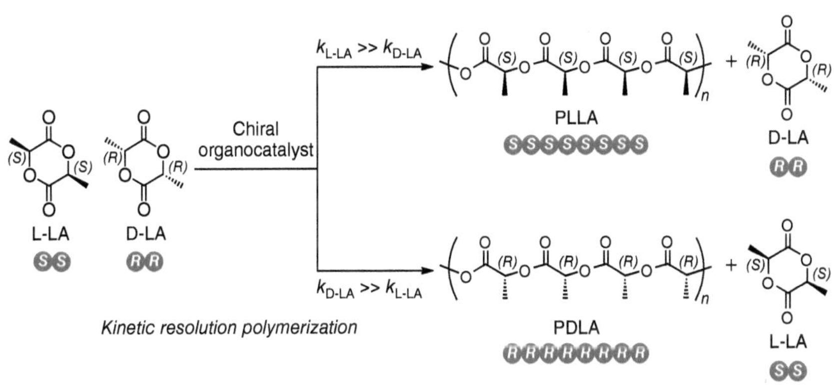

Figure 1.17 ROP of *rac*-LA using chiral organocatalysts: enantiomer-selective polymerization.

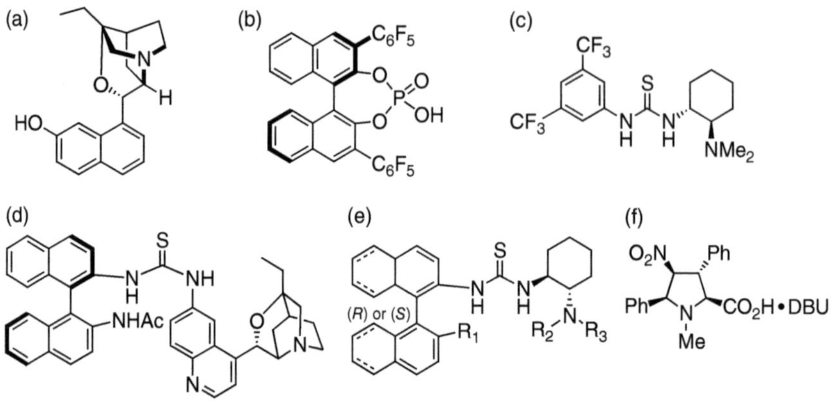

Figure 1.18 Representative chiral organocatalysts for the ROP of *rac*-LA.

are more effective in controlling the enantioselectivity of *rac*-LA. The (*R*)-catalyst polymerizes L-LA prior to D-LA, and the stereoselectivity factor is as high as 28.3 (Figure 1.18b) [71]. Takemoto's catalyst is a commonly used chiral bifunctional organocatalyst for the ROP of LA [45]. The use of enantiopure Takemoto's catalyst in catalyzing the ROP of *rac*-LA results in a relatively high isotacticity ($P_m \geq 0.80$) but low stereoselectivity ($s \leq 5.0$) (Figure 1.18c) [46]. The combination of enantiopure Takemoto's catalyst and a phosphazene base can further increase the isotacticity and reactivity [72]. Other novel chiral bifunctional organocatalysts synthesized by combining the moieties, including cinchona alkaloids, 1,1′-binaphthyl, thiourea, and *trans*-1,2-cyclohexanediamine, have also been reported for the ROP of *rac*-LA (Figure 1.18d,e) [73, 74]. Using these two catalysts shown in Figure 1.18d,e, stereoselectivity factor values of 53 and 17.5, respectively, were obtained, in addition to the high isotacticity of the PLA products. Ionic catalysts consisting of a densely substituted proline-type amino acid and DBU have also been reported to be effective for the ROP of *rac*-LA (Figure 1.18f) [75]. In the abovementioned cases, the highest stereoselectivity factor is 53. Therefore, the obtained PLA should be either L-LA-rich or D-LA-rich, instead of perfect PLLA or PDLA.

Unlike the emerging reports on the organocatalyzed ROP of *rac*-LA, there are no reports on the organocatalyzed ROP of *rac*-*β*-BL, to the best of our knowledge. It can be attributed to the intrinsic difficulties in distinguishing the enantiomers of *β*-BL. The carbonyl group activated by the catalyst is opposite to the chiral center (Figure 1.19a). Therefore, enantiomer-selective polymerization of *rac*-*β*-BL is highly challenging, even when metal-complex catalysts are used [76]. In addition, the ROP of *β*-BL can proceed via two different pathways, which further increases the difficulty in controlling the stereoselectivity in the ROP of *rac*-*β*-BL. A nucleophilic attack on the lactone carbonyl carbon results in stereoretention (Figure 1.19b), whereas that on the methine carbon next to the ester group results in stereoinversion (Figure 1.19c).

Chen *et al.* developed a novel method to circumvent the difficulties in realizing enantiomer-selective ROP of *rac*-*β*-BL using an eight-membered cyclic diolide

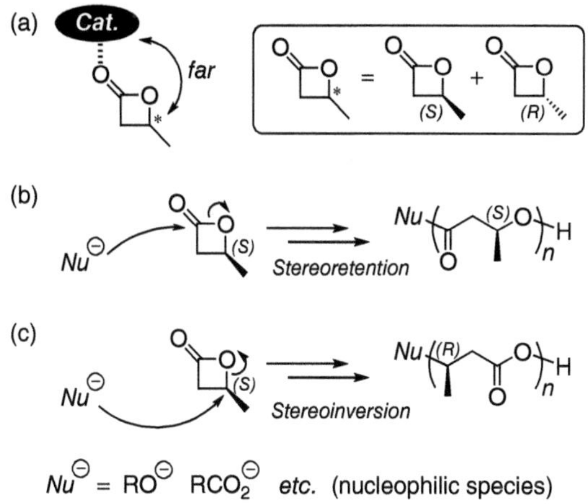

Figure 1.19 (a) Intrinsic difficulty in performing enantiomer-selective ROP of *rac*-β-BL; ROP of β-BL by nucleophilic attack on (b) carbonyl and (c) methine carbons.

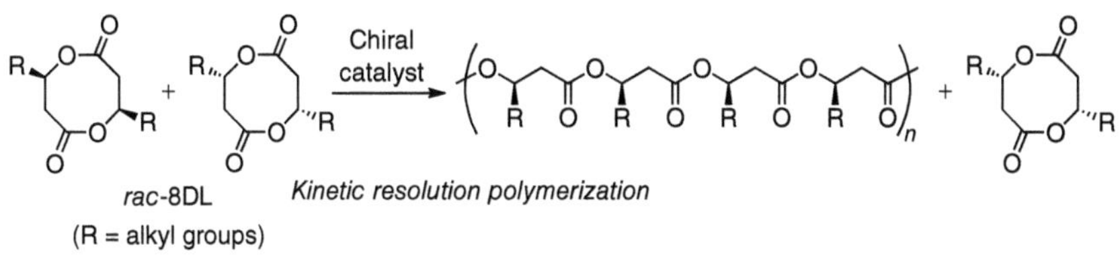

Figure 1.20 Enantiomer-selective polymerization of *rac*-8DL.

(8DL) [77–79]. Although several steps are required for monomer synthesis, near-perfect stereoselectivity in the ROP of *rac*-8DL can be achieved using sterically hindered rare-earth-metal salen complexes as catalysts. Organocatalysts such as BINOL-derived chiral phosphoric acids have also been tested for this polymerization [80]; however, the stereoselectivity factor is low, and stereogradient P(β-BL) is obtained (Figure 1.20).

In addition to enantiomer-selective polymerization, the selectivity includes other aspects such as regioselectivity and chemoselectivity. The highly regioselective ROP of methylglycolide (MG) yields an alternating copolymer of lactic acid and glycolic acid (PLGA), which is a promising biodegradable, biocompatible, and bioresorbable material for medical applications. This polymerization was achieved via a prudent selection of organobase catalysts; the phosphazene base *t*-BuP$_2$ provided a high regioselectivity. The deprotonated propagating chain ends nucleophilic attack at the lactyl carbonyl group of MG selectively (Figure 1.21) [81].

Regarding the chemoselectivity of the ROP of cyclic esters, copolymerization of different cyclic esters, especially LA and ε-CL, is of significance. Because both PLA and PCL are biodegradable polymers with complementary mechanical properties, their copolymers have been studied extensively. However, because of their drastically different polymerizabilities, statistical copolymers have conventionally been synthesized using metal-complex catalysts [82]. Recently, simple benzoic acid has been reported to catalyze this copolymerization under bulk polymerization conditions at high temperatures [83, 84] (Figure 1.22).

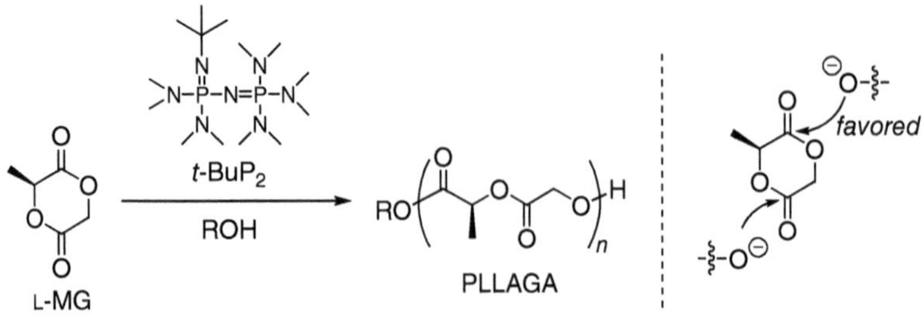

Figure 1.21 Regioselective ROP of MG by the t-BuP$_2$ catalyst.

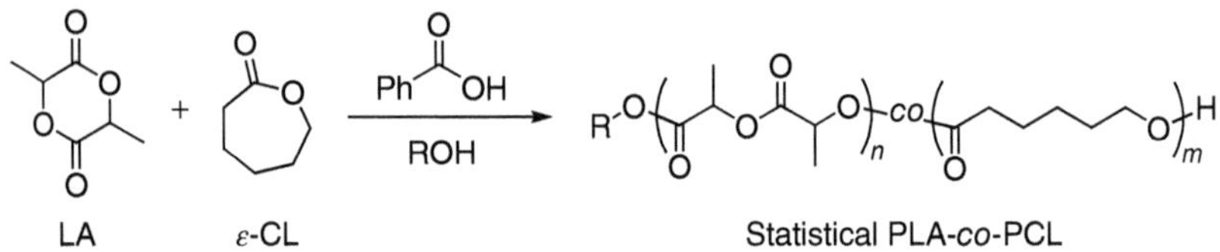

Figure 1.22 Statistical ROP of LA and ε-CL using a benzoic acid catalyst.

Integration of the ROP of cyclic esters with other types of polymerization, especially the ring-opening copolymerization (ROCOP) of cyclic anhydrides/epoxides and CO_2/epoxides, has attracted much research interest in recent years toward synthesizing degradable block-like copolymers in a one-pot, one-step manner. This type of transformation is referred to as self-switchable polymerization [85–87]. From a mechanistic perspective, self-switchable polymerization is based on catalysts that can catalyze both types of polymerization, namely ROP and ROCOP, and harnesses the intrinsic reactivity differences between the monomers in both catalytic cycles. Self-switchable polymerization was first reported to be catalyzed by metal-complex catalysts; then, organocatalysts were discovered to be effective. Considering the terpolymerization of cyclic anhydrides, epoxides, and cyclic esters as an example, the ROCOP of cyclic anhydrides/epoxides usually occurs prior to the ROP of cyclic esters, affording polyester-based block-like copolymers (Figure 1.23a). Considering the reaction mechanism, the selectivity between the ROCOP and ROP is rooted in the different reactivities of the nucleophilic attack from the hydroxy and carboxylic acid chain ends on the three monomers (Figure 1.23b). The reaction between the hydroxy group and cyclic anhydride is the most favored pathway (k_1), followed by the ring opening of the cyclic ester (k_2) and then epoxide (k_3). The carboxylic acid group preferentially reacts with the epoxide. Therefore, polyester-based block-like copolymers can be obtained by the terpolymerization of a mixture of these three monomers by following the reaction sequence shown in Figure 1.23a. Regarding the use of organocatalysts for this transformation, alkali-metal carboxylates (AMCs) [88], strong organobase catalysts [89, 90], and acid–base catalyst pairs containing triethylborane (TEB) [91] or (thio)ureas [92, 93] are representative examples (Figure 1.23c).

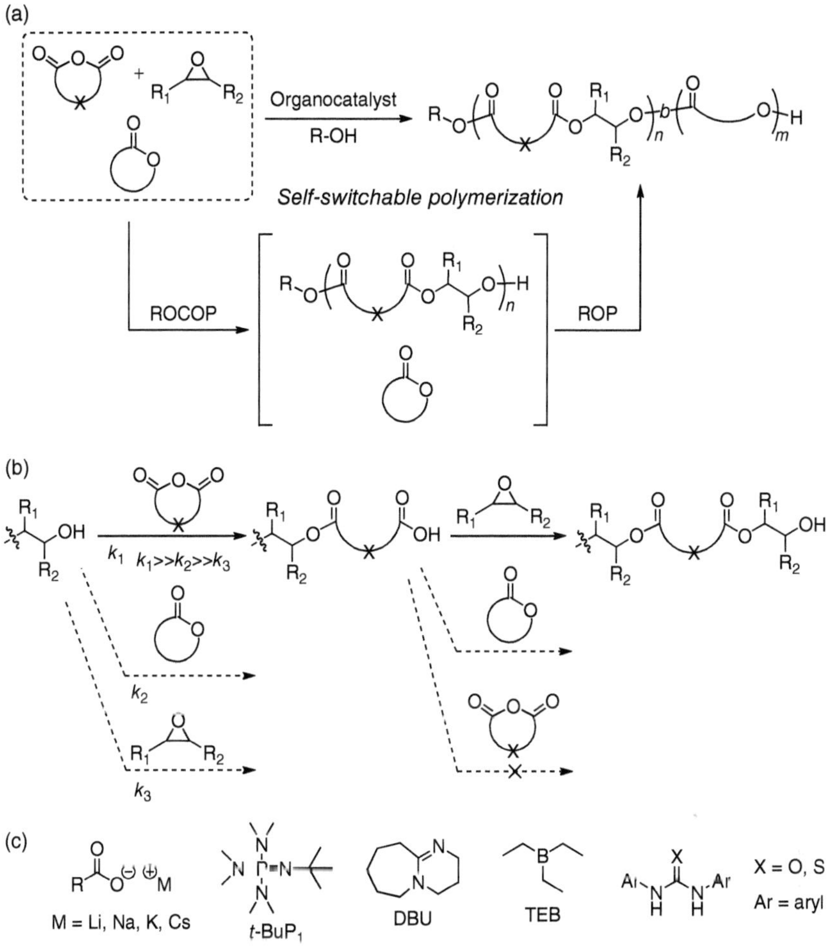

Figure 1.23 (a) Self-switchable polymerization of cyclic anhydrides, epoxides, and cyclic esters; (b) mechanistic consideration of the selectivity; (c) representative organocatalysts.

1.3.3 Higher-Temperature Tolerance

Poor thermal stability is a common feature of most organic compounds. Organocatalyst degradation usually occurs during high-temperature polymerization. Considering the ROP of LA as an example, although numerous environmentally benign organocatalytic systems have been developed, the industrial production of PLA still employs organometallic catalysts such as $Sn(Oct)_2$.

Controlling the ROP of LA using bulk organocatalysts is recognized as a significant challenge because the transesterification and racemization of PLA can easily occur at high reaction temperatures [94]. The melting point of optically pure PLA is approximately 180 °C. Hence, the industrial processing conditions for PLA are set to even higher temperatures [95]. The bulk polymerization of LA can be conducted

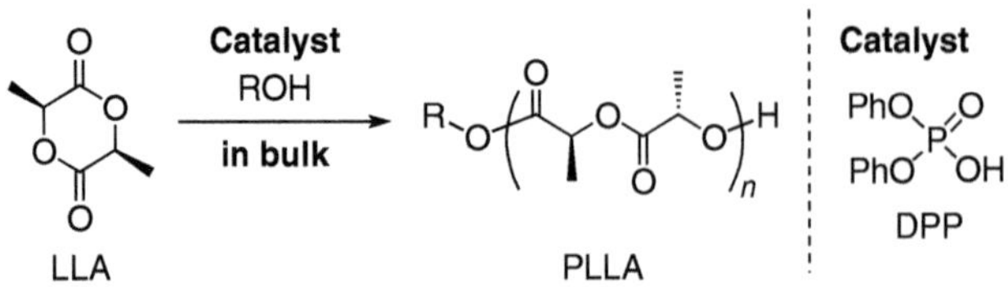

Figure 1.24 ROP of LLA in bulk catalyzed by DPP.

at a temperature above the melting point of LA (95–97 °C). Therefore, the initial organocatalyzed bulk polymerization of LA was commenced from ~100 °C.

Lactide can be catalyzed efficiently using organobase catalysts, including Brønsted bases and Lewis bases (nucleophilic catalysts). However, the bulk polymerization of LA using base catalysts is hindered by base-catalyzed racemization, polymer degradation, and the relatively low thermal stability of the catalysts [94]. Some of these drawbacks can be overcome using organoacid catalysts, although their catalytic activity in the ROP of LA is much lower than that of base catalysts. For example, DPP is an efficient catalyst for the ROP of CL, VL, and TMC in solution, but not for the ROP of LA [30, 96]. When the bulk reaction is performed at 130 °C, well-controlled PLA is obtained without racemization; however, the polymerization warrants a relatively higher catalyst loading than the ROP of other cyclic ester monomers (Figure 1.24) [97].

Acid–base binary catalysts have been extensively studied for the bulk polymerization of LA. Salts usually exhibit higher thermal stabilities than acids or bases alone. For instance, the mixing of TBD and MSA in a ratio of 1 : 1 produces a TBD:MSA (1 : 1) salt, which is calculated to be 37.4 kcal mol^{-1} more stable than free TBD and MSA (Figure 1.25) [98]. Even when the acid–base mixture cannot form crystalline salts because of ionic and hydrogen-bonding interactions, these mixtures still demonstrate an improved thermal stability [99].

A stoichiometric mixture of biogenic creatinine and carboxylic acids, including glycolic acid and acetic acid, was the first reported acid–base mixed organocatalyst for the ROP of LA in bulk (Figure 1.26a) [100]. Acid–base mixtures comprising imidazole and DMAP as the base components and various acid counterparts are also effective catalysts for this polymerization (Figure 1.26b–e) [35, 40, 101, 102]. In addition to 1 : 1 acid–base mixtures, nonstoichiometric mixtures can fine-tune the acidity/basicity, catalytic reactivity, and thermal stability. They have been described as deep eutectic solvents (DES) [103, 104]. Recently, they have been referred to as noneutectic mixture organocatalysts (NEMO), in the context of catalytic reactions [105, 106].

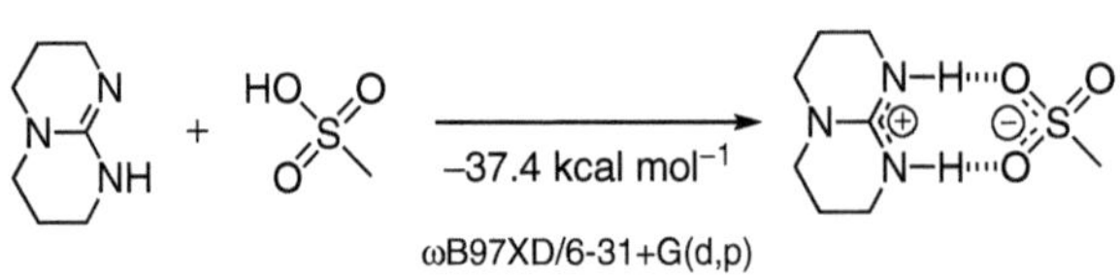

Figure 1.25 Density functional theory calculations of TBD, MSA, and the TBD:MSA (1 : 1) complex.

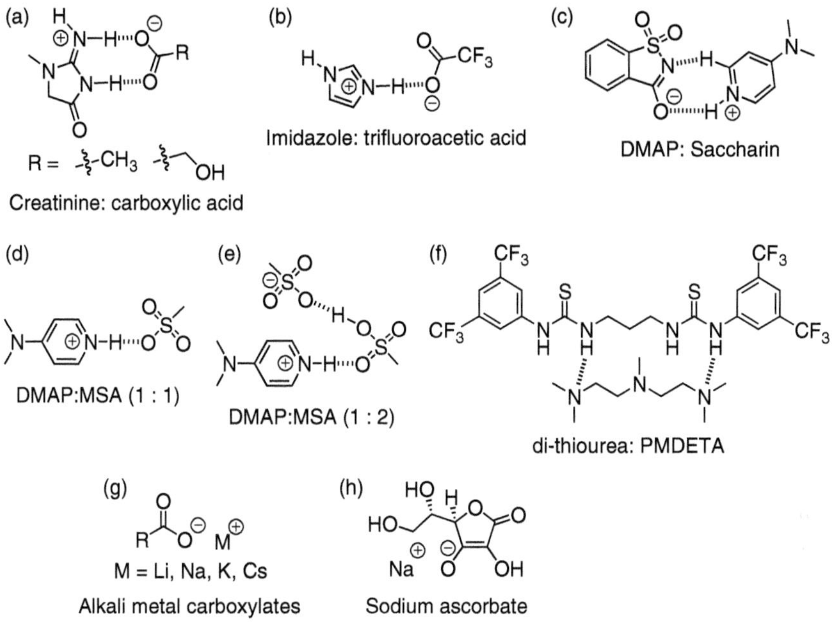

Figure 1.26 Representative acid–base mixed or salt organocatalysts for the ROP of LA in bulk.

Replacement of the acid component with a hydrogen-bonding donor, such as urea or thiourea, also affords tolerance to elevated temperatures, when suitable hydrogen-bonding donors and bases are selected, thus catalyzing the ROP of LA and other cyclic esters in bulk (Figure 1.26f) [107]. The food sweetener saccharin can form an ionic binary organocatalyst with DMAP toward the ROP of LA in bulk (Figure 1.26c) [35]. In addition to the mixtures of organic acids and organic bases, inorganic alkali metal salts (e.g. AMCs and sodium ascorbate) have been reported for the ROP of LA (Figure 1.26g,h) [39, 108]. These reports indicate the potential of safe food additives as catalysts for polymer synthesis. This discussion is detailed in Section 1.3.4.

Ionic liquids, which typically tolerate high temperatures, can also facilitate the ROP of cyclic esters. However, because they are employed in an excess amount as a solvent or they exhibited low catalytic activity when utilized in the catalytic amounts [109, 110], they are not discussed herein.

1.3.4 Safety Considerations

In numerous works on organocatalyzed polymerization, keywords such as "organic," "metal-free," "heavy-metal-free," or "transition-metal-free" are commonly mentioned to emphasize the nontoxic, safe, and environmentally benign nature of organocatalysts. However, organocatalysts need not mean safe catalysts. For instance, DMAP, the first organocatalyst reported by Hedrick *et al.* is categorized as "Danger" in its safety data sheet [5]. It is highly toxic when ingested or absorbed by the skin. Phosphazene bases have been reported as effective catalysts for the ROP of cyclic esters. However, the phosphazene-based residue is suspected to be cytotoxic [111]. In contrast, the commonly used organometallic catalyst in industry, $Sn(Oct)_2$, has been approved by the United States Food and Drug Administration (FDA), if the $Sn(Oct)_2$ content does not exceed 1 wt% of the polymer product when applied in coatings that can come into contact with food. However, the regulation for the content of stannous compounds in food is much stricter. The concentration of tin should be lower than 15 ppm, in the case of stannous chloride. The strict regulation of tin also applies to polymers for biomedical applications such as drug delivery carriers, bioresorbable sutures, and implants. To reach this high standard, Evonik Industries AG launched the RESOMER® Zero series of bioresorbable polymer materials, in which the tin concentration was controlled to be less than 1 ppm by the clean removal of tin catalyst residue from the obtained polymers. Therefore, the development of truly safe organocatalysts that meet industrial requirements is significant, especially for polymers related to food and biomedical applications, as it could eliminate the high-cost tin catalyst removal process.

Poly(lactic acid) and poly(ε-caprolactone) are two of the most important biodegradable, biocompatible, and bioresorbable polymers. Here, the ROP of LA and ε-CL are considered as examples to introduce and discuss recent developments in safe organocatalysts for the ROP of cyclic esters. In the early investigation stages of organocatalysts for polymerization, numerous carboxylic acids and amino acids were examined as catalysts for the ROP of ε-CL and δ-VL (Figure 1.27). Many of the tested acids were safe and edible, although their catalytic performances necessitated improvements [112]. For instance, tartaric acid and citric acid are widely used food additives in our daily lives. Benzoic acid, another safe carboxylic acid (whose sodium salt is commonly used as a food preservative) has recently been demonstrated to catalyze the copolymerization of LA and CL in bulk to yield the statistical P(LA-*co*-CL) copolymer, which is a commercially available bioresorbable polymer [83, 84]. In addition to safe carboxylic acids, their alkali-metal salts, such as sodium and potassium salts, are often used as food additives. For example, potassium sorbate and sodium acetate are often used as additives for food preservation and

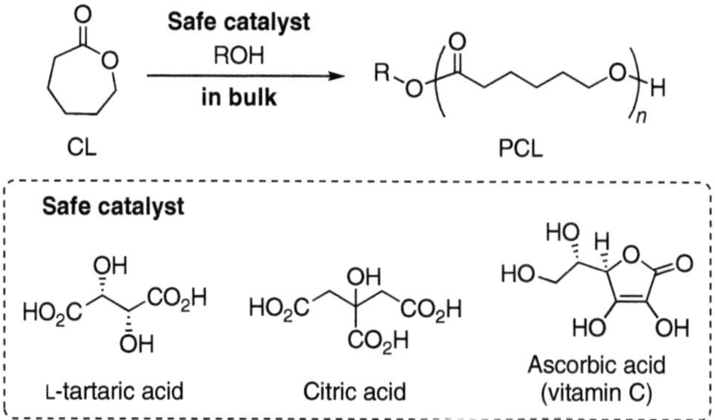

Figure 1.27 ROP of CL and representative truly safe catalysts.

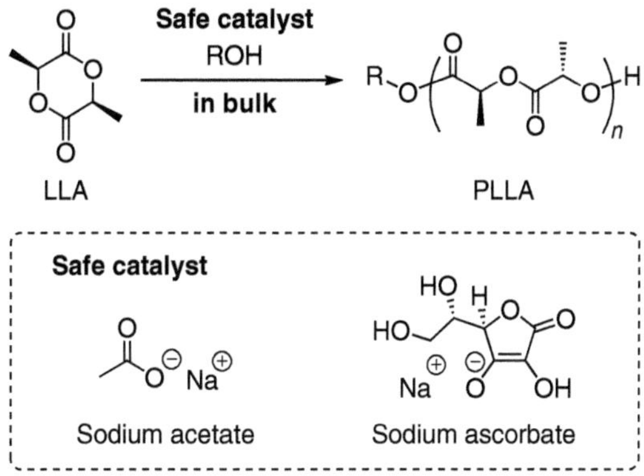

Figure 1.28 ROP of LA and representative truly safe catalysts.

flavor tuning [113]. Satoh *et al.* reported that AMCs, including the abovementioned food additives, can function as simple, effective catalysts for the ROP of cyclic esters under industrially relevant conditions (Figure 1.28) [39, 114]. In addition to the ROP of cyclic esters, the ROP of cyclic carbonates and epoxides and the ROCOP of cyclic anhydrides and epoxides can be catalyzed by AMCs [88, 115–121]. Alteration of alkali-metal cations with organic onium cations can provide new opportunities. Trimethyl glycine, a natural betaine derived from sugar beets, effectively catalyzes the ROP of TMC [43]. In addition to carboxylates, other safe alkali-metal salts have been investigated. Alkali-metal carbonates such as sodium carbonate can catalyze the ROCOP of cyclic anhydrides and epoxides [122]. Ascorbic acid and sodium ascorbate (vitamin C and its sodium salt, respectively) were discovered to be effective catalysts for the ROP of CL and LA (Figures 1.27 and 1.28) [108]. The copolymerization of CL and LA was also facilitated by sodium ascorbate. In addition to polymerization, depolymerization and chemical recycling of PLA can be conducted using sodium ascorbate.

In industrial production, the catalyst residue is commonly not removed to reduce production costs. Therefore, catalyst safety is an important factor when considering their applications. Numerous excellent catalyst designs have been reported after the investigation of organocatalysts for polymer synthesis. However, simple and safe compounds such as sodium acetate and sodium ascorbate have long been overlooked. The aforementioned studies demonstrate the potential of safe and unassuming substances in our daily lives.

1.4 Toward Higher Degradability and Recyclability

The 3R initiative (reduce, reuse, and recycle) aims to build a sound material-cycle society for sustainable development. Global plastic waste is a pressing issue that needs to be addressed. Redesigning current polymer materials toward improved degradability and recyclability is considered an important approach for polymer researchers. Regarding the polyesters synthesized via ROP, significant advancements have been achieved through the development of novel polyesters with higher degradability and recyclability. In this section, recent advancements are introduced, particularly examples involving organocatalysts.

1.4.1 Incorporation of Chemically Labile Moieties into Polymer Chains for Higher Degradability

Generally, ester bonds undergo hydrolysis. At elevated temperatures, ester hydrolysis reactions can easily occur under strongly acidic and basic conditions in the laboratory. However, under natural environmental conditions, the reaction rate of ester hydrolysis is much lower. In addition, unlike fundamental organic chemistry studies using small molecules, the functional groups on polymers often exhibit different reaction properties because of their high molecular weights. The crystallization or aggregation status of polymers can significantly influence the reaction rate of their functional groups. Regarding the degradation of polyesters via hydrolysis, the hydrolysis of the ester group is typically favored from the chain ends assisted by the free hydroxyl or carboxylic acid groups via backbiting (Figure 1.29a) [123–125].

Incorporation of highly labile moieties into polymer chains, ideally in a random manner, is a representative strategy to facilitate degradation and has been investigated extensively in recent years (Figure 1.29b). After the degradation of the highly active units, long polyester chains are cleaved into short-chain oligomers. After degradation to oligomers, the crystallization/aggregation ability decreases, and the number of reactive chain ends increases drastically, thus facilitating hydrolysis. For example, PLA is one of the most widely used biosourced, biodegradable polymer. Although it can undergo facile biodegradation in compost, its degradation in natural environments, especially in water, can take years. This is ascribed to its semicrystalline nature and the predominant depolymerization mechanism from the chain ends [124, 125]. Introduction of labile moieties randomly into the polymer chain of PLA without altering its material properties is highly desirable. The introduction of highly acid-labile hemiacetal ester functional groups is a widely studied strategy. 1,3-Dioxolan-4-one (DOX), which is synthesized from glycolic acid and formaldehyde, is a representative cyclic hemiacetal ester. The copolymerization of DOX and LLA yields the copolymer P(LLA-*co*-DOX) (Figure 1.30a) [126]. This copolymer has

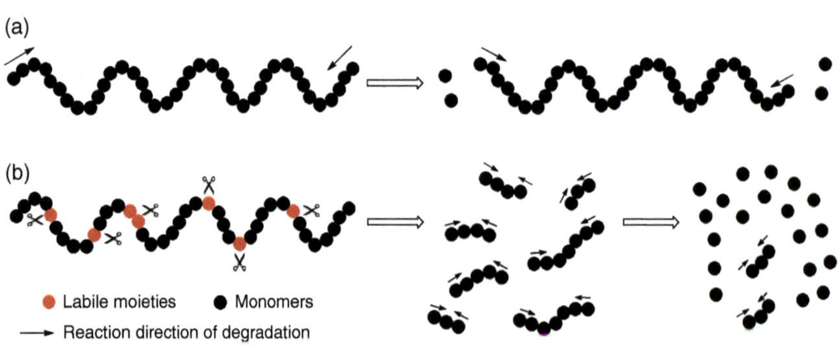

Figure 1.29 Degradation of polymers (a) from the chain ends; (b) from the labile moieties.

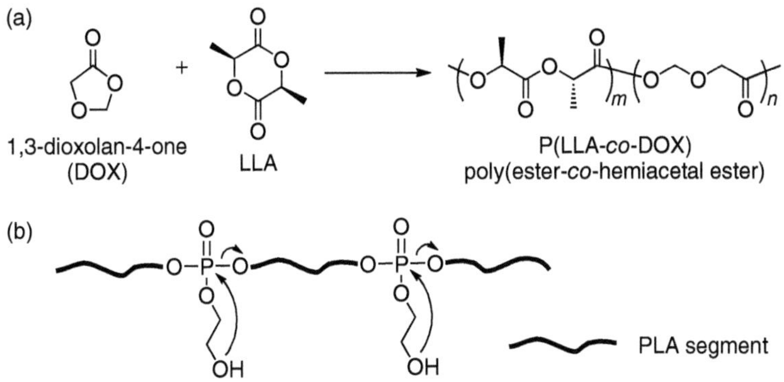

Figure 1.30 (a) Copolymerization of hemiacetal ester DOX with LLA; (b) incorporation of phosphate moiety into PLA.

thermal properties similar to LLA, namely, T_m and T_g. However, the degradability in aqueous environments increases noticeably. Upon 10% incorporation of DOX, the mass loss and reduction in molecular weight of the P(LLA-*co*-DOX) film in seawater are 2% and 35%, respectively, after 45 days; no changes are observed in PLLA during the same period. Another design for a marine-degradable PLA involves the introduction of phosphate moieties into the polymer chain (Figure 1.30b) [127]. The PLA chains undergo rapid cleavage via an RNA-inspired intramolecular attack by the free hydroxyl group, thus transforming into easily degradable lactide oligomers. However, the iterative addition of LA and cyclic phosphate monomers is required for the synthesis. Cyclic acetals, such as 1,3-dioxolane (DXL) and its derivatives, can also copolymerize with LLA [128–130]. Although the degradability, especially under acidic conditions, is significantly enhanced, the material properties differ substantially from PLA. Most of the copolymers become amorphous, with their T_g values at approximately room temperature.

In addition to copolymerization, development of novel cyclic ester monomers is an alternative approach for enhancing the degradability of polyesters. Cyclic hemiacetal esters have been studied extensively [131–133]. Although poly(hemiacetal ester)s can be synthesized via condensation polymerization [134], the ROP of cyclic hemiacetal esters has been developed as a more well-controlled method [135]. The resulting poly(hemiacetal ester)s are acid-labile and potentially biodegradable polymers. Therefore, they hold promise as degradable alternatives to current polymer materials in daily applications and for high-end applications such as drug delivery systems [136]. Regarding the polymerization conditions, poly(hemiacetal ester)s can be obtained only via cationic ring-opening polymerization (CROP) using Brønsted acid or Lewis acid catalysts (Figure 1.31a). Through anionic ring-opening polymerization (AROP), polyesters are obtained by eliminating aldehyde (Figure 1.31b). The CROP of exocyclic hemiacetal esters can also produce acid-labile poly(hemiacetal ester)s with different molecular structures (Figure 1.32) [137].

Cationic copolymerization of cyclic hemiacetal esters with other monomers to synthesize degradable polymers has also been widely investigated [138, 139]. The

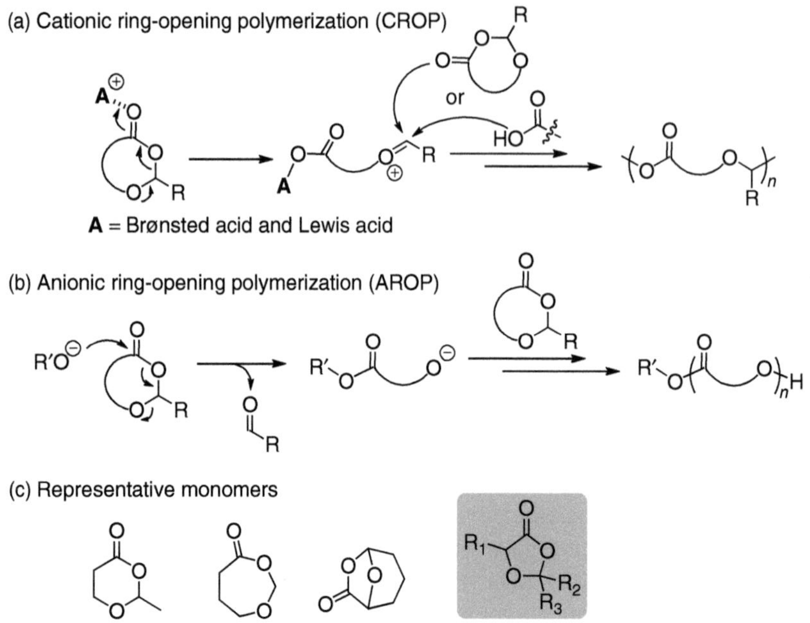

Figure 1.31 Ring-opening polymerization of cyclic hemiacetal esters via (a) cationic mechanism, (b) anionic mechanism; and (c) representative monomers (the monomer marked in gray was reported in copolymerization only).

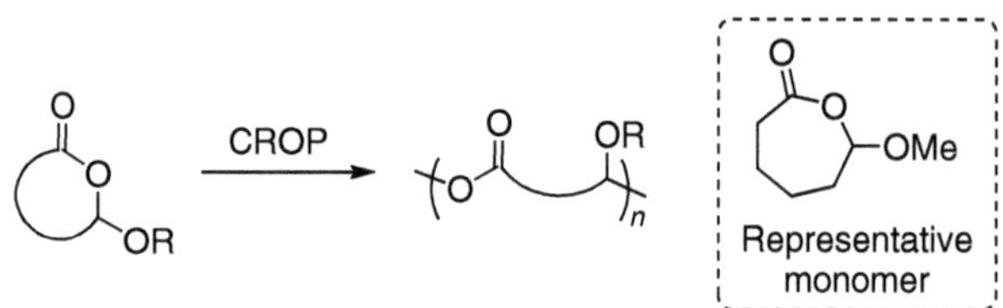

Figure 1.32 CROP of exocyclic hemiacetal esters.

synthesis of poly(hemiacetal ester)s via the copolymerization of cyclic acetals or aldehydes with cyclic anhydrides has been reported recently [140, 141]. However, these examples are not introduced in detail herein.

Other functionalities have also been introduced into the cyclic ester monomers to enhance the degradability of the resulting polymers. 4-carbomethoxyvalerolactone (CMVL), which bears a carbomethoxy group at the 4-position of δ-VL, undergoes ROP to afford poly(4-carbomethoxyvalerolactone) (PCMVL), upon catalysis by DPP (Figure 1.33a) [142, 143]. The resulting PCMVL is tough, easily degradable, and recyclable. Regarding its degradation mechanism, especially under basic conditions, the retro oxa-Michael reaction is the main reaction pathway (Figure 1.33a, right). When a ketone group is introduced at the 4-position of δ-VL, the cyclic ester 4-ketovalerolactone (KVL) can also undergo ROP to yield PKVL in a manner similar to that effected by the DPP catalyst [144]. The polymer structure is similar to that of poly(glycolide) (PGA). The synthesized PKVL shows $T_g = 7\,°C$ and $T_m = 132$

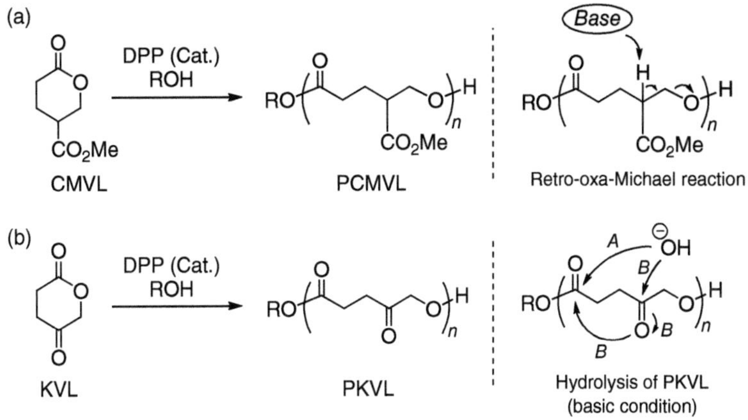

Figure 1.33 ROP of novel cyclic esters.

and 148 °C. The PKVL undergoes facile depolymerization, especially under basic conditions. The hydroxy anion can nucleophilically attack the ester carbonyl (Figure 1.33b, right, path A) and ketone groups, followed by an intramolecular attack on the ester carbonyl group (path B). This mechanism explains the rationale behind PKVL undergoing a faster hydrolysis than PVL, under basic conditions.

In addition to the aforementioned examples, amine-containing polyesters or poly(amino esters) are an important category of degradable polyesters for biomedical applications [136, 145]. This type of polyester is discussed in detail in Chapter 6.

1.4.2 Development of Novel Chemically Recyclable Polyesters

Chemical recycling of polymer materials converts plastic waste into their starting monomers or other value-added chemicals. This process is considered an important approach for achieving a circular economy of polymer materials. Among these, closed-loop recycling, i.e. converting back into monomers, has attracted considerable research attention in recent years. Among all chemically recyclable polymers, polyesters synthesized via the ROP approach are the most widely studied. Organocatalysts have been used in many reported applications.

According to the fundamental principle of polymerization, most of the polymerization process is enthalpy-driven, i.e. the enthalpy change $\Delta H < 0$. The entropy change of polymerization is always negative ($\Delta S < 0$) because the degrees of freedom decrease during the formation of polymers from monomers. The Gibbs free energy change of polymerization $\Delta G = \Delta H - T\Delta S$ should be negative ($\Delta G < 0$) to allow for the polymerization to occur. When $\Delta G > 0$, the reverse reaction, i.e. the depolymerization, occurs (Figure 1.34). When $\Delta G = 0$, the polymerization and depolymerization reach an equilibrium, and the temperature thereat is called ceiling temperature T_c. In principle, all polymers should have a T_c that can allow for reverse depolymerization to occur. However, most vinyl polymers such as polyolefins decompose via

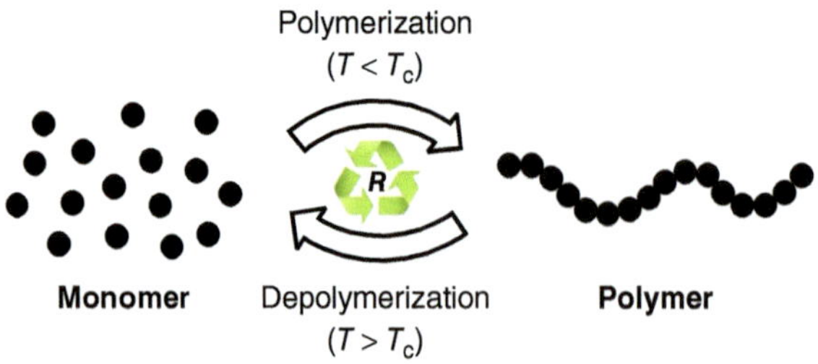

Figure 1.34 Schematic of polymerization and depolymerization.

other reaction pathways before attaining their high theoretical T_c values. Therefore, from a practical viewpoint, T_c needs to be within an appropriate range to facilitate convenient synthesis, use, and chemical recycling of polymer materials. Aside from polymers that can directly depolymerize back to monomers, numerous polymers first degrade to different low-molecular-weight compounds or oligomers, which can then be further transferred back to monomers or polymers. These cases are excluded to narrow the scope of this section.

Five-membered γ-butyrolactone (γ-BL) was considered a nonpolymerizable lactone for a long time because of its low ring strain and small ΔH value. In 2016, Chen and Hong reported the first successful preparation of high molecular weight poly(γ-butyrolactone) (P(γ-BL)) at low temperatures [1, 146]. The isolated P(γ-BL) could withstand high temperatures after the catalyst removal. The polymer could depolymerize back to γ-BL monomers via thermal degradation (Figure 1.35). Although lanthanide complexes were used as the first catalyst, base catalysts, including the organic phosphazene base and the inorganic base *t*-BuOK, also demonstrate excellent catalytic performances for this polymerization [2, 147, 148].

Limited by the theoretical T_c value, the polymerization of γ-BL has to be conducted at low temperatures, which restricts its practical application. Therefore, after Chen and Hong's seminal report, the T_c value and material properties were considered by changing the molecular design of the monomers. Numerous successful examples have been reported to date [149–156]. Here, instead of introducing the novel cyclic ester monomers individually, they are classified into three major categories: (i) bicyclic structures, (ii) substitutions, and (iii) heteroatoms. These three categories are discussed in the following paragraphs.

In the first category, bicyclic structures are used. Following the successful polymerization of γ-BL, the structure of five-membered lactone has been integrated

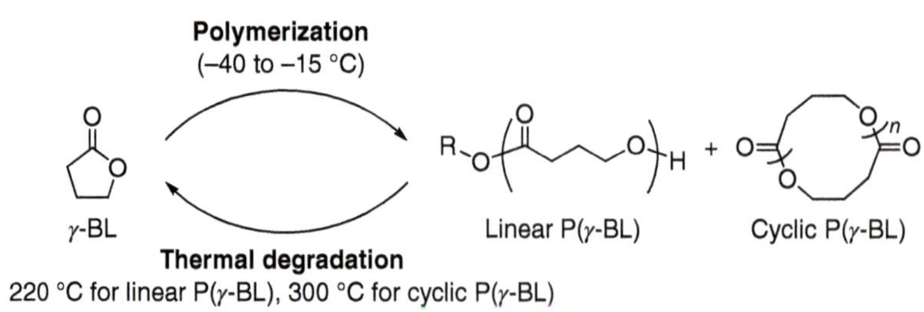

Figure 1.35 Polymerization of γ-BL and depolymerization of P(γ-BL).

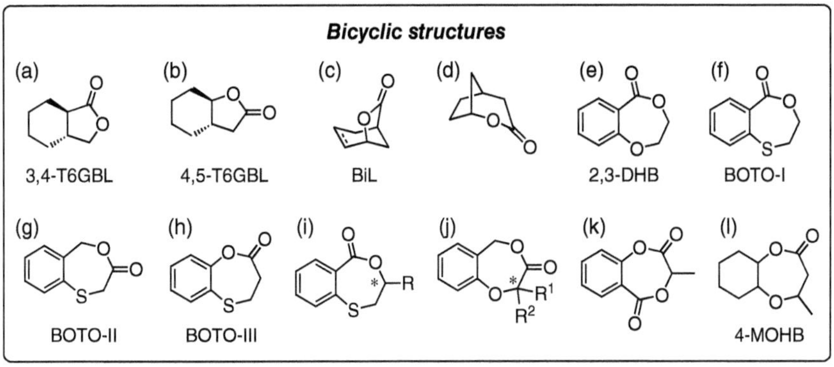

Figure 1.36 Cyclic ester monomers with bicyclic structures.

into bicyclic compounds to tune the polymerizability and material properties of the polymers. As shown in Figure 1.36a–c, the *trans*-fused bicyclic and bridged lactones contain the substructure of the five-membered γ-BL [157–159]. Compared to the normal γ-BL, the bicyclic structure endows the polymer with a higher ring strain; thus, the polymerization can be conducted at room temperature while ensuring the recyclability back to monomers. In addition to the increased polymerizability, the introduction of rigid cyclic structures into the polymer chain drastically increases the glass transition temperature and melting point of the obtained polyesters. This strategy of introducing bicyclic structures has been proven successful for seven-membered lactones (Figure 1.36d–l) [41, 160–166]. Unlike the γ-BL, the seven-membered ε-CL possess good polymerizability; however, the depolymerization process usually requires harsh conditions owing to the less favored ring-closing reaction [167–169]. The development of mild and highly selective PCL depolymerization method remains an important research topic. By introducing bicyclic structures together with heteroatoms, such as oxygen and sulfur, into the seven-membered ring, the ring-closing depolymerization process becomes more favorable, thus allowing for the chemical recycling back to the monomer under milder conditions ($\leq$ 200 °C). In addition to lowering the T_c value to enable facile chemical recycling, the incorporation of cyclic structures and heteroatoms into the polymer chain can also tune the thermal and mechanical properties of the polyester products. Furthermore, side-chain substituents, such as atactic, isotactic, and stereocomplexes, have been introduced to further fine-tune the material properties by altering the functional group and tacticity [163, 165].

The second category involves the introduction of substituents into the cyclic monomer. In particular, gem-disubstitution can significantly improve the chemical recyclability of cyclic esters through the Thorpe–Ingold effect [170]. Similar to ε-CL, δ-VL exhibits a high polymerizability and high T_c of 298 °C at $[M]_0 = 1\,\text{mol}\,\text{l}^{-1}$ [171]. Reduction in the T_c value and chemical recycling of the monomer at 140–170 °C have been demonstrated via single substitution at the β-, γ-, and δ-positions (Figure 1.37a–c) [38, 59, 142, 143, 172, 173]. When di-alkyl substitution is performed at the α-position, the T_c value decreases further to 67–115 °C at $[M]_0 = 1\,\text{mol}\cdot\text{l}^{-1}$, depending on the alkyl group (Figure 1.37d) [171]. Aside from

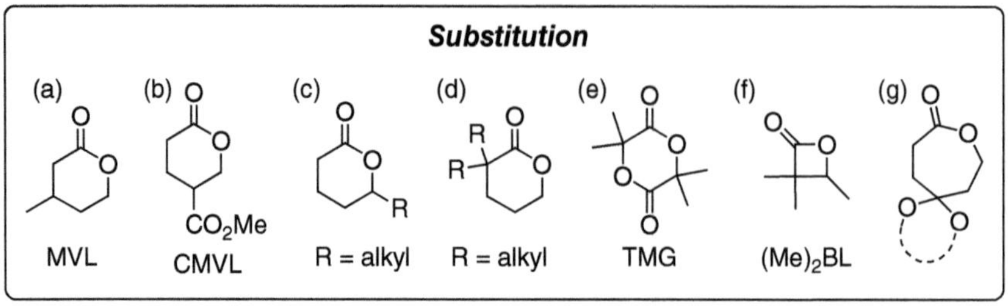

Figure 1.37 Cyclic ester monomers with various substituents.

tuning the T_c, the introduction of different substituents can alter the material properties of the resulting substituted PVLs to adapt to various applications, including plastics [143, 171], polyurethane foams [172], elastomers [38, 173], and pressure-sensitive adhesives [59]. Lactide is a representative six-membered cyclic ester monomer, and PLA is one of the most widely used biopolymers. The T_c of PLA is as high as 343 °C in bulk. The effective depolymerization of PLA to LA by zinc catalyst at approximately 200 °C has been reported. However, racemization occurs partially [174], which can be overcome by dissolving in a dilute, highly polar aprotic solvent, such as DMF and γ-valerolactone (GVL), to reduce the T_c value and suppress racemization [175]. Tetramethyl glycolide (TMG), a cyclic dimer of α-hydroxyisobutyric acid, can be polymerized to PTMG, which shows a high T_m of 185–190 °C (Figure 1.37e). Although the thermal degradation of PTMG conventionally affords methacrylic acid, when $Sn(Oct)_2$ is used as the catalyst at 265 °C, TMG can be recovered with a yield of 80.8% [176]. Polyhydroxyalkanoates (PHAs), represented by poly(3-hydroxybutyrate) (P3HB), are important biorenewable and biodegradable polymers. However, PHAs exhibit a relatively low degradation temperature of $T_{d,5\%} \approx 250$ °C. Moreover, PHAs can easily undergo elimination to form α,β-unsaturated carboxylic acids under thermal conditions. This challenge has been successfully addressed by introducing α,α-disubstituents to the four-membered β-lactone structure (Figure 1.37f) [177]. The gem-disubstitution significantly promotes the ring-closing process toward recovering the monomer under thermal depolymerization at 210 °C; however, the gaseous byproduct 2-methyl-2-butene is generated. In addition to the recyclability, the α,α-disubstituted PHAs show similar melting points, enhanced toughness, and a higher degradation temperature ($T_{d,5\%}$ = 322–335 °C), compared to the normal PHAs. The gem-disubstitution strategy has also been reported to be effective in seven-membered lactones. By introducing a spirocyclic acetal-functionality at the γ-position, the chemical recyclability improves considerably along with tunable thermal and mechanical properties, depending on the chemical structures of the acetal units (Figure 1.37g) [178].

The third category involves the introduction or alteration of the heteroatoms in cyclic esters. This strategy involves replacing the methylene unit ($-CH_2-$) with heteroatoms, such as oxygen, nitrogen, and sulfur, and replacing the ester moiety with a thioester. Poly(1,4-dioxan-2-one) (PPDO) is a biodegradable, biocompatible, and bioresorbable polymer that is used as a biomedical material. Compared to δ-VL, it has a lower T_c value of 265 °C, allowing for easier chemical recycling

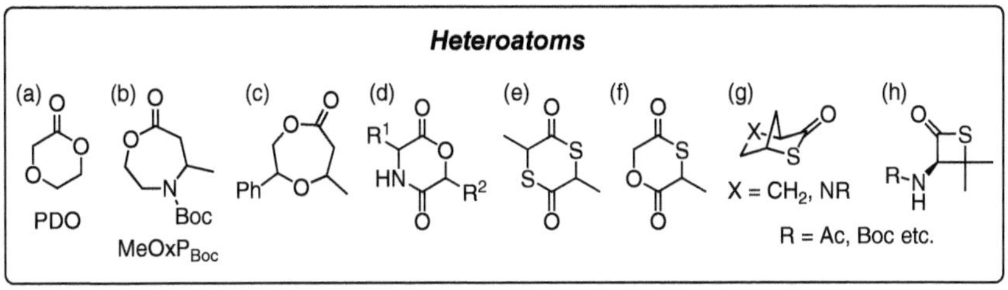

Figure 1.38 Cyclic ester monomers with heteroatoms.

(Figure 1.38a) [179, 180]. The introduction of oxygen and nitrogen atoms together with other substituents into the seven-membered lactone structure could also enable chemical recycling of the polyesters under relatively mild conditions (Figure 1.38b,c) [181, 182]. In addition, these monomers can be synthesized via the chemical upcycling of P3HB. This strategy of altering the heteroatoms can be applied to LA-like monomer structures as well. By changing one ester moiety of LA to an amide, the resulting poly(ester–amide) can be chemically recycled with ease in a closed-loop manner (Figure 1.38d) [183]. Because of the hydrogen-bonding interactions between the amide groups and the tunable side-chain functionality, the thermal and mechanical properties of poly(ester–amide)s can be adjusted over a wide range. For other novel LA-like monomers, changing the ester to a thioester group (i.e. O-to-S) is an effective approach to address the trade-off between the polymerizability of the monomer and depolymerizability of the polymer (Figure 1.38e,f) [184, 185]. More importantly, enhanced mechanical properties, particularly a higher toughness, can be achieved by a simple O-to-S substitution, without deteriorating the thermal properties. A similar O-to-S substitution strategy has been reported for five-membered cyclic thioesters fused in bridged bicycles and for four-membered β-thiolactones with a geminal dimethyl substitution. Both these monomers exhibit excellent polymerizability and recyclability (Figure 1.38g,h) [186, 187].

In summary, this section discusses different design strategies for cyclic esters, especially novel compounds, whose corresponding polyesters can be chemically recycled with ease. Further, for some cyclic ester monomers, more than one strategy has been employed to adjust the polymerizability of the monomer and depolymerizability of the polymer, while considering the material properties.

Closed-loop chemical recycling is a promising solution to the current global plastic problem. In real-world applications, wherein different types of plastic waste are collected together, closed-loop chemical recycling may encounter challenges associated with separating the polymers or recycled monomers.

1.5 Summary and Outlook

Among all types of organocatalyzed polymerization, the ROP of cyclic esters has been studied most extensively. Over the past two decades, a large variety of organocatalysts has been developed to meet the requirements of various cyclic ester monomers under different polymerization conditions. Among the novel organocatalysts, facets related to industrial applications are being increasingly considered.

This chapter discusses the aspects of reactivity, selectivity, thermal stability, and safety. In addition, the ROP of cyclic esters based on external-stimuli-responsive organocatalysts is an emerging research area [188–191]. Several representative examples, particularly those controlled by light, have been reported [192–200]. However, this research area is still in its infancy. We look forward to further improvements, especially at the levels of spatial and temporal control, to enable their application in highly sophisticated material production. In addition to the novel organocatalysts, the recent development of polyesters with excellent chemical recyclability and degradability is introduced in this chapter. In this part of the research, several aspects, including monomer synthesis, polymerization, material properties, and recyclability/degradability should be considered. With the advancements in catalyst development and monomer design, the organocatalyzed ROP of cyclic esters holds great promise for the environmentally benign production of degradable and recyclable polymeric materials.

References

1. Hong, M. and Chen, E.Y.-X. (2016). Completely recyclable biopolymers with linear and cyclic topologies via ring-opening polymerization of γ-butyrolactone. *Nat. Chem.* 8: 42–49.
2. Hong, M. and Chen, E.Y.-X. (2016). Towards truly sustainable polymers: a metal-free recyclable polyester from biorenewable non-strained γ-butyrolactone. *Angew. Chem. Int. Ed.* 55: 4188–4193.
3. Bouyahyi, M., Pepels, M.P.F., Heise, A., and Duchateau, R. (2012). ω-pentandecalactone polymerization and ω-pentadecalactone/ε-caprolactone copolymerization reactions using organic catalysts. *Macromolecules* 45: 3356–3366.
4. Ladelta, V., Bilalis, P., Gnanou, Y., and Hadjichristidis, N. (2017). Ring-opening polymerization of ω-pentadecalactone catalyzed by phosphazene superbases. *Polym. Chem.* 8: 511–515.
5. Nederberg, F., Connor, E.F., Möller, M. et al. (2001). New paradigms for organic catalysts: the first organocatalytic living polymerization. *Angew. Chem. Int. Ed.* 40: 2712–2715.
6. Myers, M., Connor, E.F., Glauser, T. et al. (2002). Phosphines: nucleophilic organic catalysts for the controlled ring-opening polymerization of lactides. *J. Polym. Sci. A Polym. Chem.* 40: 844–851.
7. Brown, H.A., De Crisci, A.G., Hedrick, J.L., and Waymouth, R.M. (2012). Amidine-mediated zwitterionic polymerization of lactide. *ACS Macro Lett.* 1: 1113–1115.
8. Coulembier, O., Dove, A.P., Pratt, R.C. et al. (2005). Latent, thermally activated organic catalysts for the on-demand living polymerization of lactide. *Angew. Chem. Int. Ed.* 44: 4964–4968.
9. Naumann, S. and Dove, A.P. (2015). *N*-heterocyclic carbenes as organocatalysts for polymerizations: trends and frontiers. *Polym. Chem.* 6: 3185–3200.

10 Maji, B., Breugst, M., and Mayr, H. (2011). *N*-heterocyclic carbenes: organocatalysts with moderate nucleophilicity but extraordinarily high Lewis basicity. *Angew. Chem. Int. Ed.* 50: 6915–6919.

11 Jaffredo, C.G., Carpentier, J.-F., and Guillaume, S.M. (2012). Controlled ROP of β-butyrolactone simply mediated by amidine, guanidine, and phosphazene organocatalysts. *Macromol. Rapid Commun.* 33: 1938–1944.

12 Moins, S., Henoumont, C., De Winter, J. et al. (2018). Reinvestigation of the mechanism of polymerization of β-butyrolactone from 1,5,7-triazabicyclo[4.4.0]dec-5-ene. *Polym. Chem.* 9: 1840–1847.

13 Lohmeijer, B.G.G., Pratt, R.C., Leibfarth, F. et al. (2006). Guanidine and amidine organocatalysts for ring-opening polymerization of cyclic esters. *Macromolecules* 39: 8574–8583.

14 Zhang, L., Nederberg, F., Pratt, R.C. et al. (2007). Phosphazene bases: a new category of organocatalysts for the living ring-opening polymerization of cyclic esters. *Macromolecules* 40: 4154–4158.

15 Liu, S., Ren, C., Zhao, N. et al. (2018). Phosphazene bases as organocatalysts for ring-opening polymerization of cyclic esters. *Macromol. Rapid Commun.* 39: 1800485.

16 Kaljurand, I., Saame, J., Rodima, T. et al. (2016). Experimental basicities of phosphazene, guanidinophosphazene, and proton sponge superbases in the gas phase and solution. *J. Phys. Chem. A* 120: 2591–2604.

17 Tshepelevitsh, S., Kütt, A., Lõkov, M. et al. (2019). On the basicity of organic bases in different media. *Eur. J. Org. Chem.* 2019: 6735–6748.

18 Vazdar, K., Margetić, D., Kovačević, B. et al. (2021). Design of novel uncharged organic superbases: merging basicity and functionality. *Acc. Chem. Res.* 54: 3108–3123.

19 Nederberg, F., Lohmeijer, B.G.G., Leibfarth, F. et al. (2007). Organocatalytic ring opening polymerization of trimethylene carbonate. *Biomacromolecules* 8: 153–160.

20 Chuma, A., Horn, H.W., Swope, W.C. et al. (2008). The reaction mechanism for the organocatalytic ring-opening polymerization of L-lactide using a guanidine-based catalyst: hydrogen-bonded or covalently bound? *J. Am. Chem. Soc.* 130: 6749–6754.

21 Kricheldorf, H.R. and Dunsing, R. (1986). Mechanism of the cationic polymerization of L,L-dilactide. *Makromol. Chem.* 187: 1611–1625.

22 Kricheldorf, H.R. and Kreiser, I. (1987). Polylactones, 11. Cationic copolymerization of glycolide with L,L-dilactide. *Makromol. Chem.* 188: 1861–1873.

23 Shibasaki, Y., Sanda, F., and Endo, T. (1999). Activated monomer cationic polymerization of 1, 3-dioxepan-2-one initiated by water-hydrogen chloride. *Macromol. Rapid Commun.* 20: 532–535.

24 Shibasaki, Y., Sanada, H., Yokoi, M. et al. (2000). Activated monomer cationic polymerization of lactones and the application to well-defined block copolymer synthesis with seven-membered cyclic carbonate. *Macromolecules* 33: 4316–4320.

25 Lou, X., Detrembleur, C., and Jérôme, R. (2002). Living cationic polymerization of δ-valerolactone and synthesis of high molecular weight homopolymer and asymmetric telechelic and block copolymer. *Macromolecules* 35: 1190–1195.

26 Bordwell, F.G. (1988). Equilibrium acidities in dimethyl sulfoxide solution. *Acc. Chem. Res.* 21 (12): 456–463.

27 Christ, P., Lindsay, A.G., Vormittag, S.S. et al. (2011). p*K*a values of chiral Brønsted acid catalysts: phosphoric acids/amides, sulfonyl/sulfuryl imides, and perfluorinated TADDOLs (TEFDDOLs). *Chem. Eur. J.* 17: 8524–8528.

28 Baśko, M. and Kubisa, P. (2008). Mechanism of propagation in the cationic polymerization of L,L-lactide. *J. Polym. Sci. A Polym. Chem.* 46: 7919–7923.

29 Baśko, M. and Kubisa, P. (2010). Cationic polymerization of L,L-lactide. *J. Polym. Sci. A Polym. Chem.* 48: 2650–2658.

30 Makiguchi, K., Satoh, T., and Kakuchi, T. (2011). Diphenyl phosphate as an efficient cationic organocatalyst for controlled/living ring-opening polymerization of δ-valerolactone and ε-caprolactone. *Macromolecules* 44: 1999–2005.

31 Delcroix, D., Couffin, A., Susperregui, N. et al. (2011). Phosphoric and phosphoramidic acids as bifunctional catalysts for the ring-opening polymerization of ε-caprolactone: a combined experimental and theoretical study. *Polym. Chem.* 2: 2249–2256.

32 Xu, J., Wang, X., Liu, J. et al. (2022). Ionic H-bonding organocatalysts for the ring-opening polymerization of cyclic esters and cyclic carbonates. *Prog. Polym. Sci.* 125: 101484.

33 Coady, D.J., Fukushima, K., Horn, H.W. et al. (2011). Catalytic insights into acid/base conjugates: highly selective bifunctional catalysts for the ring-opening polymerization of lactide. *Chem. Commun.* 47: 3105–3107.

34 Makiguchi, K., Kikuchi, S., Yanai, K. et al. (2014). Diphenyl phosphate/4-dimethylaminopyridine as an efficient binary organocatalyst system for controlled/living ring-opening polymerization of L-lactide leading to diblock and end-functionalized poly(L-lactide)s. *J. Polym. Sci. A Polym. Chem.* 52: 1047–1054.

35 Wei, F., Zhu, H., Li, Z. et al. (2019). Food sweetener saccharin in binary organocatalyst for bulk ring-opening polymerization of lactide. *Adv. Synth. Catal.* 361: 1335–1347.

36 Gontard, G., Amgoune, A., and Bourissou, D. (2016). Ring-opening polymerization of ε-caprolactone catalyzed by ionic hydrogen bond activation with bis-pyridiniums. *J. Polym. Sci. A Polym. Chem.* 54: 3253–3256.

37 Zhang, X., Jones, G.O., Hedrick, J.L., and Waymouth, R.M. (2016). Fast and selective ring-opening polymerizations by alkoxides and thioureas. *Nat. Chem.* 8: 1047–1053.

38 Li, C., Wang, L., Yan, Q. et al. (2022). Rapid and controlled polymerization of bio-sourced δ-caprolactone toward fully recyclable polyesters and thermoplastic elastomers. *Angew. Chem. Int. Ed.* 61: e202201407.

39 Saito, T., Aizawa, Y., Yamamoto, T. et al. (2018). Alkali metal carboxylate as an efficient and simple catalyst for ring-opening polymerization of cyclic esters. *Macromolecules* 51: 689–696.

40 Basterretxea, A., Gabirondo, E., Jehanno, C. et al. (2021). Stereoretention in the bulk ROP of L-lactide guided by a thermally stable organocatalyst. *Macromolecules* 54: 6214–6225.

41 Ren, F., Xian, J., Jia, Z. et al. (2023). Tetrabutylammonium halides as selectively bifunctional catalysts enabling the syntheses of recyclable high molecular weight salicylic acid-based copolyesters. *Angew. Chem. Int. Ed.* 62: e202306759.

42 Zhang, Q., Hu, C., Pang, X., and Chen, X. (2023). Multi-functional organofluoride catalysts for polyesters production and upcycling degradation. *ChemSusChem* 17: e202300907.

43 Saito, T., Takojima, K., Oyama, T. et al. (2019). Trimethyl glycine as an environmentally benign and biocompatible organocatalyst for ring-opening polymerization of cyclic carbonate. *ACS Sustain. Chem. Eng.* 7: 8868–8875.

44 Okino, T., Hoashi, Y., and Takemoto, Y. (2003). Enantioselective Michael reaction of malonates to nitroolefins catalyzed by bifunctional organocatalysts. *J. Am. Chem. Soc.* 125: 12672–12673.

45 Dove, A.P., Pratt, R.C., Lohmeijer, B.G.G. et al. (2005). Thiourea-based bifunctional organocatalysis: supramolecular recognition for living polymerization. *J. Am. Chem. Soc.* 127: 13798–13799.

46 Orhan, B., Tschan, M.J.-L., Wirotius, A.-L. et al. (2018). Isoselective ring-opening polymerization of *rac*-lactide from chiral Takemoto's organocatalysts: elucidation of stereocontrol. *ACS Macro Lett.* 7: 1413–1419.

47 Goldys, A.M. and Dixon, D.J. (2014). Organocatalytic ring-opening polymerization of cyclic esters mediated by highly active bifunctional iminophosphorane catalysts. *Macromolecules* 47: 1277–1284.

48 Specklin, D., Hild, F., Chen, L. et al. (2017). Bifunctional squaramides as organocatalysts for lactide polymerization: catalytic performance and comparison with monofunctional analogues. *ChemCatChem* 9: 3041–3046.

49 Geng, X., Liu, Z., Zhang, C., and Zhang, X. (2023). Toward stereo-and sequence-defined block copolymers via a three-site organocatalyst. *Macromolecules* 56: 4649–4657.

50 Fastnacht, K.V., Spink, S.S., Dharmaratne, N.U. et al. (2016). Bis-and tris-urea H-bond donors for ring-opening polymerization: unprecedented activity and control from an organocatalyst. *ACS Macro Lett.* 5: 982–986.

51 Pothupitiya, J.U., Dharmaratne, N.U., Marie, T. et al. (2017). H-bonding organocatalysts for the living, solvent-free ring-opening polymerization of lactones: toward an all-lactones, all-conditions approach. *Macromolecules* 50: 8948–8954.

52 Bhaw-Luximon, A., Jhurry, D., Spassky, N. et al. (2001). Anionic polymerization of D, L-lactide initiated by lithium diisopropylamide. *Polymer* 42: 9651–9656.

53 Lin, B. and Waymouth, R.M. (2017). Urea anions: simple, fast, and selective catalysts for ring-opening polymerizations. *J. Am. Chem. Soc.* 139: 1645–1652.

54 Morodo, R., Dumas, D.M., Zhang, J. et al. (2024). Ring-opening polymerization of cyclic esters and carbonates with (thio) urea/cyclopropenimine organocatalytic systems. *ACS Macro Lett.* 13: 181–188.

55 Lin, B. and Waymouth, R.M. (2018). Organic ring-opening polymerization catalysts: reactivity control by balancing acidity. *Macromolecules* 51: 2932–2938.

56 Jiang, Z., Zhao, J., and Zhang, G. (2019). Ionic organocatalyst with a urea anion and tetra-*n*-butyl ammonium cation for rapid, selective, and versatile ring-opening polymerization of lactide. *ACS Macro Lett.* 8: 759–765.

57 Shen, Y., Zhao, Z., Li, Y. et al. (2019). A facile method to prepare high molecular weight bio-renewable poly(γ-butyrolactone) using a strong base/urea binary synergistic catalytic system. *Polym. Chem.* 10: 1231–1237.

58 Shen, Y., Li, D., Kou, X. et al. (2022). Ultrafast ring-opening copolymerization of lactide with glycolide toward random poly(lactic-*co*-glycolic acid) copolymers by an organophosphazene base and urea binary catalysts. *Polym. Chem.* 13: 1861–1868.

59 Xu, C., Wang, L., Liu, Y. et al. (2023). Rapid and controlled ring-opening (co) polymerization of bio-sourced alkyl-δ-lactones to produce recyclable (co) polyesters and their application as pressure-sensitive adhesives. *Macromolecules* 56: 6117–6125.

60 Jadrich, C.N., Pane, V.E., Lin, B. et al. (2022). A cation-dependent dual activation motif for anionic ring-opening polymerization of cyclic esters. *J. Am. Chem. Soc.* 144: 8439–8443.

61 Lin, B., Hedrick, J.L., Park, N.H., and Waymouth, R.M. (2019). Programmable high-throughput platform for the rapid and scalable synthesis of polyester and polycarbonate libraries. *J. Am. Chem. Soc.* 141: 8921–8927.

62 Lin, B., Jadrich, C.N., Pane, V.E. et al. (2020). Ultrafast and controlled ring-opening polymerization with sterically hindered strong bases. *Macromolecules* 53: 9000–9007.

63 Thomas, C.M. (2010). Stereocontrolled ring-opening polymerization of cyclic esters: synthesis of new polyester microstructures. *Chem. Soc. Rev.* 39: 165–173.

64 Stanford, M.J. and Dove, A.P. (2010). Stereocontrolled ring-opening polymerisation of lactide. *Chem. Soc. Rev.* 39: 486–494.

65 Carpentier, J.-F. (2010). Discrete metal catalysts for stereoselective ring-opening polymerization of chiral racemic β-lactones. *Macromol. Rapid Commun.* 31: 1696–1705.

66 Tschan, M.J.-L., Gauvin, R.M., and Thomas, C.M. (2021). Controlling polymer stereochemistry in ring-opening polymerization: a decade of advances shaping the future of biodegradable polyesters. *Chem. Soc. Rev.* 50: 13587–13608.

67 Dove, A.P., Li, H., Pratt, R.C. et al. (2006). Stereoselective polymerization of rac-and meso-lactide catalyzed by sterically encumbered *N*-heterocyclic carbenes. *Chem. Commun.* 2881–2883.

68 Zhang, L., Nederberg, F., Messman, J.M. et al. (2007). Organocatalytic stereoselective ring-opening polymerization of lactide with dimeric phosphazene bases. *J. Am. Chem. Soc.* 129: 12610–12611.

69 Liu, S., Li, H., Zhao, N., and Li, Z. (2018). Stereoselective ring-opening polymerization of *rac*-lactide using organocatalytic cyclic trimeric phosphazene base. *ACS Macro Lett.* 7: 624–628.

70 Miyake, G.M. and Chen, E.Y.-X. (2011). Cinchona alkaloids as stereoselective organocatalysts for the partial kinetic resolution polymerization of *rac*-lactide. *Macromolecules* 44: 4116–4124.

71 Makiguchi, K., Yamanaka, T., Kakuchi, T. et al. (2014). Binaphthol-derived phosphoric acids as efficient chiral organocatalysts for the enantiomer-selective polymerization of *rac*-lactide. *Chem. Commun.* 50: 2883–2885.

72 Zaky, M.S., Wirotius, A.-L., Coulembier, O. et al. (2021). A chiral thiourea and a phosphazene for fast and stereoselective organocatalytic ring-opening-polymerization of racemic lactide. *Chem. Commun.* 57: 3777–3780.

73 Zhu, J.-B. and Chen, E.Y.-X. (2015). From meso-lactide to isotactic polylactide: epimerization by B/N Lewis pairs and kinetic resolution by organic catalysts. *J. Am. Chem. Soc.* 137: 12506–12509.

74 Dai, J., Xiong, W., Li, D.-Y. et al. (2023). Bifunctional thiourea-based organocatalyst promoted kinetic resolution polymerization of racemic lactide to isotactic polylactide. *Chem. Commun.* 59: 12731–12734.

75 Sanchez-Sanchez, A., Rivilla, I., Agirre, M. et al. (2017). Enantioselective ring-opening polymerization of rac-lactide dictated by densely substituted amino acids. *J. Am. Chem. Soc.* 139: 4805–4814.

76 Huang, H.-Y., Xiong, W., Huang, Y.-T. et al. (2023). Spiro-salen catalysts enable the chemical synthesis of stereoregular polyhydroxyalkanoates. *Nat. Catal.* 6: 720–728.

77 Tang, X., Westlie, A.H., Watson, E.M., and Chen, E.Y.-X. (2019). Stereosequenced crystalline polyhydroxyalkanoates from diastereomeric monomer mixtures. *Science* 366 (6466): 754–758.

78 Zhang, Z., Shi, C., Scoti, M. et al. (2022). Alternating isotactic polyhydroxyalkanoates via site-and stereoselective polymerization of unsymmetrical diolides. *J. Am. Chem. Soc.* 144: 20016–20024.

79 Westlie, A.H., Quinn, E.C., Parker, C.R., and Chen, E.Y.-X. (2022). Synthetic biodegradable polyhydroxyalkanoates (PHAs): recent advances and future challenges. *Prog. Polym. Sci.* 134: 101608.

80 Liu, J., Wang, J., Li, M., and Tao, Y. (2022). Gradient isoselective ring-opening polymerization of racemic cyclic diolide driven by chiral phosphoric acid catalysis. *Polym. Chem.* 13: 6328–6334.

81 Takojima, K., Makino, H., Saito, T. et al. (2020). An organocatalytic ring-opening polymerization approach to highly alternating copolymers of lactic acid and glycolic acid. *Polym. Chem.* 11: 6365–6373.

82 Nomura, N., Akita, A., Ishii, R., and Mizuno, M. (2010). Random copolymerization of ε-caprolactone with lactide using a homosalen–Al complex. *J. Am. Chem. Soc.* 132: 1750–1751.

83 Mezzasalma, L., Harrisson, S., Saba, S. et al. (2019). Bulk organocatalytic synthetic access to statistical copolyesters from L-lactide and ε-caprolactone using benzoic acid. *Biomacromolecules* 20: 1965–1974.

84 Jehanno, C., Mezzasalma, L., Sardon, H. et al. (2019). Benzoic acid as an efficient organocatalyst for the statistical ring-opening copolymerization of ε-caprolactone and L-lactide: a computational investigation. *Macromolecules* 52: 9238–9247.

85 Stößer, T., Chen, T.T.D., Zhu, Y., and Williams, C.K. (2017). 'Switch'catalysis: from monomer mixtures to sequence-controlled block copolymers. *Phil. Trans. R. Soc. A* 376: 20170066.

86 Hu, C., Pang, X., and Chen, X. (2022). Self-switchable polymerization: a smart approach to sequence-controlled degradable copolymers. *Macromolecules* 55: 1879–1893.

87 Jia, Y., Sun, Z., Hu, C., and Pang, X. (2022). Switchable polymerization: a practicable strategy to produce biodegradable block copolymers with diverse properties. *ChemPlusChem* 87: e202200220.

88 Xia, X., Suzuki, R., Takojima, K. et al. (2021). Smart access to sequentially and architecturally controlled block polymers via a simple catalytic polymerization system. *ACS Catal.* 11: 5999–6009.

89 Ji, H.-Y., Wang, B., Pan, L., and Li, Y.-S. (2018). Access to sequence-controlled block copolymers by self-switchable organocatalytic multicomponent polymerization. *Angew. Chem. Int. Ed.* 57: 16888–16892.

90 Li, H., Luo, H., Zhao, J., and Zhang, G. (2018). Sequence-selective terpolymerization from monomer mixtures using a simple organocatalyst. *ACS Macro Lett.* 7: 1420–1425.

91 Zhu, S., Wang, Y., Ding, W. et al. (2020). Lewis pair catalyzed highly selective polymerization for the one-step synthesis of A z C y (AB) x C y A z pentablock terpolymers. *Polym. Chem.* 11: 1691–1695.

92 Lin, L., Chen, X., Xiang, H. et al. (2022). Construction of triblock copolyesters via one-step switchable terpolymerization of epoxides, phthalic anhydride and ε-caprolactone using dual urea/organic base catalysts. *Polym. Chem.* 13: 801–807.

93 Liang, X., Wang, W., Zhao, D. et al. (2023). Self-switchable polymerization catalysis with monomer mixtures: using a metal-free commercial thiourea catalyst to deliver block polyesters. *Polym. Chem.* 14: 4918–4926.

94 Mezzasalma, L., Dove, A.P., and Coulembier, O. (2017). Organocatalytic ring-opening polymerization of L-lactide in bulk: a long standing challenge. *Eur. Polym. J.* 95: 628–634.

95 Lim, L.-T., Auras, R., and Rubino, M. (2008). Processing technologies for poly(lactic acid). *Prog. Polym. Sci.* 33: 820–852.

96 Makiguchi, K., Ogasawara, Y., Kikuchi, S. et al. (2013). Diphenyl phosphate as an efficient acidic organocatalyst for controlled/living ring-opening polymerization of trimethylene carbonates leading to block, end-functionalized, and macrocyclic polycarbonates. *Macromolecules* 46: 1772–1782.

97 Saito, T., Aizawa, Y., Tajima, K. et al. (2015). Organophosphate-catalyzed bulk ring-opening polymerization as an environmentally benign route leading to block copolyesters, end-functionalized polyesters, and polyester-based polyurethane. *Polym. Chem.* 6: 4374–4384.

98 Jehanno, C., Flores, I., Dove, A.P. et al. (2018). Organocatalysed depolymerisation of PET in a fully sustainable cycle using thermally stable protic ionic salt. *Green Chem.* 20: 1205–1212.

99 Basterretxea, A., Jehanno, C., Mecerreyes, D., and Sardon, H. (2019). *ACS Macro Lett.* 8: 1055–1062.

100 Li, H., Zhang, S., Jiao, J. et al. (2009). Controlled synthesis of polylactides using biogenic creatinine carboxylate initiators. *Biomacromolecules* 10: 1311–1314.

101 Coulembier, O., Josse, T., Guillerm, B. et al. (2012). An imidazole-based organocatalyst designed for bulk polymerization of lactide isomers: inspiration from nature. *Chem. Commun.* 48: 11695–11697.

102 Kadota, J., Pavlović, D., Hirano, H. et al. (2014). Controlled bulk polymerization of L-lactide and lactones by dual activation with organo-catalytic systems. *RSC Adv.* 4: 14725–14732.

103 Smith, E.L., Abbott, A.P., and Ryder, K.S. (2014). Deep eutectic solvents (DESs) and their applications. *Chem. Rev.* 114: 11060–11082.

104 del Monte, F., Carriazo, D., Serrano, M.C. et al. (2014). Deep eutectic solvents in polymerizations: a greener alternative to conventional syntheses. *ChemSusChem* 7: 999–1009.

105 Basterretxea, A., Gabirondo, E., Jehanno, C. et al. (2019). Polyether synthesis by bulk self-condensation of diols catalyzed by non-eutectic acid–base organocatalysts. *ACS Sustain. Chem. Eng.* 7: 4103–4111.

106 Moins, S., Loyer, P., Odent, J., and Coulembier, O. (2019). Preparation of a mimetic and degradable poly(ethylene glycol) by a non-eutectic mixture of organocatalysts (NEMO) via a one-pot two-step process. *RSC Adv.* 9: 40013–40016.

107 Coderre, D.N., Fastnacht, K.V., Wright, T.J. et al. (2018). H-bonding organocatalysts for ring-opening polymerization at elevated temperatures. *Macromolecules* 51: 10121–10126.

108 Zhang, Q., Hu, C., Duan, R. et al. (2022). A recyclable process between a monomer and polyester with a natural catalyst. *Green Chem.* 24: 9282–9289.

109 Kaoukabi, A., Guillen, F., Qayouh, H. et al. (2015). The use of ionic liquids as an organocatalyst for controlled ring-opening polymerization of ε-caprolactone. *Ind. Crop. Prod.* 72: 16–23.

110 Cheechana, N., Benchaphanthawee, W., Akkravijitkul, N. et al. (2021). Organocatalytic ring-opening polymerization of ε-caprolactone using *bis*(*N*-(*N*′-butylimidazolium)) alkane dicationic ionic liquids as the metal-free catalysts: polymer synthesis, kinetics and DFT mechanistic study. *Polymers* 13: 4290.

111 Xia, Y., Shen, J., Alamri, H. et al. (2017). Revealing the cytotoxicity of residues of phosphazene catalysts used for the synthesis of poly(ethylene oxide). *Biomacromolecules* 18: 3233–3237.

112 Casas, J., Persson, P.V., Iversen, T., and Córdova, A. (2004). Direct organocatalytic ring-opening polymerizations of lactones. *Adv. Synth. Catal.* 346: 1087–1089.

113 Mohammadzadeh-Aghdash, H., Sohrabi, Y., Mohammadi, A. et al. (2018). Safety assessment of sodium acetate, sodium diacetate and potassium sorbate food additives. *Food Chem.* 257: 211–215.

114 Li, F., Suzuki, R., Gao, T. et al. (2023). Alkali metal carboxylates: simple, efficient, and industrial relevant catalysts for controlled polymer synthesis. *Bull. Chem. Soc. Jpn.* 96: 1003–1018.

115 Takojima, K., Saito, T., Vevert, C. et al. (2020). Facile synthesis of poly(trimethylene carbonate) by alkali metal carboxylate-catalyzed ring-opening polymerization. *Polym. J.* 52: 103–110.

116 Xia, X., Suzuki, R., Gao, T. et al. (2022). One-step synthesis of sequence-controlled multiblock polymers with up to 11 segments from monomer mixture. *Nat. Commun.* 13: 163.

117 Gao, T., Xia, X., Tajima, K. et al. (2022). Polyether/polythioether synthesis via ring-opening polymerization of epoxides and episulfides catalyzed by alkali metal carboxylates. *Macromolecules* 55: 9373–9383.

118 Xia, X., Gao, T., Li, F. et al. (2022). Multidimensional control of repeating unit/sequence/topology for one-step synthesis of block polymers from monomer mixtures. *J. Am. Chem. Soc.* 144: 17905–17915.

119 Suzuki, R., Xia, X., Gao, T. et al. (2022). Synthesis of hyperbranched polyesters via the ring-opening alternating copolymerisation of epoxides with a cyclic anhydride having a carboxyl group. *Polym. Chem.* 13: 5469–5477.

120 Xia, X., Gao, T., Li, F. et al. (2023). Sequential polymerization from complex monomer mixtures: access to multiblock copolymers with adjustable sequence, topology, and gradient strength. *Macromolecules* 56: 92–103.

121 Ozen, C., Satoh, T., and Maeda, S. (2020). A theoretical study on the alkali metal carboxylate-promoted L-lactide polymerization. *J. Comput. Chem.* 41: 2197–2202.

122 Xie, H., Zheng, L., Feng, J. et al. (2023). Alkali metal carbonate catalyzed copolymerization of anhydrides and epoxides: a simple, efficient and versatile approach to well-defined alternating polyesters. *Polym. Chem.* 14: 1630–1638.

123 Bruice, T.C. and Fife, T.H. (1962). Hydroxyl group catalysis. III. 1 The nature of neighboring hydroxyl group assistance in the alkaline hydrolysis of the ester bond. *J. Am. Chem. Soc.* 84: 1973–1979.

124 Shih, C. (1995). Chain-end scission in acid catalyzed hydrolysis of poly(D,L-lactide) in solution. *J. Control. Release* 34: 9–15.

125 Karamanlioglu, M., Preziosi, R., and Robson, G.D. (2017). Abiotic and biotic environmental degradation of the bioplastic polymer poly(lactic acid): a review. *Polym. Degrad. Stab.* 137: 122–130.

126 Martin, R.T., Camargo, L.P., and Miller, S.A. (2014). Marine-degradable polylactic acid. *Green Chem.* 16: 1768–1773.

127 Rheinberger, T., Wolfs, J., Paneth, A. et al. (2021). RNA-inspired and accelerated degradation of polylactide in seawater. *J. Am. Chem. Soc.* 143: 16673–16681.

128 Kost, B. and Basko, M. (2021). Synthesis and properties of L-lactide/1,3-dioxolane copolymers: preparation of polyesters with enhanced acid sensitivity. *Polym. Chem.* 12: 2551–2562.

129 Cichoń, K., Bak-Sypien, I.I., Basko, M., and Kost, B. (2023). Synthesis and characterization of functionalized polylactides containing acetal units. *Macromolecules* 56: 6951–6967.

130 Shen, T., Chen, K., Chen, Y., and Ling, J. (2023). Ring-opening polymerization of cyclic acetals: strategy for both recyclable and degradable materials. *Macromol. Rapid Commun.* 44: 2300099.

131 Okada, M., Sumitomo, H., and Yamamoto, Y. (1974). Cationic polymerization of 6, 8-dioxabicyclo [3.2.1] octan-7-one. *Makromol. Chem.* 175: 3023–3028.

132 Okada, M., Sumitomo, H., and Atsumi, M. (1984). Specific formation of a polymer containing five-membered oxalactone rings in the main chain in the cationic ring-opening polymerization of 6, 8-dioxabicyclo [3.2.1] octan-7-one. *Macromolecules* 17: 1840–1843.

133 Boucher, D., Laviéville, S., Ladmiral, V. et al. (2024). Hemiacetal esters: synthesis, properties, and applications of a versatile functional group. *Macromolecules* 57: 810–829.

134 Otsuka, H. and Endo, T. (1999). Poly(hemiacetal ester)s: new class of polymers with thermally dissociative units in the main chain. *Macromolecules* 32: 9059–9061.

135 Neitzel, A.E., Haversang, T.J., and Hillmyer, M.A. (2016). Organocatalytic cationic ring-opening polymerization of a cyclic hemiacetal ester. *Ind. Eng. Chem. Res.* 55: 11747–11755.

136 Dirauf, M., Muljajew, I., Weber, C., and Schubert, U.S. (2022). Recent advances in degradable synthetic polymers for biomedical applications-beyond polyesters. *Prog. Polym. Sci.* 129: 101547.

137 Neitzel, A.E., Barreda, L., Trotta, J.T. et al. (2019). Hydrolytically-degradable homo-and copolymers of a strained exocyclic hemiacetal ester. *Polym. Chem.* 10: 4573–4583.

138 Hyoi, K., Kanazawa, A., and Aoshima, S. (2019). Cationic ring-opening co-and terpolymerizations of lactic acid-derived 1,3-dioxolan-4-ones with oxiranes and vinyl ethers: nonhomopolymerizable monomer for degradable co-and terpolymers. *ACS Macro Lett.* 8: 128–133.

139 Takebayashi, K., Kanazawa, A., and Aoshima, S. (2023). Cationic ring-opening copolymerization of cyclic acetals and 1,3-dioxolan-4-ones via the activated monomer mechanism and transacetalization reaction. *Macromolecules* 56: 5524–5533.

140 Zhang, X., Zhang, C., and Zhang, X. (2022). A facile and unprecedented route to a library of thermostable formaldehyde-derived polyesters: highly active and selective copolymerization of cyclic acetals and anhydrides. *Angew. Chem. Int. Ed.* 61: e202117316.

141 Zhang, X., Guo, W., Zhang, C., and Zhang, X. (2023). A recyclable polyester library from reversible alternating copolymerization of aldehyde and cyclic anhydride. *Nat. Commun.* 14: 5423.

142 Fahnhorst, G.W. and Hoye, T.R. (2018). A carbomethoxylated polyvalerolactone from malic acid: synthesis and divergent chemical recycling. *ACS Macro Lett.* 7: 143–147.

143 Fahnhorst, G.W., De Hoe, G.X., Hillmyer, M.A., and Hoye, T.R. (2020). 4-Carboalkoxylated polyvalerolactones from malic acid: tough and degradable polyesters. *Macromolecules* 53: 3194–3201.

144 Xu, S., Wang, Y., and Hoye, T.R. (2020). Poly(4-ketovalerolactone) from levulinic acid: synthesis and hydrolytic degradation. *Macromolecules* 53: 4952–4959.

145 Wang, X., Zhang, Z., and Hadjichristidis, N. (2023). Poly(amino ester)s as an emerging synthetic biodegradable polymer platform: recent developments and future trends. *Prog. Polym. Sci.* 136: 101634.

146 Myers, D., Cyriac, A., and Williams, C.K. (2016). To react the impossible ring. *Nat. Chem.* 8: 3–4.

147 Zhao, N., Ren, C., Li, H. et al. (2017). Selective ring-opening polymerization of non-strained γ-butyrolactone catalyzed by a cyclic trimeric phosphazene base. *Angew. Chem. Int. Ed.* 56 (42): 12987–12990.

148 Falconnet, A., Nicolas, M., Vollgraff, T. et al. (2023). Facile preparation of biodegradable poly(γ-butyrolactone) via base-assisted ring-opening polymerization. *Green Chem.* 25: 3624–3632.

149 Hong, M. and Chen, E.Y.-X. (2017). Chemically recyclable polymers: a circular economy approach to sustainability. *Green Chem.* 19: 3692–3706.

150 Tang, X. and Chen, E.Y.-X. (2019). Toward infinitely recyclable plastics derived from renewable cyclic esters. *Chem* 5: 284–312.

151 Coates, G.W. and Getzler, Y.D.Y.L. (2020). Chemical recycling to monomer for an ideal, circular polymer economy. *Nat. Rev. Mater.* 5: 501–516.

152 Shi, C., Reilly, L.T., Phani Kumar, V.S. et al. (2021). Design principles for intrinsically circular polymers with tunable properties. *Chem* 7: 2896–2912.

153 Xu, G. and Wang, Q. (2022). Chemically recyclable polymer materials: polymerization and depolymerization cycles. *Green Chem.* 24: 2321–2346.

154 Liu, Y. and Lu, X.-B. (2023). Emerging trends in closed-loop recycling polymers: Monomer design and catalytic bulk depolymerization. *Chem. Eur. J.* 29: e202203635.

155 Plummer, C.M., Li, L., and Chen, Y. (2023). Ring-opening polymerization for the goal of chemically recyclable polymers. *Macromolecules* 56: 731–750.

156 Li, Z., Shen, Y., and Li, Z. (2024). Ring-opening polymerization of lactones to prepare closed-loop recyclable polyesters. *Macromolecules* https://doi.org/10.1021/acs.macromol.3c01912.

157 Zhu, J.-B., Watson, E.M., Tang, J., and Chen, E.Y.-X. (2018). A synthetic polymer system with repeatable chemical recyclability. *Science* 360: 398–403.

158 Shi, C., Li, Z.-C., Caporaso, L. et al. (2021). Hybrid monomer design for unifying conflicting polymerizability, recyclability, and performance properties. *Chem* 7: 670–685.

159 Shi, C., Clarke, R.W., McGraw, M.L., and Chen, E.Y.-X. (2022). Closing the "one monomer–two polymers–one monomer" loop via orthogonal (de) polymerization of a lactone/olefin hybrid. *J. Am. Chem. Soc.* 144: 2264–2275.

160 MacDonald, J.P. and Shaver, M.P. (2016). An aromatic/aliphatic polyester prepared via ring-opening polymerisation and its remarkably selective and cyclable depolymerisation to monomer. *Polym. Chem.* 7: 553–559.

161 Lizundia, E., Makwana, V.A., Larrañaga, A. et al. (2017). Thermal, structural and degradation properties of an aromatic–aliphatic polyester built through ring-opening polymerisation. *Polym. Chem.* 8: 3530–3538.

162 Li, L.-G., Wang, Q.-Y., Zheng, Q.-Y. et al. (2021). Tough and thermally recyclable semiaromatic polyesters by ring-opening polymerization of benzo-thia-caprolactones. *Macromolecules* 54: 6745–6752.

163 Tu, Y.-M., Wang, X.-M., Yang, X. et al. (2021). Biobased high-performance aromatic–aliphatic polyesters with complete recyclability. *J. Am. Chem. Soc.* 143: 20591–20597.

164 Li, Z., Shen, Y., and Li, Z. (2022). Chemical upcycling of poly(3-hydroxybutyrate) into bicyclic ether–ester monomers toward value-added, degradable, and recyclable poly(ether ester). *ACS Sustain. Chem. Eng.* 10: 8228–8238.

165 Fan, H.-Z., Yang, X., Chen, J.-H. et al. (2022). Advancing the development of recyclable aromatic polyesters by functionalization and stereocomplexation. *Angew. Chem. Int. Ed.* 61: e202117639.

166 Bruckmoser, J., Remke, S., and Rieger, B. (2022). Ring-opening polymerization of a bicyclic lactone: polyesters derived from norcamphor with complete chemical recyclability. *ACS Macro Lett.* 11: 1162–1166.

167 Nelißen, M., Keul, H., and Höcker, H. (1995). Ring-closing depolymerization of poly(ε-caprolactone). *Macromol. Chem. Phys.* 196: 1645–1661.

168 Persenaire, O., Alexandre, M., Degée, P., and Dubois, P. (2001). Mechanisms and kinetics of thermal degradation of poly(ε-caprolactone). *Biomacromolecules* 2: 288–294.

169 Abe, H., Takahashi, N., Ju Kim, K. et al. (2004). Effects of residual zinc compounds and chain-end structure on thermal degradation of poly(ε-caprolactone). *Biomacromolecules* 5: 1480–1488.

170 Jung, M.E. and Piizzi, G. (2005). gem-Disubstituent effect: theoretical basis and synthetic applications. *Chem. Rev.* 105: 1735–1766.

171 Li, X.-L., Clarke, R.W., Jiang, J.-Y. et al. (2023). A circular polyester platform based on simple gem-disubstituted valerolactones. *Nat. Chem.* 15: 278–285.

172 Schneiderman, D.K., Vanderlaan, M.E., Mannion, A.M. et al. (2016). Chemically recyclable biobased polyurethanes. *ACS Macro Lett.* 5: 515–518.

173 Brutman, J.P., De Hoe, G.X., Schneiderman, D.K. et al. (2016). Renewable, degradable, and chemically recyclable cross-linked elastomers. *Ind. Eng. Chem. Res.* 55: 11097–11106.

174 Alberti, C. and Enthaler, S. (2020). Depolymerization of end-of-life poly(lactide) to lactide via zinc-catalysis. *ChemistrySelect* 5: 14759–14763.

175 Cederholm, L., Wohlert, J., Olsén, P. et al. (2022). "Like recycles like": selective ring-closing depolymerization of poly(L-lactic acid) to L-lactide. *Angew. Chem. Int. Ed.* 61: e202204531.

176 Nishida, H., Andou, Y., Watanabe, K. et al. (2010). Poly(tetramethyl glycolide) from renewable carbon, a racemization-free and controlled depolymerizable polyester. *Macromolecules* 44: 12–13.

177 Zhou, L., Zhang, Z., Shi, C. et al. (2023). Chemically circular, mechanically tough, and melt-processable polyhydroxyalkanoates. *Science* 380: 64–69.

178 Tu, Y.-M., Gong, F.-L., Wu, Y.-C. et al. (2023). Insights into substitution strategy towards thermodynamic and property regulation of chemically recyclable polymers. *Nat. Commun.* 14: 3198.

179 Nishida, H., Yamashita, M., Hattori, N. et al. (2000). Thermal decomposition of poly(1,4-dioxan-2-one). *Polym. Degrad. Stab.* 70: 485–496.

180 Nishida, H., Yamashita, M., Endo, T., and Tokiwa, Y. (2000). Equilibrium polymerization behavior of 1,4-dioxan-2-one in bulk. *Macromolecules* 33: 6982–6986.

181 Li, Z., Zhao, D., Huang, B. et al. (2022). Chemical upcycling of poly(3-hydroxybutyrate)(P3HB) toward functional poly(amine-alt-ester) via tandem degradation and ring-opening polymerization. *Macromolecules* 55: 9697–9704.

182 Zhao, D., Li, Z., Shen, Y., and Li, Z. (2023). Crystalline stereoregular poly(ether-ester) via MeAl [Salen]-catalyzed well-controlled ring-opening polymerization of enantiopure cyclic ether-ester monomer. *Macromolecules* 56: 6019–6026.

183 Guo, Y.-T., Shi, C., Du, T.-Y. et al. (2022). Closed-loop recyclable aliphatic poly(ester-amide)s with tunable mechanical properties. *Macromolecules* 55: 4000–4010.

184 Wang, Y., Li, M., Chen, J. et al. (2021). O-to-S substitution enables dovetailing conflicting cyclizability, polymerizability, and recyclability: dithiolactone vs. dilactone. *Angew. Chem. Int. Ed.* 60: 22547–22553.

185 Wang, Y., Zhu, Y., Lv, W. et al. (2023). Tough while recyclable plastics enabled by monothiodilactone monomers. *J. Am. Chem. Soc.* 145 (3): 1877–1885.

186 Yuan, J., Xiong, W., Zhou, X. et al. (2019). Tough while recyclable plastics enabled by monothiodilactone monomers. *J. Am. Chem. Soc.* 141: 4928–4935.

187 Shi, C., McGraw, M.L., Li, Z.-C. et al. (2020). High-performance pan-tactic polythioesters with intrinsic crystallinity and chemical recyclability. *Sci. Adv.* 6 (34): eabc0495.

188 Lai, H., Zhang, J., Xing, F., and Xiao, P. (2020). Recent advances in light-regulated non-radical polymerisations. *Chem. Soc. Rev.* 49: 1867–1886.

189 Doerr, A.M., Burroughs, J.M., Gitter, S.R. et al. (2020). Advances in polymerizations modulated by external stimuli. *ACS Catal.* 10: 14457–14515.

190 Chen, M. and Chen, C. (2020). Controlling the ring-opening polymerization process using external stimuli. *Chin. J. Chem.* 38: 282–286.

191 Yadav, N. and Chundawat, T.S. (2023). Recent developments in the UV-visible light-intervened ring-opening polymerization of cyclic esters. *Polym. Chem.* 14: 2435–2451.

192 Osaki, M., Takashima, Y., Yamaguchi, H., and Harada, A. (2009). Switching of polymerization activity of cinnamoyl-α-cyclodextrin. *Org. Biomol. Chem.* 7: 1646–1651.

193 Neilson, B.M. and Bielawski, C.W. (2013). Photoswitchable NHC-promoted ring-opening polymerizations. *Chem. Commun.* 49: 5453–5455.

194 Fu, C., Xu, J., and Boyer, C. (2016). Photoacid-mediated ring opening polymerization driven by visible light. *Chem. Commun.* 52: 7126–7129.

195 Dai, Z., Cui, Y., Chen, C., and Wu, J. (2016). Photoswitchable ring-opening polymerization of lactide catalyzed by azobenzene-based thiourea. *Chem. Commun.* 52: 8826–8829.

196 Eisenreich, F., Kathan, M., Dallmann, A. et al. (2018). A photoswitchable catalyst system for remote-controlled (co) polymerization in situ. *Nat. Catal.* 1: 516–522.

197 De Roover, Q., Vucko, T., Vincent, S.P. et al. (2021). Photocontrolled lactide ROP by the light-regulated release of potassium acetate from an azobenzene-bridged crown ether. *Cat. Sci. Technol.* 11: 6048–6052.

198 Zhang, X., Ma, Q., Jiang, Y. et al. (2021). Visible light-regulated organocatalytic ring-opening polymerization of lactones by harnessing excited state acidity. *Polym. Chem.* 12: 885–892.

199 Hu, S., Chen, J., Zhang, X. et al. (2022). Visible light-regulated organocatalytic ring-opening polymerization of lactones using hydroxybenzophenones as photocatalyst. *ACS Appl. Polym. Mater.* 4: 3361–3368.

200 Liu, Z., Zhang, X., Sun, P. et al. (2023). Harnessing the photo-acidity of organic dyes for the development of ring-opening polymerization of lactones under visible light. *Polym. Chem.* 14: 4414–4418.

2

Organocatalyzed Copolymerization of CO_2 with Epoxide Toward Polycarbonate Synthesis

Xiaoshuang Feng and Yves Gnanou

King Abdullah University of Science and Technology (KAUST), Physical Sciences and Engineering Division, Thuwal 23955, Saudi Arabia

2.1 Introduction

Since the first report by Inoue on CO_2/epoxide copolymerization [1], considerable efforts have been devoted to this research area in order to make use of CO_2, a thermodynamically stable and over-released greenhouse gas that can generate valuable polymeric materials upon copolymerization [2–4]. Many catalysts have been developed that are either hetero- or homogeneous organometallic complexes mostly based on transition metals or main group metals such as Zn [5–7], Cr [8], Co [9], Mg [10, 11], Al [12], etc.; the most efficient among them brought about CO_2/epoxides copolymerizations with very high activities through a coordination-insertion mechanism. The related works on organometallic copolymerization of CO_2 with epoxides were well-reviewed by Coates, Darensbourg, Williams, *et al.* [13–16] and recently by Zhang, Meng, Kozak, *et al.* [17–20]. Despite their success in CO_2/epoxide copolymerization, these metal-based catalysts presented some disadvantages: (i) the complexity of their structures with ligands which necessitated multistep synthesis; (ii) the contamination of the produced polycarbonates by residual metals, especially in the case of homogeneous transition metal-based complexes, which required a post-polymerization metal/removal step due to their color and often their toxicity [21–24]. In this context, achieving CO_2/epoxides copolymerization under metal-free conditions was a breakthrough as it addressed the above issues. In 2016, relying on mere oxyanionic growing chains and not on coordinated metal-based complexes, we reported the copolymerization of CO_2/epoxides carried out in the presence of triethylborane (TEB) associated with onium salts [25]. Such a totally different copolymerization approach resorting to metal-free conditions triggered the interests of many researchers, entailing attention to this research topic. Up to now, almost all organocatalytic CO_2/epoxides copolymerizations that were developed are based on boron systems. As shown in Figure 2.1, many epoxide monomers could be copolymerized with CO_2 catalyzed by trialkylborane compounds. Borane-mediated (co)polymerization was recently completely reviewed by Zhang, Nauman, and us

Organocatalysts in Polymer Chemistry: Synthesis and Applications, First Edition. Edited by Zhibo Li.

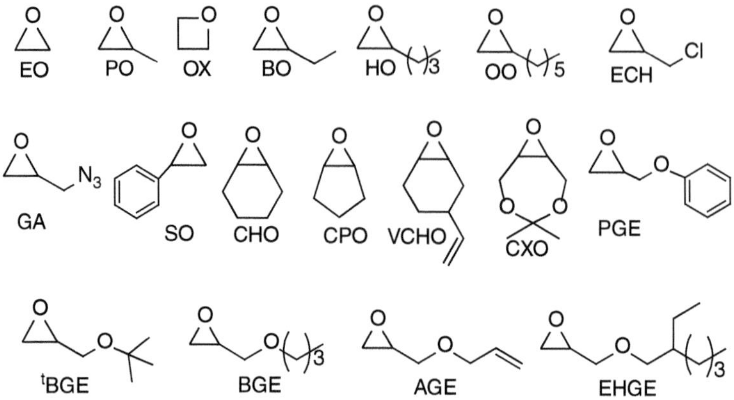

Figure 2.1 Reported organocatalytically copolymerizable epoxide monomers with CO_2.

[26, 27]; some of the previous works on organocatalytic preparation of polycarbonates and polyesters were also reviewed by Li and coworkers [28]. In this chapter, we focus on the recent progress of organocatalytic CO_2/epoxides copolymerization based on alkylborane, including the discovery and development of CO_2/epoxides copolymerization catalyzed by TEB, the synthesis of polycarbonate-based block copolymers. Lastly, the most recent works related to bifunctional organoboron catalysts are also discussed.

2.2 Discovery of TEB Catalyzed CO_2/Epoxides Copolymerization

In the pursuit of a new and convenient CO_2/epoxide copolymerization method, the first thought that came to our minds was the anionic ring-opening copolymerization approach, which is widely utilized for polyether synthesis [29]. Some epoxides such as PO are prone to side reactions such as transfer and, therefore, cannot be polymerized under living conditions, preventing the formation of high molar mass samples. The solution to this issue has been to resort to Lewis acid, such as tributylaluminum (TiBA), which activates epoxide on the one hand, and forms ate complexes with the growing oxyanions on the other hand, thus resulting in an enhanced propagation and elimination of transfer reactions [30]. When attempting to anionically copolymerize CO_2 with epoxides, besides the two cross-addition reactions of CO_2 to alkoxide and epoxide to carbonate anions, the system is actually mainly dominated by backbiting reactions that produce cyclic carbonates. Upon using Lewis acid that can cap the growing anions – both alkoxides and carbonates – and thus generate ate complexes of lower reactivity, backbiting reactions could be prevented. To favor propagation, an excess of Lewis acid is necessary to activate specifically epoxide monomer (Figure 2.2). Under these conditions, reactions (5) and (6) can be averted. Upon using a judiciously chosen Lewis acid to activate epoxide monomers, the ate complexes $RO^-(LA)Z^+$ and $RCO_3^-(LA)Z^+$ can cross propagate ($k_{p1,2}$, $k_{p2,1}$) and even

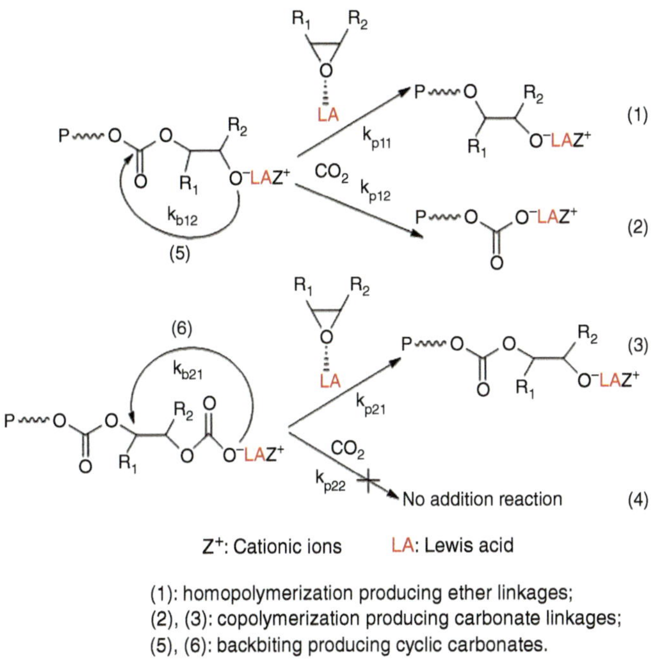

Figure 2.2 Reactions involved in anionic copolymerization of CO_2 and epoxides [25].

propagate for $RO^-(LA)Z^+$ ($k_{p1,1}$), but would be prevented from backbiting ($k_{b1,2}$ and $k_{b2,1}$). To favor the formation of polycarbonates over polyether, conditions and, in particular, the amount of Lewis acid should be adjusted to favor $k_{p1,2}$ over $k_{p1,1}$.

Following this idea, we first reported a novel CO_2/epoxide copolymerization system that occurred through classical anionic polymerization in contrast to earlier works, which were all based on conventional coordinated mechanisms. In this work, using lithium alkoxides or lithium salts as an initiator associated with a commercially available Lewis acid, TiBA, the alternating copolymerization of CO_2 with CHO could be achieved with a high linear vs. cyclic selectivity [31]. Based on the characterization results of the polycarbonates obtained, the copolymerization proceeded through an anionic mechanism and exhibited a "living" character. In addition, since the counterion used in such anionic copolymerization was lithium, block copolymers of polystyrene-*b*-polycarbonate type could be easily obtained by one-pot sequential polymerization of the respective monomers starting from sec-BuLi as initiator. In the presence of a strong Lewis acid such as TiBA, which is known to activate the epoxide monomers, a tight ate complex of low reactivities formed between the growing oxyanions and TiBA which is associated with a small size of lithium cation. Such ate complexes exhibited a high preference for copolymerization ($k_{p1,2}$) over homopolymerization ($k_{p1,1}$), ensuring the formation of linear poly(cyclohexene carbonate)s ($k_{p1,2} > k_{p1,1}$ and $k_{p1,2} > k_{-p1,2}$) with low content in ether linkages and high alternating structures. Although low turnover numbers (TON) in the range of around 50 were observed, the successful alternating

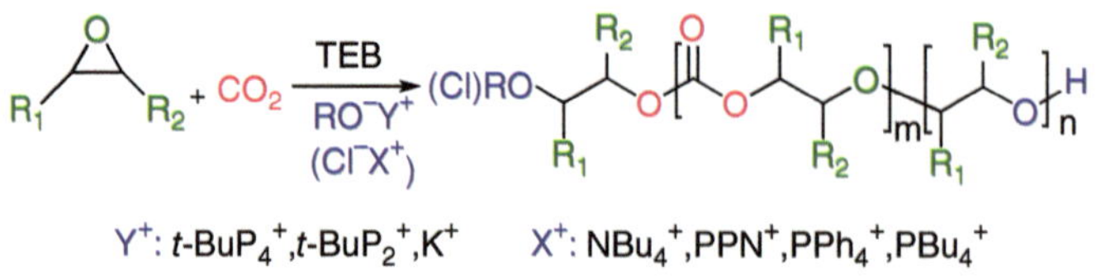

Figure 2.3 The anionic copolymerization of epoxides and CO_2 using the catalyst system of TEB combined with an onium salt or organic base. Source: Zhang et al. [25]/American Chemical Society.

copolymerization of CO_2 with CHO when initiated by the Li^+/TiBA system proves that the concept shown in Figure 2.2 works. In other words, the anionic copolymerization of CO_2 and epoxides carried out in the presence of Lewis acid produced alternating polycarbonates in a similar way to coordinating organometallic systems. To address the low copolymerization turnover number of such lithium-TiBA system and its inability to bring about the copolymerization of CO_2 with PO, other initiating systems comprising much bulkier cations than lithium were screened and other Lewis acids were tried. When bulkier cations were associated with oxyanions, the reactivity of the latter propagating species was dramatically enhanced due to the weak interaction between the growing anions and the cations. Concomitantly, a milder Lewis acid than TiBA capable of activating epoxides and at the same time forming loose ate complexes with the growing alkoxides was chosen. Following this reasoning (Figure 2.2), in 2016 we discovered a new system for the copolymerization of CO_2 with two common epoxides of PO and CHO (Figure 2.3) [25]. As bulky cations, those of onium salt were selected, and milder Lewis acid TEB was associated with the growing alkoxides and served as an epoxide activator through boron–oxygen interaction. The copolymerization was conducted under mild conditions (2 eq. TEB to the initiator, 10 bars of CO_2, 60 ~ 80 °C) and exhibited high activity and selectivity. In the case of PO, the TON values were as high as 490. The molar masses (M_n) of the copolymers were in the range of 4.7 ~ 50.0 kDa with narrow dispersity ($Đ = M_w/M_n < 1.20$). The linear/cyclic selectivity was >90%, and the carbonate contents were up to 99% in the produced polycarbonates. In the case of CHO, TON values jumped close to 4000. Perfectly alternating PCHC with M_n up to 76.4 kDa were obtained with negligible formation of cyclic carbonates. Later on, with purified CO_2 by TiBA, the catalytic activity could be significantly increased, the TON values under optimized conditions being equal to 9280 in the case of CHO, and 3840 in the case of PO [32].

The key roles played by TEB were elucidated by density functional theory (DFT) calculations [33]. Four steps were studied: step I corresponds to epoxide ring-opening by the chloride anion in the presence of TEB; step II is related to the subsequent insertion of CO_2; step III shows the alternating insertion of an epoxide facilitated by TEB; step IV characterizes the leaving of TEB followed by a new round of polymerization. As an example of the copolymerization of CO_2 and CHO initiated by a chloride anion whose theoretical profile is shown in Figure 2.4, CHO is activated by TEB, followed by its ring-opening by the chloride anion. Then, the insertion

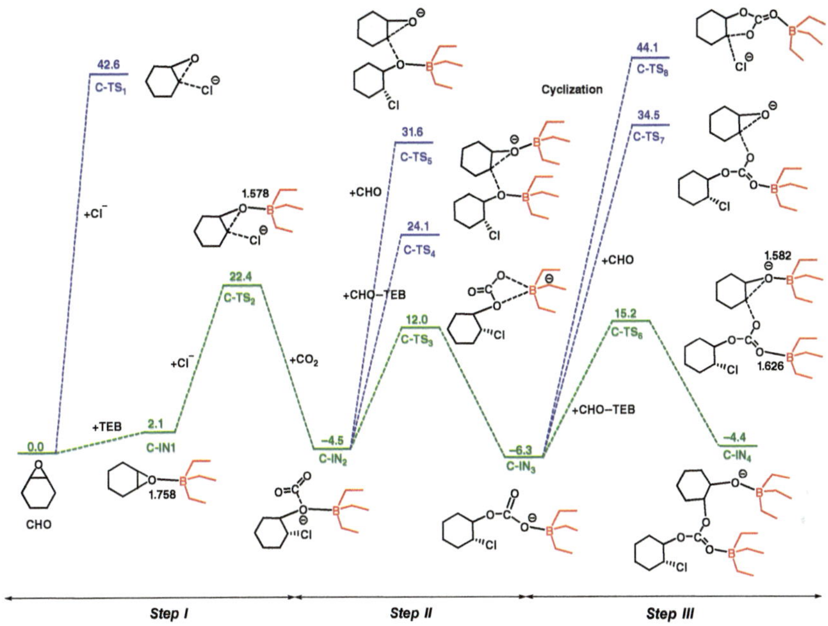

Figure 2.4 DFT calculation model of CHO/CO_2 copolymerization (kcal mol^{-1}). Source: Zhang et al. [33]/American Chemical Society.

of CO_2 occurs, giving rise to an alternating structure. Another TEB-activated CHO is then ring-opened and inserted at the chain end. After the anion migrates to the just ring-opened CHO, the initial TEB then moves and activates another CHO for the next catalytic round. Possible competitive reactions (blue lines) are found to be unfavorable because of the high energy barriers. The calculations also suggest that 2 eqs of TEB with respect to the initiator are needed while less of TEB will result in higher TS energies and thus hinder the copolymerization steps. The CHO model under similar reaction conditions shows lower energy barriers than those of the PO model, in agreement with the enhanced reactivity for the CHO and CO_2 copolymerization. The remarkable decrease of barriers in the presence of TEB confirms the epoxide activation role played by the Lewis acid.

It was the first report describing the preparation of polycarbonates by copolymerization of CO_2 with epoxides under metal-free conditions and with comparable activities to the established organometallic systems. In comparison with the latter systems, TEB is commercially available and does not need multistep synthesis, unlike organometallic systems. In addition, the produced polycarbonates are colorless, free of metal residues, and thus could well be used without further purification in areas such as food package, coating, biological and electronic fields. In addition, the copolymerization is carried out through "living" anionic approach which could be applied to the design and synthesis of CO_2-based copolymers with different structures in combination with copolymerization of other types of monomers and polymerization methods.

2.3 Development of TEB Catalyzed CO_2/Epoxides Copolymerization

After succeeding in copolymerizing CO_2 with two common epoxides, PO and CHO, we investigated the feasibility of TEB-mediated copolymerizations of CO_2 with other oxirane monomers including alkyl ether or allyl substituted ones such as BO, HO, OO, BGE, AGE, and EHGE [34]. Due to the steric hindrance of their substituents, the reactivities of the monomers such as BO, HO, and OO were lower than that of PO; the presence of ether function or double bonds further lowers the reactivity of monomers such as BGE, EHGE, and AGE, because of possible competitive interactions between TEB with their ether oxygen and double bonds. However, with initiators including bulkier cations such as phosphoniums from PPNCl or from t-BuP_4, alternating polycarbonates could be obtained for all the above oxirane monomers in the presence of 3 eq. TEB to the initiator, the M_n of these polycarbonates reached up to 152 kDa and the linear/cyclic selectivities were above 95%.

The case of EO must be distinguished from all other oxirane monomers whose poly(ethylene oxide) (PEO) is hydrophilic, but the copolymerization of EO produces poly(ethylene carbonate) (PEC) which is hydrophobic. When investigating the copolymerization of EO with CO_2 we found that it proceeded very fast; such copolymerization exhibits the highest activity among all the epoxide monomers [35]. Instead of using 2 eq. of TEB with respect to the initiator as in the case of PO, poly[(ethylene carbonate)-co(ethylene oxide)] (PECEO), samples with carbonate contents above 90% could be obtained under a CO_2 pressure of 10–30 bar with just 1–1.2 eqs. TEB with respect to the initiator in tetrahydrofuran or hexane. Upon varying the pressure of CO_2 or manipulating the amount of TEB respected to the initiator, the carbonate content could be fine-tuned from above 90% to below 2%, affording an array of obtained copolymer samples with hydrophobic, amphiphilic, and hydrophilic properties. Importantly, when carbonate content decreases to below 10 mol%, PECEO exhibits "PEO" like properties such as high crystallinity, wettability, and also degradability, unlike regular PEO. One PEOEC sample of molar mass 208 kg mol^{-1} with 4.4 mol% of carbonate content was selected for degradation evaluation under different conditions: (i) pH = 8.5 at 25 °C (commonly used in surface modification, cell culture, enzyme assay, and electrophoretic applications); (ii) pH value = 7.4 at 37 °C (mimic conditions in body fluid); (iii) pH = 6.5 at 37 °C (mimic the slightly acidic microenvironment of tumor). The degradation rates are following in this order: basic > neutral > acidic, in all cases, after one month the molar masses of the initial PEOEC sample after degradation could be remarkably degraded into thousands of g/mol which can be excreted from the body without any concerns (the renal threshold for a complete excretion is below 40 kg mol^{-1}) [36].

Zhang et al. tested the copolymerization behavior of epoxides bearing a phenyl group, i.e. phenyl glycidyl ether (PGE) and styrene oxide (SO) with CO_2 under similar experimental polymerization conditions. Due to the large size of their electron-withdrawing phenyl substituent, PGE and SO lend themselves to the formation of cyclic carbonates by backbiting. Upon using more TEB (6 and 8 eq. of TEB to the initiator for PGE and SO), propagation could be favored over backbiting and a high linear vs. cyclic selectivity could be achieved. Such phenyl-contained polycarbonates were obtained as colorless solids with a high refractive index (n_d) of 1.55 ~ 1.56 (590 nm, 20 °C, cast film), which are promising materials for optical applications [37]. In the meantime, we investigated the polymerization of another challenging monomer, glycidyl azide (GA), in the presence of TEB [38]. GA, long regarded as a nonpolymerizable monomer either cationically or anionically as being prone to transfer and elimination due to the acidic character of its hydrogens carried by its β-carbon [39–41], could be homopolymerized in the presence of TEB. TEB not only helped homopolymerize GA to form poly(glycidyl azide) ether but also to copolymerize it with CO_2 and generate alternating poly(glycidyl azide carbonate)s. However, high molar mass PGAC could not be obtained because of transfer side reactions that could not totally be suppressed. Nevertheless, a controlled copolymerization could be accomplished up to 11 kDa with the synthesis of well-defined samples of poly(glycidyl azide carbonate). The terpolymerization of GA, PO (or CHO), and CO_2 was also investigated, and these polycarbonate copolymers carrying reactive azido groups could be easily derived by click chemistry [38].

In addition to its applicability to the copolymerization of CO_2 with most epoxide monomers as described above, such TEB-mediated copolymerization system also works for four-membered cyclic ethers and oxetane. Indeed, oxetane whose ring strain is lower than that of oxiranes (112 vs. 106 kJ mol^{-1} for oxirane and oxetane) [42] exhibits a rather low reactivity. Trimethylene carbonate (TMC), a six-membered cyclic carbonate monomer, is widely used for the preparation of biodegradable poly(trimethylene carbonate) (PTMC) through ring-opening polymerization (ROP) [43]. One generally resorts to phosgene or its derivatives to synthesize TMC. In contrast, the direct copolymerization of oxetane and CO_2 is seldom used to produce PTMC due to the low reactivity of oxetanes, especially in the presence of organocatalysts [42, 44–48]. Alves, et al. first reported a metal-free oxetane/CO_2 coupling using tetrabutylammonium iodide as a catalyst and a hydrogen bond donor (HBD), namely 1,3-bis-(hydroxyhexafluoroisopropyl) benzene (1,3-bis-HFAB) as an activator of oxetane. These authors could obtain an oligomeric α,ω-dihydroxyl PTMC at 130 °C with 98% linear vs. cyclic selectivity [47]. On the other hand, Dove, Coulembier, and their colleagues used iodine as an activator in combination with organic superbases, such as 1,5,7-triazabicyclo[4.4.0]dec-5-ene (TBD), for direct copolymerization of oxetane with CO_2 at 105 °C, which afforded PTMC with molar mass around 6.0 kg mol^{-1} and 89–99% linear vs. cyclic selectivity [48]. Shortly after, the same group used an $I_2/CH_3CO_2(NBu_4)$ binary catalyst system for the coupling of CO_2 with oxetane and observed that either TMC or PTMC could be selectively produced upon varying the temperature. As shown in Figure 2.5a, cyclic carbonate was obtained at 55 °C, whereas changing the reaction temperature to 105 °C yielded PTMC with a molar mass of up to 6.4 kg mol^{-1} [42].

Instead of the abovementioned HBD or iodine used as an activator of oxetane, TEB or tributyl borane (TBB) was found to promote high activity when associated with an onium salt for the selective coupling of CO_2 with oxetanes [44]. In contrast

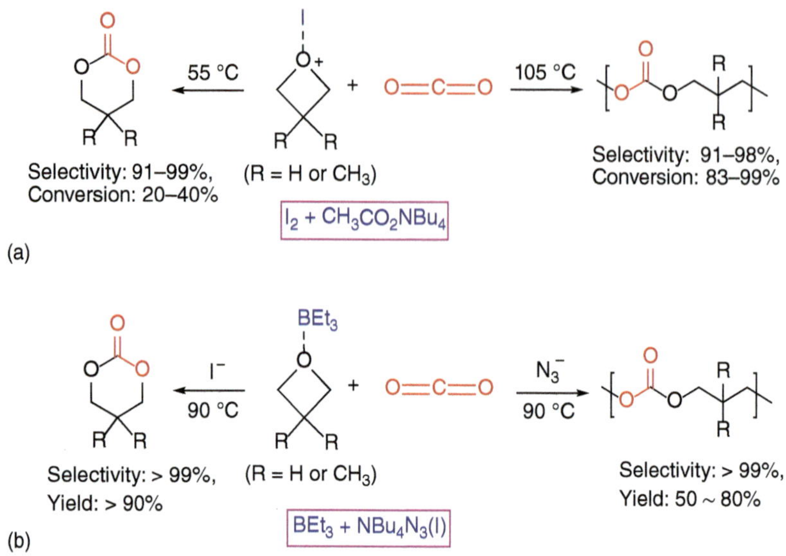

Figure 2.5 Organocatalytical selective coupling of oxetanes with CO_2 using (a) I_2/tetrabutylammonium acetate. Source: Huang et al. [42]/American Chemical Society.; (b) triethylborane and halide or azido tetrabutylammonium salts. Source: Zhang et al. [44]/American Chemical Society.

with onium iodide used as an initiator in the presence of TEB, only cyclic TMC was obtained at 90 °C as shown in Figure 2.5b; on the other hand, with onium chloride or onium azide as initiator, only linear polycarbonates were generated at 90 °C in the presence of TEB. The selectivity was determined by the type of anions of the onium salts used. As a good leaving group, iodide favors backbiting reaction and the formation of TMC, whereas the poor leaving groups such as chloride and azide allow the propagation to occur and the growth of PTMC with molar mass up to 8.0 kg mol^{-1}. Such processes also work for the coupling of substituted oxetane, namely 3,3-dimethyl oxetane with CO_2 which respectively afforded 5,5-dimethyl-1,3-dioxane-2-one and polycarbonates with high selectivity. Upon monitoring the kinetics of such reaction by *in situ* IR, this boron-mediated coupling process was shown to occur through direct copolymerization of oxetane with CO_2, unlike reported salen metallic complexes [45, 46, 49], or other organocatalytic systems [42, 47, 48] for which linear PTMCs were obtained as major products through the concomitant ROP of in situ formed TMC.

When onium chloride salts were used as initiators, the polycarbonates obtained exhibited only one terminal hydroxyl group. It is known that one promising application for the production of polyurethanes is low molar mass polycarbonate diols, whose performances are deemed superior compared to conventional polyether and polyester diols [50–54]. Unlike onium chloride salts or alkoxide salts, carboxylic acids are very common compounds either as natural or synthetic products. After observing that onium carboxylates behave very similar to onium chloride when used as initiators for the copolymerization of CO_2 with epoxides in the presence of

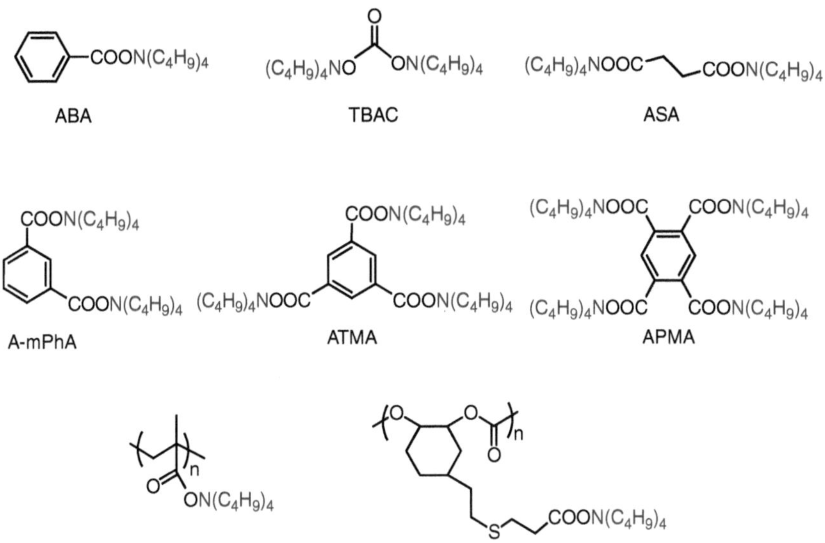

Figure 2.6 Multifunctional tetrabutylammonium carboxylates as initiators for the preparation of polycarbonate polyols. Source: Patil et al. [55] and Feng and coworkers [56].

TEB, we took advantage of the availability of carboxylic acids with functionalities varying from 1 to 2 to 3 and more to prepare hydroxyl-terminated polycarbonates with various functionalities [55].

Upon neutralization with tetrabutylammonium hydroxide, these carboxylic acids turned out to be excellent initiators for the abovementioned copolymerizations (Figure 2.6). Well-defined polycarbonate polyols were successfully synthesized using multifunctional initiators, the M_n varying from 1 to 10 kDa, and the terminal hydroxyl functionalities from 1 to 4. In addition, by changing the pressure of CO_2 from 10 to 2 bar, the carbonate content could be tuned from 95% to 50%. Considering the large amounts of initiators and TEB that were necessary for the preparation of low molar mass polycarbonate polyols, a recycling process of TEB and the cation was thus developed as an answer to the question of the high cost of polyol production [57].

Zhang and coworkers also disclosed a zwitterionic mechanism for the copolymerization of CO_2 with PO [58]. In this system, the ammonium cations were generated *in situ* by ring-opening of epoxide by tertiary amine TEA, and the *in situ* formed cations functioning like tetrabutylammonium phosphonium halide or alkoxide salts used as the initiator for the copolymerization of PO with CO_2, generating an end-to-end zwitterion made of a TEB-masked anion and an onium cation (Figure 2.7). This *in situ* generated zwitterion catalyst system is highly effective, quite similar to bifunctional borane systems discussed in subsequent sections, and does not need 2 eq. of borane as required in previous reports, affording alternating PPC of molar mass up to 56.0 kg mol^{-1} with feeding ratio of TEA/TEB/PO (1/1/400) and productivity of 171 g PPC per gram catalyst. This mechanism was supported by DFT

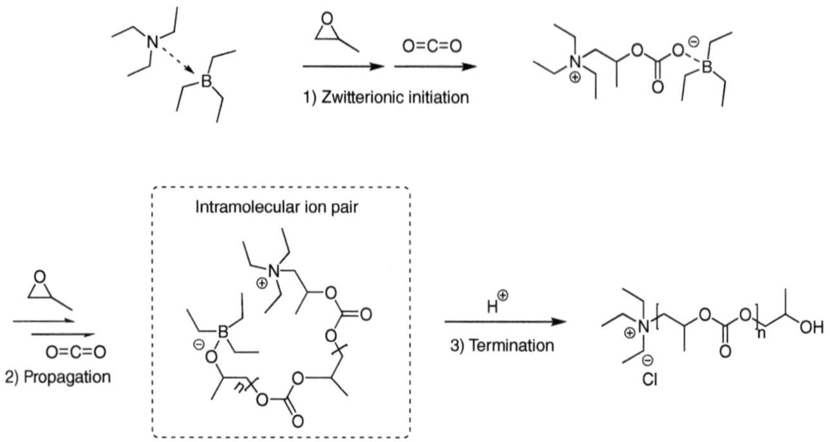

Figure 2.7 Proposed mechanism for the TEB/TEA catalyzed zwitterionic CO_2/PO copolymerization. Source: Wang et al. [58]/American Chemical Society.

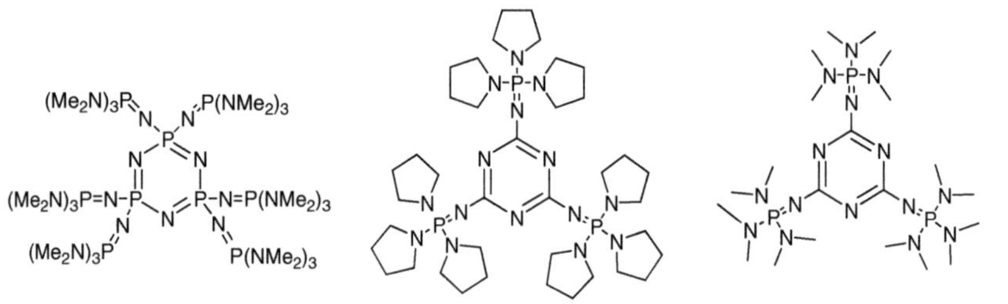

CTPB | C_3N_3-Py-P_3 | C_3N_3-Me-P_3

(p*K*a = 33.3 in CD3CN, φ = 1.4 nm) (p*K*a = 26.5 in CD3CN, φ = 1.3 nm) (p*K*a = 25.7 in CD3CN, φ = 1.1 nm)

Figure 2.8 Cyclic phosphazene organobases used for copolymerization of CO_2 with epoxides with TEB. Source: Zhang et al. [60]/John Wiley & Sons.

calculations and the characterization results of the obtained PPCs which featured terminal triethylammonium and hydroxyl groups.

Besides traditional onium salts such as tetrabutylammonium, bis(triphenylphosphine)iminium (PPN) phosphonium, and *t*-BuP_4 phosphazenium which served as an initiator in the presence of TEB, Li, Liu, and coworkers synthesized cyclic trimeric phosphazene bases [59], which they used as initiators along with TEB for the copolymerization of CO_2 and epoxides [60]. Among the phosphazenes (CTPB, C_3N_3-Py-P_3, and C_3N_3-Me-P_3), as shown in Figure 2.8, it was found that C_3N_3-Py-P_3 with a relatively low basicity (pK_a = 26.5 in CD_3CN) and a large molecular size (φ = 1.3 nm) exhibited the highest catalytic activity and selectivity. In sharp contrast, the basicity of CTPB was too strong, which resulted in a tighter adduct accompanied by neutralization or reduction of their individual parent precursors. When tested in CO_2/CHO copolymerization under 1 MPa and 80 °C, a TON value of 12,240 was obtained with the C_3N_3-Py-P_3/TEB system. The highly active C_3N_3-Py-P_3/TEB system could efficiently copolymerize CO_2 with CHO under 1 atm pressure and at room temperature with TOF up to 95 h^{-1}.

Overall, metal-free TEB-based systems including onium salts exhibit wide applicability and high selectivity and activity with respect to the copolymerization of epoxides with CO_2. In comparison to the state of art organometallic systems used to copolymerize CO_2/PO (Table 2.1), commercially available reagents of NBu_4Cl/TEB afford PPC with a productivity of 768 g polycarbonate per gram catalyst, exhibiting high industrialization potential, especially for high molar mass polycarbonate preparation.

2.4 Synthesis of CO_2-Based Block Copolymers

In comparison to traditional bisphenol A-based aromatic polycarbonates, aliphatic CO_2-based polycarbonates generally suffer from poor thermal and mechanical properties. To address this issue, attempts were made to copolymerize epoxides with CO_2 along with other monomers to produce both random and blocky CO_2-based copolymers. TEB-based Lewis pair systems were demonstrated to work not only for CO_2/epoxides copolymerization but also for the "living" (co)polymerizations of epoxides/COS [62], epoxides/anhydrides [63–65], epoxides/isocyanates [66], cyclic esters [67], and epoxides/dialdehyde [68]. The livingness, versatility, and compatibility of TEB based systems with oxygenated monomers allowed the synthesis of block copolymers with different structures in combination with other polymerization methods.

2.4.1 One-Step Ter-, Quater-Polymerization

Taking advantage of the versatility of TEB-based systems, other monomers, as shown in Figure 2.9, could be ter- or quater polymerized along with CO_2/epoxides, affording copolymers including polycarbonates, polyesters, and others. Upon feeding the reaction medium on monomers, including CO_2 either sequentially, gradually, or at once, CO_2-based random, gradient, or quasi-block copolymers could be obtained.

Meng and coworkers systematically investigated the effects of used Lewis pairs on the microstructures of poly(carbonate-ester) copolymers obtained upon ter- or quater-polymerization of CO_2 with epoxides and anhydrides. Through quarter-copolymerization of CO_2/CHO/PO/PA initiated by TEB and bis(triphenylphosphine)iminium chloride (PPNCl) Lewis pair, diblock, and triblock copolymers with little tapering could be obtained. Quadripolymers exhibiting high T_g (86–115 °C) and tensile strength (up to 54.8 MPa) were shown to possess thermal, mechanical properties, and transparency similar to polystyrene [69]. With Lewis pair made of TEB and tertiary amine triethylamine (TEA), random copolymers of molar mass up to 130 kg mol^{-1} were obtained with high selectivity and a productivity of 1.2 kg polymer per gram catalyst. Such TEB/TEA Lewis pair was shown to work for the random copolymerization of CO_2 with EO, BO, CHO epoxides, and PA, NBA anhydrides [70]. On the other hand, an

Table 2.1 The productivity comparison of typical catalyst systems for CO_2/PO copolymerization.

Catalyst system	Structure	Mn (kg mol^{-1})	Reaction conditions	Productivity (g/g)a	Refs.
Zinc β-diiminate	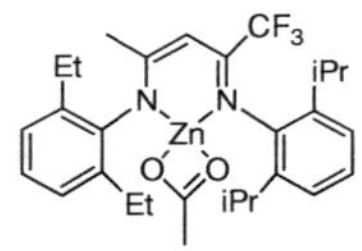	35.0	25 °C 0.7 MPa	84	2002 [5]
Bifunctional (salen)Co complex	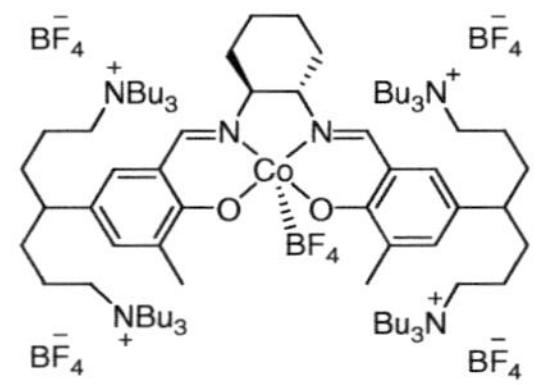	285	80 °C 2.0–1.7 MPa	761	2008 [60]
Bifunctional (porphyrin) Al complex	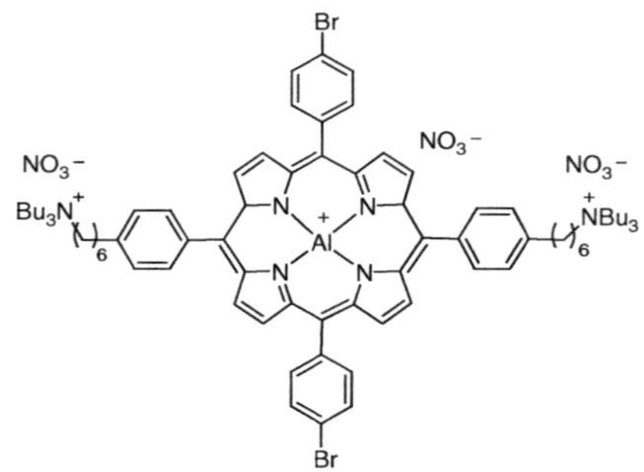	96	80 °C 3.0 MPa	376	2014 [61]
NBu_4Cl/TEB	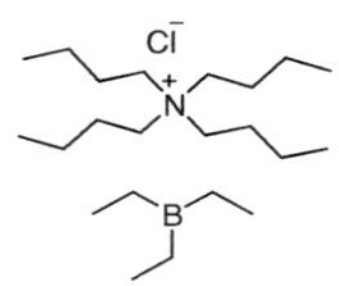	50.0 203	60 °C 1.0 MPa	105 768	2016 [25] 2019 [32]

a) Productivity = g polymer per gram catalyst.

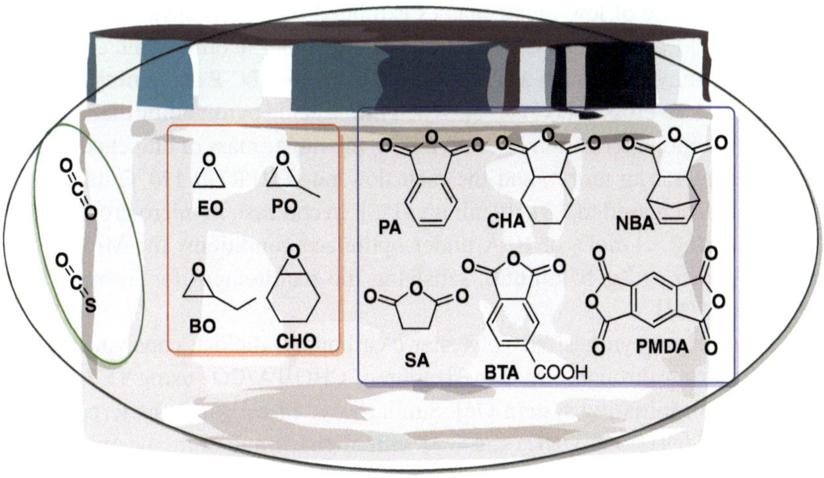

Figure 2.9 Reported comonomers: COS, epoxides, anhydrides in organocatalytic ter- or quater-polymerization with CO_2 in one step.

organic base such as 7-methyl-1,5,7-triazabicyclo[4.4.0] decane-5-ene (MTBD) or 1,8-diazabicyclo[5.4.0]undec-7-ene (DBU) coupled with TEB, gave rise to poly(ester-*co*-carbonate) tapered copolymers upon copolymerizing CO_2, PO, and PA with properties similar to those produced by the TEB/PPNCl pair [71]. The incorporation of PA units was found to efficiently enhance the thermal properties as shown by the higher T_g value (47 °C) of the copolymer compared to commercial PPC with 43 mol% of polyester incorporation; a faster rate of biodegradation was observed compared to commercial PBAT. The same group used TEB/tetrabutyl ammonium m-phthalate (A-mPhA) pair to prepare various copolymers with random/gradient/block structures, such as random CHO/EO/CO_2 and PO/EO/CO_2 terpolymers, gradient PA/EO/CO_2 terpolymers and triblock PEC-PCHC-PEC copolymers including poly(ethylene carbonate) segments by a tandem strategy [72].

The chemoselectivity of the Lewis pair made of TEB/DBU (MTBD) for PO/anhydrides over CO_2/PO copolymerization was also confirmed by Zhao et al. [73]. In this more recent work, the authors screened different organic bases and found that the chemoselectivity negatively correlated with the strength of the bases used. The content of ester-carbonate linkage decreased from 42% in the case of *t*-BuP_4 (p*K*a = 42.7 in MeCN), to 41% in the case of *t*-BuP_2 (p*K*a = 33.5 in MeCN) and 31% in the case of *t*-BuP_1 (p*K*a = 26.9 in MeCN). In the case of DBU and MTBD (p*K*a = 24.3 in MeCN) associated with TEB, no junction of ester and carbonate linkages was detected, and the copolymer produced upon terpolymerizing of PO/PA/CO_2 was blocky in nature. Notably, the same authors derived hyperbranched topologies with more terminal functionalities upon using 1,3,4-benzene tricarboxylic anhydride (TBA) as inimer and terpolymerizing CO_2 with a library of epoxides and anhydrides. In the presence of TBA, the molar mass of obtained copolymers was moderate in the range of 32–86 kg mol^{-1} due to its low reactivities. The mechanical properties were not provided in this report.

To address the issue of low molar masses exhibited by poly(propylene carbonate phthalate) (PPC-P) copolymers, Meng and coworkers on the other hand employed cross-linkable anhydrides such as PMDA, NBA to create PPC-P containing branching points or micro-cross-links [74, 75]. When introducing pyromellitic dianhydride (PMDA) into terpolymerization of PO/PA/CO_2, the molar mass of branched PPC-P increased up to 156 kg mol^{-1}, and the melt flow rate (MFR) at 170 °C under the load of 2.16 kg decreased to 7.3 g/10 minutes [75]. In contrast, in micro-cross-linked PPC-P containing <1 mol% of NBA under optimized conditions, the MFR could be tailored down to 2 g/10 minutes, satisfying the requirement for foaming and blowing PPC-P [74].

Li and coworkers synthesized poly(ester-*b*-carbonate) diblock copolymers with very little tapering through terpolymerization of CHO/PA/CO_2 using TEB/PPNCl Lewis pair as an initiating system [76]. Similarly, we investigated the terpolymerization of epoxides (PO or CHO) and CO_2 with aliphatic and aromatic anhydrides (SA, PA) respectively [77]. The aliphatic anhydride (SA) exhibited a much higher preference for epoxides than the aromatic anhydride (PA) during terpolymerization, leading in the former case to the formation of block copolymers with little tapering. In the case of PA, random to-block copolymers could be obtained by varying the feeding ratio of PA to epoxide. In addition, pure diblock poly(ester-*b*-carbonate)s could be easily generated through sequential copolymerization of epoxides with anhydrides first and epoxides with CO_2 afterward [76, 77].

After the successful PO/COS copolymerization mediated a metal-free Lewis pair including TEB [62], Zhang and his coworkers investigated terpolymerization of CO_2/COS/PO to achieve the incorporation of sulfur atoms into polycarbonate chains [78]. The authors screened different Lewis pair combinations and found that *N*,*N*-dimethylcyclohexylamine/TEB can effectively catalyze the CO_2/COS/PO terpolymerization, affording a sulfur-containing polycarbonate with high molar mass up to 58.4 kg mol^{-1} and narrow dispersity ($Đ$ = 1.24–1.56). Due to the higher reactivity of COS compared to CO_2, the obtained terpolymers displayed a gradient nature with tunable thiocarbonate units (27–81%) upon varying the feed ratio of COS to other monomers. The terpolymers exhibited a high refractive index of 1.55, suggesting applications in the field of optical materials.

2.4.2 Polycarbonate Block Copolymers from Orthogonal Polymerization

Mechanistically distinct polymerization systems can afford unique polycarbonate-based block copolymers that include other blocks with different properties, the prerequisite for the successful preparation of such copolymers being the compatibility between the two distinct polymerizations, thus allowing the orthogonal growth of their respective blocks.

Using TEB/PPNCl Lewis pair organocatalyst, Tao and his coworkers described a tandem strategy to synthesize poly(ester-*b*-carbonate) block copolymers through sequence-selective polymerization from a mixture of *O*-carboxyanhydride (OCA) and epoxides (Figure 2.10) [79]. The higher reactivity of OCA monomers compared

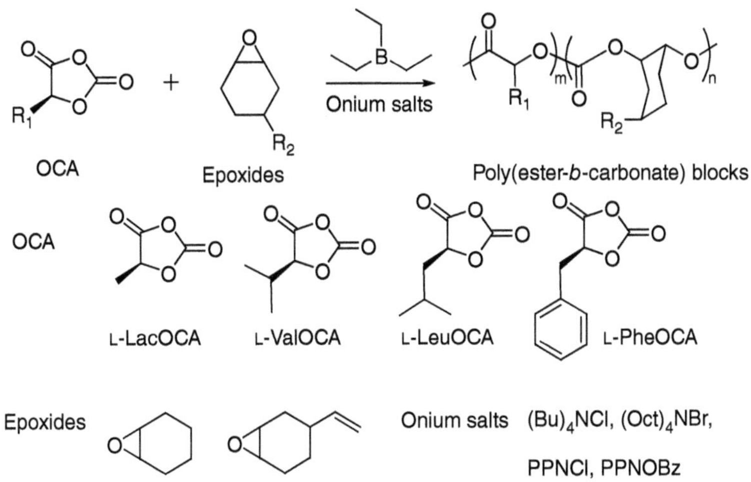

Figure 2.10 Synthesis of poly(ester-*b*-carbonate) block copolymers through tandem ROP of OCA and ROCOP of epoxides with CO_2. Source: Tang et al. [79]/John Wiley & Sons.

to that of CHO/CO_2 guaranteed the selective formation of polyester as the first block; and the CO_2 which was *in situ* liberated upon ROP of OCA used for CHO/CO_2, the copolymerization with CHO to form the second polycarbonate block. The orthogonality or extremely high selectivity of these two polymerizations was confirmed by experimental results and further supported by DFT calculations. Not only could a well-defined diblock copolymer of polyester and polycarbonate be generated from a mixture of OCA and CHO, but an ABCB tetrablock copolymer could also be easily produced in one-pot through the sequential addition of another OCA monomer. Indeed, in comparison with fewer available cyclic ester monomers, OCA monomers provide a wide availability of various aliphatic and aromatic options to produce polyesters. Different polyester and polycarbonate block copolymers can be readily obtained by selective copolymerization of epoxides and OCAs carrying different substituents.

At about the same time, we described the synthesis of di- and triblock copolymers composed of polycarbonates and polyvinyl blocks using a bifunctional initiator-CTA system, TTC-COOH, for the ROCOP of epoxide/CO_2 from its carboxylic acid group and its trithiocarbonate function serving transfer agent for the RAFT polymerization of vinyl monomers (Figure 2.11) [80]. To overcome the instability of TTC-CO_2(NBu_4), TTC-COOH was neutralized by mild organobase tributylamine or NHC *in situ*. The compatibility of these two distinct polymerizations was confirmed by control experiments. This organocatalytic system could easily be used to synthesize diverse AB diblock copolymers from a library of epoxides and vinyl monomers. Surely, ABC triblock copolymers could also be generated through the sequential addition of a third monomer such as ethylene oxide, cyclohexene oxide/CO_2, or vinyl monomers. Considering the versatility of RAFT polymerization for vinyl monomers and TEB/onium initiating system for the ROCOP of epoxides

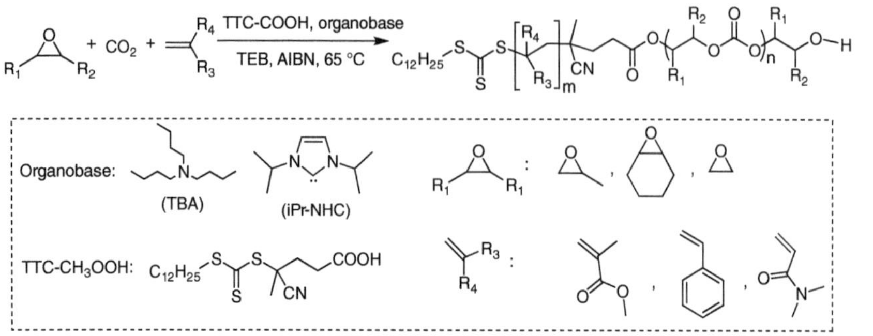

Figure 2.11 Synthesis of AB diblock copolymers using bifunctional initiator TTC-COOH for the orthogonal ROCOP of epoxide/CO_2 and RAFT polymerization of vinyl monomers. Source: Patil et al. [80]/Royal Society of Chemistry.

with CO_2, the described process can actually be used to produce block copolymers with a wide range of block compositions and structures.

2.4.3 All-Polycarbonate Block Copolymers

The versatility of TEB-mediated copolymerization of CO_2 with oxirane monomers prompted the authors of this work to synthesize all-polycarbonate triblock copolymers by sequential copolymerization of CO_2 with two different epoxides in Figure 2.12a, for instance of OO and CHO, using a difunctional initiator [34]. Such triblock copolymers were composed of a central soft poly(octene carbonate) (T_g −24 °C), flanked by two hard PCHC (T_g 120 °C) blocks, exhibiting elastomeric properties. Upon varying the ratio of OO to CHO and their respective ratios to the initiator, three all-polycarbonate triblock samples were prepared with M_n of around 350 kDa and 22, 26, and 29 mol% hard block content, respectively. The AFM analysis clearly showed a microphase-separated structure with the hard PCHC block being dispersed in the continued soft block phases (Figure 2.13). All three samples demonstrated typical elastomeric behavior characterized by a high elongation at break from 1052% to 331% and ultimate tensile strength in the range of 1.4 ~ 2.5 MPa. Such materials that are totally made of polycarbonate and are entirely degradable may find applications in tissue engineering.

Taking advantage of the ability of tetrabutylammnonium carboxylates like onium halide salts to serve as initiators in the previous section [55], all-polycarbonate graft copolymers were synthesized in two steps using a “grafting from” approach (Figure 2.12b) [56]. In the first step, poly(vinyl cyclohexene carbonate) backbones of two different sizes ($PVCHC_{35}$, $PVCHC_{283}$) were prepared by copolymerization of vinyl cyclohexene oxide (VCHO) with CO_2 in the presence of TEB, using tetrabutyl ammonium succinate (TBAS) as initiator. In a second step, the dangling vinyl double bonds of PVCHC were transformed into carboxylic acid groups. After partial neutralization of the latter by tetrabutylammonium hydroxide, the PPC grafts could be grown from carboxylic sites carried by the backbone by copolymerization of PO with CO_2. Through this example, it was demonstrated that

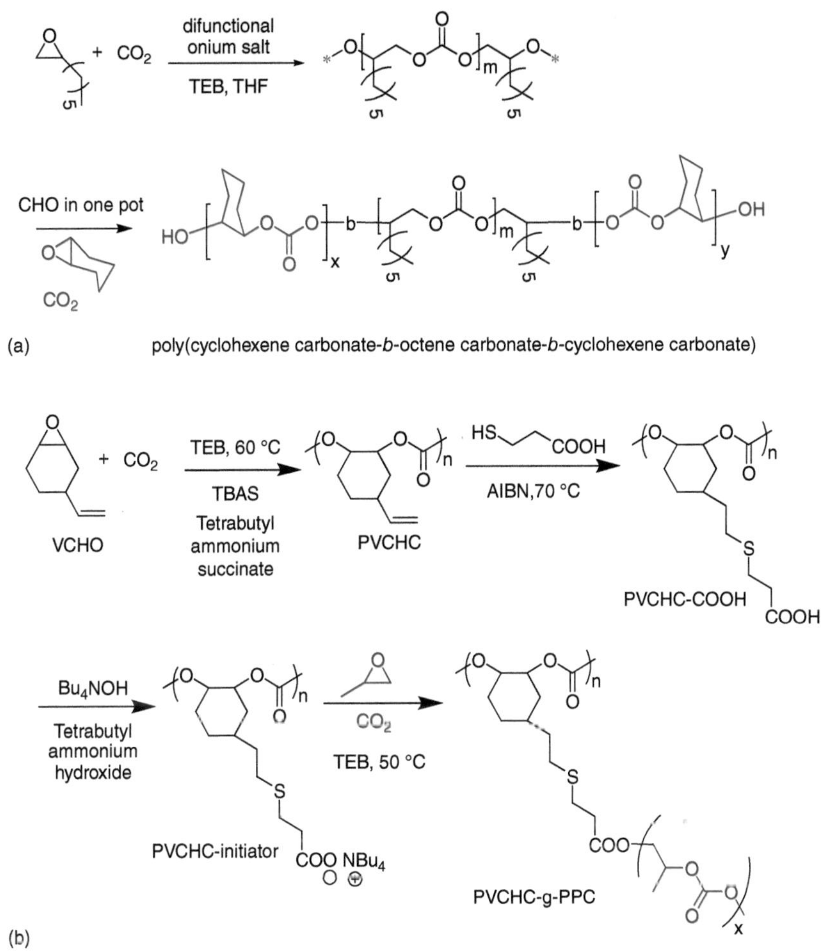

Figure 2.12 All-polycarbonate block copolymers prepared under metal-free conditions (a) PCHC-*b*-POC-*b*-PCHC triblock copolymers. Source: Jia et al. [34]/American Chemical Society.; (b) graft copolymers with PVCHC as backbone and PPC as a side chain. Source: Alagi et al. [56]/American Chemical Society.

this chemistry also works for other polymeric carboxylate ammonium initiators. For instance, when poly(methacrylic acid) (PMAA) generated by reversible addition–fragmentation chain transfer (RAFT) polymerization of methacrylic acid was used as backbone followed by CO_2/PO copolymerization, PPC grafted copolymer on PMAA (PMAA-g-PPC) could similarly be prepared. Through such "grafting from" approach, graft copolymers with well-defined structures and molar masses up to millions g/mol could be obtained without any difficulty. In comparison to polycarbonate graft copolymers synthesized by macrotransfer strategy, the transfer reactions were never complete and therefore the obtained grafted copolymers were often contaminated with an excess of nongrafted polycarbonate linear chains which had to be removed by fractionation [81]. The overall structure and thus the

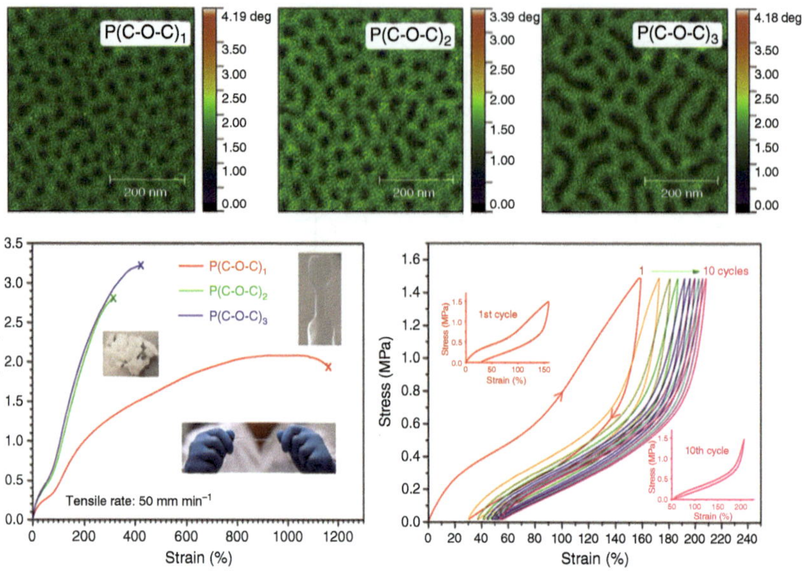

Figure 2.13 AFM phase separation images (upper), and stress-strain curves for the tensile bars made from PCHC-*b*-POC-*b*-PCHC triblock copolymers and cyclic tensile curve for P(C-O-C)$_3$ (bottom). Source: Reproduced with permission from [34]/American Chemical Society/CC BY 4.0.

morphology of these graft copolymers were shown to dramatically impact both their thermal and mechanical properties.

2.4.4 CO_2-Based Triblock Copolymers

Because CO_2-based polycarbonates such as poly(propylene carbonate) (PPC) or poly(cyclohexene carbonate) (PCHC) suffer from low tensile strength and brittle properties, a series of poly(propylene carbonate-*co*-propylene oxide) (PPPC)-based ABA triblock copolymers were prepared with a view of generating polycarbonate-based materials with better physical properties (Figure 2.14) [82]. Generally, using a difunctional initiator, PPNBSK, the copolymerization of PO with CO_2 under different pressure of CO_2 afforded a central PPPC block with tunable softness, then subsequent growth of a hard block at both sides, generated well-defined high molar mass ABA triblock copolymers.

It is known that TEB-based ate complexes cannot be used to initiate ROP of cyclic esters such as lactone and lactide. In a previous report, we showed that the addition of hydrogen-bonding donors such as amines and (thio)ureas in the same TEB-based initiating system as the one applied for polycarbonate synthesis can promote the ROP of cyclic ester efficiently in a controlled way [67]. Therefore, after generating the middle PPPC block, and releasing unreacted CO_2, PLA block could be grown upon concomitant addition of lactide and 1,3-diphenylurea (UPh) to afford PLA-PPPC-PLA triblock copolymers (Figure 2.14) [82]. Some other epoxides such as EO, BO, and OO were also employed for central polycarbonate

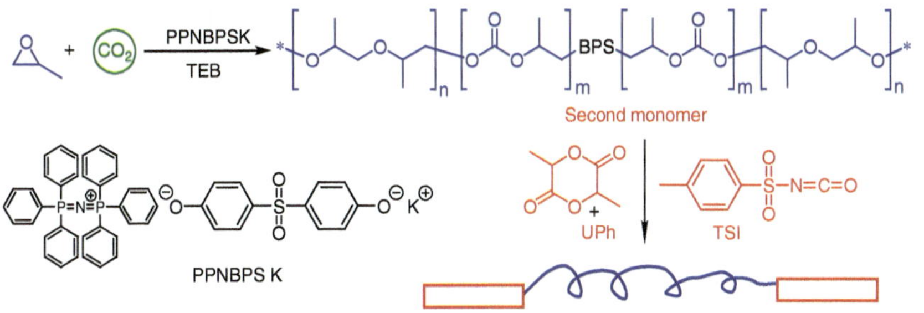

Figure 2.14 CO_2-based triblock copolymers prepared through sequential copolymerization of CO_2/epoxide, and LA or TSI/epoxide. Source: Liu et al. [82] and Jia and coworkers [83].

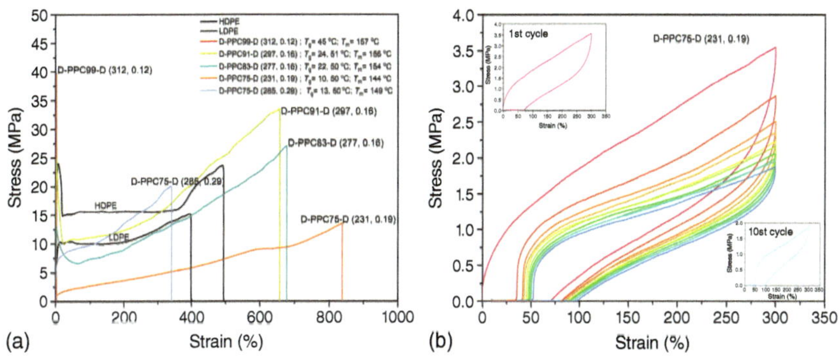

Figure 2.15 (a) Representative stress-strain curves of PDLA-*b*-PPPC-*b*-PDLA triblock copolymers comparing with HDPE and LDPE, (b) Cyclic stress-strain curve of D-PPC75-D (231, 0.19). Source: Liu et al. [82]/American Chemical Society/CC BY 4.0.

formation. Following such a process, many ABA samples associating 12–29 wt% of PDLA, PLLA, and PRLA as hard blocks were prepared; the molar masses of all the samples were above 150 kg mol^{-1}, ensuring the phase separation of polycarbonate and PLA blocks in such ABA triblock copolymers. Glass transition temperatures corresponding to PPPC and PLA blocks along with melting/crystallization points of PDLA/PLLA were clearly observed. As shown in Figure 2.15, after arranging PPPC and PLA blocks into triblocks, the formed copolymers exhibited different properties varying with the composition of the central blocks chosen. Mechanical properties were also affected by the stereo-structures of PLAs. Samples with similar composition and molar masses including racemic PLA blocks and enantiomeric PLLA block showed more modest mechanical characteristics than those fitted with PDLA blocks due to the presence of *rac*-LA and LLA of lower purity than DLA. When the central block was composed of PPC with more than 83% carbonate content and $T_g > 22$ °C, the copolymers exhibited plastic properties, while samples composed of 75% of carbonate content displayed elastomeric properties. A total of 10 cycles were run with sample D-PPC75-D (231, 0.19) with no delay time between cycles in this case, the recovery rate $[(\varepsilon_{300}-\varepsilon_{residual})/\varepsilon_{300}]$ was found to be 77%. Sample D-PPC83-D (277, 0.17) and D-PPC9-D (297, 0.16) exhibited comparable or better mechanical properties to the widely used HDPE and LDPE polyolefin materials. With 33 wt% of CO_2 content, the former two samples showed degradability unlike HDPE and LDPE.

Similarly, taking advantage of the alternating copolymerization occurring between *p*-tosyl isocyanate (TSI) and epoxides under the same conditions as those observed for CO_2/epoxides polycarbonate synthesis [66], we also reported the synthesis of PTPU-*b*-PPPC-*b*-PTPU (U-PC-U) ABA triblock copolymers. In this work, two PTPU hard blocks were associated with a central PPPC block just through sequential copolymerization of PO with TSI. The second block PTPU formed exhibited a glass transition temperature of 107 °C, a value comparable to that of PCHC, due to its different chemical nature phase separation with low T_g polycarbonates (PPPC) that was more favorable than PCHC hard blocks [83]. As for the preparation of triblock PLA-PPPC-PLA, the carbonate content in the central PPPC block and the compositions were varied upon modifying the polymerization conditions. The molar masses of the obtained triblock U-PC-U copolymers could reach up to 300 kg mol^{-1}. Samples with central PPPC block and rigid blocks less than 30 wt% showed significantly higher toughness, higher tensile strength, and higher elongation at break than commercial HDPE and LDPE (Figure 2.16).

It should be noted that in these types of CO_2-based triblock copolymers which exhibited high molar masses and well-defined structure, PO was used as monomer and solvent. So, the excess PO present in the second step of ROP of LA and PO/TSI copolymerization did not give rise to homopolymerization to the formation of ether linkages, which was supported by control experiments and DFT calculations. In addition, the productivity of TEB in such triblock copolymer syntheses is above 500 g polymer per gram catalyst. The easy, cost-effective process has great

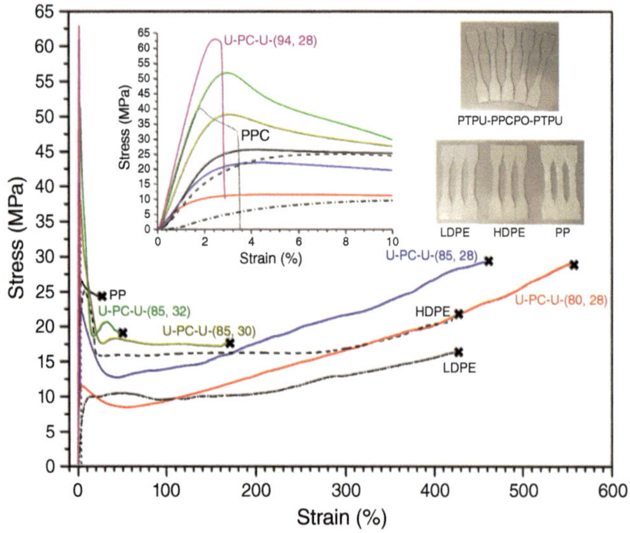

Figure 2.16 Stress-strain curves of U-PC-U triblock copolymers and reference HDPE and LDPE polymers. Source: Reproduced with permission from [83]/American Chemical Society.

industrialization potential to give access to CO_2-based triblock products and replace today's commodity plastics that harm the environment due to their chemical inertness.

2.5 Bifunctional Organoboron Catalysts

In 2006, Nozaki and coworkers reported the design of a new bifunctional catalyst in which cobalt salen complex was covalently linked to piperidinyl and piperidinium arms; such a complex was found to copolymerize epoxides with CO_2 with high activity and selectivity in comparison to bicomponent (salcy)Co(III)/cocatalyst system [84]. Other research groups, Lee and coworkers [85] and Lu and coworkers [86], resorted to similar approaches to design and synthesize multifunctional catalysts that would maintain proximity between propagating chains and the Lewis acid, thus catalyzing more efficiently copolymerization at low loadings [87].

Wu and coworkers reported bifunctional boron-based catalysts following a similar approach to that unveiled by Nozaki, Lee, and Lu for their Co-based complex bicomponent system. In these bifunctional catalysts, 9-borabicyclo(3.3.1)nonane (9-BBN) is tethered to an ammonium cation [88]. In comparison to a bicomponent system such as hexyl-9-BBN/butyltriethylammonium bromide, bifunctional organocatalyst Et_3N-C_5-BBNBr (Figure 2.17) demonstrates a remarkably high activity for the copolymerization of CHO with CO_2 with turnover frequency up to 4900 h^{-1}. The authors proposed that such a high activity was due to the intramolecular synergistic effect between the electrophilic boron center and the quaternary ammonium salt, which was experimentally confirmed by reaction kinetics studies, multiple control experiments, ^{11}B NMR investigation, and its crystal structure. DFT calculations further supported experimental results that the Gibbs free energy of the ring-opening of CHO by Br^- is as low as 16.46 kcal mol^{-1}. As a comparison, the DFT calculation results for CO_2/CHO copolymerization catalyzed by TEB/Cl indicated an energy barrier of 20.3 kcal mol^{-1} for the ROP of CHO activated by TEB as shown in Figure 2.4 [33].

These results have attracted broad attention and triggered the design and synthesis of other bifunctional boron-based catalysts and their application in the (co)polymerization of different monomers and monomer combinations. Many bifunctional organocatalysts carrying 9-BBN, borinane (BRN) moieties respectively associated with ammonium, phosphonium were designed and prepared. The structures of some representative efficient bifunctional organocatalysts are shown in Figure 2.17. In this section, we only summarize recent results related to CO_2/epoxides copolymerization for polycarbonate synthesis.

2.5.1 9-BBN Tethered with Ammonium Cations

After their success at efficiently copolymerizing CO_2/CHO, Wu and coworkers checked the polymerization activity of CO_2 with another internal epoxide, cyclopentene oxide (CPO) using the same catalyst Et_3N-C_5-BBNBr. Unfortunately, only cyclic

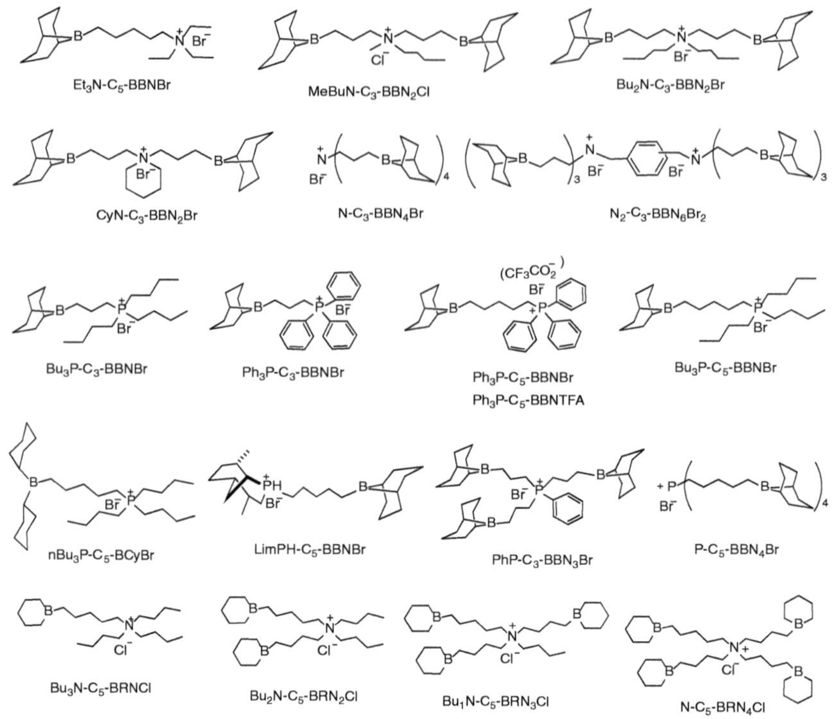

Figure 2.17 Representative effective bifunctional organoboron-based catalyst examples for CO_2/epoxides copolymerization. Nomenclature follows the rules proposed by Lin. Source: Schaefer et al. [89]/American Chemical Society.

carbonates *cis*-cPC were produced in lieu of the generation of PCPC polycarbonate. To this end, they designed and prepared a series of bifunctional organocatalysts that carry two BBN moieties, i.e. Bu_2N-C_5-BBN_2Br, CyN-C_3-BBN_2Br. Under optimal conditions, CyN-C_3-BBN_2Br could produce alternating PCPC with >99% linear vs. cyclic selectivity and a production efficiency of 1.0 kg of PCPC per gram of catalyst. In addition, in the presence of CyN-C_3-BBN_2Br dinuclear organoboron catalysts and KOH (molar ratio of 1 : 1), the produced PCPC was shown capable of depolymerization into CPO monomer in 95% of isolated yield without any side diol or cPC products [90]. Continuing in the same direction and taking advantage of different reactivity of terminal epoxides and internal epoxides toward ROP and copolymerization with CO_2, the same group used dinuclear organoboron catalyst MeBuN-C_3-BBN_2Cl to selectively sequentially ROP of PO and CO_2/CHO reversibly from an epoxide mixture of PO and CHO under different atmospheres (switch between N_2 and CO_2) to obtain well-defined polyether-*b*-polycarbonate block copolymer [91]. This approach was also suitable for various combinations of terminal epoxides and internal epoxides to deliver corresponding well-defined polyether-block-polycarbonate copolymers, including internal epoxides such as VCHO and 4,4-dimethyl-3,5,8-trioxa-bicyclo[5.1.0]octane (CXO); and terminal epoxides such as EO, BO, and AGE.

Epichlorohydrin (ECH) is an industrially available monomer, which is challenging to copolymerize with CO_2 in comparison to other epoxide monomers because of its electron-withdrawing chloromethyl group carried by its oxirane ring favoring back-biting reaction which in turn generates cyclic carbonates, especially at high temperature. To address this, Wu and coworkers prepared a pinwheel-shaped organoboron catalysts carrying four 9-BBN moieties N-C_3-BBN_4Br and tested the copolymerization of ECH with CO_2 with the latter catalyst. The highly active N-C_3-BBN_4Br was shown to actually bring about CO_2/ECH copolymerization under mild conditions (25–40 °C, 25 bar of CO_2), producing alternating poly(chloropropylene carbonate) with >99% linear vs. cyclic selectivity and molar masses up to 36.5 kg mol^{-1} and highest T_g of 45.4 °C up to date [92]. The authors explain that the high density of intramolecular tetranuclear boron centers provides a high local concentration of Lewis acidic centers to efficiently trap the dissociated carbonate anion, thus preventing the carbonate backbiting reaction for the formation of cyclic carbonates.

Cheng and Liu designed and synthesized a series of bifunctional multinuclear organoboron catalysts featuring six boron centers and two quaternary ammonium bromides linked by *o*-, *m*-, or *p*-benzenedimethylenyl group, N_2-C_3-BBN_6Br [93]. Their catalytic performances were investigated in the copolymerizations of CO_2 respectively with epoxide PO, CHO, and ECH. The obtained experiment results demonstrate that the structures linking the two ammonium cations control the copolymerization CO_2 with different epoxides. N_2-C_3-BBN_6Br with ortho-disubstituted linker generates 45% of ether linkages during CO_2/PO copolymerization but exhibits its highest activity in CO_2/CHO copolymerization, allowing the formation of fully alternating PCHCs with >99% linear vs. cyclic selectivity. On the other hand, N_2-C_3-BBN_6Br with para-disubstituted linker exhibits high activity in CO_2/PO and CO_2/ECH copolymerizations, producing PPC with the highest carbonate linkages (88%). However, in comparison to CO_2/ECH copolymerization catalyzed by tetranuclear N-C_3-BBN_4Br, the molar masses of poly(chloropropylene carbonate)s were quite modest with values of 10 kg mol^{-1}. Based on the DFT calculations, boron pairs (up to seven pairs) found in the N_2-C_3-BBN_6Br with ortho-disubstituted linker were demonstrated to be more conducive to the continuous insertion of PO and the formation of ether linkages in CO_2/PO copolymerization; it also exhibited higher activity in CHO/CO_2 copolymerization.

2.5.2 9-BBN Tethered with Phosphonium Cations

In TEB-mediated CO_2/epoxide copolymerization, the polymerization activity varies with the size of the cation of the onium salt used: phosphonium > ammonium > imidazolium > potassium as previously reported [25]. Indeed, the bulkier the size of phosphonium cation compared to the ammonium one leads to weaker interactions with oxyanions, which in turn favor the ring-opening of epoxides, as also observed in TiBA-assisted ROP of PO for high molar mass PPO synthesis [94]. Therefore, bifunctional organoboron catalysts featuring bulkier phosphonium cations are expected to exhibit higher activities. The Wu and coworkers [95] and Lin and coworkers [89] groups independently reported the design and synthesis of a series of 9-BBN-based bifunctional catalysts tethered to different phosphonium cations. In comparison to their ammonium analogs, Bu_3P-C_5-BBN_1Br indeed exhibits a double higher activity in CO_2/CHO copolymerization, and PhP-C_3-BBN_3Br a higher linear vs. cyclic selectivity (86% vs. 60%) along with a slightly higher carbonate content in the obtained polycarbonates (96% vs. 90%) from CO_2/ECH copolymerization [95]. Lin's group at the same time designed and prepared an array of 35 distinct bifunctional catalysts featuring different borane moieties, with phosphonium cations, and linkers between these phosphoniums and boron atoms [89]. The performances of these catalysts were systematically evaluated for CO_2/CHO copolymerization by in situ Raman spectroscopy, and factors such as Lewis acidity of boranes, phosphonium substituents, and linker length were investigated. Similar to Wu's observations, it was found that the copolymerization of CHO and CO_2 proceeded faster $k_{obs} = 2.1 \times 10^{-4}\ s^{-1}$ than with nitrogen analog Bu_3N-C5-BBNBr ($k_{obs} = 1.2 \times 10^{-4}\ s^{-1}$). The authors further used P-C_5-BBN_4Br or tertiary phosphonium borane catalysts such as LimPH-C_5-BBNBr in copolymerization reactions of CO_2 respectively with BO, BGE, VCHO CHO, and prepared telechelic polyalkylene carbonate polyols using dialcohols as chain transfer agents.

Li and coworkers prepared Ph_3P-C_n-BBNX bifunctional phosphonium 9-BBN catalysts with different linker spacer associated with anion Cl^-, Br^-, I^-, CF_3COO^- and tested in the copolymerization of CO_2 with CHO; especially, telechelic PCHC diols ($1.5\ kg\ mol^{-1}$ to $7.5\ kg\ mol^{-1}$) were generated with Ph_3P-C_5-BBNTFA using water as chain transfer agent [96]. Similarly, Darensbourg et al. chose Ph_3P-C_3-BBNBr among all the prepared catalysts for catalyzing the copolymerization of CO_2 with a range of other commonly employed epoxides. Only copolymerizing with CHO, VCHO gave polycarbonates, CHO resembling epoxides such as monoterpenes, α-pinene oxide, and limonene oxide were unsuccessful, and aliphatic epoxides such as PO, BO, and SO lead only to cyclic carbonates. The authors tested the moisture stability and found it much less sensitive to atmospheric moisture than their ammonium analogs [97].

2.5.3 Borinane Tethered with Ammonium Cations

There is obviously a substantial improvement as described above when using 9-BBN-based bifunctional catalyst for copolymerization of CO_2 with CHO, CPO, VCHO, and ECH, but the improvement was only moderate when applied to PO/CO_2 copolymerization [93, 97, 98]. We reasoned that bifunctional organoboron carrying more accessible boron centers than in the BBN case can potentially exhibit an even higher reactivity. We learned from our experience in alkylborane-mediated polymerization that steric hindrance around boron centers is an important factor affecting the polymerization of epoxides. For instance, bulky triphenylborane has a higher Lewis acidity than TEB but is unfit toward PO and CO_2 copolymerization [99]. Steric hindrance around boron centers and Lewis acidity are two criteria influencing the overall performance of boron-based catalysts. We thus designed and synthesized a series of bifunctional organoboron catalysts based on borinane (BRN). Indeed, the six-membered cyclic borinane features the same steric characteristics as that exhibited by TEB. Pure borinane was synthesized via Brown's method [100] with a yield of up to 95%. Once a small quantity of borinane was obtained, it was thus used as seeds for amplification from 2 mol of borinane affording 3 mol of borinane without any purification procedure [101]. Beyond our expectation, the borinane-based catalyst Bu_3N-C_5-BRNCl enabled the copolymerization of CO_2 with PO and generated alternating PPC of molar mass up to 50 kg mol^{-1} with >90% selectivity [102]. The productivity of this catalyst was 271.5 g of PPC per gram catalyst and 5.7 kg of PCHC per gram catalyst, respectively. This catalyst also works for other epoxides such as BO, OO, and AGE. Moreover, it enabled an efficient copolymerization of CO_2 with oxetane, affording alternating poly(oxetane carbonate) with unprecedented molar mass up to 16.1 kg mol^{-1} in comparison to TEB/$PPNN_3$ bicomponent system [44]. In addition, Bu_3N-C_5-BRNCl could be easily recycled after being neutralized by succinic acid through a simple precipitation process when it was used to prepare telechelic PPC-diols. Continuing our efforts to prepare low molar mass polycarbonate polyols cost-effectively, we attempted the preparation of these polyols with N-C_5-BRN_4Cl in the presence of chain transfer agents [103]. In the case of PCHC polyol, the productivity was 18.2 kg per gram catalyst, comparable to state-of-the-art metallic catalysts [104]; when copolymerizing CO_2 with PO in the presence of chain transfer agents, N-C_5-BRN_4Cl could be used to produce poly(ether propylene carbonate) polyols with 20 wt% of CO_2 incorporation and a productivity of 1.1 kg per gram catalyst. The very high performance of these catalysts which required a low loading shows a great potential for application in large-scale preparation of CO_2-based polyols.

2.6 Conclusions

Since our first report on the preparation of polycarbonates in 2016 under metal-free conditions, much progress has been made in organocatalytic CO_2/epoxide copolymerization during the past eight years. The easy access to this copolymerization provides a convenient and powerful tool to copolymerize a wide range of oxirane monomers with CO_2, avoiding complex ligand design and multistep synthesis in comparison to traditional organometallic catalytic systems. The copolymerizations are generally carried out with an anionic ring-opening approach under "living" conditions, allowing precise control of the obtained polycarbonates of their molar masses, the carbonate contents, compositions, and functionalities. The catalytic activities of the TEB/Lewis base system and those of the recently developed bifunctional boron-based organocatalysts are comparable in CO_2/epoxide copolymerization, and even better in some cases than that of state-of-the-art organometallic systems. Despite the great progress made during the past two decades in CO_2/epoxides copolymerization, the applications of obtained aliphatic polycarbonates are very limited due to their poor properties, which must be addressed by whatever metallic or metal-free copolymerization systems are used. Telechelic polycarbonate polyols are found to exhibit better performance than conventional polyether polyols in applications such as polyurethanes. Some works related to catalyst recycling and CO_2/epoxide copolymerization in the presence of transfer agents using bifunctional boron-based organocatalysts are reported to decrease the cost. However, the production of poly(propylene carbonate) or poly(ether propylene carbonate) polyols still cannot compete with low-cost benchmark polyether polyols. Ter- or block-copolymerization with other monomers to form random or block copolymers stands out to be a good option to improve the properties of high molar mass PPC or PCHC. Yet taking advantage of versatility and compatibility and metal-free produced products, boron-based organocatalytic system exhibits high potential to promote the commercialization application of CO_2-based polycarbonates.

References

1 Inoue, S., Koinuma, H., and Tsuruta, T. (1969). Copolymerization of carbon dioxide and epoxide. *J. Polym. Sci. Part B: Polym. Lett.* 7: 287–292.

2 von der Assen, N., Jung, J., and Bardow, A. (2013). Life-cycle assessment of carbon dioxide capture and utilization: avoiding the pitfalls. *Energy Environ. Sci.* 6: 2721–2734.

3 Artz, J., Müller, T.E., Thenert, K. et al. (2018). Sustainable conversion of carbon dioxide: an integrated review of catalysis and life cycle assessment. *Chem. Rev.* 118: 434–504.

4 Martín, C., Fiorani, G., and Kleij, A.W. (2015). Recent advances in the catalytic preparation of cyclic organic carbonates. *ACS Catal.* 5: 1353–1370.

5 Allen, S.D., Moore, D.R., Lobkovsky, E.B., and Coates, G.W. (2002). High-activity, single-site catalysts for the alternating copolymerization of CO_2 and propylene oxide. *J. Am. Chem. Soc.* 124: 14284–14285.

6 Lee, B.Y., Kwon, H.Y., Lee, S.Y. et al. (2005). Bimetallic anilido-aldimine zinc complexes for epoxide/CO_2 copolymerization. *J. Am. Chem. Soc.* 127: 3031–3037.

7 Li, Y., Zhang, Y.-Y., Liu, B., and Zhang, X.-H. (2018). HCAII-inspired catalysts for making carbon dioxide-based copolymers: The role of metal-hydroxide bond. *Chin. J. Polym. Sci.* 36: 139–148.

8 Darensbourg, D.J. and Mackiewicz, R.M. (2005). Role of the cocatalyst in the copolymerization of CO_2 and cyclohexene oxide utilizing chromium salen complexes. *J. Am. Chem. Soc.* 127: 14026–14038.

9 Cohen, C.T., Chu, T., and Coates, G.W. (2005). Cobalt catalysts for the alternating copolymerization of propylene oxide and carbon dioxide: combining high activity and selectivity. *J. Am. Chem. Soc.* 127: 10869–10878.

10 Kember, M.R. and Williams, C.K. (2012). Efficient magnesium catalysts for the copolymerization of epoxides and CO_2; using water to synthesize polycarbonate polyols. *J. Am. Chem. Soc.* 134: 15676–15679.

11 Xiao, Y., Wang, Z., and Ding, K. (2006). Intramolecularly dinuclear magnesium complex catalyzed copolymerization of cyclohexene oxide with CO_2 under ambient CO_2 pressure: kinetics and mechanism. *Macromolecules* 39: 128–137.

12 Ikpo, N., Flogeras, J.C., and Kerton, F.M. (2013). Aluminium coordination complexes in copolymerization reactions of carbon dioxide and epoxides. *Dalton Trans.* 42: 8998–9006.

13 Coates, G.W. and Moore, D.R. (2004). Discrete metal-based catalysts for the copolymerization of CO_2 and epoxides: discovery, reactivity, optimization, and mechanism. *Angew. Chem. Int. Ed.* 43: 6618–6639.

14 Darensbourg, D.J. (2007). Making plastics from carbon dioxide: salen metal complexes as catalysts for the production of polycarbonates from epoxides and CO_2. *Chem. Rev.* 107: 2388–2410.

15 Kember, M.R., Buchard, A., and Williams, C.K. (2011). Catalysts for CO_2/epoxide copolymerisation. *Chem. Commun.* 47: 141–163.

16 Lu, X.-B., Ren, W.-M., and Wu, G.-P. (2012). CO_2 copolymers from epoxides: catalyst activity, product selectivity, and stereochemistry control. *Acc. Chem. Res.* 45: 1721–1735.

17 Li, Y., Zhang, Y.-Y., Hu, L.-F. et al. (2018). Carbon dioxide-based copolymers with various architectures. *Prog. Polym. Sci.* 82: 120–157.

18 Wang, Y. and Darensbourg, D.J. (2018). Carbon dioxide-based functional polycarbonates: metal catalyzed copolymerization of CO_2 and epoxides. *Coord. Chem. Rev.* 372: 85–100.

19 Kozak, C.M., Ambrose, K., and Anderson, T.S. (2018). Copolymerization of carbon dioxide and epoxides by metal coordination complexes. *Coord. Chem. Rev.* 376: 565–587.

20 Xu, Y., Lin, L., Xiao, M. et al. (2018). Synthesis and properties of CO_2-based plastics: environmentally-friendly, energy-saving and biomedical polymeric materials. *Prog. Polym. Sci.* 80: 163–182.

21 Sujith, S., Min, J.K., Seong, J.E. et al. (2008). A highly active and recyclable catalytic system for CO_2/propylene oxide copolymerization. *Angew. Chem. Int. Ed.* 47: 7306–7309.

22 Hongfa, C., Tian, J., Andreatta, J. et al. (2008). A phase separable polycarbonate polymerization catalyst. *Chem. Commun.* 8: 975–977.

23 Hauenstein, O., Reiter, M., Agarwal, S. et al. (2016). Bio-based polycarbonate from limonene oxide and CO_2 with high molecular weight, excellent thermal resistance, hardness and transparency. *Green Chem.* 18: 760–770.

24 Bahramian, B., Ma, Y., Rohanizadeh, R. et al. (2016). A new solution for removing metal-based catalyst residues from a biodegradable polymer. *Green Chem.* 18: 3740–3748.

25 Zhang, D., Boopathi, S.K., Hadjichristidis, N. et al. (2016). Metal-free alternating copolymerization of CO_2 with epoxides: fulfilling "green" synthesis and activity. *J. Am. Chem. Soc.* 138: 11117–11120.

26 Zhang, C., Geng, X., Zhang, X. et al. (2023). Alkyl borane-mediated metal-free ring-opening (co)polymerizations of oxygenated monomers. *Prog. Polym. Sci.* 136: 101644.

27 Naumann, S. (2023). Borane catalysis for epoxide (co)polymerization. *Polym. Chem.* 14: 1834–1862.

28 Ma, Y.K., Liu, S.F., and Li, Z.B. (2022). Organocatalytic alternating copolymerization of CO_2/epoxide/anhydride for synthesis of polycarbonate/polyester copolymers. *Acta Polym. Sin.* 53: 1041–1056.

29 Herzberger, J., Niederer, K., Pohlit, H. et al. (2016). Polymerization of ethylene oxide, propylene oxide, and other alkylene oxides: synthesis, novel polymer architectures, and bioconjugation. *Chem. Rev.* 116: 2170–2243.

30 Billouard, C., Carlotti, S., Desbois, P., and Deffieux, A. (2004). "Controlled" high-speed anionic polymerization of propylene oxide initiated by alkali metal alkoxide/trialkylaluminum systems. *Macromolecules* 37: 4038–4043.

31 Zhang, D., Zhang, H., Hadjichristidis, N. et al. (2016). Lithium-assisted copolymerization of CO_2/cyclohexene oxide: a novel and straightforward route to polycarbonates and related block copolymers. *Macromolecules* 49: 2484–2492.

32 Jia, M., Hadjichristidis, N., Gnanou, Y., and Feng, X. (2019). Monomodal ultrahigh-molar-mass polycarbonate homopolymers and diblock copolymers by anionic copolymerization of epoxides with CO_2. *ACS Macro Lett.* 8: 1594–1598.

33 Zhang, D.-D., Feng, X., Gnanou, Y., and Huang, K.-W. (2018). Theoretical mechanistic investigation into metal-free alternating copolymerization of CO_2 and epoxides: the key role of triethylborane. *Macromolecules* 51: 5600–5607.

34 Jia, M., Zhang, D., de Kort, G.W. et al. (2020). All-polycarbonate thermoplastic elastomers based on triblock copolymers derived from triethylborane-mediated sequential copolymerization of CO_2 with various epoxides. *Macromolecules* 53: 5297–5307.

35 Jia, M., Zhang, D., Gnanou, Y., and Feng, X. (2021). Surfactant-emulating amphiphilic polycarbonates and other functional polycarbonates through metal-free copolymerization of CO_2 with ethylene oxide. *ACS Sustain. Chem. Eng.* 9: 10370–10380.

36 Webster, R., Didier, E., Harris, P. et al. (2007). PEGylated proteins: evaluation of their safety in the absence of definitive metabolism studies. *Drug Metab. Dispos.* 35: 9–16.

37 Chen, Z., Yang, J.-L., Lu, X.-Y. et al. (2019). Triethyl borane-regulated selective production of polycarbonates and cyclic carbonates for the coupling reaction of CO_2 with epoxides. *Polym. Chem.* 10: 3621–3628.

38 Boopathi, S.K., Hadjichristidis, N., Gnanou, Y., and Feng, X. (2019). Direct access to poly(glycidyl azide) and its copolymers through anionic (co-)polymerization of glycidyl azide. *Nat. Commun.* 10: 293.

39 Nagahama, S., Tanaka, T., and Matsumoto, A. (2004). Supramolecular control over the stereochemistry of diene polymers. *Angew. Chem. Int. Ed.* 43: 3811–3814.

40 Scadeng, O., Ferguson, M.J., and West, F.G. (2011). Intermolecular [3 + 3]-cycloadditions of azides with the nazarov intermediate. *Org. Lett.* 13: 114–117.

41 Brown, H.C., Midland, M.M., Levy, A.B. et al. (1987). Organoboranes for synthesis. *Tetrahedron* 43: 4079–4088.

42 Huang, J., Jehanno, C., Worch, J.C. et al. (2020). Selective organocatalytic preparation of trimethylene carbonate from oxetane and carbon dioxide. *ACS Catal.* 10: 5399–5404.

43 Fukushima, K. (2016). Poly(trimethylene carbonate)-based polymers engineered for biodegradable functional biomaterials. *Biomater. Sci.* 4: 9–24.

44 Zhang, C.-J., Wu, S.-Q., Boopathi, S. et al. (2020). Versatility of boron-mediated coupling reaction of oxetanes and epoxides with CO_2: selective synthesis of cyclic carbonates or linear polycarbonates. *ACS Sustain. Chem. Eng.* 8: 13056–13063.

45 Darensbourg, D.J., Moncada, A.I., Choi, W., and Reibenspies, J.H. (2008). Mechanistic studies of the copolymerization reaction of oxetane and carbon dioxide to provide aliphatic polycarbonates catalyzed by (salen)CrX complexes. *J. Am. Chem. Soc.* 130: 6523–6533.

46 Darensbourg, D.J. and Moncada, A.I. (2009). (Salen)Co(II)/*n*-Bu4NX catalysts for the coupling of CO_2 and oxetane: selectivity for cyclic carbonate formation in the production of poly-(trimethylene carbonate). *Macromolecules* 42: 4063–4070.

47 Alves, M., Grignard, B., Boyaval, A. et al. (2017). Organocatalytic coupling of CO_2 with oxetane. *ChemSusChem* 10: 1128–1138.

48 Huang, J., De Winter, J., Dove, A.P., and Coulembier, O. (2019). Metal-free synthesis of poly(trimethylene carbonate) by efficient valorization of carbon dioxide. *Green Chem.* 21: 472–477.

49 Darensbourg, D.J. and Moncada, A.I. (2010). Tuning the selectivity of the oxetane and CO_2 coupling process catalyzed by (salen)CrCl/*n*-Bu4NX: cyclic carbonate formation vs aliphatic polycarbonate production. *Macromolecules* 43: 5996–6003.

50 Alagi, P., Ghorpade, R., Choi, Y.J. et al. (2017). Carbon dioxide-based polyols as sustainable feedstock of thermoplastic polyurethane for corrosion resistant metal coating. *ACS Sustain. Chem. Eng.* 5: 3871–3881.

51 Roudsari, G.M., Mohanty, A.K., and Misra, M. (2017). Exploring the effect of poly(propylene carbonate) polyol in a biobased epoxy interpenetrating network. *ACS Omega* 2: 611–617.

52 Wang, J., Zhang, H., Miao, Y. et al. (2016). UV-curable waterborne polyurethane from CO_2-polyol with high hydrolysis resistance. *Polymer* 100: 219–226.

53 Pohl, M., Danieli, E., Leven, M. et al. (2016). Dynamics of polyether polyols and polyether carbonate polyols. *Macromolecules* 49: 8995–9003.

54 Wang, J., Zhang, H., Miao, Y. et al. (2016). Waterborne polyurethanes from CO_2 based polyols with comprehensive hydrolysis/oxidation resistance. *Green Chem.* 18: 524–530.

55 Patil, N.G., Boopathi, S.K., Alagi, P. et al. (2019). Carboxylate salts as ideal initiators for the metal-free copolymerization of CO_2 with epoxides: synthesis of well-defined polycarbonates diols and polyols. *Macromolecules* 52: 2431–2438.

56 Alagi, P., Zapsas, G., Hadjichristidis, N. et al. (2021). All-polycarbonate graft copolymers with tunable morphologies by metal-free copolymerization of CO_2 with epoxides. *Macromolecules* 54: 6144–6152.

57 Patil, N., Bhoopathi, S., Chidara, V. et al. (2020). Recycling a borate complex for synthesis of polycarbonate polyols: towards an environmentally friendly and cost-effective process. *ChemSusChem* 13: 5080–5087.

58 Wang, Y., Zhang, J.-Y., Yang, J.-L. et al. (2021). Highly selective and productive synthesis of a carbon dioxide-based copolymer upon zwitterionic growth. *Macromolecules* 54: 2178–2186.

59 Liu, S., Ren, C., Zhao, N. et al. (2018). Phosphazene bases as organocatalysts for ring-opening polymerization of cyclic esters. *Macromol. Rapid Commun.* 39: 1800485.

60 Zhang, J., Wang, L., Liu, S., and Li, Z. (2022). Synthesis of diverse polycarbonates by organocatalytic copolymerization of CO_2 and epoxides: from high pressure and temperature to ambient conditions. *Angew. Chem. Int. Ed. Engl.* 61: e202111197.

61 Sheng, X., Wang, Y., Qin, Y. et al. (2014). Aluminum porphyrin complexes via delicate ligand design: emerging efficient catalysts for high molecular weight poly(propylene carbonate). *RSC Adv.* 4: 54043–54050.

62 Yang, J.-L., Wu, H.-L., Li, Y. et al. (2017). Perfectly alternating and regioselective copolymerization of carbonyl sulfide and epoxides by metal-free Lewis pairs. *Angew. Chem. Int. Ed.* 56: 5774–5779.

63 Kummari, A., Pappuru, S., and Chakraborty, D. (2018). Fully alternating and regioselective ring-opening copolymerization of phthalic anhydride with epoxides using highly active metal-free Lewis pairs as a catalyst. *Polym. Chem.* 9: 4052–4062.

64 Ji, H.-Y., Chen, X.-L., Wang, B. et al. (2018). Metal-free, regioselective and stereoregular alternating copolymerization of monosubstituted epoxides and tricyclic anhydrides. *Green Chem.* 20: 3963–3973.

65 Hu, L.-F., Zhang, C.-J., Wu, H.-L. et al. (2018). Highly active organic Lewis pairs for the copolymerization of epoxides with cyclic anhydrides: metal-free access to well-defined aliphatic polyesters. *Macromolecules* 51: 3126–3134.

66 Jia, M., Hadjichristidis, N., Gnanou, Y., and Feng, X. (2021). Polyurethanes from direct organocatalytic copolymerization of *p*-tosyl isocyanate with epoxides. *Angew. Chem. Int. Ed.* 60: 1593–1598.

67 Liu, J., Gnanou, Y., and Feng, X. (2022). Expanding the scope of boron-based ate complexes by manipulating their reactivity: the case of cyclic esters and their (co)polymers. *Macromolecules* 55: 1800–1810.

68 Patil, N., Gnanou, Y., and Feng, X.S. (2022). Anionic copolymerization of *o*-phthalaldehyde with epoxides: facile access to degradable polyacetals and their copolymers under ambient conditions. *Macromolecules* 55: 7817–7826.

69 Ye, S., Wang, W., Liang, J. et al. (2020). Metal-free approach for a one-pot construction of biodegradable block copolymers from epoxides, phthalic anhydride, and CO_2. *ACS Sustain. Chem. Eng.* 8: 17860–17867.

70 Liang, J., Ye, S., Wang, S. et al. (2022). Biodegradable copolymers from CO_2, epoxides, and anhydrides catalyzed by organoborane/tertiary amine pairs: high selectivity and productivity. *Macromolecules* 55: 6120–6130.

71 Liang, J., Ye, S., Wang, W. et al. (2021). Performance tailorable terpolymers synthesized from carbon dioxide, phthalic anhydride and propylene oxide using Lewis acid-base dual catalysts. *J. CO2 Util.* 49: 101558.

72 Ye, S., Ren, Y., Liang, J. et al. (2022). One-pot construction of random, gradient and triblock copolymers from CO_2, epoxides and phthalic anhydride by metal-free catalyst. *J. CO2 Util.* 65: 102223.

73 Zhao, M.J., Zhu, S.S., Zhang, G.C. et al. (2023). One-step synthesis of linear and hyperbranched CO_2-based block copolymers via organocatalytic switchable polymerization. *Macromolecules* 56: 2379–2387.

74 Liang, J.X., Wang, S.Y., Wu, C.E. et al. (2023). A new biodegradable CO_2-based poly(ester-*co*-carbonate): molecular chain building up with crosslinkable domain. *J. CO2 Util.* 69: 102403.

75 Wang, W.J., Ye, S.X., Liang, J.X. et al. (2022). Architecting branch structure in terpolymer of CO_2, propylene oxide and phthalic anhydride: an enhancement in thermal and mechanical performances. *Chin. J. Polym. Sci.* 40: 462–468.

76 Zhang, J., Wang, L., Liu, S. et al. (2021). A Lewis pair as organocatalyst for one-pot synthesis of block copolymers from a mixture of epoxide, anhydride, and CO_2. *Macromolecules* 54: 763–772.

77 Chidara, V.K., Boopathi, S.K., Hadjichristidis, N. et al. (2021). Triethylborane assisted synthesis of random and block poly(ester-carbonate)s through one-pot terpolymerization of epoxides, CO_2, and cyclic anhydrides. *Macromolecules* 54: 2711–2719.

78 Kiriratnikom, J., Guo, J., Cao, X. et al. (2022). Metal-free terpolymerization of propylene oxide, carbon dioxide, and carbonyl sulfide: a facile route to sulfur-containing polycarbonates with gradient sequences. *J. Polym. Sci.* 60: 3414–3419.

79 Tang, J., Li, M., Wang, X., and Tao, Y. (2022). Switchable polymerization organocatalysis: from monomer mixtures to block copolymers. *Angew. Chem. Int. Ed.* 61: e202115465.

80 Patil, N., Gnanou, Y., and Feng, X.S. (2022). Orthogonally grown polycarbonate and polyvinyl block copolymers from mechanistically distinct (co)polymerizations. *Polym. Chem.* 13: 2988–2998.

81 Honda, S. and Sugimoto, H. (2016). Carbon dioxide-derived immortal brush macromolecules with poly(propylene carbonate) side chains. *Macromolecules* 49: 6810–6816.

82 Liu, J., Jia, M., Gnanou, Y., and Feng, X. (2023). One-pot synthesis of CO_2-based polylactide-*b*-poly(ether carbonate)-*b*-polylactide triblock copolymers and their mechanical properties. *Macromolecules* 56: 1615–1624.

83 Jia, M., Liu, J., Gnanou, Y., and Feng, X. (2023). Triblock copolymers from CO_2, propylene oxide, and *p*-tosyl isocyanate of higher toughness than polyethylenes. *Macromolecules* 56: 3631–3640.

84 Nakano, K., Kamada, T., and Nozaki, K. (2006). Selective formation of polycarbonate over cyclic carbonate: copolymerization of epoxides with carbon dioxide catalyzed by a cobalt(III) complex with a piperidinium end-capping arm. *Angew. Chem.* 118: 7432–7435.

85 Noh, E.K., Na, S.J., S, S. et al. (2007). Two components in a molecule: highly efficient and thermally robust catalytic system for CO_2/epoxide copolymerization. *J. Am. Chem. Soc.* 129: 8082–8083.

86 Ren, W.-M., Liu, Z.-W., Wen, Y.-Q. et al. (2009). Mechanistic aspects of the copolymerization of CO_2 with epoxides using a thermally stable single-site cobalt(III) catalyst. *J. Am. Chem. Soc.* 131: 11509–11518.

87 Lidston, C.A.L., Severson, S.M., Abel, B.A., and Coates, G.W. (2022). Multifunctional catalysts for ring-opening copolymerizations. *ACS Catal.* 12: 11037–11070.

88 Yang, G.-W., Zhang, Y.-Y., Xie, R., and Wu, G.-P. (2020). Scalable bifunctional organoboron catalysts for copolymerization of CO_2 and epoxides with unprecedented efficiency. *J. Am. Chem. Soc.* 142: 12245–12255.

89 Schaefer, J., Zhou, H., Lee, E. et al. (2022). Tertiary and quaternary phosphonium borane bifunctional catalysts for CO_2/epoxide copolymerization: a mechanistic investigation using in situ Raman spectroscopy. *ACS Catal.* 12: 11870–11885.

90 Yang, G.W., Wang, Y., Qi, H. et al. (2022). Highly selective preparation and depolymerization of chemically recyclable poly(cyclopentene carbonate) enabled by organoboron catalysts. *Angew. Chem. Int. Ed.* 61: e202210243.

91 Zhang, Y.-Y., Yang, G.-W., Xie, R. et al. (2022). Sequence-reversible construction of oxygen-rich block copolymers from epoxide mixtures by organoboron catalysts. *J. Am. Chem. Soc.* 144: 19896–19909.

92 Yang, G.W., Xu, C.K., Xie, R. et al. (2021). Pinwheel-shaped tetranuclear organoboron catalysts for perfectly alternating copolymerization of CO_2 and epichlorohydrin. *J. Am. Chem. Soc.* 143: 3455–3465.

93 Tong, Y.T., Cheng, R.H., Dong, H. et al. (2022). Highly active bifunctional dual-arm organoboron catalysts bearing cooperative intramolecular structures for the copolymerization of CO_2 and epoxides. *J. CO2 Util.* 60: 101979.

94 Labbé, A., Carlotti, S., Billouard, C. et al. (2007). Controlled high-speed anionic polymerization of propylene oxide initiated by onium salts in the presence of triisobutylaluminum. *Macromolecules* 40: 7842–7847.

95 Zhang, Y.Y., Lu, C., Yang, G.W. et al. (2022). Mechanism-inspired upgradation of phosphonium-containing organoboron catalysts for epoxide-involved copolymerization and homopolymerization. *Macromolecules* 55: 6443–6452.

96 Wang, X.-W., Hui, J.-W., Li, Y.-T. et al. (2023). Facile synthesis of polycarbonate diol via copolymerization of CO_2 and cyclohexene oxide catalysed by a combination of one-component phosphonium borane Lewis pair and water. *Chin. J. Polym. Sci.* 41: 735–744.

97 Sengoden, M., Bhat, G.A., and Darensbourg, D.J. (2022). Bifunctional organoboron-phosphonium catalysts for coupling reactions of CO_2 and epoxides. *RSC Adv.* 12: 32440–32447.

98 Yang, G.-W., Xu, C.-K., Xie, R. et al. (2022). Precision copolymerization of CO_2 and epoxides enabled by organoboron catalysts. *Nat. Synth.* 1: 892–901.

99 Andrea, K.A. and Kerton, F.M. (2019). Triarylborane-catalyzed formation of cyclic organic carbonates and polycarbonates. *ACS Catal.* 9: 1799–1809.

100 Brown, H.C. and Pai, G.G. (1983). Organoboranes: XXVIII. Convenient procedures for the synthesis of borinane. *J. Organomet. Chem.* 250: 13–22.

101 Chen, C., Gnanou, Y., and Feng, X. (2022). Borinane boosted bifunctional organocatalysts for ultrafast ring-opening polymerization of cyclic ethers. *Macromolecules* 55: 10662–10669.

102 Chen, C., Gnanou, Y., and Feng, X.S. (2022). Borinane-based organoboron catalysts for alternating copolymerization of CO_2 with cyclic ethers: improved productivity and facile recovery. *Polym. Chem.* 13: 6312–6321.

103 Chen, C., Gnanou, Y., and Feng, X. (2023). Ultra-productive upcycling CO_2 into polycarbonate polyols via borinane-based bifunctional organocatalysts. *Macromolecules* 56: 892–898.

104 Chapman, A.M., Keyworth, C., Kember, M.R. et al. (2015). Adding value to power station captured CO_2: tolerant Zn and Mg homogeneous catalysts for polycarbonate polyol production. *ACS Catal.* 5: 1581–1588.

3

Organocatalyzed Ring-Opening Copolymerization of Cyclic Anhydride and Cyclic Ether (Acetal)

Xun Zhang, Chengjian Zhang, and Xinghong Zhang

Zhejiang University, State Key Laboratory of Biobased Transportation Fuel Technology, International Research Center for X Polymers, Department of Polymer Science and Engineering, No. 866, Yuhangtang Road, Hangzhou, 310058, China

3.1 Introduction

With the rapid evolution of our era, demand for polymer materials continues to rise. Polyesters, an important type of polymers, have a wide range of applications in various fields. Polyesters can be roughly divided into aliphatic, semi-aromatic and aromatic polyesters. For semi-aromatic polyesters, such as poly(ethylene terephthalate) (PET), one of the earliest industrialized polyesters, exhibit excellent mechanical strength, fatigue resistance, and other distinctive features due to the incorporation of large aromatic structure into polymer chains [1]. PET is extensively used as fibers, blow-molded bottles, and other packaging materials. Produced annually on a multibillion-pound scale, PET plays a pivotal role in meeting the growing demand for polymer applications, and significantly can be easily recovered for reuse because PET is usually used without any additives. In contrast, aliphatic polyesters initially received less attention due to their inferior physical and chemical properties, stemming from the absence of aromatic or rigid functionalities in the polymer backbone. In very recently years, aliphatic polyesters are highly concerned because they have abundant renewable monomer sources [2, 3], ease of degradation into useful products [4, 5], and high biocompatibility [6] and thus a potentially sustainable or recyclable alternatives to petroleum-based polymers. Currently, the development and industrial production of poly(lactic acid) (PLA) and poly(hydroxyl alkanoates) (PHAs) stand out as two successful examples. Consequently, these typical biodegradable polymers have found diverse applications, ranging from specialized biomedical devices to packaging materials [7, 8].

The most common synthesis route for polyesters is through step-growth polymerization of diacids or diesters with diols (Figure 3.1) [1]. For example, poly(butylene succinate) and poly(butyleneadipate-co-terephthalate) prepared by this method, exhibit good mechanical performance, heat resistance, and biodegradability, which render them highly favored and one of the best-performing materials in both biodegradable plastic research and market applications. However, this approach

Organocatalysts in Polymer Chemistry: Synthesis and Applications, First Edition. Edited by Zhibo Li.

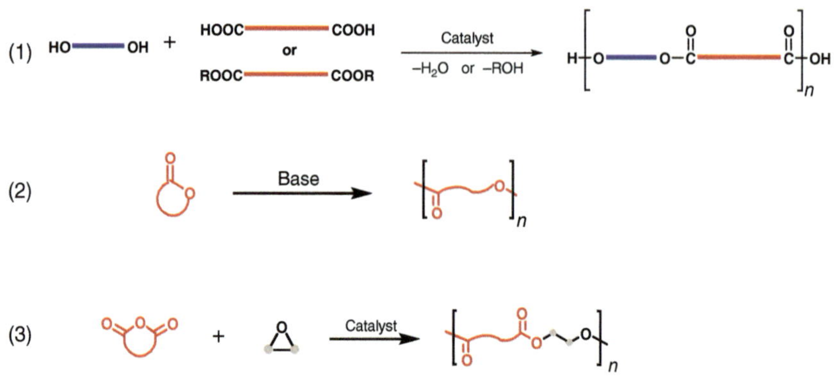

Figure 3.1 Illustration of the general reactions to prepare polyesters.

generates small-molecule by-products, such as water or alcohol during synthesis. The elimination of these by-products is energy-intensive. Furthermore, as with all step-growth polymerizations, obtaining products with a high number-average molecular weight (M_n) requires polymerization to achieve very high monomer conversions, usually at elevated temperatures for extended period of time. This extended process inevitably results in a broad molecular weight distribution, with the dispersity ($Đ$ or M_w/M_n, where M_w is the weight-average molecular weight) of the polymers typically reaching around 2 [1]. In general, low molecular weight polyester diols are produced in industry and used as the reactive prepolymers through the reaction of end –OH groups with isocyanates for making foams, elastomers, adhesives, coatings and so on.

In contrast to step-growth polymerization, chain-growth polymerization can afford high-molecular-weight polymers with controlled dispersity. The synthesis of polyesters via chain-growth manner has attracted great interest. For example, ring-opening polymerization (ROP) of cyclic lactones (such as PLA, poly(ε-caprolactone) (PCL), Figure 3.1) is a practical method to synthesize polyesters, which has been studied extensively with a wide variety of newly developed initiators/catalysts, including organocatalysts, metal alkoxides, and various metal complexes, which is intensively discussed elsewhere. However, the ability to synthesize polyesters via this method is limited in obtaining low-cost monomers, high production cost and poor overall performances compared with traditional non-degradable polymers such as polyolefins [9, 10]. An alternative chain-growth route to both aliphatic and semi-aromatic polyesters is the ring opening alternating copolymerization (ROAC) of epoxides and cyclic anhydrides (Figure 3.1). The use of two distinct monomers allows for facile tuning of properties and even post-polymerization modification [9, 11]. Up to date, by adopting different combinations of cyclic anhydrides and epoxides, over 500 polyesters can be successfully synthesized. The key role of catalysts is underscored in achieving efficient and highly selective copolymerization to yield high molecular weight products. Therefore, this copolymerization route is a very promising way to produce new polyesters.

In this chapter, we will introduce the ROAC of cyclic anhydrides and epoxides, with a special focus on the pivotal role of organocatalysts in the copolymerization, highlighting their unique advantages. The use of organocatalysts greatly expanded the application fields of polyesters. We also specifically highlight the copolymerization of cyclic anhydrides with other monomers, including tetrahydrofuran (THF), cyclic acetals and even aldehydes, to synthesize a new family of polyesters. With the introduction of these innovative methods and the use of new low-cost monomers, we detail the expanded classification of polyesters.

3.2 The Copolymerization of Epoxide with Cyclic Anhydride

3.2.1 The Scope of Epoxides and Cyclic Anhydrides

The copolymerization of epoxides and cyclic anhydrides is now an attractive route to produce aliphatic and semi-aromatic polyesters owing to the availability of those two distinct families of monomers, as shown in Figure 3.2.

As the downstream products of the petrochemical industry, almost all the epoxides can be prepared from the oxidation of the corresponding olefins. Some of them can be produced in large scale with low cost or have the great potential for large scale production, such as ethylene oxide (EO), propylene oxide (PO), 1,2-butylene oxide (BO), epichlorohydrin (ECH), isobutylene oxide (IBO) and so on. Some epoxides can be converted into polymers with good thermal properties benefited the inherent aromatic ring or aliphatic ring such as allyl glycidyl ether (AGE), phenyl glycidyl ether (PGE) and cyclohexene oxide (CHO), as seen in Figure 3.2. Some epoxides with the functional group are also important for post-polymerization so that it can enhance the performance further, such as AGE, vinyl cyclohexene oxide (VCHO). Bio-based epoxides such as *trans*-(*R*)-limonene oxide (LO), PO and EO, can be directly obtained from biomass to show the potential to replace the traditional polymers.

Cyclic anhydrides, as the oxygen-rich monomers, can be easily prepared by the dehydration of the corresponding diacids. Most of the diacids are very cheap products and even can be extracted from some natural product. For example, succinic acid, a raw material of succinic anhydride (SA), can be abstracted from amber. Some cyclic anhydrides can be produced in large scale with the low cost such as maleic anhydride (MA), glutaric anhydride (GA), diglycolic anhydride (DGA), thiodiglycolic anhydride (TDGA) and phthalic anhydride (PA), as shown in Figure 3.2. The cyclic anhydrides with the aromatic or aliphatic ring, such as PA, hexahydrophthalic anhydride (HHPA), tetrahydrophthalic anhydride (THPA), *endo*-bicyclo[2.2.2]oct-5-ene-2,3-dicarboxylic anhydride (BCODA) and *cis*-5-norbornene-*endo*-2,3-dicarboxylic anhydride (NA), can endow the copolymers with the improved thermal properties. Especially, the NA can be prepared from bio-based norbornene and MA via a *Diels-Alder* reaction. In addition to NA, the cyclic anhydrides like MA, methylsuccinic anhydride (MSA), citraconic anhydride (CA) can also be obtained from biomass. Of special note, the introduction of

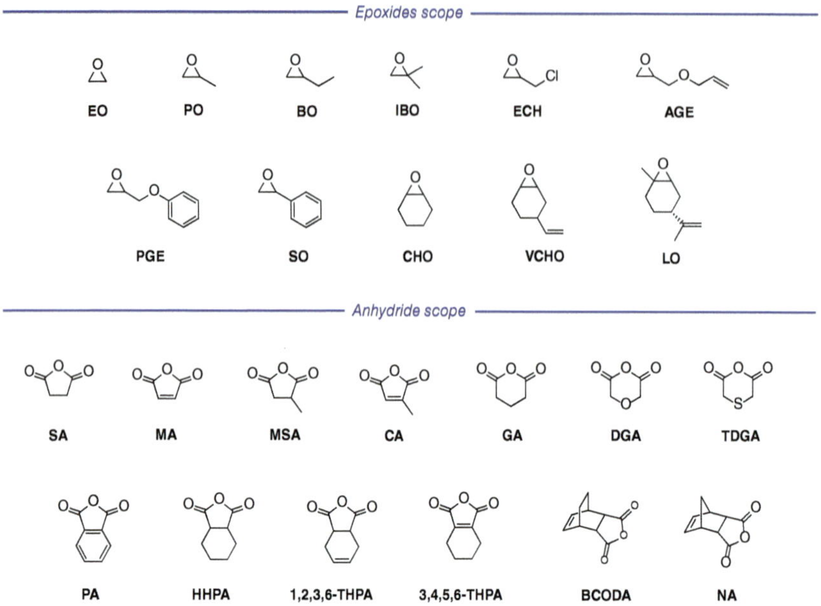

Figure 3.2 The scope of representative epoxides and anhydrides for copolymerization.

hetero-atoms like O, S into the cyclic anhydrides like DGA, TDGA can reduce the carbon content of the obtained polymers, i.e. low-carbon polymers [12], and endow these polymers with some special characters like high refractive index.

The first report of the copolymerization of epoxides with cyclic anhydrides for polyesters were disclosed in 1960s [13]. In early, many catalyst systems such as inorganic salts, tertiary amines, and metal alkoxides were be applied for this copolymerization. However, the copolymerization reactions by these catalysts often suffer from poor controllability, low molecular weights, and broad molecular weight distribution accompanied by polyether contamination from epoxides homopolymerization [14]. Next, we will briefly discuss the recent development of metal-catalyzed copolymerization, which facilitates the reader's comparison with organocatalysis.

3.2.2 Metal Catalysts

Significant progress in metal catalyzed copolymerization of epoxide and cyclic anhydride has been made with high activity and selectivity over the past few decades. In 1985, Aida and Inoue reported the first example of controlled copolymerization of PO with PA, catalyzed by an aluminum porphyrin catalyst with tetralkyl ammonium halide as the catalyst promoter (Figure 3.3, cat. 1) [15, 16]. A diverse array of metal complexes were then developed for this route based on porphyrin ligand by changing the center metals, such as Mn [17], Fe [17], Co [18], Cr [18–20] and Al [16, 18] complexes (Figure 3.3, cat. 1–2). In addition to porphyrin ligands, catalysts composed of other ligand skeletons and metal ions have been widely used in the copolymerization

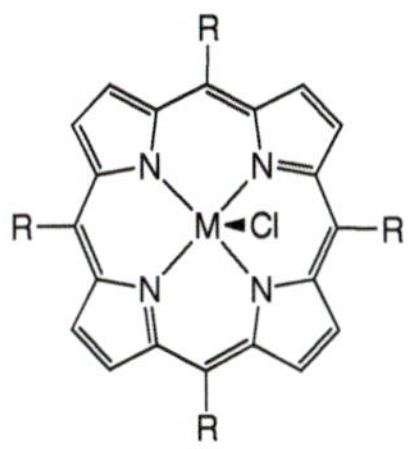

R = Ph, C_6F_5
M = Al, Cr, Co

cat. 1

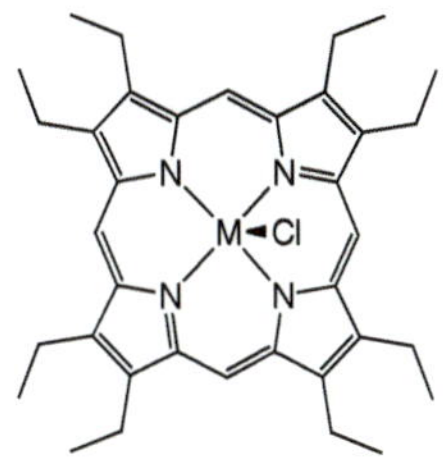

M = Al, Cr, Co

cat. 2

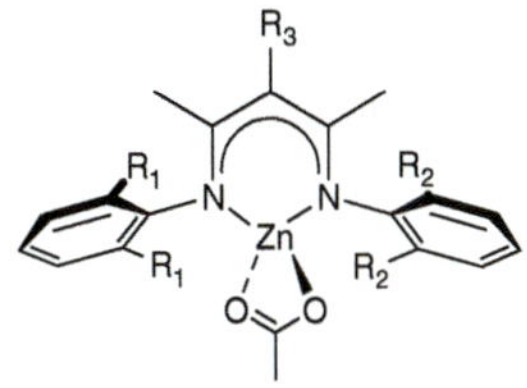

R_1, R_2 = Et, ^{i}Pr
R_3 = H, Me, CN

cat. 3

Figure 3.3 The metalloporphyrin catalysts and diimine catalysts.

of epoxides and cyclic anhydrides. The more famous ones are catalysts composed of a diiminate ligand skeletons and salicylaldimine skeletons complexed with metals.

In 2007, Coates and coworkers found that a β-diiminate zinc acetate [(BDI)ZnOAc] catalyst (Figure 3.3, cat. 3), which had been used for epoxides/CO_2 copolymerization, is the first highly active catalyst for the alternating copolymerization of a number of epoxides and cyclic anhydrides [21, 22]. Four years later, Duchateau and coworkers reported a chromium salophen complex (Figure 3.4, cat. 4) accompanied with 4-dimethylaminopyridine (DMAP) for the copolymerization of CHO and cyclic anhydrides [23]. The bio-based monomer such as LO was also used to copolymerized with PA to yield partially renewable polyesters with better activity in bulk (TOF: $64\,h^{-1}$), low molar mass ($M_n < 1.8\,kg\,mol^{-1}$) and moderate ester linkage content of less than 73%.

Similar to the salophen complex, the *N,N'*-bis(salicylidene)cyclohexanediamine chromium (III) chloride[(salen)CrCl] (Figure 3.4, cat. 5) was also applied to the copolymerization of epoxides with cyclic anhydrides. Coates and coworkers

N,N'-Bis(salicylidene)ethylenediamine (salen) catalysts.and related metal catalysts

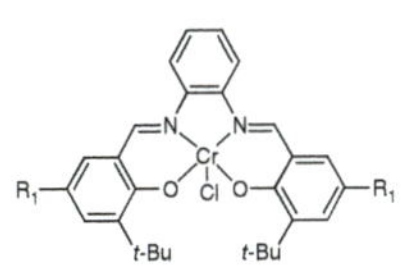

cat. 4

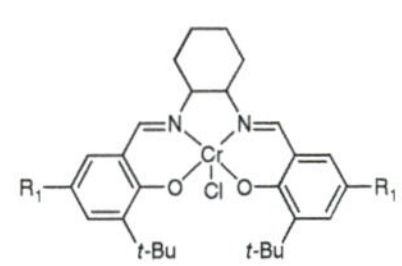

cat. 5

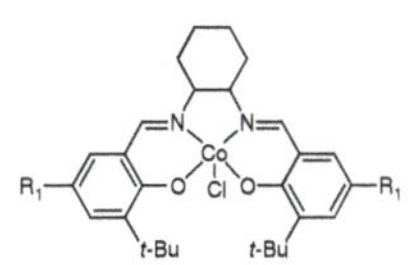

cat. 6

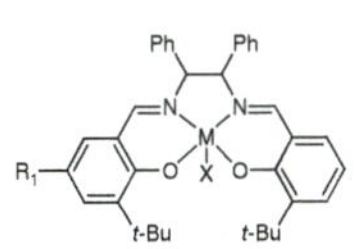

cat. 7

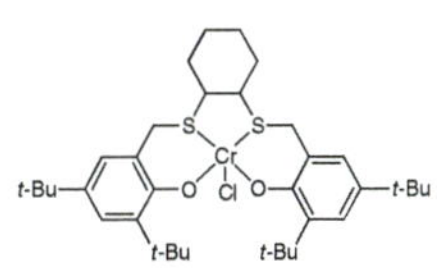

cat. 8

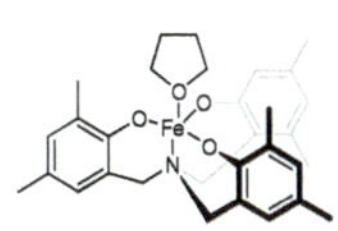

cat. 9

Figure 3.4 Representative *N, N'*-Bis(salicylidene)ethylenediamine catalysts and related catalysts.

reported the successful use of these chromium or cobalt complexes (Figure 3.4, cat. 6, 7) respectively for PO/MA copolymerizations, showing high activity [24]. With a nucleophilic co-catalyst, such as bis(triphenylphosphine)iminium salts ([PPN]X, X = Cl^-, N_3^-, DNP^-), phosphines, ammonium salts, and others, Duchateau and colleagues observed the binary catalysts exhibit significantly enhanced activity [25]. Shortly after, other ligand such as tetradentate sulfur ligands (Figure 3.4, cat. 8) [26] and geometrically more flexible iron aminotriphenolate metal complexes (Figure 3.4, cat. 9) [11] also were synthesized and then used for copolymerization of epoxides with cyclic anhydrides.

Williams and coworkers reported a series of effective dinuclear or multinuclear catalysts based on light metals or transition metals with group **1** metal such as Al, Cd, Ca, Mg, Zn, Fe, Li, Na, K and so on, for the copolymerization of epoxides with anhydrides. Among them, the heterodinuclear Zn/Mg catalyst (Figure 3.5, cat. 10) shows improved copolymerization selectivity, controllability, and significantly greater activity as compared to either of the homodinuclear analogues or any combinations of them, suggesting a synergic relationship between the di- or multi-metals [27–30]. In addition, heterodinuclear catalysts with dinucleating Schiff-base orthovanillin ligand [27], such as Fe/K (Figure 3.5, cat. 11) and Al/K (Figure 3.5, cat. 12) show significantly improved catalytic activities (TOF: 1152 h^{-1} for Fe/K, 1072 h^{-1} for Al/K) for PA/CHO copolymerization [30].

Lu and coworkers also reported a series of dinuclear chromium-salan catalysts (Figure 3.5, cat. 13) for the copolymerization of terminal epoxides and cyclic anhydrides [31]. The dinuclear metal-salen catalysts such as Cr [32, 33], Al [34, 35] and Co [36] bearing a binaphthol linker (Figure 3.5, cat. 14, 15) also showed significantly higher activity than the mononuclear salan Cr complexes. In addition, the trinuclear complexes catalysts (Figure 3.6 cat. 16) and dinuclear bifunctional catalysts (Figure 3.6, cat. 17) show the higher activity (TOF: 142 h^{-1} for Mg/Mg/Na

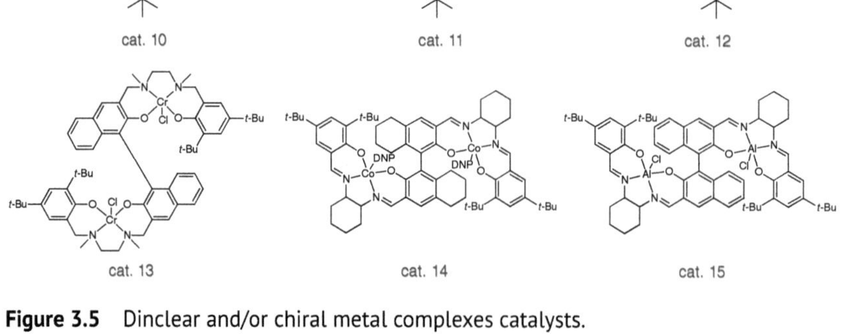

Figure 3.5 Dinclear and/or chiral metal complexes catalysts.

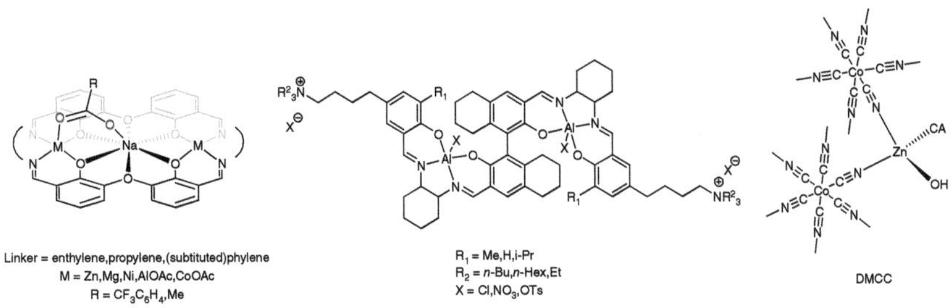

cat. 16 cat. 17 cat. 18

Figure 3.6 Bifunctional and multinuclear metal complexes catalysts.

trinuclear catalyst in PA/CHO copolymerization) than the single nuclear catalyst or combination of them, which show a clear synergic effect between the metals in either dinuclear/trinuclear multi-chirality catalysts, or between metal Lewis acid center with arching Lewis base in catalyst molecules. Very recently, Zhang and coworkers reported a heterogeneous zinc–cobalt (III) double metal cyanide complex (DMCC) catalyst with the proposed active center as cat. 18 in Figure 3.6 that is very effective to the copolymerization of MA with IBO for preparing semicrystalline polyesters [37]. In fact, DMCC is highly active catalyst for a variety of cyclic anhydrides and epoxides and can be used to synthesize many polyesters with high activity, selectivity, and high molecular weight. It is proposed that the collaborative catalysis of surface metals contributes to the copolymerization.

Overall, advancements in metal catalysts have been substantial, showcasing significant improvements in both activity and polymerization control. These catalysts demonstrate the ability to produce copolymers with high molecular weights and exceptional catalyst activity (up to 1152 h^{-1}). Of special note, the prevalent use of the binary metal catalyst systems, i.e. metal Lewis acid plus organic base, is a common strategy for the copolymerization. However, the challenges arise from the toxicity of metal ions by heteroatoms, the presence of metal residues in the final polymer, leading to toxicity and a dark color, significantly restricting the widespread application of these polymers on a large scale. Consequently, organocatalysts emerge as a promising solution, offering significant potential to address these challenges and find applications in biomedical, food packaging, and other areas.

3.2.3 Organocatalysts

Organocatalytic copolymerization of epoxides with cyclic anhydrides is a metal-free route to the synthesis of polyesters without metal ion residues. It is a highly desired synthetic pathway but remains a huge challenge. In very recent years, the organocatalytic polymerization is a hot topic in the field of polymer chemistry and is rapidly developed with the discoveries of many effective organocatalysts, from the simple organic base to the Lewis pairs for many types of polymerizations [38–49]. Herein, we discuss the polymer selectivity, catalytic activity, alternating

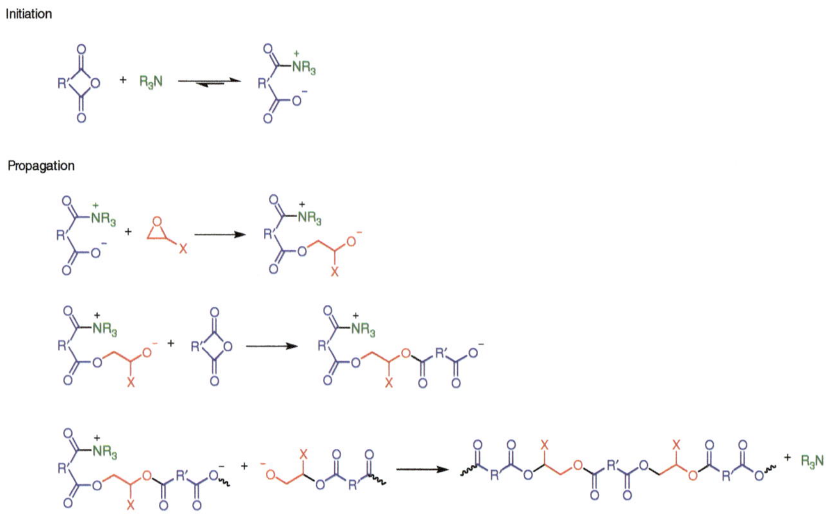

Figure 3.7 The proposed mechanism catalyzed by tertiary amines. (X, R and R′ are substituents). Source: Fischer [13]/John Wiley & Sons.

degree, and molecular weight control for the copolymerization of epoxides with cyclic anhydrides via organocatalysts.

3.2.3.1 Organic Bases

In 1960, Fischer and coworkers used the tertiary amines as the catalysts (note that the concept of organocatalyst was proposed in 2000s) for the copolymerization of epoxides with cyclic anhydrides (Figure 3.7). These tertiary amines showed rather low activity and selectivity for the copolymerization [13]. In this reaction, the cyclic anhydrides were activated by tertiary amines firstly, which causes deactivation of the amine. The ring opening polymerization of epoxides is suppressed by the anhydride-amine combination. Then, the carboxyl anion reacted with the epoxides to afford alkoxide anion, which attacks the cyclic anhydride for chain growth. Finally, these alternating steps repeated would give rise to the polyester. Of course, the displacement of the amine by alkoxide ion can also give rise to the polyester. Then the metal complex catalysts were developed and well-studied as the above discussion. In 2015, a highly stereoregular polyester was synthesized by the organocatalysts mediated alternating copolymerization of CHO and NA stereoisomers [50]. The geometrical structure of polyester can be tuned simply by modulating the type of NA isomers (*endo*- or *exo*-NA), monomer feed ratio and reaction temperature. The *cis*-(>99%) and *trans*-polyesters (>99%) exhibit high glass transition temperature up to 129.8 and 115.9 °C, respectively. Shortly after, the Hošt'álek et al. [51] conducted a comprehensive study of experiments conditions and refer to that PPNCl proved to be the most efficient initiator, which produced highly alternating polyesters with molar mass up to 2.1 kg mol^{-1}.

Organophosphazene bases [52], as a class of non-ionic, weakly nucleophilic organic strong bases, have achieved great success in the organocatalytic

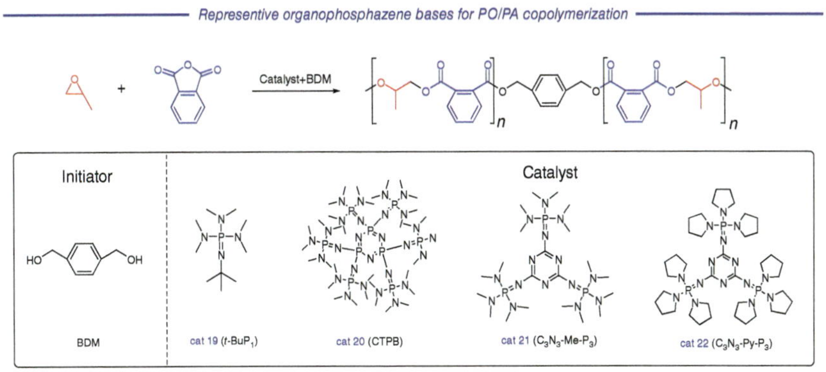

Figure 3.8 The representative organophosphazene bases for PO/PA copolymerization.

ring-opening polymerization of cyclic lactones. In 2017, Zhao and coworkers [53] employed the phosphazene base for the ring-opening copolymerization of epoxides and cyclic anhydrides. In the copolymerization of EO and PA, the non-nucleophilic character and appropriate basicity of t-BuP_1 (Figure 3.8, cat. 19) has engendered a unique self-buffering mechanism. They pointed out that compared to t-BuP_4 which can deprotonate both EO-terminals (aliphatic hydroxyl) and PA-terminals (benzoic carboxyl), the weakly basic t-BuP_1 can deprotonate only PA-terminals (benzoic carboxyl). The t-BuP_1 alternately and repeatedly acts as the activator of PA-terminals (by abstracting proton) and deactivator of EO terminals (by returning proton) so that carboxylate species are active enough to react with EO and the activity of hydroxyl species is kept low enough to avoid side reactions, even at full conversion of PA. By this self-buffering mechanism, the t-BuP_1 rendered the carboxyl and hydroxyl terminals appropriately active to allow chain growth in a strictly alternating manner with the completely inhibited side reactions such as epoxide self-propagation, transesterification. Based on this mechanism, the copolymerizations of PA with other substituted epoxides such as PO, BO, SO, AGE, CHO, VCHO and LO were achieved. A variety of polyesters with diverse structures were obtained with the molecular weights up to 12.4 kg mol^{-1} [54].

Organophosphazenes are performed very well in the field of organocatalytic synthesis of polymers, and their activity and selectivity are largely affected by their basicity and molecular size. Li and coworkers designed and synthesized three types of organophosphazenes with different molecular sizes and basicity, i.e. cyclic trimeric phosphazene base (CTPB) (Figure 3.8, cat. 20), 2,4,6-tris[tri(dimethylamino)iminophosphorane]-1,3,5-triazine (C_3N_3-Me-P_3) (Figure 3.8, cat. 21) and 2,4,6-tris[tri(1-pyrrolidinyl)-iminophosphorane]-1,3,5-triazine (C_3N_3-Py-P_3) (Figure 3.8, cat. 22) [55]. There catalysts were effective to the alternating copolymerization of PO and PA. The catalytic activity of cat. 22 was higher than that of t-BuP_1 under the same reaction conditions owing to large molecular size of cat. 22, although cat. 22 is less basic than t-BuP_1. In addition, the basicity of cat. 22 is lower than that of CTPB, however, the polymerization controllability of cat. 22 is better than that of the CTPB. This is because that the large shape and size of cat.

22 are conducive to the formation of loose ion pairs. The resultant polyester has a controllable molecular weight with narrow molecular weight distribution. Note that the application of cats. 20–22 might be difficult because they need complex synthetic steps and high cost.

3.2.3.2 Bicomponent Lewis Pair

Bicomponent Lewis pairs that are composed of Lewis acid and Lewis base (also known as cocatalyst or catalyst promoter, the two are often used confusedly in the literature), have cooperative catalysis effect for polymerization, and are widely investigated in very recent years. A simple metal Lewis pairs, zinc alkyls (or aryls) and amines as Lewis pairs, as reported by Li and coworkers, succeeded in the copolymerization of epoxides and cyclic anhydrides with a high catalytic activity (TOF: $210\ h^{-1}$) and perfectly alternating degree (>99%) [56]. This kind of metal Lewis pairs have almost the same catalyst effect as traditional large complex metal catalysts, which provides a good idea for the design of simple and practical metal-free polymerization catalyst. In the field of organocatalytic polymerization, people continue to study the application of organic Lewis pairs in polymerization, hoping that they will have similar catalytic activity to metal catalysts.

In 2016, Gnanou, Feng and their coworkers firstly disclosed triethylborane (TEB) as the Lewis acid and quaternary ammonium salt or phosphate salt as the Lewis base for the copolymerization of epoxides (e.g.: PO, CHO) and carbon dioxide(CO_2) [39]. In 2018, Zhang and coworkers [43] firstly reported the combination of TEB with organic base (Figure 3.9) for the copolymerization of epoxides with cyclic anhydrides. They proposed a homologous tetra-coordinate TEB-bonded (i.e. B-O species) oxyanion for the chain growth, similar to the metal catalyst in Section 2.2 for forming Metal-O species, as shown in Figure 3.10a. Other epoxides (PO, PGE, and ECH) and cyclic anhydrides (MA, PA, SA, and DGA) were successfully extended. Side reactions including transesterification and polyether formation were suppressed to a minimal content.

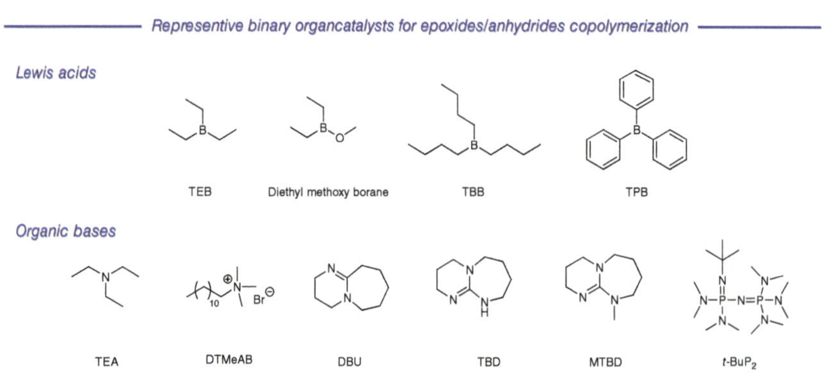

Figure 3.9 Representative binary organocatalysts for epoxides/cyclic anhydrides copolymerization.

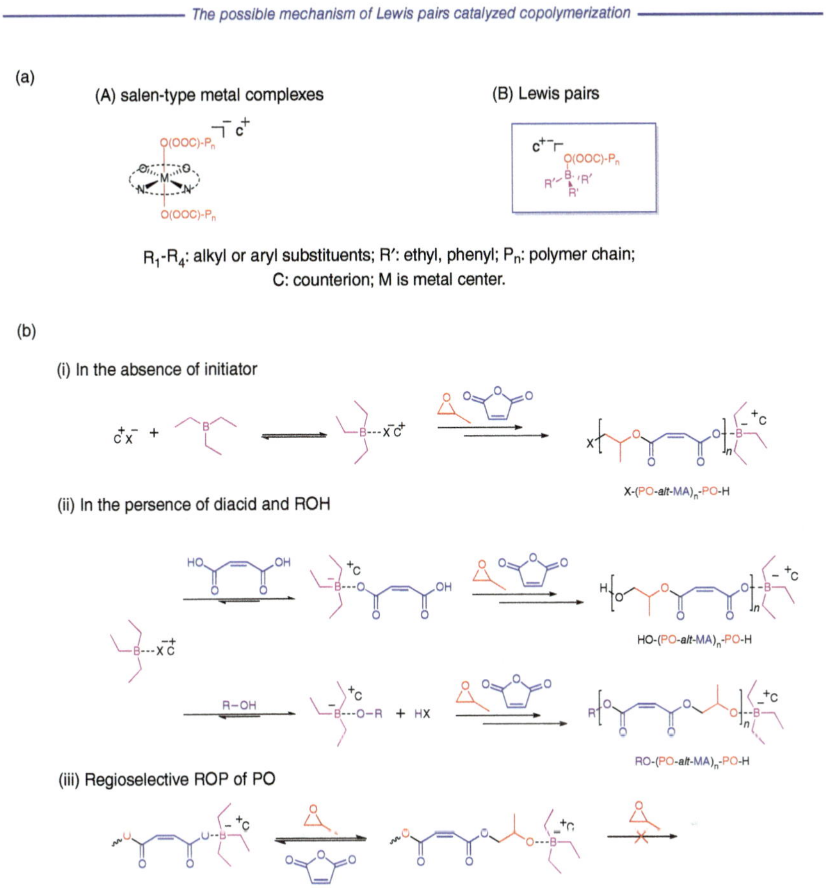

Figure 3.10 (a) hexacoordinate metal species [(salen)M complexes], and tetracoordinate boron species for the chain growth. (R_1–R_4: alkyl or aryl substituents; R′: ethyl, phenyl; P_n: polymer chain; C: counterion; M is metal center). (b) The possible mechanism of TEB/Lewis pairs for alternating copolymerization of epoxides with cyclic anhydrides. Source: Hu et al. [43]/American Chemical Society.

The obtained polyesters exhibited well-defined structures, M_ns up to 20.0 kg mol^{-1}, and $Đ$ of 1.12–1.39. Of significance, the presence of TEB promoted the regioselectivity for the ring-opening reaction of PO. The 72% head-to-tail diads were obtained by the TEB/DTMeAB which is higher than the DTMeAB alone (about 60%) for PO/MA copolymerization (Figure 3.10b). Further investigation on borane-catalyzed PO/SA copolymerization indicated that the activity order was consistent with the acidity of boranes and the basicity of organic bases, viz TEB > diethyl methoxy borane > TBB > $(C_6H_5)_3B$ and t-BuP_2 > t-BuP_1 > MTBD > DBU > TEA [57, 58].

In mechanism, the authors concluded that the improvement in catalyst activity by adding TEB may be attributed to the coordination of Br^- to boron to form Lewis pairs. When the Lewis pairs combined with the initiator, i.e. diacid, as shown in Figure 3.10b, by the reaction of TEB with the diacid, that is the deprotonation, a

TEB-bonded carboxylate (alkoxide) anions are formed and more nucleophilic for the alternating chain growth. Such tetracoordinate boron species as the chain growth center can also suppress the homopolymerization of PO even in the excess of PO and provides a moderate steric hindrance for regioselective ring-opening reaction of PO at methylene (CH_2) site. Therefore, the high activity (TOF: 102 and 303 h^{-1} for the PO/MA and PO/PA copolymerization, respectively) and moderate regioselectivity can be achieved with the suppressed side reactions.

Afterwards, Chakraborty and coworkers [59] resorted to various Lewis acids (TEB, Et_3Al, Et_2Zn, and n-Bu_2Mg) in combination with onium salts (PPNCl) or organic bases (TBD and DBU) to facilitate the copolymerization of epoxides with PA. The combination of TEB/PPNCl was shown as the most effective catalyst for the copolymerization of CHO with PA, yielding high-molar-mass copolymers with M_ns up to 57.5 kg mol^{-1} and $Đ$ as low as 1.07. Li and coworkers [60] used various types of Lewis pairs for epoxides/cyclic anhydrides copolymerization, wherein the TEB/PPNCl combination exhibited the highest catalytic activity (TOF up to 330 h^{-1}) for the copolymerization of PO with tricyclic anhydride CPMA, a *Diels–Alder* adduct of cyclopentadiene and MA. Nine mono-substituted epoxides and four tricyclic anhydrides were used as monomers to synthesize a library of polyesters [56]. Of interest, such processes exhibited high regioselectivity of up to 98% for aliphatic epoxides and glycidyl ethers and 78% for SO bearing an electron-withdrawing phenyl group. Both steric and electronic effects of the phenyl group competitively drove the attack of SO, which is agreed with the Zhang's result [43].

Zhao and coworkers, used phosphazene bases and TEB as catalysts for the copolymerization of cyclic anhydrides with large excess epoxides, no polyether formation was observed at high temperatures even after the cyclic anhydride was completely consumed [61]. They found that the protonolysis of TEB by COOHs occurs during the copolymerization of cyclic anhydrides and epoxides when conducted at high reaction temperatures. They revealed the generated acyloxyborane species, which can be further protonolyzed by OHs at relatively late stages of the copolymerization to generate boronic ester (alkoxyborane) species. Kim and coworkers exploited the organocatalytic copolymerizations of cyclic anhydrides and a functionalized epoxide, *t*-butyl glycidoxy acetate, followed by acidic deprotection to access degradable polyesters with carboxylic acid pendants [62]. DGA and GA were also used as cyclic anhydrides to prepare two types of polyesters. They demonstrated that the introduction of the oxygen heteroatom into the cyclic anhydrides significantly influenced their thermal properties and pH-dependent self-association behavior in an aqueous solution.

The synthesis of high molecular weight polyesters via ROAC of cyclic anhydrides and epoxides using organocatalysts remain a major challenge. In 2020, Li and coworkers [63], reported a cyclic trimeric phosphazene base (CTPB) combined with TEB was used as an efficient catalyst for the copolymerization of PA with epoxides. It was found that with the addition of TEB as a cocatalyst, even a small amount of TEB, such as 5 mol% relative to CTPB, significantly improved the activity and PA conversion. By reducing the loading of CTPB and TEB, high molecular weight polyesters were facilely synthesized (M_n up to 118.5 kg mol^{-1}) with perfectly alternating structure and narrow polydispersities, the copolymerization were fast with a TOF of up to 500 h^{-1}. In addition, Gnanou and coworkers used various amounts of TEB to regulate the reactivity of EO in its copolymerization with cyclic anhydrides. The reactivity of EO could be varied on demand, either fully alternating copolymer or poly(ester-*co*-ether) with random sequence could be prepared, paving the way for the synthesis of ester-containing degradable poly(ethylene oxide)s [64].

Later, on the basis of the observation of the simplest Lewis pair composed of TEB and TEA for fully alternating copolymerization of PO and CO_2 [65], a route to prepare biodegradable poly(ester-*co*-carbonate)s via a terpolymerization of PO, CO_2, and PA has been developed by Meng et al. With using the TEB/TEA Lewis pair at a very low catalyst loadings (PO/PA/TEB/TEA:16000/2000/4/1), the productivity was 1.2 kg polymer/g catalyst with low cyclic carbonate content (< 5%) at 65 °C [66]. Moreover, the molecular weight of the resultant copolymer was up to 131 kg mol^{-1} with a relatively narrow $Đ$ of 1.31. The TEB/TEA Lewis pair is also applicable for various epoxides and cyclic anhydrides with CO_2 to provide random copolymers and the glass transition temperatures of the resultant polymers can be adjusted from 34 to 131 °C. These poly(ester-*co*-carbonate)s with random or block structures can be simply tuned by selecting different organic bases for coupling with TEB [67]. The incorporation of polyester segments effectively modifies the thermal and mechanical properties of poly(propylene carbonate) with 43% polyester incorporated showing a higher T_g value (47 °C) than commercial PPC.

In 2018, Zhang and coworkers firstly reported a metal-free, living copolymerization of epoxides and carbonyl sulfide (COS) through the combination of Lewis bases (e.g.: phosphazene, amidine, and guanidine) and thioureas as hydrogen-bond donors, a poly(monothiocarbonate) was synthesized with 100% alternating degree, >99% tail-to-head content, controlled M_ns up to 98.4 kg mol^{-1}, and narrow polydispersities (1.13–1.23) [68]. They proposed the mechanism of supramolecular specific recognition of thiourea to epoxide and base to COS promote the copolymerization cooperatively. In 2019, Meng and coworkers [69] reported the combination of (thio)ureas and organic bases, as shown in Figure 3.11, as a cooperative metal-free catalyst for the preparation of fully alternating copolymer of epoxides and cyclic anhydrides. They used a library of commercial (thio)ureas and organic bases for various polyester syntheses. In particular, the U1/PPNCl pair (Figure 3.11) stood out as the most active and effective Lewis pair for the ROCOP of epoxides and cyclic anhydrides. A high TOF (456 h^{-1}) was achieved for the CHO/PA copolymerization, which is comparable to the results using metal catalysts. Polyesters with medium to high molecular weights up to 34.6 kg mol^{-1} were obtained, and most exhibited narrow dispersity as low as 1.07. A mechanism with the two initiating routes by (thio)urea/PPNCl Lewis pair was proposed for copolymerization, as presented in Figure 3.12. In the pathway A, Cl^- initiates the polymerization at one direction, while in the pathway B, the polymer chain propagates at two directions by the carboxylate anion. Cl^- and carboxylate anion can be generated by the activation of the (thio)urea/PPNCl Lewis pair. The Lewis pair can also activate the monomers during the reaction as the previous report [70, 71] and the ring-opening of epoxides

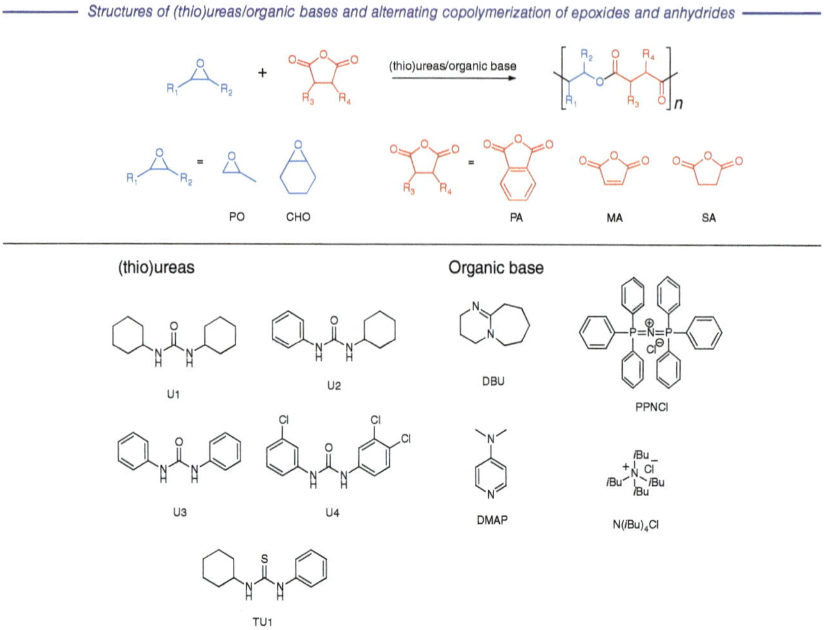

Figure 3.11 Represented (thio)ureas catalysts as the Lewis acid for epoxides/cyclic anhydrides copolymerization.

is proved as the rate-determining step. Then as the monomer continue to insert the polymer chain, the polyester segments are formed.

Apart from TEB and (thio)urea, Zhu and Kou [72] used the protic acids as Lewis acid for copolymerization of epoxides and cyclic anhydrides with organic base in 2022. The combination of organic base and protic acid were demonstrated as effective binary catalysts for fully alternating copolymerization of CHO and PA. A series of commercial organic bases and protic acids were employed and the effects of different types and ratios of organic bases and protic acids were performed, indicating that DMAP/diphenyl phosphate showed higher activity (TOF = $574\,h^{-1}$) and perfectly alternating selectivity (>99%), producing polymers with narrow polydispersity (1.08–1.22). It should be noted that the (thio)urea/organic base by Meng et al. and the protic acid/organic base are metal-free, borane-free catalysts for the copolymerization of epoxides and cyclic anhydrides.

3.2.3.3 Bifunctional Lewis Pair

It is a general strategy to improve the catalyst performance by covalently linking Lewis acid and base in one molecule, that is, bifunctional Lewis pair. Recently, by analogy of attaching a co-catalyst (sometimes, catalyst promoter) to a metal ligand framework [48, 49], the organocatalyst pending quaternary ammonium salt (QAS) shows the great catalytic activity for copolymerization of epoxides with cyclic anhydrides. In 2021, Wu and coworkers [73] reported highly effective copolymerization of epoxides with cyclic anhydrides, using a series of borane

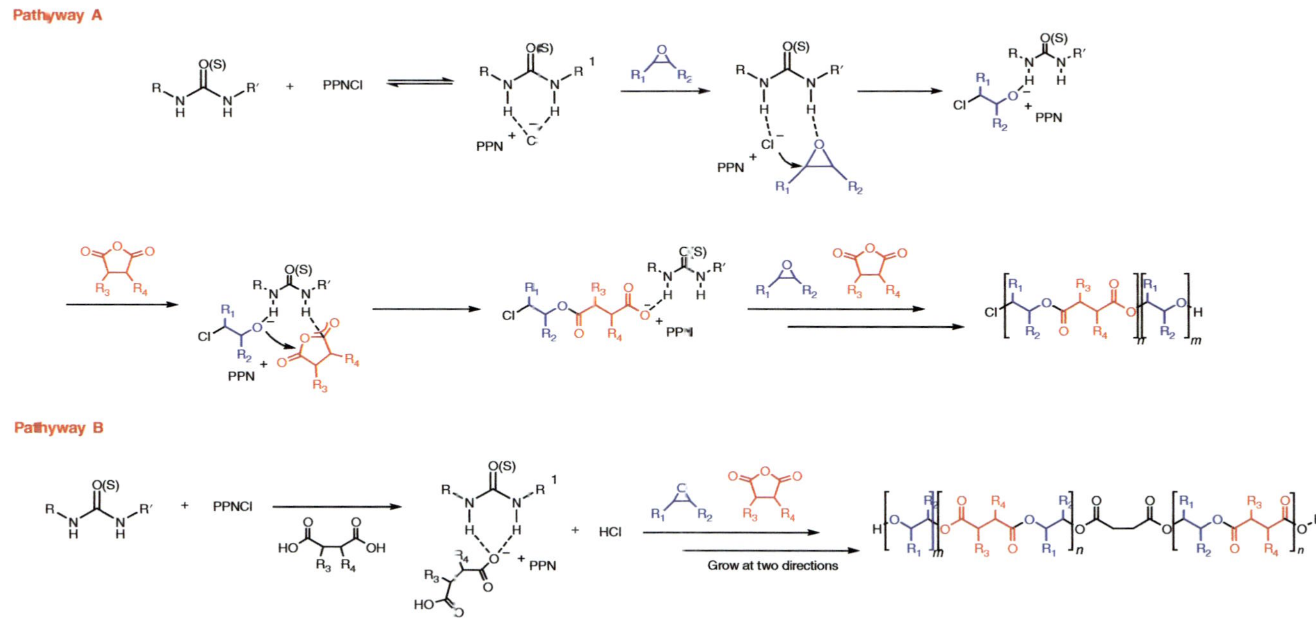

Figure 3.12 Proposed mechanism for the alternating polymerization of epoxides with cyclic anhydrides catalyzed by the (thio)urea/PPNCl pair. Source: Lin et al. [69]/Royal Society of Chemistry.

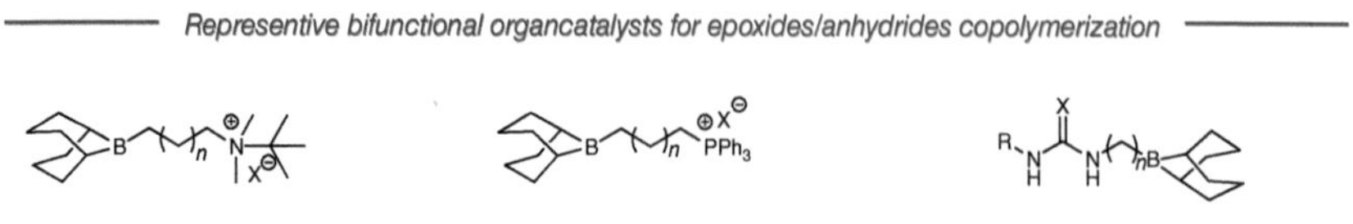

n = 1,2,3
X = Cl,Br,I
cat. 23

n = 1,2,3,4
X = Cl,Br,I
cat. 24

n = 3,4,5,6 X = O,S
R = Ph,3-CF_3Ph,3,5-$(CF_3)_2$Ph,C_6H_{11},$(CH_3)_2$CH
cat. 25

Figure 3.13 Representative bifunctional organocatalysts for the copolymerization of epoxides and cyclic anhydrides.

compounds tethered with QAS groups (Figure 3.13, cat. 23). This organoboron catalysts could withstand reaction temperature as high as 180 °C in the copolymerization of CHO and PA without any side reactions being observed. The M_ns of the produced poly(CHO-*alt*-PA) arrived 94.5 kg mol^{-1} with a polydispersity of 1.19. The authors reported an unprecedented turnover number (TON) of up to 9900, equivalent to the productivity of 7.4 kg of polyester/g of catalyst, at a low monomer/catalyst feeding ratio (CHO/PA/catalyst = 20000/10000/1) at 150 °C. Li and coworkers reported a bifunctional phosphonium borane Lewis pair, which serves as either initiator or catalyst for the copolymerization [74]. This catalyst featuring pentamethylene –$(CH_2)_5$– linked triphenylphosphonium bromide (Lewis base) and a borabicyclo[3.3.1]nonane moiety (Figure 3.13, cat. 24) showed a high TOF of 1920 h^{-1}, thermal robustness (120 °C) and a good alternating selectivity (>92%) for the copolymerization of PA and CHO. Furthermore, the cyclic anhydrides were extended to bio-renewable SA and MA, as well as industrially relevant epoxide PO. The presence of two types of initiating species derived from Br$^-$ and cyclohexane-1,2-diol (CHD) in the polymerization setup were verified and a higher molecular weight fraction (51.8 kg mol^{-1}) of the produced poly(PA-*alt*-CHO) were obtained with a narrow polydispersity ($Đ$ = 1.10).

Tao and coworkers [75] reported a new organocatalyst combined a (thio)urea moiety and a Lewis acidic boron center (Figure 3.13, cat. 25). The power of these catalysts was illustrated in the copolymerization of epoxides with cyclic anhydrides, giving excellent selectivity and activity. The ring-opening copolymerization of epoxides with cyclic anhydrides mediated by these catalysts achieved >99% polyester selectivity, and higher activity with TOFs ranging from 162 to 408 h^{-1}, as compared to either of the di-thiourea analogues or any combinations of them. This strategy reveals that tug-of-war between two distinct catalytic sites can enhance the catalytic performance. The improvement is achieved by using H-bond donors to pull off the propagating chain-end, simultaneously enhancing the role of monomer activation and nucleophilicity of the propagation intermediates. This innovative approach developed by the Tao group points to a novel design strategy for this area of catalysis.

Since the use of binary TEB/organobase catalyst for the copolymerization of epoxide and cyclic anhydride was first reported by Zhang and coworkers [43], many research groups have developed effective organocatalysts for this copolymerization. The developed organocatalysts, from nucleophilic tertiary amines to superbase organophosphazenes, from single component catalysts (initiators) to binary Lewis pairs, and bifunctional Lewis pairs, show satisfied polymerization control, selectivity and activity. For the PO/MA copolymerization, TEB combined with PPh_4Br show high activity with a turnover frequency of $102\,h^{-1}$. TEB with CTPB show a TOF of up to $500\,h^{-1}$ and afford a high molecular weight polyester ($118.5\,kg\,mol^{-1}$) for PO/PA copolymerization. Of significance, the (thio)ureas as the Lewis acid show the great potential to replace the TEB because (thio)ureas are easy to obtain, low-cost and more safer during use. The U1/PPNCl pair show a high TOF of $456\,h^{-1}$ toward the copolymerization of CHO with PA. Up to date, the copolymerizations of several epoxides with cyclic anhydrides catalyzed by organocatalysts shows comparable activity and selectivity to metal catalysts. In particular, organocatalysts exhibit stronger ability to the oxygenated monomers than metal catalysts [12].

3.3 The Copolymerization of Oxetane with Cyclic Anhydride

In addition to high-ring tension epoxides (i.e. three-membered ethers), four-membered ethers, especially oxetane (OX), have also attracted more attention as key monomers for polymerizations, because it is a derivative of biomass chemical 1,3-propanediol. Oxetane is an active unit in the molecular structure of many natural and synthetic compounds with various biological or pharmaceutical activities such as anti-cancer, anti-HIV, and inhibition of glutamine synthase. Considering the ability to polymerize, oxetane has low ring strain and therefore is less even unreactive upon nucleophilic attack. There are few reports on the copolymerization of oxetane with cyclic anhydride including the use of metal catalysts [76, 77].

In 1999, Endo and coworkers reported the first example of oxetane copolymerization with cyclic anhydrides by a titanium complex with large steric hindrance (Figure 3.14a) [79]. In this example, the titanium complex was effective to the copolymerization of oxetane with different cyclic anhydrides having 5, 6 and 7-membered ring at 120 °C. Unfortunately, the resultant copolymers had low molecular weights, low polyester selectivity and small amounts of undesired polyether linkage (34–54%).

A few years later, Nishikubo and coworkers reported copolymerization of oxetanes with cyclic anhydrides catalyzed by quaternary onium salts (Figure 3.14b) [78]. The copolymerization of sugar derivative 3-(benzyloxymethyl)-3-ethyloxetane (BMEO) with PA was carried out very well, afforded polyesters with fully alternating sequences. By changing the substitutes of oxetanes and cyclic anhydrides, the resultant copolymer was obtained with moderate molecular weight up to $11.1\,kg\,mol^{-1}$.

Recently, Buchard and coworkers used metal salen complexes and PPNCl as catalysts to successfully copolymerize sugar-derived oxetane **1** and cyclic anhydrides, forming a new type of fully alternating sugar-derived polyesters (Figure 3.15) [80]. The obtained polyesters exhibit the good thermal stability with the initial thermal

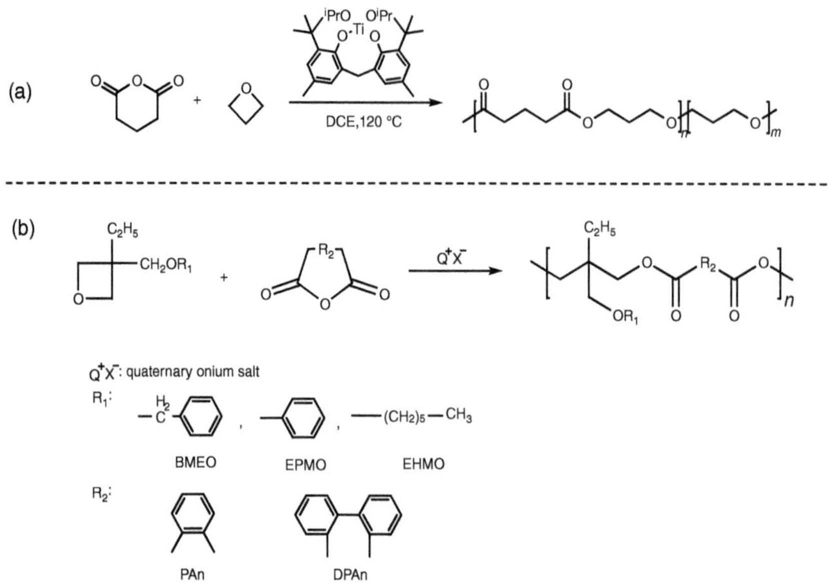

Figure 3.14 (a) Copolymerization of OX with GA by titanium complexes as initiators. (b) The copolymerization of related oxetane with cyclic anhydrides. Source: Kameyama [78]/American Chemical Society.

decomposition temperature of >212 °C, and a board range of glass transition temperatures from 60 to 145 °C. These polyesters have the 100% renewable content prepared by copolymerizing various bio-based monocyclic, bicyclic, and tricyclic anhydrides (as shown in Figure 3.16) with the sugar-derived oxetane **1**. However, the copolymerization catalyzed by metal systems need a long reaction time (over 20 h) with low TOF ($< 4.3\ h^{-1}$) and also produce a small amount of polyether linkages (<5%), which limited the production of these polyesters in an industry scale.

Xu and coworkers recently successfully performed the organocatalytic copolymerization of this bio-based oxetane with cyclic anhydrides in a controlled manner using the TEB/t-BuP_1 catalyst and alcohol as initiator [81]. This metal-free catalyst can make the copolymerization to carry out in a shorter reaction time (6–9 h) with a higher TOF ($8.4 \sim 11.5\ h^{-1}$) and >99% selectivity. Taking the **1**/SA copolymerization as an example, they proposed a Lewis pair cooperative mechanism (Figure 3.16). At the initial stage, the hydroxyl is deprotonated by superbase t-BuP_1 to form an alkoxide anion, and followed by the attack of the alkoxide anion to SA, a succinate salt is formed. In equilibrium, the proton of the t-BuP_1 cation [t-BuP_1H] can be back to the succinate salt to form dormant species. As TEB activates oxetane, nucleophilic attack on the activated oxetane rapidly forms an alkoxide intermediate (species B in Figure 3.16) fast. At this stage, proton exchange equilibrium still exists. The alkoxide then reacts repeatedly with SA and **1** to form t hydroxyl-terminated or carboxyl-terminated copolymers.

As discussed above, the copolymerization of oxetanes with cyclic anhydrides is promising because the partially or fully biomass-derived, oxygen-rich copolymers

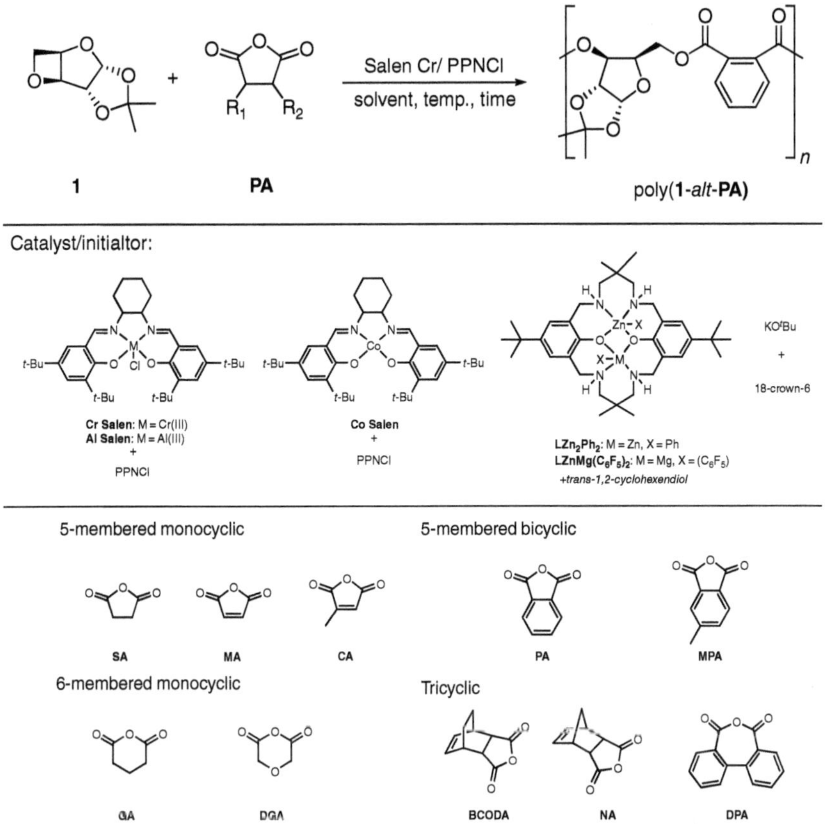

Figure 3.15 The initial and catalytic systems for the copolymerization of **1** and **PA**.

can be synthesized. Overall, due to the low ring strain of oxetane derivatives and poor catalyst efficiency, it is highly desired to develop active metal-free catalyst, and low cost oxetane-based monomers in future.

3.4 The Copolymerization of Tetrahydrofuran with Cyclic Anhydride

The copolymerization of cyclic anhydrides with 3 and 4-membered cyclic ethers can be carried out in an anionic manner, as the above discussion. 5 or larger membered cyclic ethers have nearly zero enthalpy changes that inhibited their ring-opening polymerization via anionic way. For example, tetrahydrofuran (THF), as a typical five-membered cyclic ether, is unable to undergo anionic ring-opening polymerization. In fact, THF can only be carried out by cationic polymerization manner through the activation of ether bond, as discussed below.

There are few reports on copolymerization of cyclic anhydrides with THF. Takasu and coworkers [82] reported the copolymerization THF with cyclic anhydrides

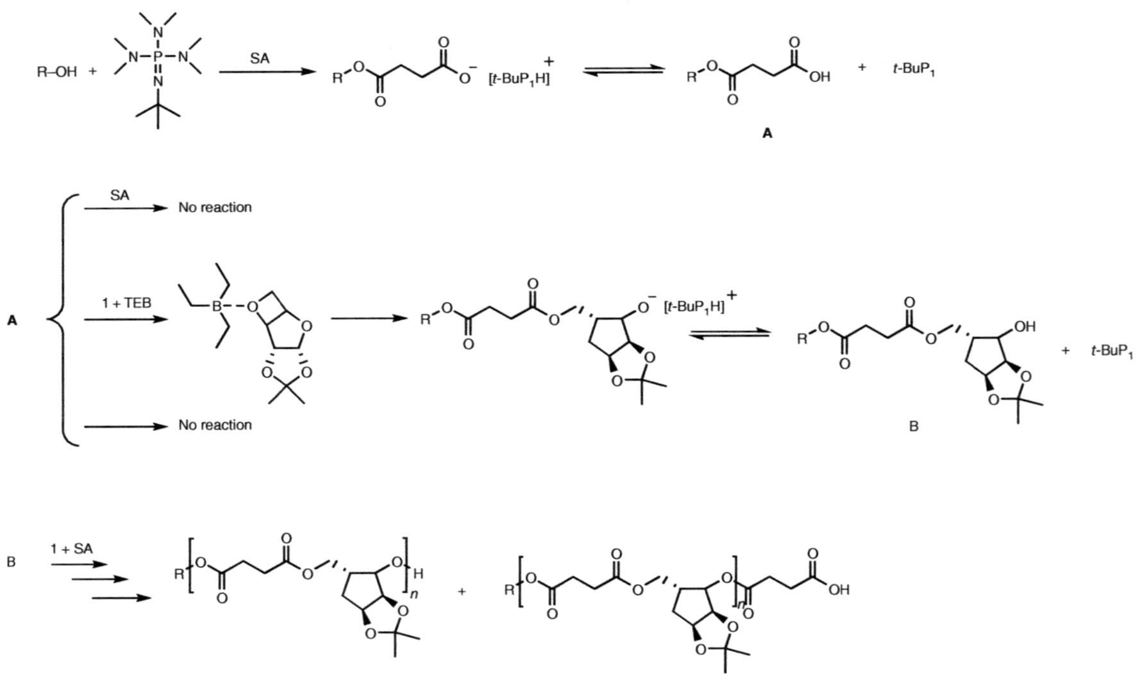

Figure 3.16 Plausible mechanistic pathways for the copolymerization of **1** and **SA**. Source: Zheng et al. [81]/John Wiley & Sons.

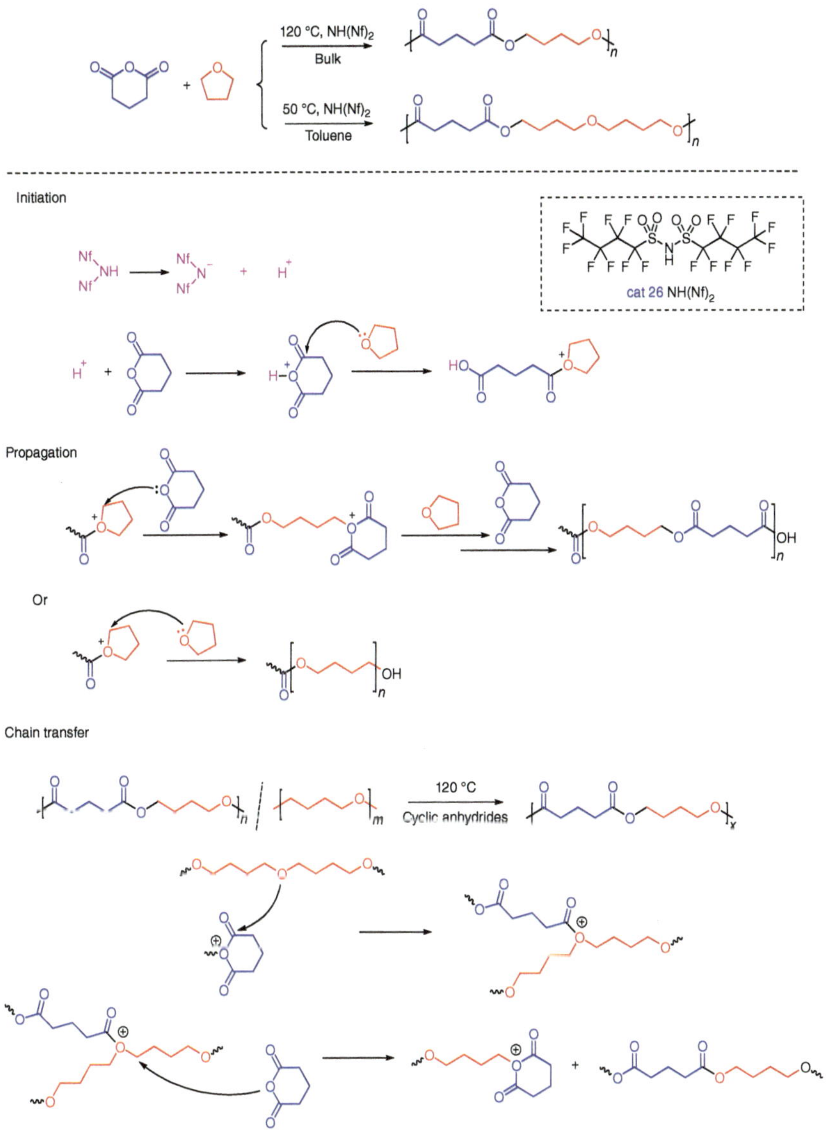

Figure 3.17 Initiation and chain growth of $(Nf)_2NH$-catalyzed copolymerization of THF and cyclic anhydrides. Source: Tang et al. [82]/John Wiley & Sons.

in 2012. They used nonafluorobutanesulfonimide (Nf_2NH) as a cationic catalyst (Figure 3.17) for the copolymerization at 50–120 °C. Through the copolymerization of THF with several cyclic anhydrides, they found that ether units were temporarily formed at high temperatures (100–120 °C), and the ether units were subsequently converted into ester units via chain transfer reactions. They pointed out that at the propagating stage, the THF can homopolymerize as ether units and cyclic anhydrides remained in the mixture. Then the ether oxygen can nucleophilic attack the propagating center of the cyclic anhydride and subsequent insertion of the remaining cyclic anhydrides monomer. After the chain transfer reaction, a new oxonium ion of cyclic anhydride was regenerated and finally resulting in completely alternating polyester. In contrast, at the low temperatures of 25–50 °C, the ether units produced first are not converted into polyester, thus obtaining poly(ester-*r*-ether).

Currently, THF/cyclic anhydride copolymerization can only be conducted catatonically. THF, as the product of complete hydrogenation of bio-based furan, is a very cheap bio-based chemical, and thus its copolymerization with cyclic anhydride is worthy of study. Note that, other low strain cyclic ether with more than six-membered rings cannot be homopolymerized by either anionic polymerization or cationic polymerization.

3.5 The Copolymerization of Cyclic Acetal (Aldehyde) with Cyclic Anhydride

Cyclic acetals and aldehydes are very common chemicals, most of which can be obtained from biomass (Figure 3.18). Cyclic acetal (1–5 in Figure 3.18), known as low-tension heterocyclic compounds, has a ring larger than five-membered and can be (co)polymerized catatonically. In view of thermodynamics, cyclic

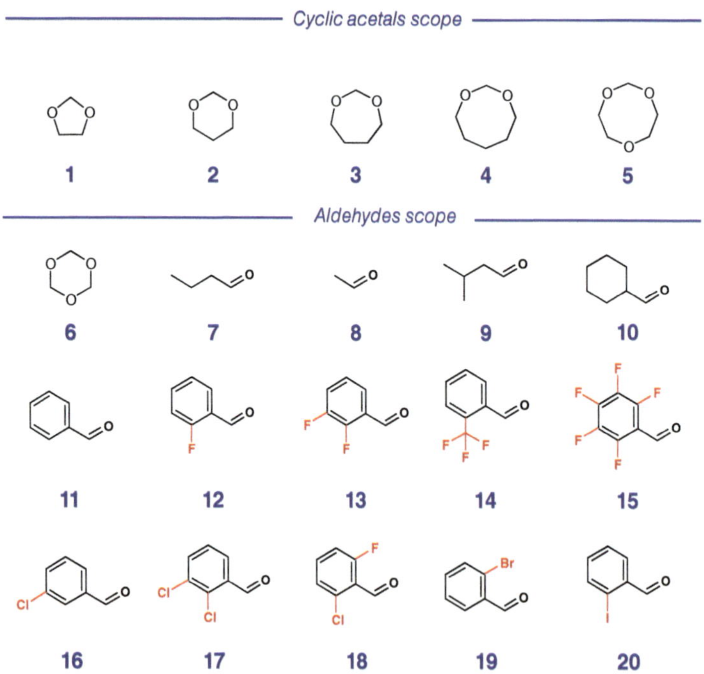

Figure 3.18 The representative acetals and aldehydes monomer for copolymerization with anhydrides.

acetal tends to polymerize more easily than THF. In addition, cyclic acetals can be synthesized from the simple condensation reaction small molecule diol and aldehyde that are common industrial feedstocks. Plenty of aldehyde compounds (7–10 in Figure 3.18) can be extracted directly from natural plants or synthesized by the dehydrogenation of renewable alcohols on a large scale. Aromatic aldehyde with diverse structures can also be easily obtained (11–20 in Figure 3.18), which can provide heat-resistant structure or functional groups for the potential polymers. However, these mono-substituted (except formaldehyde) aldehydes are quite difficult to (co)polymerization due to their low enthalpy of polymerization. These monomers are the ideal renewable and cheap raw material to produce polyesters.

Cyclic acetals are a class of low ring tension monomers. No reports on their (co)polymerization through anionic mechanism. Only a small part of them can polymerization via cationic mechanism by themselves or with other vinyl monomers [83]. In 2021, Zhang and coworkers reported the first example of copolymerization of cyclic anhydrides with cyclic acetals [84], using commercially available Lewis acids or Brønsted acids as the cationic catalysts under mild conditions, a new family of poly(acetal-*alt*-ester)s were obtained with high activity within short reaction time. The copolymerization of 5 cyclic acetals and 9 six-membered cyclic anhydrides afforded 45 fully alternating poly(acetal-*alt*-ester)s. They proposed a copolymerization mechanism involved three steps, as shown in Figure 3.19. First, the cyclic acetal polymerizes very rapidly, and then the polyacetal formed degrades into acetal monomers until an equilibrium of polyacetal and cyclic acetal. Finally, the cyclic acetal is copolymerized with the cyclic anhydrides to form poly(acetal-*alt*-ester)s, shifting the equilibrium to the left. As a newly developed polyester, the acetal group was introduced into main chain contribute to the polyesters with quite low glass transition temperatures from −64 to −27 °C. In addition, the obtained polyesters with a dicarboxylic structure and high thermal decomposition temperature (275–324 °C) show fluid state at room temperature. This newly developed method is robust and efficient and a wide range of catalysts

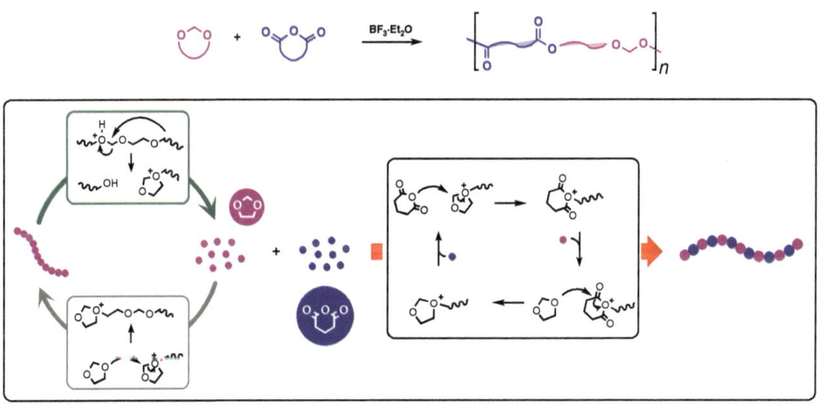

Figure 3.19 The copolymerization of epoxides with acetals. Source: Zhang et al. [84]/John Wiley & Sons.

such as $BF_3{\cdot}Et_2O$, $B(C_6F_5)_3$, MeOTf, and $NH(OTf)_2$ are effective at wide reaction temperatures (25–140 °C).

On the basis of understanding the copolymerization of cyclic acetals with cyclic anhydrides, Zhang and colleagues pointed out that the one-step upcycling of polytetrahydrofuran (PTHF) to polyesters can be completed through a cationic route [85]. The complete recycle of PTHF to THF is quite difficult because of the existence of equilibrium of "PTHF ⇌ THF" during depolymerization process. By introducing cyclic anhydrides into the PTHF depolymerization process, the equilibrium of "PTHF ⇌ THF" will be shift to the formation of THF that was rapidly consumed by its copolymerization with cyclic anhydride. Simple reaction conditions and the use of commonly used catalysts were suitable and successful in providing 18 polyesters with extended 18 cyclic anhydrides. This strategy provides an upcycling method to reuse waste PTHF.

Different from the large membered cyclic ethers and cyclic acetals, mono-substituted aldehydes (as shown in Figure 3.18) are a class of linear molecules. Due to the low polymerization enthalpy change, most mono-substituted aldehydes are not used as the monomers for polymerization reactions, and only a very small number of aldehydes, such as formaldehyde (in the form of paraformaldehyde, **6** in Figure 3.18), can be easily polymerized. The oxygen-rich aldehydes are one-carbon feedstocks (like CO_2, COS) and good candidates to prepare low-carbon polymer. In 2023, Zhang and coworkers realized the copolymerization of mono-substituted aldehydes and cyclic anhydrides for the first time (Figure 3.20) [86]. The commercially available 56 aldehydes and 5 cyclic anhydrides were employed successfully to prepare 60 fully alternating polyesters within several minutes or seconds. The resultant polyesters

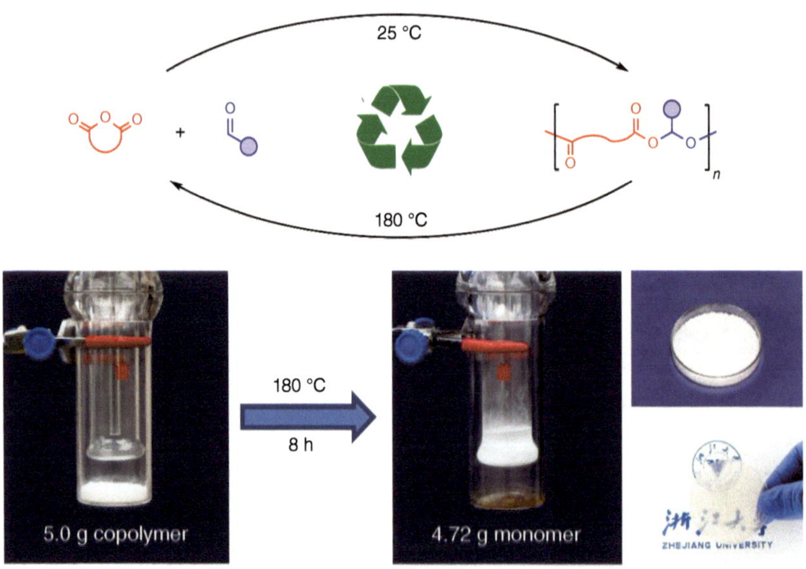

Figure 3.20 The copolymerization of aldehydes with cyclic anhydrides. Source: Reproduced with permission from [86]/Springer Nature/CC BY 4.0.

show a broad range of glass transition temperatures from −57 to 73 °C because the formed acetal linkage makes the polyester flexible. All the polyesters exhibit the thermal degradation temperatures from 170 to 230 °C.

Of great interest, the copolymerization exhibits the chemical reversibility attributed to its relatively low enthalpy, which makes the resulting polyesters performed a closed-loop recycling to monomers at high temperatures, and thus can be recycled. For example, in the copolymerization of acetaldehyde with GA, the ceiling temperature of the resultant polymer is estimated to be ca. 141.9 °C at the monomer concentration of 4.4 M. In addition, take the copolymer of 3,5-dichlorobenzaldehyde and GA as an instance, 5.0 g copolyesters can be recycled to 4.72 g monomers in quantitative yield (up to 94%) under 180 °C by a simple sublimation process, as shown in Figure 3.20. The obtained monomers can be repolymerized to polyesters with the comparable properties to the start polyester.

3.6 Polyester-Based Block Copolymer

As discussed above, the copolymerization of cyclic anhydrides with the above-mentioned oxygenated monomers, e.g. epoxides, oxetanes, THF, cyclic acetals and aldehydes, is a powerful tool to the synthesis of new polyesters with diverse structures. Of course, it is highly desired to introduce other polymer blocks into these polyesters, to improve the overall performances. Previous studies have shown that borane compounds are effective to either epoxides/cyclic anhydrides copolymerization or epoxide/COS (or CO_2) copolymerization and can accommodate the shift of different growing anions [39, 41, 43]. Therefore, these catalysts are potential for the preparation of block copolymers. Because of the versatility of TEB-based catalyst system and diversity of epoxides and cyclic anhydrides, terpolymerization through kinetic determination, or sequential polymerizations of monomers have been reported for the synthesis of block copolymers with different compositions and structures.

Herein, we introduce the synthesis of block copolymers containing polyester and polyether blocks using organocatalysts. As the example shown in Figure 3.21a, the copolymerization of epoxide and cyclic anhydride affords a block copolymer of polyester-*b*-polyether [87, 88]. Zhao and coworkers reported the successful synthesis of block copolymers with strict $(AB)_nB_m$ type sequence from a mixture of two easily available PA and PO or EO (Figure 3.21a) [87], using TEB as the catalyst and phosphazene base (t-BuP_1) as the cocatalyst. Block copolymers with controlled high molecular weights (up to 56.6 kg mol^{-1}) can be formed in one step, and the block architectures can be enriched by the use of mono-, di-, or tetra-hydroxyl initiators.

PLA, as a crystalline polyester, is often used as the rigid segments in making block copolymers. The binary TEB/organic base catalyst for PO ROP or copolymerization with other monomer such as CO_2 make the diverse polycarbonates segments incorporated and tune the performance of the final block copolymers. According to the activation mechanism of TEB, binding to TEB can weaken the nucleophilicity

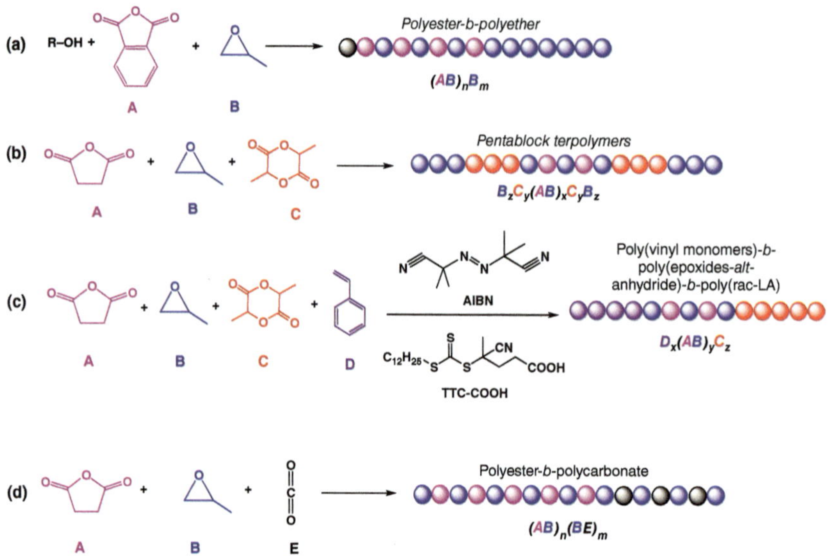

Figure 3.21 (a) One-step approach to ester-ether block copolymers, (b) $A_zC_y(AB)_xC_yA_z$ pentablock terpolymers, (c) poly(vinyl monomers)-*b*-poly(epoxides-*alt*-anhydride)-*b*-poly (*rac*-LA) triblock copolymers, and (d) polyester-*b*-polycarbonate block copolymers.

and basicity of the growing oxyanion, thereby slowing down the ring-opening reaction of cyclic esters. By this method, Wang and coworkers [89] reported the one-step synthesis of ester-ether block copolymers via kinetic control. From a mixture of epoxides, cyclic anhydrides and LA, the copolymerization of epoxides with cyclic anhydrides can be successfully performed first through the Lewis pair (TEB/DBU = 2 : 1) in the mixed monomers with water as an initiator. Then the ROP of *rac*-LA, and finally the ROP of epoxides. $A_zC_y(AB)_xC_yA_z$ pentablock terpolymers can be obtained with M_n of 3.4–13.4 kg mol^{-1} and $Đ$ of < 1.3 without transesterification reaction (Figure 3.21b).

Furthermore, the shifting of growing chain center is also a useful method to prepare block copolymers. For example, using a trithiocarbonate compound (TTC–COOH) bearing a carboxylic group as a chain transfer agent, simultaneous chain propagation of ROCOP/RAFT and ROP/RAFT sequentially occurred from a mixture of vinyl monomers, epoxides, cyclic anhydrides, and LA in the presence of TEB/DBU (3/1) and AIBN (Figure 3.21c) [90]. A poly(vinyl monomers)-*b*-poly(epoxides-*alt*-anhydride)-*b*-poly(*rac*-LA) triblock copolymer was successfully generated with M_n of 8.9–29.2 kg mol^{-1} and $Đ$ of <1.2. In this protocol, three polymerization mechanisms simultaneously occur for the formation of the final block copolymers.

The block copolymers with polyester and polycarbonate blocks have also been widely investigated. Terpolymerization of CO_2/epoxides/cyclic anhydrides can afford such block copolymers, as shown in Figure 3.21d. Through polymerization of four monomers, i.e. CO_2, CHO, PO and PA using the TEB/PPNCl catalyst, Meng and coworkers [91] successfully synthesized a quadri-block copolymer with high T_g

(>90 °C) and tensile strength (up to 54.8 MPa). This block copolymer has nearly the same thermal properties, mechanical properties, and transparency as polystyrene. Recently, a cost-effective dual urea/organic base catalyst also was used to synthesize triblock copolyesters via a switchable terpolymerization of mixed monomers of epoxides, PA and CL by the same group [92]. By the combination of ureas with organic base to evaluate the catalytic activity and switchable selectivity, the U1/PPNCl Lewis pair was demonstrated with the highest catalytic activity to synthesize the block copolymer. The formation of these block copolymers from mixed monomers is decided by polymerization kinetics and the existence/exhaustion of PA is the switch between ROAC and ROP reaction cycles. By using the diol as the initiator, the terpolymerization of CHO/PA/CL started with the copolymerization of PA with CHO until PA was completely consumed, followed by the ring opening polymerization of CL. Finally the PCL-*b*-P (PA-*alt*-CHO)-*b*-PCL well-defined triblock copolyesters were obtained.

Phosphazene bases also show great potential in the synthesis of block copolymers. Li and coworkers synthesized polyester-*b*-polycarbonate di-block copolymers with very little tapering through terpolymerization of CHO, PA, and CO_2 by the PPNCl/TEB Lewis pair [93]. The binary TEB/C_3N_3-Py-P_3 (Figure 3.8) system also was reported by the same group for the terpolymerization of CO_2/epoxide/anhydride [94]. They revealed the activity and chemoselectivity were highly dependent upon the molar ratio of C_3N_3-Py-P_3 to TEB, and sequence-controlled poly(ester-carbonate) were synthesized through one-pot/one-step manner by tuning the phosphazene/TEB ratios. At molar ratio of 1/0.5 (C_3N_3-Py-P_3 to TEB) with BDM as the initiator, the terpolymerization of CHO, PA, and CO_2 exhibited an unprecedentedly high chemoselectivity and well-defined triblock PCHC-*b*-poly(PA-*alt*-CHO)-*b*-PCHC copolymers were obtained. By increasing the C_3N_3-Py-P_3/TEB molar ratio to 1, a tapered copolymer could be obtained. With further increasing the amount of TEB, the random copolymers with high content of polycarbonate were synthesized. Similarly, Gnanou, Feng and coworkers used the TEB with PPNCl or Bu_4NN_3 as the catalyst for the terpolymerization of epoxides (PO or CHO), CO_2 and aliphatic or aromatic anhydrides (SA, PA), afforded a tapering poly(ester-*b*-carbonate) [95].

3.7 Conclusions and Outlook

To sum up, in this chapter, we introduce the research and development of organocatalytic copolymerization of cyclic anhydrides with epoxides, oxetanes, THF, cyclic acetals and aldehydes. These monomers are clearly rich of oxygen, and thus the elemental composition of the resultant polymers has a low carbon content, which opens up a new path for low-carbon production of polymer materials [12].

In methodology, it is powerful to the synthesis of new family of polyesters with diverse structures. First, diverse oxygenated monomers can be used for the copolymerization. EO, PO are low cost and currently produced in industry in large quantities, which is the basis for competition with traditional polymers. Cyclic anhydrides, for example, MA, SA are usually prepared from C4 products either petrochemical industry or biomass derivatives. Oxetane can be derived from 1,3-propanediol, THF can be produced from 1,4-butanediol that is currently produced in large scale and low cost, cyclic acetals are synthesized from small molecule diol and aldehyde, providing unlimited chances to the synthesis of poly(acetal-ester) with excellent thermal properties and tunable mechanical properties.

Second, organocatalyst is proved to be very suitable to the polymerization of oxygenated monomers [45], without any metal ion residues in final products. In addition, the advantages of organocatalyst for the polymerization are: (i) thousands of low-cost small molecules can be directly used. (ii) either binary Lewis acid-base system or covalent-linked Lewis acid-base system with subtle design, based on supramolecular interactions with monomers/initiators, provides infinite combinations for polymerizations, and new understandings on polymerization catalysis. (iii) compared with metal catalysts, organocatalysts with lower molecular weight or fewer atomic numbers can achieve efficient and highly selective polymerizations.

Third, the resultant polyesters have diverse structures, and thus tunable performances, (bio) degradability and recyclability. For example, the copolymerization of aldehydes with cyclic anhydrides can produce up to 60 unprecedented polyesters with new structures. These new structures endow these polyesters with broaden range in thermal properties. Importantly, the new properties of this copolymer enable circular recycling of these polyesters.

However, the formation of polyesters and their blocks using above monomers and organocatalysts are an emerging technology in polymer chemistry, there are some challenges that should be kept in mind.

(1) We need low-cost monomers that can be produced at large-scale. Polyesters are a new family of oxygen-rich polymers, the first principle of wide application of these polymers is the price. This is not a science question but a true one that we must keep in mind when looking for or designing new monomers. That is, how can we find low-cost monomers for developing useful, sustainable polymers? The authors believe that the monomers in this chapter provide choices to make such polymers.
(2) We need thermally robust, highly active low-cost organocatalysts. The organocatalytic polymerizations have great gap with traditional metal catalyst systems. The improvements of catalyst activity and the category expansion of metal-free catalysts are still challenges. At this moment, the borane related catalyst is the focus of attention, because of their good catalyst activity, but the air- and moisture- sensitive character limited their applications. Bifunctional catalysts exhibit improved catalytic activities, as most of previous works [73–75], but the complex synthesis, high cost and the removal from the products are still the challenge. The catalyst design of new structure, borane-free and halogen-free is an important development direction in the future.
(3) We need oxygenated polyesters with similar properties as traditional polymers while is sustainable. The introduction of oxygen atoms in the form of ester, acetal groups into the polymer backbone is for potential sustainability including

(bio)degradability and recyclability. While the usability (basically, comparable mechanical and thermal properties similar to traditional polyolefins) of a polymer is the first need. Therefore, the subtle control over the sequence, regioselectivity and even stereoselectivity are particularly important during synthesis. However, the polyesters herein show rather limited properties.

Finally, this chapter discusses the organocatalytic copolymerizations of cyclic anhydrides with various oxygen-containing monomers. These innovative approaches hold great promise for expanding the classification of polyester and creating new materials that satisfy usability and sustainability. The hope of is that these oxygenated polymers, low-carbon alternatives to fully carbon-rich polymers, will make our planet a better place.

References

1 Odian, G. (2004). *Principles of Polymerization*, 4e. Hoboken, NJ: Wiley.

2 Hillmyer, M.A. and Tolman, W.B. (2014). Aliphatic polyester block polymers: renewable, degradable, and sustainable. *Acc. Chem. Res.* 47 (8): 2390–2396.

3 Gross, R.A. and Kalra, B. (2002). Biodegradable polymers for the environment. *Science* 297 (5582): 803–807.

4 Olson, D.A., Gratton, S.E., DeSimone, J.M., and Sheares, V.V. (2006). Amorphous linear aliphatic polyesters for the facile preparation of tunable rapidly degrading elastomeric devices and delivery vectors. *J. Am. Chem. Soc.* 128 (41): 13625–13633.

5 Brown, A.H. and Sheares, V.V. (2007). Amorphous unsaturated aliphatic polyesters derived from dicarboxylic monomers synthesized by *Diels–Alder* chemistry. *Macromolecules* 40 (14): 4848–4853.

6 Lecomte, P. and Jérôme, C. (2012). *Recent Developments in Ring-Opening Polymerization of Lactones*, 245. Berlin, Heidelberg: Springer-Verlag.

7 Ikada, Y. and Tsuji, H. (2000). Biodegradable polyesters for medical and ecological applications. *Macromol. Rapid Commun.* 21 (3): 117–132.

8 Engels, H.W., Pirkl, H.G., Albers, R. et al. (2013). Polyurethanes: versatile materials and sustainable problem solvers for today's challenges. *Angew. Chem. Int. Ed. Engl.* 52 (36): 9422–9441.

9 Paul, S., Zhu, Y., Romain, C. et al. (2015). Ring-opening copolymerization (ROCOP): synthesis and properties of polyesters and polycarbonates. *Chem. Commun. (Camb.)* 51 (30): 6459–6479.

10 Longo, J.M., Sanford, M.J., and Coates, G.W. (2016). Ring-opening copolymerization of epoxides and cyclic anhydrides with discrete metal complexes: structure-property relationships. *Chem. Rev.* 116 (24): 15167–15197.

11 Sanford, M.J., Peña Carrodeguas, L., Van Zee, N.J. et al. (2016). Alternating copolymerization of propylene oxide and cyclohexene oxide with tricyclic anhydrides: access to partially renewable aliphatic polyesters with high glass transition temperatures. *Macromolecules* 49 (17): 6394–6400.

12 Xia, Y., Zhang, C., Wang, Y. et al. (2023). From oxygenated monomers to well-defined low-carbon polymers. *Chin. Chem. Lett.* 35: 108860.

13 Fischer, R.F. (1960). Polyesters from expoxides and anhydrides. *J. Polym. Sci.* 44 (143): 155–172.

14 Lustoň, J. and Vašš, F. (1984). *Anionic Copolymerization of Cyclic Ethers with Cyclic Anhydrides*, vol. 56. Berlin, Heidelberg: Springer.

15 Aida, T. and Inoue, S. (1985). Catalytic reaction on both sides of a metalloporphyrin plane: alternating copolymerization of phthalic anhydride and epoxypropane with an aluminum porphyrin-quaternary salt system. *J. Am. Chem. Soc.* 107 (5): 1358–1364.

16 Aida, T., Sanuki, K., and Inoue, S. (1985). Well-controlled polymerization by metalloporphyrin: synthesis of copolymer with alternating sequence and regulated molecular weight from cyclic acid anhydride and epoxide catalyzed by the system of aluminum porphyrin coupled with quaternary organic salt. *Macromolecules* 18 (6): 1049–1055.

17 Robert, C., Ohkawara, T., and Nozaki, K. (2014). Manganese-corrole complexes as versatile catalysts for the ring-opening homo- and co-polymerization of epoxide. *Chemistry* 20 (16): 4789–4795.

18 Chatterjee, C. and Chisholm, M.H. (2011). The influence of the metal (Al, Cr, and Co) and the substituents of the porphyrin in controlling the reactions involved in the copolymerization of propylene oxide and carbon dioxide by porphyrin metal(III) complexes. 1. Aluminum chemistry. *Inorg. Chem.* 50 (10): 4481–4492.

19 Huijser, S., HosseiniNejad, E., Sablong, R. et al. (2011). Ring-opening Co- and terpolymerization of an alicyclic oxirane with carboxylic acid anhydrides and CO_2 in the presence of chromium porphyrinato and salen catalysts. *Macromolecules* 44 (5): 1132–1139.

20 Harrold, N.D., Li, Y., and Chisholm, M.H. (2013). Studies of ring-opening reactions of styrene oxide by chromium tetraphenylporphyrin initiators: mechanistic and stereochemical considerations. *Macromolecules* 46 (3): 692–698.

21 Jeske, R.C., DiCiccio, A.M., and Coates, G.W. (2007). Alternating copolymerization of epoxides and cyclic anhydrides: an improved route to aliphatic polyesters. *J. Am. Chem. Soc.* 129 (37): 11330–11331.

22 Jeske, R.C., Rowley, J.M., and Coates, G.W. (2008). Pre-rate-determining selectivity in the terpolymerization of epoxides, cyclic anhydrides, and CO_2: a one-step route to diblock copolymers. *Angew. Chem. Int. Ed. Engl.* 47 (32): 6041–6044.

23 Hosseini Nejad, E., Paoniasari, A., Koning, C.E., and Duchateau, R. (2012). Semi-aromatic polyesters by alternating ring-opening copolymerisation of styrene oxide and anhydrides. *Polym. Chem.* 3 (5).

24 DiCiccio, A.M. and Coates, G.W. (2011). Ring-opening copolymerization of maleic anhydride with epoxides: a chain-growth approach to unsaturated polyesters. *J. Am. Chem. Soc.* 133 (28): 10724–10727.

25 Hosseini Nejad, E., van Melis, C.G.W., Vermeer, T.J. et al. (2012). Alternating ring-opening polymerization of cyclohexene oxide and anhydrides: effect of catalyst, cocatalyst, and anhydride structure. *Macromolecules* 45 (4): 1770–1776.

26 Si, G., Zhang, L., Han, B. et al. (2015). Novel chromium complexes with a [OSSO]-type bis(phenolato) dianionic ligand mediate the alternating ring-opening copolymerization of epoxides and phthalic anhydride. *Polym. Chem.* 6 (35): 6372–6377.

27 Deacy, A.C., Durr, C.B., Kerr, R.W.F., and Williams, C.K. (2021). Heterodinuclear catalysts Zn(II)/M and Mg(II)/M, where M = Na(I), Ca(II) or Cd(II), for phthalic anhydride/cyclohexene oxide ring opening copolymerisation. *Cat. Sci. Technol.* 11 (9): 3109–3118.

28 Diment, W.T., Gregory, G.L., Kerr, R.W.F. et al. (2021). Catalytic synergy using Al(III) and group 1 metals to accelerate epoxide and anhydride ring-opening copolymerizations. *ACS Catal.* 11 (20): 12532–12542.

29 Diment, W.T. and Williams, C.K. (2022). Chain end-group selectivity using an organometallic Al(III)/K(I) ring-opening copolymerization catalyst delivers high molar mass, monodisperse polyesters. *Chem. Sci.* 13 (29): 8543–8549.

30 Diment, W.T., Rosetto, G., Ezaz-Nikpay, N. et al. (2023). A highly active, thermally robust iron(III)/potassium(I) heterodinuclear catalyst for bio-derived epoxide/anhydride ring-opening copolymerizations. *Green Chem.* 25 (6): 2262–2267.

31 Liu, J., Bao, Y.-Y., Liu, Y. et al. (2013). Binuclear chromium–salan complex catalyzed alternating copolymerization of epoxides and cyclic anhydrides. *Polym. Chem.* 4 (5): 1439–1444.

32 Liu, F.-P., Li, J., Liu, Y. et al. (2019). Alternating copolymerization of trans-internal epoxides and cyclic anhydrides mediated by dinuclear chromium catalyst systems. *Macromolecules* 52 (15): 5652–5657.

33 He, G.-H., Liu, Y.-L., Liu, Y., and Lu, X.-B. (2022). Enantioselective resolution copolymerization of racemic cis-epoxides and cyclic anhydrides mediated by multichiral bimetallic chromium complexes. *Macromolecules* 55 (10): 3869–3876.

34 Li, Y.N., Liu, Y., Yang, H.H. et al. (2022). Intramolecular partners in asymmetric catalysis copolymerization: highly enantioselective and controllable at enhanced temperatures and low loadings. *Angew. Chem. Int. Ed. Engl.* 61 (22): e202202585.

35 Li, J., Liu, Y., Ren, W.M., and Lu, X.B. (2016). Asymmetric alternating copolymerization of meso-epoxides and cyclic anhydrides: efficient access to enantiopure polyesters. *J. Am. Chem. Soc.* 138 (36): 11493–11496.

36 Li, J., Ren, B.H., Wan, Z.Q. et al. (2019). Enantioselective resolution copolymerization of racemic epoxides and anhydrides: efficient approach for stereoregular polyesters and chiral epoxides. *J. Am. Chem. Soc.* 141 (22): 8937–8942.

37 Hu, L., Zhang, X., Cao, X. et al. (2021). Alternating copolymerization of isobutylene oxide and cyclic anhydrides: a new route to semicrystalline polyesters. *Macromolecules* 54 (13): 6182–6190.

38 Dove, A.P., Pratt, R.C., Lohmeijer, B.G. et al. (2005). Thiourea-based bifunctional organocatalysis: supramolecular recognition for living polymerization. *J. Am. Chem. Soc.* 127 (40): 13798–13799.

39 Zhang, D., Boopathi, S.K., Hadjichristidis, N. et al. (2016). Metal-free alternating copolymerization of CO_2 with epoxides: fulfilling "green" synthesis and activity. *J. Am. Chem. Soc.* 138 (35): 11117–11120.

40 Hong, M. and Chen, E.Y. (2016). Towards truly sustainable polymers: a metal-free recyclable polyester from biorenewable non-strained gamma-butyrolactone. *Angew. Chem. Int. Ed. Engl.* 55 (13): 4188–4193.

41 Yang, J.L., Wu, H.L., Li, Y. et al. (2017). Perfectly alternating and regioselective copolymerization of carbonyl sulfide and epoxides by metal-free Lewis pairs. *Angew. Chem. Int. Ed. Engl.* 56 (21): 5774–5779.

42 Hu, S., Zhao, J., Zhang, G., and Schlaad, H. (2017). Macromolecular architectures through organocatalysis. *Prog. Polym. Sci.* 74: 34–77.

43 Hu, L.-F., Zhang, C.-J., Wu, H.-L. et al. (2018). Highly active organic Lewis pairs for the copolymerization of epoxides with cyclic anhydrides: metal-free access to well-defined aliphatic polyesters. *Macromolecules* 51 (8): 3126–3134.

44 Li, M., Tao, Y., Tang, J. et al. (2019). Synergetic organocatalysis for eliminating epimerization in ring-opening polymerizations enables synthesis of stereoregular isotactic polyester. *J. Am. Chem. Soc.* 141 (1): 281–289.

45 Zhang, C.-J. and Zhang, X.-H. (2019). Organocatalytic polymerization. *Science China Chem.* 62 (9): 1087–1089.

46 McGraw, M.L. and Chen, E.Y.X. (2020). Lewis pair polymerization: perspective on a ten-year journey. *Macromolecules* 53 (15): 6102–6122.

47 Fukushima, K. and Nozaki, K. (2020). Organocatalysis: a paradigm shift in the synthesis of aliphatic polyesters and polycarbonates. *Macromolecules* 53 (13): 5018–5022.

48 Zhang, C., Geng, X., Zhang, X. et al. (2023). Alkyl borane-mediated metal-free ring-opening (co)polymerizations of oxygenated monomers. *Prog. Polym. Sci.* 136: 101644.

49 Naumann, S. (2023). Borane catalysis for epoxide (co)polymerization. *Polym. Chem.* 14: 1834–1862.

50 Han, B., Zhang, L., Liu, B. et al. (2015). Controllable synthesis of stereoregular polyesters by organocatalytic alternating copolymerizations of cyclohexene oxide and norbornene anhydrides. *Macromolecules* 48 (11): 3431–3437.

51 Hošťálek, Z., Trhlíková, O., Walterová, Z. et al. (2017). Alternating copolymerization of epoxides with anhydrides initiated by organic bases. *Eur. Polym. J.* 88: 433–447.

52 Liu, S., Ren, C., Zhao, N. et al. (2018). Phosphazene bases as organocatalysts for ring-opening polymerization of cyclic esters. *Macromol. Rapid Commun.* 39 (24): e1800485.

53 Li, H., Zhao, J., and Zhang, G. (2017). Self-buffering organocatalysis tailoring alternating polyester. *ACS Macro Lett.* 6 (10): 1094–1098.

54 Li, H., Luo, H., Zhao, J., and Zhang, G. (2018). Well-defined and structurally diverse aromatic alternating polyesters synthesized by simple phosphazene catalysis. *Macromolecules* 51 (6): 2247–2257.

55 Wang, L., Zhang, J., Zhao, N. et al. (2020). Synthesis of tris-phosphazene bases with triazine as core and their applications for efficient ring-opening alternating copolymerization of epoxide and anhydride: notable effect of basicity and molecular size. *ACS Macro Lett.* 9 (9): 1398–1402.

56 Ji, H.-Y., Chen, X.-L., Wang, B. et al. (2018). Metal-free, regioselective and stereoregular alternating copolymerization of monosubstituted epoxides and tricyclic anhydrides. *Green Chem.* 20 (17): 3963–3973.

57 Hu, L.F., Chen, D.J., Yang, J.L., and Zhang, X.H. (2020). An investigation of the organoborane/Lewis base pairs on the copolymerization of propylene oxide with succinic anhydride. *Molecules* 25 (2): 253.

58 Lanfang, H. (2019). Zhang Xinghong, recent progress in the copolymerization of cyclic ethers and cyclic anhydrides via organocatalysts. *J. Funct. Polym.* 32 (3): 259–270.

59 Kummari, A., Pappuru, S., Gupta, P.K. et al. (2019). Metal-free Lewis pair catalyst synergy for fully alternating copolymerization of norbornene anhydride and epoxides: biocompatible tests for derived polymers. *Mater. Today Commun.* 19: 306–314.

60 Ji, H.-Y., Wang, B., Pan, L., and Li, Y.-S. (2018). Lewis pairs for ring-opening alternating copolymerization of cyclic anhydrides and epoxides. *Green Chem.* 20 (3): 641–648.

61 Luo, H., Zhou, Y., Li, Q. et al. (2023). Oxygenated Boron species generated in situ by protonolysis enables precision synthesis of alternating polyesters. *Macromolecules* 56 (5): 1907–1920.

62 Lee, S., Jung, W.H., Kim, M. et al. (2023). Tailoring the interplay between two monomers in the properties of degradable polyesters synthesized via ring-opening alternating copolymerization. *ACS Macro Lett.* 12 (5): 590–597.

63 Zhang, J., Wang, L., Liu, S., and Li, Z. (2020). Phosphazene/Lewis acids as highly efficient cooperative catalyst for synthesis of high-molecular-weight polyesters by ring-opening alternating copolymerization of epoxide and anhydride. *J. Polym. Sci.* 58 (6): 803–810.

64 Chidara, V.K., Gnanou, Y., and Feng, X. (2022). Using triethylborane to manipulate reactivity ratios in epoxide-anhydride copolymerization: application to the synthesis of polyethers with degradable ester functions. *Molecules* 27 (2).

65 Wang, Y., Zhang, J.-Y., Yang, J.-L. et al. (2021). Highly selective and productive synthesis of a carbon dioxide-based copolymer upon zwitterionic growth. *Macromolecules* 54 (5): 2178–2186.

66 Liang, J., Ye, S., Wang, S. et al. (2022). Biodegradable copolymers from CO_2, epoxides, and anhydrides catalyzed by organoborane/tertiary amine pairs: high selectivity and productivity. *Macromolecules* 55 (14): 6120–6130.

67 Liang, J., Ye, S., Wang, W. et al. (2021). Performance tailorable terpolymers synthesized from carbon dioxide, phthalic anhydride and propylene oxide using Lewis acid-base dual catalysts. *J. CO2 Util.* 49.

68 Zhang, C.J., Wu, H.L., Li, Y. et al. (2018). Precise synthesis of sulfur-containing polymers via cooperative dual organocatalysts with high activity. *Nat. Commun.* 9 (1): 2137.

69 Lin, L., Liang, J., Xu, Y. et al. (2019). Fully alternating sustainable polyesters from epoxides and cyclic anhydrides: economical and metal-free dual catalysis. *Green Chem.* 21 (9): 2469–2477.

70 Zhang, X., Jones, G.O., Hedrick, J.L., and Waymouth, R.M. (2016). Fast and selective ring-opening polymerizations by alkoxides and thioureas. *Nat. Chem.* 8 (11): 1047–1053.

71 Pothupitiya, J.U., Dharmaratne, N.U., Jouaneh, T.M.M. et al. (2017). H-bonding organocatalysts for the living, solvent-free ring-opening polymerization of lactones: toward an all-lactones, all-conditions approach. *Macromolecules* 50 (22): 8948–8954.

72 Zhu, X. and Kou, X. (2022). Organic bases and protic acids as binary catalysts for ring-opening alternating copolymerization of epoxides and cycle anhydrides. *Chem. Pap.* 76 (4): 2145–2152.

73 Xie, R., Zhang, Y.Y., Yang, G.W. et al. (2021). Record productivity and unprecedented molecular weight for ring-opening copolymerization of epoxides and cyclic anhydrides enabled by organoboron catalysts. *Angew. Chem. Int. Ed. Engl.* 60 (35): 19253–19261.

74 Hui, J., Wang, X., Yao, X., and Li, Z. (2022). A one-component phosphonium borane Lewis pair serves as a dual initiator and catalyst in the ring-opening alternating copolymerization of anhydrides and epoxides. *Polym. Chem.* 13 (47): 6551–6563.

75 Wang, J., Zhu, Y., Li, M. et al. (2022). Tug-of-war between two distinct catalytic sites enables fast and selective ring-opening copolymerizations. *Angew. Chem. Int. Ed. Engl.* 61 (36): e202208525.

76 Ahmad, S., Yousaf, M., Mansha, A. et al. (2016). Ring-opening reactions of oxetanes: a review of methodology development and synthetic applications. *Synth. Commun.* 46 (17): 1397–1416.

77 Bull, J.A., Croft, R.A., Davis, O.A. et al. (2016). Oxetanes: recent advances in synthesis, reactivity, and medicinal chemistry. *Chem. Rev.* 116 (19): 12150–12233.

78 Kameyama, A., Ueda, K., Kudo, H., and Nishikubo, T. (2002). The first synthesis of alternating copolymers of oxetanes with cyclic carboxylic anhydrides using quaternary onium salts. *Macromolecules* 35: 3792–3794.

79 Takeuchi, D., Aida, T., and Endo, T. (1999). The first example of the copolymerization of cyclic acid anhydrides with oxetane by bulky titanium bisphenolates. *Macromol. Rapid Commun.* 20 (12): 646–649.

80 McGuire, T.M., Clark, E.F., and Buchard, A. (2021). Polymers from sugars and cyclic anhydrides: ring-opening copolymerization of a D-xylose anhydrosugar oxetane. *Macromolecules* 54 (11): 5094–5105.

81 Zheng, L., Feng, J., Xie, H., and Xu, Z. (2023). Organocatalytic synthesis of sugar-derived polyesters via ring-opening alternating copolymerization of anhydrosugar oxetane and anhydrides. *J. Polym. Sci.* 61 (14): 1430–1438.

82 Tang, T., Oshimura, M., Yamada, S. et al. (2012). Synthesis of periodic copolymers via ring-opening copolymerizations of cyclic anhydrides with tetrahydrofuran using nonafluorobutanesulfonimide as an organic catalyst and subsequent transformation to aliphatic polyesters. *J. Polym. Sci. A Polym. Chem.* 50 (15): 3171–3183.

83 Abel, B.A., Snyder, R.L., and Coates, G.W. (2021). Chemically recyclable thermoplastics from reversible-deactivation polymerization of cyclic acetals. *Science* 373 (6556): 783–789.

84 Zhang, X., Zhang, C., and Zhang, X. (2022). A facile and unprecedented route to a library of thermostable formaldehyde-derived polyesters: highly active and selective copolymerization of cyclic acetals and anhydrides. *Angew. Chem. Int. Ed. Engl.* 61 (13): e202117316.

85 Zhang, X., Sun, Y., Zhang, C., and Zhang, X. (2023). Upcycling polytetrahydrofuran to polyester. *CCS Chem.* 5 (5): 1233–1241.

86 Zhang, X., Guo, W., Zhang, C., and Zhang, X. (2023). A recyclable polyester library from reversible alternating copolymerization of aldehyde and cyclic anhydride. *Nat. Commun.* 14 (1): 5423.

87 Li, H., He, G., Chen, Y. et al. (2019). One-step approach to polyester–polyether block copolymers using highly tunable bicomponent catalyst. *ACS Macro Lett.* 8 (8): 973–978.

88 Ji, H.-Y., Song, D.-P., Wang, B. et al. (2019). Organic Lewis pairs for selective copolymerization of epoxides with anhydrides to access sequence-controlled block copolymers. *Green Chem.* 21 (22): 6123–6132.

89 Zhu, S., Wang, Y., Ding, W. et al. (2020). Lewis pair catalyzed highly selective polymerization for the one-step synthesis of $A_zC_y(AB)_xC_yA_z$ pentablock terpolymers. *Polym. Chem.* 11 (10): 1691–1695.

90 Zhu, S., Zhao, Y., Ni, M. et al. (2020). One-step and metal-free synthesis of triblock quaterpolymers by concurrent and switchable polymerization. *ACS Macro Lett.* 9 (2): 204–209.

91 Ye, S., Wang, W., Liang, J. et al. (2020). Metal-free approach for a one-pot construction of biodegradable block copolymers from epoxides, phthalic anhydride, and CO_2. *ACS Sustain. Chem. Eng.* 8 (48): 17860–17867.

92 Lin, L., Chen, X., Xiang, H. et al. (2022). Construction of triblock copolyesters via one step switchable terpolymerization of epoxides, phthalic anhydride and ε-caprolactone using dual urea/organic base catalysts. *Polym. Chem.* 13 (6): 801–807.

93 Zhang, J., Wang, L., Liu, S. et al. (2021). A lewis pair as organocatalyst for one-pot synthesis of block copolymers from a mixture of epoxide, anhydride, and CO_2. *Macromolecules* 54 (2): 763–772.

94 Ma, Y., You, X., Zhang, J. et al. (2023). Synthesis of sequence-specific poly(ester-carbonate) copolymers via chemoselective terpolymerization controlled by the stoichiometric ratio of phosphazene/triethylborane. *Angew. Chem. Int. Ed. Engl.* 62 (25): e202303315.

95 Chidara, V.K., Boopathi, S.K., Hadjichristidis, N. et al. (2021). Triethylborane-assisted synthesis of random and block poly(ester-carbonate)s through one-pot terpolymerization of epoxides, CO_2, and cyclic anhydrides. *Macromolecules* 54 (6): 2711–2719.

4

Organocatalysts for the Preparation of Degradable and Closed-Loop Recyclable Polyesters

Yong Shen[1] *and Zhibo Li*[1,2]

[1]*Qingdao University of Science and Technology, College of Chemical Engineering, No.53 Zhengzhou Road, Qingdao, 266042, China*
[2]*Qingdao University of Science and Technology, Key Laboratory of Biobased Polymer Materials, Shandong Provincial Education Department, College of Polymer Science and Engineering, No.53 Zhengzhou Road, Qingdao, 260042, China*

4.1 Introduction

The large production and indiscriminate disposal of plastics have resulted in serious resource and environment crisis, which motivate efforts to establish a circular plastic economy [1–4]. Different strategies have been proposed to address the end-of-life issue of plastics, including the development of bio-derivative polymers, biodegradable polymers, closed-loop recyclable polymers as well as more efficient methods to upcycle the commodity plastics [5–9]. Aliphatic polyesters are of particular interest and have drawn increased attentions in recent years due to their potential bio-renewable feedstocks, biodegradability and closed-loop recyclability as well as good thermal and mechanical properties. Aliphatic polyesters have been used in a wide range of applications from biotechnology to commodity plastics, such as surgical suture, drug delivery carriers, tissue engineering scaffolds, food packaging, straws and disposable meal boxes. Aliphatic polyesters are a class of hetero-chain macromolecules containing repeating aliphatic ester groups in their main chain, which can be prepared by step-growth polycondensation or ring-opening polymerization (ROP). The step-growth polycondensation generally requires strictly equal stoichiometry of diol and dicarboxylic acid as well as high temperature and vacuum to remove the small molecular side products so as to reach high-molecular-weight polyesters. Although it is very effective for the synthesis of aromatic polyesters, step-growth polycondensation suffers from low molecular weights of resultant polyesters, poor control over the chain-end groups as well as undesired side reactions under harsh conditions for the preparation of aliphatic polyesters. Alternatively, ROP of corresponding lactones or lactides has been proven as an effective strategy to prepare aliphatic polyesters considering the advantages of atomic economy, good control over molecular weights and narrow dispersities, well-defined chain-end groups and tunable topology of the macromolecules as

Organocatalysts in Polymer Chemistry: Synthesis and Applications, First Edition. Edited by Zhibo Li.

well as the capability to prepare copolymers with controlled monomer sequences. ROP proceeds via chain-growth polymerization mechanism and may exhibit "living"/controlled polymerization behaviors in the presence of suitable catalysts. Catalysts play an important role for the ROPs of cyclic esters not only to improve the polymerization rate but also to control the selectivity of chain propagation over other competing side reactions (inter- or intra- transesterification, etc). Despite the dominant position of organometallic catalysts in the past decades, organocatalysts have attracted increased interests in recent years given their advantages of low cost, easy handle, enhanced stability to oxygen and water as well as facile removal from the resultant polymers [10–12]. Therefore, organocatalysts are highly desirable for the microelectronic and biomedical applications where the presence of metal residues is deleterious to the material performances.

In this chapter, we will first introduce the different catalytic mechanism mediated by various organocatalysts, including electrophilic monomer activation mechanism, nucleophilic monomer activation mechanism, chain-end activation mechanisms as well as the bifunctional activation of both propagating chain-end and monomer. The organocatalytic ROPs of cyclic esters will then be discussed according to the ring size of the monomers. Specially, the recent achievements for the closed-loop recyclable polyesters will be highlighted. Finally, we will simply discuss the current challenges and possible future directions for the organocatalytic ROP.

4.2 Catalytic Mechanism of Various Organocatalysts

4.2.1 Electrophilic Monomer Activation Mechanism

Organic acids have been developed as efficient catalysts for the ROPs of lactides and lactones. The carbonyl group of lactides or lactones can be electrophilic activation by protonation. The activated monomers are then nucleophilically attacked by the alcohol initiator or propagating chain-end to undergo ring-opening to produce hydroxyl terminated polymer chain (Scheme 4.1). Strong acids, such as trifluoromethanesulfonic acid, are required for the effective solution ROP of lactide. In contrast, much milder organic acids can be used as catalysts for the ROPs of lactones. Despite the wide availability of organic acids, the ROPs mediated by organic acids still suffer from slow polymerization rate, high catalyst loading, poor controllability as well as low molecular weights of obtained polyesters.

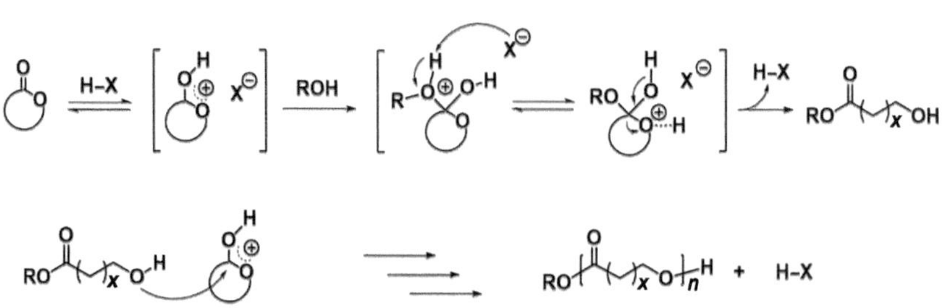

Scheme 4.1 Electrophilic monomer activation mechanism for ROP of cyclic esters.

4.2.2 Nucleophilic Monomer Activation Mechanism

When a nucleophile is used as the catalyst for the ROPs of lactides and lactones, including 4-(dimethylamino)pyridine (DMAP), phosphines and *N*-heterocyclic carbenes (NHCs), a nucleophilic monomer activation mechanism is proposed for the polymerization. The nucleophile can directly attack the carbonyl group of lactides or lactones to form an active zwitterionic intermediate, which is protonation by the alcohol initiator or propagating alcohol chain-end. The deprotonated alkoxide then attacks the protonated intermediate to produce hydroxyl terminated polymer chain and regenerate the nucleophile (Scheme 4.2). Note that zwitterionic ROPs of lactones or lactides mediated by a nucleophile are possible in the absence of alcohol initiator.

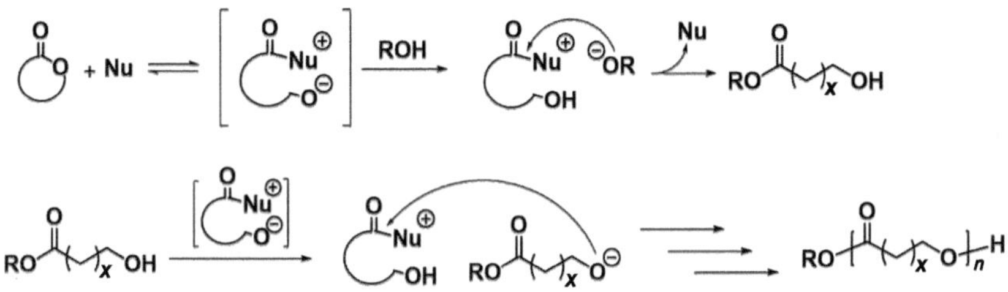

Scheme 4.2 Nucleophilic monomer activation mechanism for ROP of cyclic esters.

4.2.3 Chain-End Activation Mechanism

When a non-nucleophilic organobase is used as the catalyst, the ROPs of lactides and lactones generally proceed via a chain-end activation mechanism. Instead of directly activating the monomer, the organobase can activate the alcohol initiator or propagating chain-end by deprotonation or hydrogen bond. The nucleophilic attack to carbonyl of the monomer is facilitated by the enhanced nucleophilicity of alcohol initiator or propagating chain-end (Scheme 4.3). One challenge for the chain-end activation mechanism is to achieve a high catalytic activity while simultaneously remaining good control over the polymerization since the enhanced nucleophilicity of alcohol initiator or propagating chain-end also leads to undesired transesterification or back-biting side reactions.

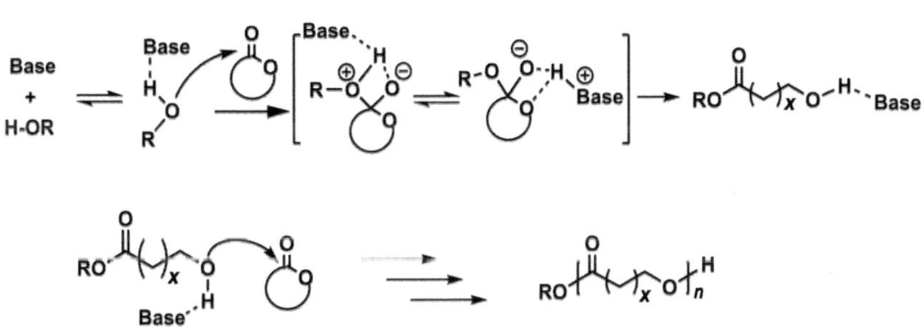

Scheme 4.3 Chain-end activation mechanism for ROP of cyclic esters.

4.2.4 Bifunctional Activation Mechanism

Compared to the single monomer activation or chain-end activation mechanism, bifunctional activation of both the monomer and propagating chain-end can simultaneously enhance the polymerization rate and selectivity over the ROP. An electrophile can activate the monomer while a base is used to activate the alcohol initiator or propagating chain-end (Scheme 4.4). It is reported that ROPs of lactides and lactones mediated by 1,5,7-triazabicyclo[4.4.0]dec-5-ene (TBD) proceed via the bifunctional activation mechanism, in which TBD serves as both hydrogen bonding donor to monomer and hydrogen bonding acceptor to alcohol/propagating chain-end.

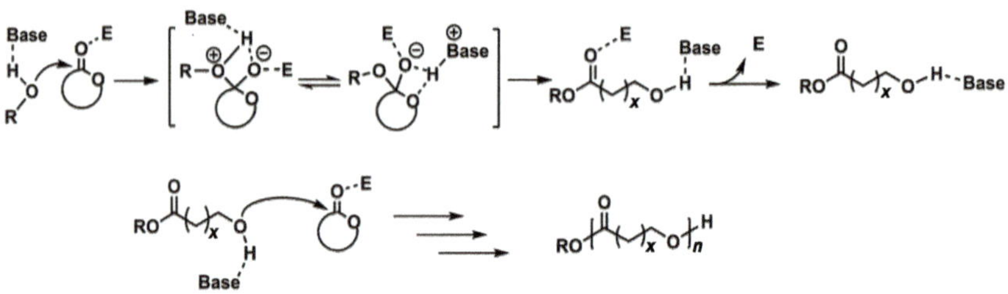

Scheme 4.4 Bifunctional activation mechanism for ROP of cyclic esters.

Another widely reported catalytic system that undergoes bifunctional activation mechanism is the base/(thio)urea binary catalyst. Based on the relative pK_a values of the (thio)urea and the base, the bifunctional activation mechanism can be further classified as cooperative hydrogen bond activation mechanism and (thio)urea anion activation mechanism [13]. When the base is incapable of deprotonating the (thio)urea, the ROPs undergo the cooperative hydrogen bond activation mechanism in which the base activates the alcohol initiator or propagating chain-end and the (thio)urea activates the monomer by hydrogen bonding, respectively. When a (thio)urea is combined with a base that strong enough to deprotonate the (thio)urea, the anionic portion of the (thio)urea anion can nucleophilically activate the alcohol initiator or propagating chain-end through hydrogen bond while the remaining N–H moiety can electrophilically activate the carbonyl of monomer through hydrogen bond (Scheme 4.5) [14, 15]. Generally speaking, the base/urea binary catalyst exhibits higher catalytic activity compared to the base/thiourea binary catalyst

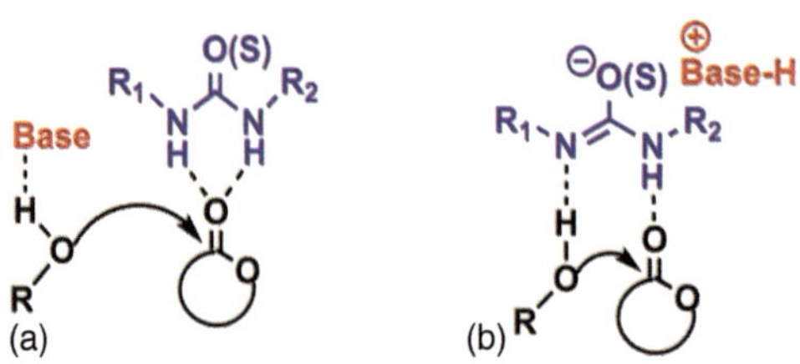

Scheme 4.5 (a) Cooperative hydrogen bond activation mechanism and (b) (thio)urea anion activation mechanism.

when ureas and thioureas of comparable acidity are used as co-catalysts. On the other hand, the relative pK_a values of the (thio)urea and the base have a big impact on the catalytic activity. Stronger base usually leads to enhanced polymerization rate for a given urea or thiourea. For a given base, the catalyst exhibits the highest activity when the pK_a value of the (thio)urea is closely matched with that of the conjugated acid of the base. In such a case, the active catalytic species, neutral (thio)urea and (thio)urea anion, are in rapid equilibrium. The solvent polarity also plays an important role on the catalytic mechanism. The cooperative hydrogen bond activation mechanism is favored in nonpolar solvents while (thio)urea anion activation mechanism is favored in polar solvents [16, 17].

4.3 Organocatalytic ROP of Lactones or Lactides

4.3.1 γ-Butyrolactone and its Derivatives

As an important member of polyhydroxyalkanoates (PHAs), poly(4-hydroxy-butyrate) (P4HB) is generally produced by bacterial fermentation and has been approved by the US Food and Drug Administration (FDA) for biomedical applications, such as surgical suture and surgical mesh. Compared to poly(L-lactide) (PLLA) and poly(L-lactic-*co*-glycolic acid) (PLGA) copolymers, P4HB exhibits suitable degradation rate and will not lead to undesirable highly acidic degradation product accumulation *in vivo* [18, 19]. These superior properties make P4HB a promising biomaterial as implanted tissue scaffolds for tissue repair and regeneration. In addition to bacterial fermentation process that generally requires tedious purification procedure, P4HB can also be produced by a chemo-synthetic procedure via ROP of γ-butyrolactone (γBL). As such, P4HB is also named as PγBL. γ-Butyrolactone (γBL) is a five-membered lactone that can be produced from succinic acid, which is a bio-derivative chemical and has been selected as the top potential bio-platform chemical by the US Department of Energy [20]. However, γBL has been recognized as a "non-polymerizable" monomer for a long time due to its low strained five-membered lactone ring. Even the polymerization conducted under a ultra-high pressure (1000 MPa) in the presence of trifluoromethane sulfonic acid as the catalyst only produced oligomers [21]. The low strain energy of five-membered lactone ring of γBL results in a small negative enthalpy change ($\Delta H_p^\ominus = -5.4\,\text{kJ mol}^{-1}$) and a large negative entropy change ($\Delta S_p^\ominus = -39.6\,\text{J mol}^{-1}\,\text{K}^{-1}$) during the ROP process. As such, ROP of γBL cannot proceed at room temperature (RT) due to the slightly positive Gibbs free energy change ($\Delta G_p^\ominus > 0$) [22, 23]. The ceiling temperature (T_c), which is defined as the highest critical polymerization temperature, can be calculated according to the equation of $T_c = \Delta H_p^\ominus/(\Delta S_p^\ominus + R\ln[\gamma BL]_0)$. When the initial monomer concentration $[\gamma BL]_0 = 1$ or $10\,\text{mol L}^{-1}$, the polymerization can occur at temperature below −136 and −9 °C, respectively. In addition to fulfill the thermodynamic requirements that conducted the polymerization at low temperature (typical at −40 ~ −60 °C), it is crucial to choose suitable catalyst to achieve moderate monomer conversion and produce high-molecular-weight PγBL.

Chen and coworkers made a breakthrough and realized the ROP of γBL under ambient pressure to produce linear or cyclic PγBL at low temperature (−40 °C) in the presence of $La[N(SiMe_3)_2]_3$ or yttrium amide as the catalysts [22]. Notably, the linear or cyclic PγBL can be depolymerized back to pristine γBL monomer with high yield by heating the bulk materials at 220 or 300 °C, respectively, thus establishing the "monomer-polymer-monomer" closed-loop life cycle for γBL. In addition to rare-earth metal based catalysts, a series of organobases has been employed for the ROP of γBL. The commonly used organobases that showed high activities for the ROP of other lactones only produce PγBL oligomers. For example, the ROP of γBL catalyzed by TBD achieved a low γBL conversion of 33% within 24 h, producing PγBL with a low M_n = 5.5 kDa. Excitingly, the strong organophosphazene base t-BuP_4 shows high catalytic activity toward the ROP of γBL. t-BuP_4 can directly initiated the polymerization by deprotonation of γBL to generate reactive enolate species, followed by ROP of γBL (pathway a, Scheme 4.6). However, this pathway achieved a low conversion of γBL. As alternative, when t-BuP_4 was first mixed with an alcohol as the initiator, the polymerization proceeded via the chain-end activation mechanism and achieved a high γBL conversion up to 90%, producing PγBL with a relative high molecular weight (M_n = 26.7 kg mol^{-1}) (pathway b, Scheme 4.6) [24]. Nevertheless, the strong basicity of t-BuP_4 and weak hydrogen bonding between t-BuP_4-H^+ and RO^- resulted in the strong nucleophilicity of propagating alkoxide chain-end and undesirable back-biting side reaction as evidenced by the formation of cyclic PγBL. In contrast, the relative weaker organophosphazene base CTPB exhibited better control over the ROP of γBL in the presence of an alcohol as the initiator. The reduced nucleophilicity of propagating alkoxide chain-end can suppress the transesterification side reaction, thus selectively producing linear PγBL at a low monomer conversion [25]. The superbase CTPB was also used for the preparation of PγBL containing block copolymers. PγBL-b-PLLA was successfully prepared by a one-pot sequential ROP of γBL and L-LA [26]. A series of amphiphilic PEG-b-PγBL diblock

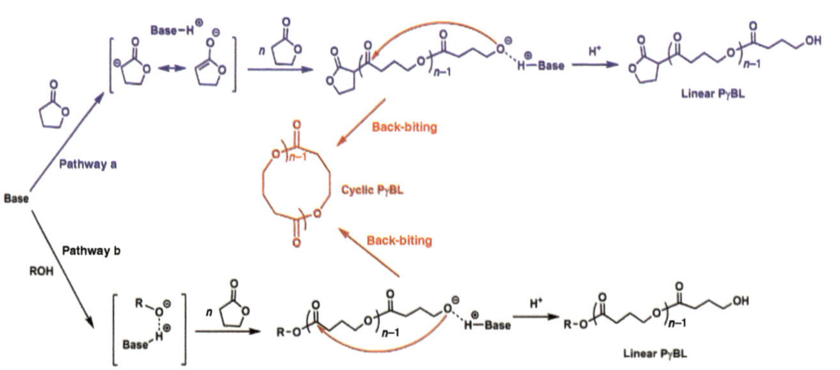

Scheme 4.6 Different pathways for the ROP of γBL in the presence of organophosphazene base. Pathway a: the polymerization was initiated by lactone enolate anion generated by deprotonation of α-H of γBL; pathway b: polymerization was initiated by an alcohol and proceeded via a chain-end activation mechanism. Source: Hong et al. [24] and Zhao et al. [25].

copolymers with different molecular weights was prepared by using poly(ethylene glycol) monomethyl ether as the macroinitiator in the presence of CTPB as the catalyst [27]. It was reported that *N*-heterocyclic olefins (NHOs) as a single organobase or in combination with lithium salts as Lewis pairs can be used for the ROP of γBL. But both catalytic systems resulted in moderate γBL conversions (<70%) and afforded PγBL oligomers ($M_n < 10\,\text{kg mol}^{-1}$) [28].

Compared to the single organobases, the strong base/(thio)urea binary catalysts exhibits higher catalytic activity and better control over the ROP of γBL. The polymerization generally proceeds via the (thio)urea anion activation mechanism. The (thio)urea anion generated by deprotonation of (thio)urea with strong base can simultaneously activate the alcohol/propagation chain-end and γBL monomer via hydrogen bond, thus accelerating the chain propagation via a intramolecular nucleophilic attack. On the other hand, the strong hydrogen bonding between propagating chain-end and (thio)urea anion depresses the transesterification or back-biting side reactions. Both the enhanced chain propagation rate and reduced chain transfer rate make it possible to synthesize linear PγBL with a high molecular weight (Scheme 4.7). For example, compared to the single CTPB that achieved a PγBL with a moderate molecular weight of $M_n = 27.2$ kDa, the CTPB/U1 (Figure 4.1) binary catalyst exhibited better control over the ROP of γBL, producing a high-molecular-weight PγBL ($M_n = 80.4$ kDa) with good mechanical properties comparable with that of P4HB obtained by bacterial fermentation [29]. Thanks to the controlled ROP of γBL mediated by CTPB/urea binary catalyst, telechelic PγBL diols with well-defined terminal hydroxyl groups were successfully prepared. These PγBL diols were then reacted with 4,4′-methylene diphenyl diisocyanate (MDI) to produce polyurethanes that behaved as thermoplastics or elastomers. Notably, the obtained polyurethanes can be chemically recycled back to γBL with high purity and yield via a tandem dissociation of urethane linkage and followed depolymerization of PγBL diols by heating under vacuum in the presence of stannous octoate ($Sn(Oct)_2$) as the catalyst [30]. Zhang et al. also reported well-controlled ROP of γBL to produce linear PγBL in the presence of *t*-BuP_4/thiourea as the binary catalyst

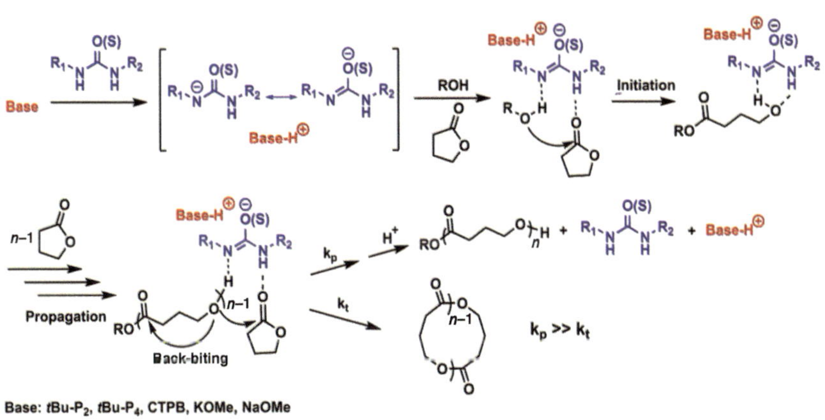

Scheme 4.7 The ROP of γBL catalyzed by the strong base/(thio)urea binary catalyst.

[31]. In contrast to the low catalytic activity of alkoxide alone, alkoxide/urea as an effective catalyst successfully achieved the bulk ROP of γBL [32]. Recently, a bisurea (U2) (Figure 4.1) in combination with organophosphazene CTPB as the binary catalyst was used for the ROP of γBL. The bisurea as a co-catalyst exhibited much higher catalytic activity compared to the monourea as supported by the kinetic experiments. A high-molecular-weight PγBL with good mechanical properties was achieved within 10 min, thus greatly reduced the polymerization time and associated high energy cost. It was supposed that the successively intramolecular hydrogen bonding activation with additional urea moiety of the bisurea greatly improved the catalytic activity of CTPB/bisurea binary catalyst [33].

α-Methylene-γ-butyrolatone (MBL) is also name as tulipalin A, which is naturally found in tulips or can be produced from bio-renewable itaconic acid [34]. MBL is a five-membered lactone bearing with an exocyclic double bond. MBL has been extensively investigated as a sustainable replacement to the petroleum-based methyl methacrylate (MMA) considering the higher glass transition temperature and better solvent resistance of its vinyl-addition polymerization (VAP) product $P(MBL)_{VAP}$ compared with PMMA [35–39]. On the other hand, MBL can proceed via ROP of γBL ring to produce biodegradable functional polyester $P(MBL)_{VAP}$. However, the chemoselective ROP of MBL is even more challenging compared to the ROP of γBL due to the unfavorable thermodynamics of ROP pathway ($\Delta H_p = -5.9\,\text{kJ mol}^{-1}$, $\Delta S_p = -40.1\,\text{J mol}^{-1}\,\text{K}^{-1}$) as well as the competitive VAP through Michael-type addition across the highly reactive exocyclic double bond. Therefore, in addition to fulfill the thermodynamic requirements to conduct the polymerization at low temperature, it is necessary to choose a suitable catalyst that chemo-selectively catalyzes the ROP pathway. Chen and coworkers realized the chemo-selective ROP of MBL to exclusively produce $P(MBL)_{ROP}$ using $La[N(SiMe_3)_2]_3$ or yttrium-based catalysts via a coordination-insertion mechanism [40]. In contrast to organometallic catalysts, organocatalysts are generally more difficult to achieve subtle region- and chemo-selectivity. The only example that achieve the organocatalytic ROP of MVL to produce exclusively $P(MBL)_{ROP}$ was recently reported by using CTPB/urea as the binary catalyst [41]. It was proposed that the VAP pathway of MBL should be proceeded via a carbanion intermediate that generally required a stronger base as the catalyst compared to the ROP pathway that proceeded via an alkoxide intermediate. In addition, the urea anions have been proven to show high catalytic activity toward the ROP of γBL due to the dual activation of propagating chain-end and γBL ring. As such, it was speculated that suitable combinations of organobases and ureas as the binary catalyst can achieve the chemo-selective ROP of MVL. These hypotheses were supported by experimental data as well as the computational calculation. A urea featured with a 4-trifluoromethylphenyl and a cyclohexyl group (U3) in combination with CTPB was used as an effective catalyst to achieve the chemo-selective ROP of MVL. Ureas with higher pK_a values as the co-catalysts resulted in the formation of $P(MBL)_{VAP}$ or the cross-linked networks. It is worth pointing out that a urea bearing with more electron-withdrawing substituents of two 3,5-bis(trifluoromethyl)phenyl groups as the co-catalyst can achieved the depolymerization of $P(MBL)_{ROP}$ to recover MBL with a high yield in dilute solution. Nevertheless, Only moderate monomer conversion (up to 50%) and $P(MBL)_{ROP}$ oligomer (M_n = 5.9 kDa) were obtained using the reported organocatalyst. More efforts are required to address this challenge. The same group later reported the chemo-selective copolymerization of MBL with δ-valerolactone (δVL) and ε-caprolactone (εCL) to produce functional copolyesters with composition-dependent properties using CTPB/urea binary catalyst [42].

Although PγBL and $P(MBL)_{ROP}$ have shown great promises as chemically recyclable polyesters with closed-loop life cycles, the low polymerization temperature (typically at or below −40 °C) and associated high energy cost still hamper their large production and practical applications. Different strategies have been proposed to manipulate the thermodynamic parameters and polymerization temperatures of five-membered γBL ring while remaining the ability to depolymerize back to monomers. Chen and co-workers presented the introduction of a *trans*-cyclohexyl ring to the α and β positions of the γBL core to afford 3,4-T6GBL that possessed improved ring strain compared to γBL. The ROP of 3,4-T6GBL was achieved in bulk at RT to produce closed-loop recyclable polymer with a high molecular weight up to 1000 kg mol^{-1} in the presence of organometallic catalysts [43]. However, the organocatalytic ROP of 3,4-T6GBL has not been reported probably due to the presence of the acidic α proton adjacent to the carbonyl, which is prone to be deprotonated by organobases to produce non-polymerizable *cis* isomer. In contrast, its constitutional isomer 4,5-T6GBL also exhibited an increased ring strain and better polymerizability compared to γBL. Moreover, the living/controlled ROP of 4,5-T6GBL was successfully achieved at RT in the presence of organobase/(thio)urea as the binary catalyst to produce a linear polymer with M_n up to 106 kg mol^{-1} and narrow dispersity ($Đ$ = 1.04) [44]. Interestingly, *N*-heterocyclic carbenes, including 1,3-di-*tert*-butylimidazol-2-ylidene (I^tBu) and 1,3-diisopropylimidazol-2-ylidene (I^iPr), were also effective catalyst to achieve the ROP of 4,5-T6GBL to selectively produce cyclic polymers with M_n = 41 ~ 72 kg mol^{-1} via a zwitterionic ROP. In addition to the introduction of a *trans*-fused ring, a bridged bicyclic moiety was also introduced to the γBL core to afford new lactone monomers with the capability of closed-loop recyclability. For example, an offspring [3.2.1] bicyclic lactone, 6-oxabicyclo[3.2.1]octan-7-one (BiL), which was featured with a high T_c εCL substructure and a low T_c γBL substructure, was recently reported by Chen and coworkers [45]. The ROP of BiL conducted at RT in toluene with an initial monomer concentration of 6.0 mol L^{-1} in the presence of TBD as the catalyst achieved a 95% monomer conversion after 48 h, affording a PBiL with a M_n of 56.7 kg mol^{-1} and a slightly broad dispersity of $Đ$ = 1.22. Due to the strong nucleophilicity and basicity of TBD, epimerization at the carbon stereogenic center adjacent to the carbonyl carbon occurred during ROP, resulting in PBiL containing both *cis*(82%) and *trans*(18%) stereoconfigurations. As such, the obtained PBiL was an amorphous material with a high glass transition temperature of T_g = 119 °C. Interestingly, TBD can also catalyze the depolymerization of obtained PBiL (including both *cis* and *trans* chains) at 120 °C to recover BiL with high yield within 30 h. The same group later reported a similar bicyclic lactone with BiL, 6-oxabicyclo[3.2.1]oct-3-en-7-one ($BiL^=$) [46]. $BiL^=$ can undergo ROP to produce polyester or ring-opening metathesis

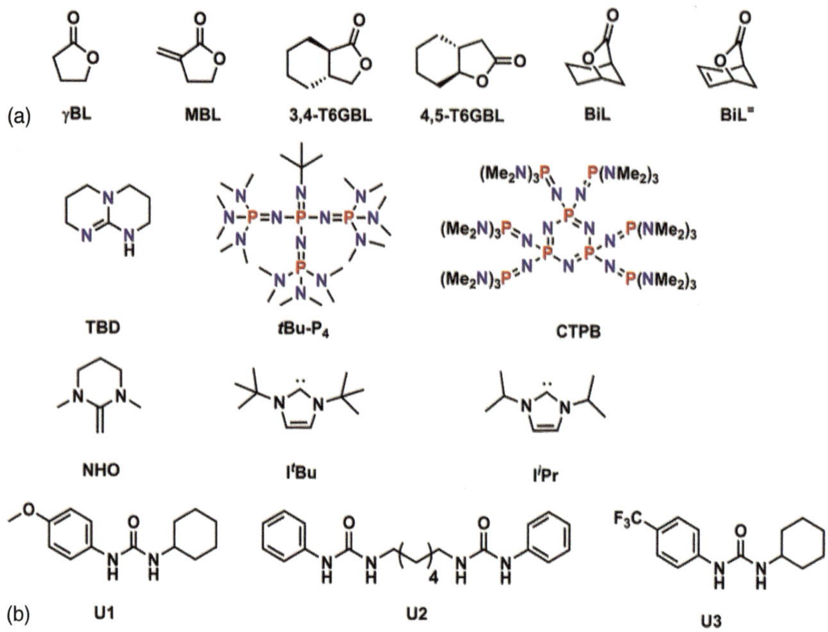

Figure 4.1 (a) The chemical structures of five-membered lactones discussed in this section as well as (b) the organocatalysts used for their ROPs.

polymerization (ROMP) to afford polyolefin. The ROP of BiL⁼ conducted at [BiL⁼]/[TBD]/[BnOH] = 500/1/1 achieved a monomer conversion of 69% after 48 h using benzyl alcohol (BnOH) as the initiator at RT in toluene, affording P(BiL⁼) with a M_n = 31.3 kg mol^{-1} and $Đ$ = 1.33. Similar to BiL, the ROP of BiL⁼ catalyzed by TBD also led to epimerization and produced amorphous P(BiL⁼). Note that for both BiL and BiL⁼, their stereoretention ROPs without epimerization were realized using La[N(SiMe$_3$)$_2$]$_3$ as the catalyst. Figure 4.1 gives the chemical structures of five-membered lactones discussed in this section and the organocatalysts used for their ROPs.

4.3.2 δ-Valerolactone and ε-Caprolactone

δ-Valerolactone (δVL) and ε-caprolactone (εCL) is a six-membered lactone and a seven-membered lactone, respectively. Despite their different ring size, both δVL and εCL are generally considered as strained lactones and investigated together. Both monomers possess a high T_c and their ROPs can be considered as non-equilibrium polymerization. As such, the organocatalytic ROPs of δVL and εCL are discussed in the same section. Compared to γBL and its derivatives that only limited organocatalysts were effective for their ROPs, the ROPs of δVL and εCL have been successfully achieved with a variety of organocatalysts, including organic acids, *N*-heterocyclic carbenes (NHCs), TBD, organophosphazene bases and base/urea binary catalysts.

Brönsted acids have been reported as effective catalysts for the ROPs of δVL and εCL via an electrophilic monomer activation mechanism. The protonated monomers are attacked by the alcohol initiator or terminal hydroxyl of propagating chain and then undergo ring-opening to form hydroxyl terminated polyesters. Carboxylic acids [47, 48], trifluoromethanesulfonic acid (HOTf) [49], methanesulfonic acid (MSA) [49], trifluoromethanesulfonimide ($HNTf_2$) [50] and diphenyl phosphate (DPP) [51, 52] have been investigated for the ROPs of δVL and εCL. Carboxylic acids exhibited low catalytic activity due to their weak acidity. For example, the bulk ROP of εCL conducted at 90 °C reached a monomer conversion of 81% within 12 h despite at a high catalyst loading (10 equivalents of initiator) in the presence of fumaric acid as the catalyst [47]. Both HOTf and MSA achieved controlled ROP of εCL at 30 °C in the presence of primary or secondary alcohols as the initiators to produce PεCL with M_n up to 11.0 kDa and narrow dispersities ($1.07 < Đ < 1.21$). Despite the large difference (more than 10 units) of their pK_a values, MSA exhibited comparable catalytic activity with HOTf. In contrast to HOTf that achieved an optimal catalytic activity at a catalyst to initiator ratio of 1, the catalytic activity of MSA increased as the catalyst to initiator ratio increased [49]. $HNTf_2$ as another strong Brönsted acid was also an effective catalyst for the ROP of δVL in the presence of an alcohol as the initiator. The polymerization exhibited "living"/controlled behavior as supported by the kinetic and chain extension experiments at a low catalyst loading (0.1 equivalent relative to initiator). Various alcohols were successfully used as initiators, such as 6-azido-1-hexanol, 2,3,4,5,6-pentafluorobenzyl alcohol, and *N*-(2-hydroxyethyl)maleimide, to prepare PδVL with terminal functional groups [50]. In addition to the strong Brönsted acids, DPP as a commercially available, low toxic organic acid was also investigated for the ROP of δVL and εCL [51, 52]. Despite its weak acidity, DPP exhibited high catalytic activity and good control over the polymerization in the presence of 3-phenyl-1-propanol (PPA) as the initiator. For example, the polymerizations conducted in toluene at RT at a feeding molar ratio of $[\delta VL \text{ or } \varepsilon CL]_0/[DPP]_0/[PPA]_0 = 50/1/1$ achieved a δVL conversion of 95% at 1 h or a εCL conversion of 97% at 8 h, respectively. In addition, the ROPs exhibited "living"/controlled polymerization behaviors. Compared to $HNTf_2$, PδVL with a higher molecular weight of $M_n = 27.5\,kg\,mol^{-1}$ was achieved using DPP as the catalyst [51]. Although different Brönsted acids have been developed as effective catalysts for the ROPs of δVL and εCL, the organic acids still face challenges, including slow polymerization rate, high catalyst loading and the low molecular weights of obtained polyesters.

N-Heterocyclic carbenes (NHCs) as a class of strong nucleophiles and neutral bases have also been developed as catalysts for the ROPs of lactones and lactides [53–56]. The catalytic activities of NHCs are strongly dependent on the nature of the substituents. Less sterically demanding carbenes are more active for the ROPs of lactones compared to their sterically encumbered analogues. For example, the sterically encumbered carbenes 1,3-bis-(2,4,6-trimethylphenyl)imidazol-2-ylidene **(1)** and 1,3-bis(2,6-diisopropylphenyl)imidazol-2-ylidene **(2)** cannot catalyze the ROP of εCL at RT in THF at an initial monomer concentration of 0.5–2.0 mol L^{-1}.

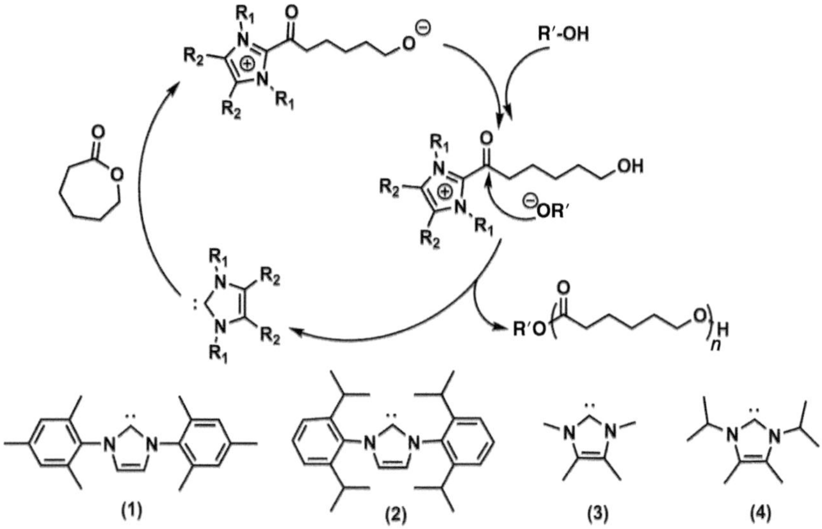

Scheme 4.8 (a) The proposed nucleophilic monomer activation mechanism for the NHC catalyzed ROP of εCL in the presence of an alcohol as the initiator as well as the chemical structures of NHCs. Source: [57]/American Chemical Society.

In contrast, controlled ROPs of εCL were achieved to produce PCL with predictable molecular weights and narrow dispersities with target DP up to 200 in the presence of the more electron-rich and less sterically hindered carbenes 1,3,4,5-tetramethylimidazol-2-ylidene (**3**) and 1,3,4,5-tetramethylimidazol-2-ylidene (**4**) [57]. Nevertheless, broaden of dispersities was observed when the monomer conversion was above 70%, indicating the existence of transesterification side reactions. It was proposed that the ROP of lactones and lactides catalyzed by NHC proceeded via a nucleophilic monomer activation mechanism, in which the NHC as a nucleophile directly attacked the carbonyl of εCL to form a zwitterionic acylimidazole intermediate. The alcohol initiator was then deprotonated by the intermediate to generate alkoxide, which then attacked the protonated intermediate to produce hydroxyl terminated polymer chain and regenerated the NHC (Scheme 4.8). It is worth to pointing out that the NHC can mediate the zwitterionic ROPs of lactones or lactides to prepare cyclic polyesters in the absence of alcohol initiator [58–60]. In particular, cyclic PCLs with high molecular weights up to $114\,\mathrm{kg\,mol^{-1}}$ but broad dispersities ($Đ = 1.29 \sim 2.10$) were achieved in the presence of 1,3-diisopropyl-4,5-dimethylimidazol-2-ylidene or 1,3,4,5-tetramethylimidazol-2-ylidene as the catalyst/initiator [60]. The kinetics investigation of zwitterionic ROP of εCL mediated by NHC suggested that less than 60% of the NHC was converted to active zwitterions due to slow and reversible initiation compared to fast propagation. These zwitterions Z_n rapidly propagated with monomer and underwent cyclization by intramolecular backbiting to produce cyclic chains C_x and lower molecular weight zwitterions Z_{n-x}. The cyclic chains C_x can be reactivated by growing zwitterions Z_n to generate high-molecular-weight zwitterions Z_{n+x} (Scheme 4.9) [61].

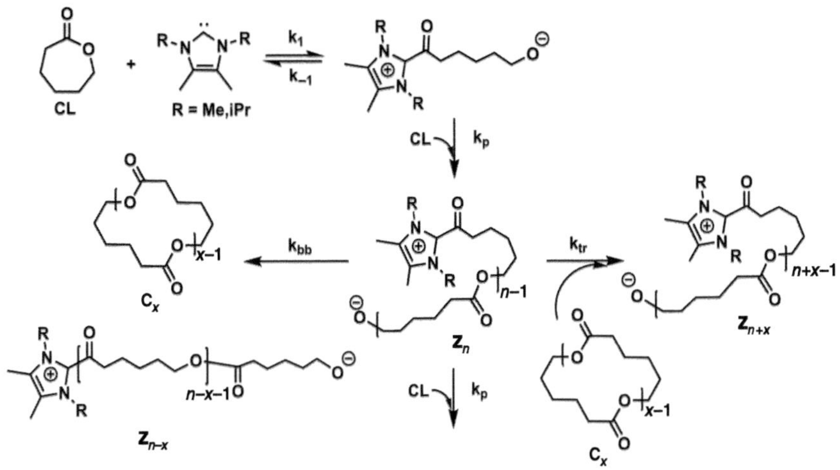

Scheme 4.9 The proposed mechanism for zwitterionic ROP of εCL to prepare cyclic PCL. Source: [61]/American Chemical Society.

Waymouth and Hedrick have developed the strong guanidine base TBD (pK_a^{MeCN} = 26.0) as an effective catalyst for the ROPs of δVL and εCL [62]. The polymerizations achieved near quantitative conversions of δVL and εCL within hours in the presence of 0.5 mol% TBD (relative to monomers), producing polymers with predictable molecular weights up to degree of polymerization (DP) of 200, high chain-end group fidelity and narrow dispersities ($Đ < 1.16$). Although the ROPs mediated by TBD exhibited living/controlled polymerization characteristics, broaden of dispersities at high monomer conversions was observed, suggesting the occurrence of transesterification when most monomers were consumed. Despite the similar basicity, MTBD (pK_a^{MeCN} = 25.4) and DBU (pK_a^{MeCN} = 24.3) alone were not active for the ROPs of δVL and εCL. The exceptional high catalytic activity of TBD was attributed to the ability of TBD to simultaneously activate monomer and alcohol/propagating chain-ends. In contrast to directly attacking the carbonyl of monomers, the theoretical investigation supported the hydrogen bonding activation mechanism as evidenced by the lower calculated activation energy barrier. Similar to the urea anion activation mechanism, TBD served as both hydrogen bonding donor to monomer and hydrogen bonding acceptor to alcohol/propagating chain-end (Scheme 4.10) [63].

Organophosphazene bases as a class of non-nucleophilic, neutral organobases are effective catalysts for the ROPs of δVL and εCL in the presence of alcohol initiators. The organophosphazene bases can activate the initiator or propagating chain-end by deprotonation or hydrogen bond, which then nucleophilically attack the carbonyl of the monomers to start the polymerization. It was reported that living/controlled ROPs of δVL and εCL were achieved in the presence of BEMP (pK_a^{MeCN} – 27.6) or tBu-P_1 (pK_a^{MeCN} = 26.9) as the catalyst [64]. In contrast to the ROP of δVL, the polymerization rate of εCL was quite slow. For example, the bulk ROP of δVL reached a monomer conversion of 69% at RT within 73 h in the presence of a BEMP loading

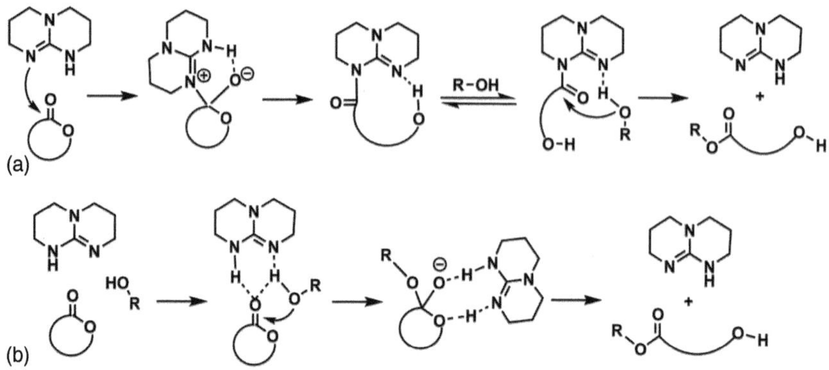

Scheme 4.10 (a) Nucleophilic monomer activation mechanism by directly attacking the carbonyl group of monomers; (b) Bifunctional activation of both monomer and alcohol by hydrogen bond. Source: [63]/American Chemical Society.

of 1 mol% relative to the monomer. In contrast, the bulk ROP of εCL only reached a low monomer conversion of 14% at 80 °C for 10 days at the same catalyst loading. Compared to BEMP and *t*Bu-P_1, *t*Bu-P_2 (pK_a^{MeCN} = 33.5) with a stronger basicity exhibited improved catalytic activity and good control over the ROPs of δVL and εCL [65]. When the ROP of εCL was conducted at an initial monomer concentration of 3 mol L^{-1} in toluene at RT, a monomer conversion of 86% was achieved in the presence of 1 mol% catalyst loading (relative to the monomer), producing PεCL with a narrow dispersity of $Đ = 1.11$. In addition to primary alcohol, secondary alcohol, (aliphatic/aromatic) amide, carboxylic acid, phenol and thiophenol were also used as initiators but achieved reduced polymerization rate and controllability. Polystyrene beads supported *t*Bu-P_2 also exhibited high catalytic activity toward the ROP of δVL but exhibited low catalytic activity for the ROP of εCL. Polystyrene beads supported *t*Bu-P_2 can be recycled by simple filtration for at least five times without obvious drop in catalytic activity [29]. When a stronger phosphazene base *t*Bu-P_4 (pK_a^{MeCN} = 42.6) was used as the catalyst, the ROP of εCL exhibited accelerated but worse controlled polymerization as evidenced by the bimodal distribution of obtained PεCL [66]. The phosphazene base CTPB (pK_a^{MeCN} = 33.3) also exhibited high catalytic activity toward the ROP of εCL in the presence of alcohol initiators. A high-molecular-weight PεCL with a M_n = 112.8 kDa was achieved at a feeding molar ratio of [εCL]/[CTPB]/[BnOH] = 2000/1/1 [67].

When a single base is used as the catalyst for the ROP, one challenge is to achieve a high catalytic activity while simultaneously remaining good control over the polymerization. This challenge is associated with the chain-end activation mechanism mediated by single base. For example, BEMP and *t*Bu-P_1 exhibits good control over the polymerization but suffers from low catalytic activity. On the other hand, *t*Bu-P_4 not only promotes the chain propagation but also catalyzes the transesterification or back-biting side reactions. In contrast, the (thio)urea anion generated by deprotonation of (thio)urea with alkoxide sodium or alkoxide potassium exhibits high selectivity for the ROP of lactones relative to transesterification of polymer chains [14, 15]. As such, rapid and controlled ROPs of δVL and εCL can be realized by choosing suitable combinations of bases and (thio)ureas [16, 17, 68–71]. Kiesewetter and co-workers investigated the use of a commercial urea triclocarban (TCC) as the co-catalyst in combination with various organobases for the ROPs of δVL and εCL [68]. When TCC was combined with MTBD (pK_a^{MeCN} = 25.4) or DBU (pK_a^{MeCN} = 24.3) as the binary catalyst, the ROPs of δVL and εCL exhibited living/controlled polymerization behaviors in benzene. The polymerizations reached high monomer conversions up to 90% within hours in the presence of 5 mol% catalysts relative to the monomers. Note that neither single MTBD or DBU was effective catalyst for the ROPs of δVL and εCL [62]. Additionally, urea as a hydrogen bonding donor remained active in polar, hydrogen bond accepting solvents, such as tetrahydrofuran (THF) and acetone, while the ROP rates considerably reduced in the presence of thioureas as hydrogen bonding donors. When a stronger base BEMP (pK_a^{MeCN} = 27.6) was combined with TCC, the polymerizations were proposed to proceed dominantly via a urea anion activation mechanism. As such, the polymerization considerably accelerated and remained controlled, achieving full conversion of δVL within 3 min in benzene at an initial monomer concentration of 1 mol L^{-1} and a 5 mol% loading of BEMP/TCC relative to δVL. Recently, Liu and Li et al. reported the ROPs of δVL and εCL catalyzed by CTPB/urea binary catalyst to produced polyesters with well-defined chain-ends, controlled and predictable molecular weights and narrow dispersities [69]. The addition of urea greatly improved the polymerization controllability while remaining high catalytic activity. In addition, PVL-*b*-PCL diblock copolyester was prepared by sequential ROP of δVL and εCL.

4.3.3 δ-Valerolactone Derivatives

Compared to the unsubstituted analogue (δVL), the alkyl substituted δ-valerolactones are thermodynamically less polymerizable. The restricted rotation of the polymer chain due to the existence of alkyl substituent leads to a more negative polymerization entropy change $\Delta S_p^\ominus$ for the ROP of alkyl substituted δ-valerolactones compared to δVL. As such, the alkyl substituted δ-valerolactones generally possess lower ceiling temperatures (T_cs) than δVL. Despite their low polymerization activities, the low T_cs and associated equilibrium polymerization characteristics make them potential candidates to prepare chemically recyclable polymers with closed-loop life cycles, which have attracted increased interests in recent years.

β-Methyl-δ-valerolactone (βMδVL) is a methyl substituted six-membered lactone at the β position of δVL, which can be produced by direct fermentation of glucose or an efficient semisynthetic approach that employs high-yielding chemical reactions to transform mevalonate. The estimated cost of βMδVL is less than 2 US dollars kg^{-1}, which makes it attractive replacement to the petroleum based chemicals. Hillmyer and Zhang *et al.* reported the bulk ROP of βMδVL at RT in the presence of 0.05 wt% TBD to achieve an equilibrium monomer conversion of ~91% within 3 h [72]. The obtained poly(βMδVL) was an amorphous aliphatic polyester with a low glass transition temperature (T_g = −51 °C). The well-defined PLLA–PβMδVL–PLLA triblock copolymers were also prepared by sequential ROPs of βMδVL and L-lactide (L-LA) in the presence of TBD as the catalyst. The obtained PLLA–PβMδVL–PLLA triblock copolymers with a specific composition and molar mass behaved as strong elastomers with comparable recoverability, tensile strength, and ultimate elongation to the commercial petroleum based elastomers.

δ-Caprolactone (δCL) is a isomer of βMδVL bearing a methyl group at the δ position of δVL. δCL is naturally found in fruits and heated milk or can be produced from bio-sourced 5-hydroxymethylfurfural [73–77]. δCL exhibits lower polymerization activity compared to βMδVL and δVL, which is probably due to the less nucleophilicity of propagating chain-end (secondary alcohol species). A series of different catalysts, including organic acids, bases and organometallic complexes have been attempted for the ROP of δCL, but the polymerization still face challenges including slow polymerization rate, high catalyst loading, poor control as well as low molecular weights of resultant polymers [78–80]. For example, when DPP was used as the catalyst, 24 h was required to reach an equilibrium monomer conversion at a catalyst loading of 1 mol% relative to the monomer [81]. Recently, Shen and Li et al. presented the living/controlled ROP of δCL in the presence of strong base/urea as the binary catalyst [82]. When a urea bearing a trifluoromethylphenyl and a cyclohexyl group was combined with potassium alkoxide (KOMe) as the catalyst, the ROP of δCL reached an equilibrium monomer conversion of 86% within 10 min in THF. As comparison, the polymerization catalyzed by single KOMe achieved a low δCL conversion of 58% within 2 h, producing a polymer with a broad distribution of $Đ = 1.42$. Notably, the combination of *t*Bu-P_2 and a bisurea exhibited even higher catalytic activity with a turnover frequency (TOF) of 2580 h^{-1}, affording PδCL with a high molecular weight M_n up to 100.9 kg mol^{-1} at a low catalyst loading of 0.1 mol% relative to the monomer. Sequential ROP of δCL and L-LA afforded well-defined PLLA-*b*-PδCL-*b*-PLLA triblock copolymers, which behaved as thermoplastics or thermoplastic elastomers depending on the compositions and molecular weights. Remarkably, PδCL was capable to depolymerize back to pristine δCL with an almost quantitative yield by heating the materials in the presence of $Sn(Oct)_2$ as the catalyst. The chemical recycling of obtained PLLA-*b*-PδCL-*b*-PLLA triblock copolymers was also achieved to recover ethyl lactate and δCL with high yield.

In addition to δCL, alkyl-δ-lactones bearing with longer alkyl substituents can also be produced from biomass on a large scale and commercially used as food additives and fragrance ingredients [83, 84]. But the ROPs of alkyl-δ-lactones, especially for those with long alkyl substituents, have been scarcely explored [80, 85]. Hillmyer and co-workers explored the effects of substituent length on their polymerization kinetics and thermodynamics for a series of alkyl-δ-lactones [80]. Although good control over the ROPs of alkyl-δ-lactones was achieved with DPP, couples of days were necessary to reach the equilibrium monomer conversion. In contrast, rapid and controlled polymerization was achieved at RT in bulk in the presence of *t*Bu-P_2/urea as the binary catalyst. For example, *t*Bu-P_2/urea achieved a much higher turnover frequency (TOF) value of 486 h^{-1} for the ROP of δ-octalactone (δOL) compared to that of DPP as the catalyst (TOF = 0.71 h^{-1}). The sequential ROPs of alkyl-δ-lactones and L-lactide was realized using *t*Bu-P_2 and a bisurea as the catalyst. The obtained triblock copolymers can be directly used as pressure-sensitive adhesives without addition of tackifiers or additives [86].

The ROPs of mono-alkyl substituted δ-lactones generally produce amorphous polyesters due to the non-stereoselective polymerization and resultant low stereoregularity of polyesters. α-Methylene-δ-valerolactone (MVL) is a bifunctional monomer that can be prepared by reacting δ-valerolactone (δVL) with ethyl formate via a simple two-step protocol [87]. Similar with MBL, MVL can also either undergo ROP to produce functional polyester or VAP to produce PMMA analogue. The chemo-selective and controlled ROP of MVL to produce high-molecular-weight polyester with M_n up to 74.6 kg mol^{-1} was recently reported by using *t*Bu-P_2 in combination with a bisurea as the catalyst. The polymerization exhibited living/controlled characteristics as supported by the good linear correlation of the molar mass of $P(MVL)_{ROP}$ as a function of monomer conversion and feeding molar ratio of monomer to initiator as well as the chain extension experiments. $P(MVL)_{ROP}$ was a semi-crystalline polyester with a degree of crystallinity X_c of 52%. The high-molecular-weight $P(MVL)_{ROP}$ behaved as a tough thermoplastic with mechanical properties comparable to commodity plastics. Moreover, $P(MVL)_{ROP}$ can be depolymerized back to pristine MVL with a 96% yield by heating at 130 °C under vacuum in the presence of $Sn(Oct)_2$ as the catalyst [88]. The chemo-selective copolymerization of MVL with εCL to produce functional copolyesters was also achieved with *t*Bu-P_2/urea binary catalyst [89].

4.3.4 Lactide

Poly(L-lactide) (PLLA) is one of the most investigated aliphatic polyesters and has achieved industrial production. A wide range of organic acids and bases has been developed as efficient catalysts for the ROP of lactide. Despite the same ring size, lactide exhibits distinct different polymerization behaviors with δVL and εCL. Compared to δVL and εCL, stonger acids, such as trifluoromethanesulfonic acid, are required for the effective solution ROP of lactide. In contrast, when an organobase is used as the catalyst, lactide exhibits higher polymerization activity than δVL and εCL. A mild organobase is enough to catalyze the ROP of lactide, such as DMAP, phosphines, DBU and TBD, but they are not efficient catalysts toward the ROPs of δVL and εCL. In contrast to δVL and εCL, lactide has two chiral centers. Most recent studies focus on the stereoselective ROP of racemic lactide (*rac*-LA) to prepare highly isotatic polylactide. Several excellent reviews have summarized the developments of organocatalytic ROP of lactide [11, 90–94]. As such, we will concentrate on the discussion of advancements achieved in past decade, especially the ROP of lactide mediated by the base/(thio)urea binary catalyst.

Waymouth and Hedrick reported the thiourea/amine binary catalyst for the living/controlled ROP of lactide in 2005 [95]. The thiourea/amine binary catalyst exhibited extraordinary selectivity toward the chain propagation relative to transesterification as supported by the predictable molecular weights and narrow dispersities ($Đ < 1.10$). It was proposed the polymerization proceeded via a bifunctional activation mechanism in which thiourea activated LA monomer and amine

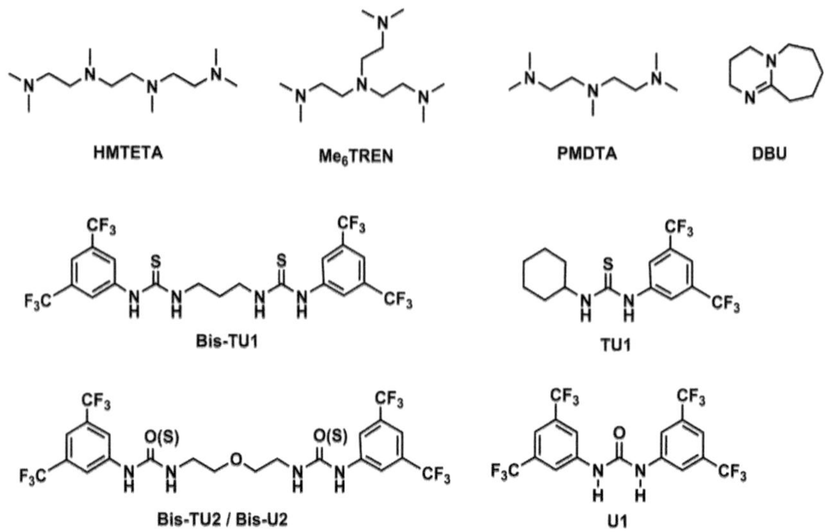

Figure 4.2 The chemical structures of organobases and (thio)ureas used for ROP of L-LA.

activated the initiator or propagating chain-end, respectively [96]. Despite the good controllability, the thiourea/amine binary catalyst exhibited low catalytic activity. For example, 48 h was required to reach a monomer conversion of 97% at a catalyst loading of 5 mol% relative to L-LA [95]. Kiesewetter and coworkers reported accelerated ROP of LA using a bis-thiourea as the hydrogen bonding donor [97]. A series of commercial available alkylamines was used as the bases. When an equivalent of alkylamine HMTETA (Figure 4.2) combined with a bis-thiourea (Bis-TU1) was used as the catalyst, the ROP of L-LA reached 90% monomer conversion within 15 min at a catalyst loading of 2.5 mol% relative to the monomer. In contrast, HMTETA/TU1 (5 mol% each relative to monomer) achieved a monomer conversion of 94% within 90 min. Notably, the alkylamine/bis-thiourea binary catalyst remained good control over the ROP of L-LA. The polymerization rate acceleration was also observed when other alkylamines were used. The reason for the accelerated polymerization rate was speculated to be the activation of a single monomer by both thiourea moieties of bis-thiourea [98]. Computation investigation indicated that LA activation via the activated TU structure is favored in energy compared to the dual thiourea activation structure (Scheme 4.11). Subsequent studies further confirmed that bis-(thio)ureas and tri-(thio)ureas bearing with two or three (thio)urea hydrogen bond donating moieties generally exhibited improved monomer activation ability and catalytic activity compared to mono-(thio)ureas [99]. The enhanced catalytic activity is proposed to originate from the successively increased intramolecular hydrogen bond activation with each additional (thio)urea moiety. The bis-urea has been proven to be more active than bis-thiourea for the ROPs of lactones and cyclic carbonate monomers. On the contrary, the bis-thiourea exhibited better catalytic performance than bis-urea toward the ROP of LA either in bulk or in solution [100]. For example, the bulk polymerization conducted at 100 °C reached a monomer conversion of

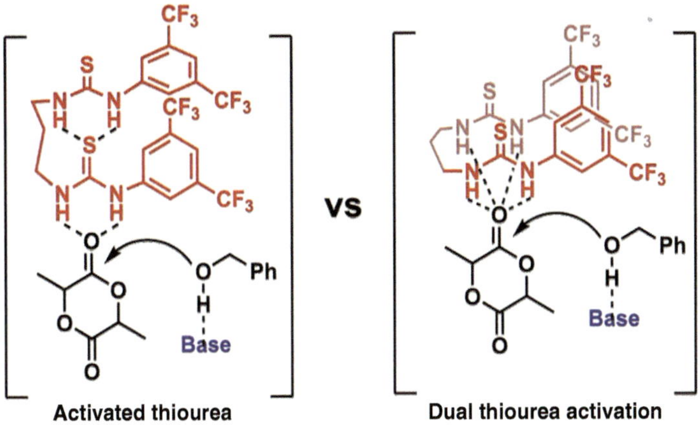

Scheme 4.11 The activated thiourea mechanism and the dual thiourea activation mechanism.

90% within 15 min in the presence of Me_6TREN/bis-thiourea (Bis-TU2) as the catalyst. In contrast, when a bis-urea (Bis-U2) was used as the co-catalyst, 105 min was required to achieve same monomer conversion (Figure 4.2). The difference in activity was attributed to the fact that Bis-TU2 can be deprotonated by Me_6TREN and the polymerization proceeded via a thiourea anion mechanism. In contrast, Me_6TREN/Bis-U2 mediated ROP of L-LA proceeded via the cooperative hydrogen bond activation mechanism.

In addition to use a bis-(thio)urea or tri-(thio)urea as the co-catalysts, the polymerization can also be accelerated when a stronger base than alkylamine was used. For example, rapid and controlled ROP of L-LA was achieved in the presence of DBU/U1 binary catalyst (Figure 4.1), achieving a 92% monomer conversion within 6 seconds. The almost infinite combinations of organobases and (thio)ureas provides opportunities to adjust catalytic activity and controllability. Shen and coworkers achieved the rapid and controlled ROP of L-LA via choosing appropriate combinations of organobases and ureas [101]. It is worth noting that a high-molecular-weight PLLA with a M_n up to 73.8 kDa was obtained at a low catalyst loading of 0.05 mol%. Zhao et al. developed a highly active and selective organocatalyst via a one-step dehydration reaction between tetra-*n*-butyl ammonium hydroxide and *N,N'*-diarylurea. The controlled ROP of L-LA was achieved to produce PLLA with widely tunable molecular weights (M_n = 4–130 kg mol^{-1}) and relatively narrow dispersities [102].

In addition to the stereoselective ROP of *rac*-LA, one challenge for the organocatalysts is to achieve the ROP of lactide at industrial relevant conditions, i.e. under solvent-free condition. The bulk polymerization of lactide requires a temperature as high as 180 °C. Besides the intensive racemization, transesterification and back biting side reactions, organocatalysts are usually unstable and become deactivated at such a high temperature. It was reported that DPP cannot catalyze the ROP of L-LA in solution. Surprisingly, Saito *et al.* reported that the bulk ROP of L-LA was achieved at 130 °C in the presence of DPP as the catalyst and

3-phenyl-1-propanol (PPA) as the initiator [52]. A polymerization conducted at [L-LA]/[DPP]/[PPA] = 100/1/2 achieved a monomer conversion of 86% within 24 h, producing a highly isotactic PLLA with a M_n of 6.8 kg mol^{-1} and a $Đ$ of 1.23. Similarly, benzoic acid (BA) has been developed as an efficient catalyst for the bulk ROP of L-LA conducted at temperature up to 180 °C by Taton and Coulembier *et al.* [103, 104] The ROP of L-LA mediated by BA exhibited living/controlled polymerization characteristics with targeted DP up to 75 as supported by the well consistence of measured molecular weights with the theoretical values as well as the narrow dispersities ($Đ < 1.29$). A bifunctional activation mechanism was proposed to explain the high catalytic activity of BA, which activated both the monomer and the propagating hydroxy chain-end through hydrogen bond (Scheme 4.12). Notably, BA can be readily recycled by simple sublimation and reused for five times without significant catalytic activity loss.

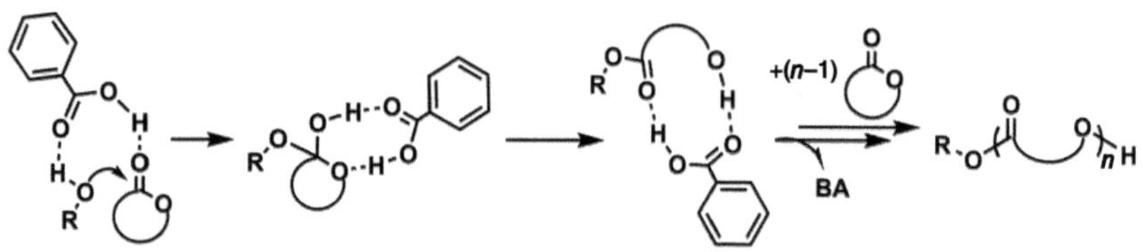

Scheme 4.12 The proposed bifunctional activation mechanism for the BA catalyzed ROP of lactide. Source: [103]/ Royal Society of Chemistry.

Guo and coworkers investigated the bulk ROP of L-lactide using pyridinium halides as the catalysts [105]. 4-(*N,N*-dimethylamino)pyridine hydrochloride (DMAP·HCl) exhibited good catalytic activity toward the bulk ROP of L-LA at 140 °C, achieving quantitative monomer conversions within 10 h at a feeding molar ratio of [catalyst]/[L-LA] = 1/120. The polymerization exhibited living/controlled behaviors, producing PLLA with M_n up to 16.6 kg mol^{-1} and narrow dispersities ($Đ$ = 1.13–1.17). Sardon and coworkers developed a nonstoichiometric acid-base organocatalyst consisting of DMAP and methanesulfonic acid (MSA) (DMAP:MSA = 1 : 2) for the bulk ROP of L-lactide [106]. Stereoretention ROP of L-LA was achieved under solvent-free condition at elevated temperature up to 180 °C, producing colorless PLLA with a M_n up to 40 kg mol^{-1} and a chiral purity in excess of 98%. Epimerization was much less favorable in the presence of DMAP:MSA = 1 : 2 as supported by the computational investigation.

The bis-(thio)urea and alkylamine base mediated ROP of L-LA can also be conducted in a solvent-free condition [107]. The bis-thiourea (Bis-TU1) combined with PMDTA were selected as the optimal catalytic system, which exhibited high catalytic activity and selectivity toward the bulk ROP of L-LA at 100 °C. Although PLLA precipitated and solidified before complete monomer conversion, the ROP still exhibited living/controlled polymerization characteristics as supported by thc lincar correlation of molecular weights with monomer conversions as well as narrow dispersities.

4.3.5 Large-Membered Lactones

As one of the most attractive biodegradable aliphatic polyesters, the ROPs of macrolactones have received extensive research in the past two decades [108, 109]. The long-chain aliphatic polyesters produced by ROPs of macrolactones possess crystalline properties similar to polyethylene. The excellent mechanical properties and high working temperature make them promising alternatives to commodity polyolefins [108, 110]. Compared with small and medium-sized lactones, the macrolactones, such as ω-pentadecalactone (PDL) and dodecanolactone (DDL), exhibit distinct different ROP behaviors. Unlike the enthalpy-driven polymerization of small and medium-sized lactones, macrolactones (≥12 ring members) are considered as non-strained monomers. Their polymerizations are entropy-driven due to the increase of conformational freedom of polymers compared to that of monomers [111].

Duchateau and coworkers attempted the metal-free ROP of PDL using a series of *N*-heterocyclic organic catalysts, including TBD, MTBD, 1,2,3-triisopropylguanidine, DBU, DMAP, imidazole, indoles and *N*-heterocyclic carbenes [112]. Among the tested catalysts, only TBD was the efficient catalyst for the polymerization of PDL conducted in bulk or in toluene solution at 100 °C in the presence of benzyl alcohol as the initiator. A series of *N*-heterocyclic olefins (NHOs) was then attempted toward the ROP of PDL [113]. The polymerization conducted in toluene at 110 °C only achieved a low monomer conversion (46%) after 240 min, producing a low-molecular-weight poly(ω-pentadecalactone) (PPDL) with a $M_n = 6.3\ \text{kg mol}^{-1}$ and a broad dispersity of $Đ = 2.36$.

Organophosphazene bases have been developed as efficient catalysts for the ROPs of lactones with small to medium size, such as γBL, δVL and εCL. In contrast, the catalytic activity of organophosphazene bases toward the ROPs of macrolactones is relatively low. Moreover, it is necessary to screen suitable catalysts to promote ROP while inhibiting the occurrence of side reactions. Hadjichristidis *et al.* attempted the bulk polymerization of PDL at 80 °C catalyzed by $t\text{Bu-P}_2$. When the catalyst loading was 2 mol% relative to the monomer, the polymerization rate was slow ($\text{TOF} = 0.08\ \text{h}^{-1}$) and the conversion of PDL only reached 20% after 5 days [114]. As the loading of $t\text{Bu-P}_2$ increased by 10 times and the polymerization temperature increased to 100 °C, faster polymerization rate ($\text{TOF} = 1\ \text{h}^{-1}$) and higher monomer conversion (93%) were observed. When the stronger base $t\text{Bu-P}_4$ ($\text{p}K_a^{\text{MeCN}} = 42.6$) was used as the catalyst, higher catalytic activity was observed for the ROP of PDL. The PDL monomers were completely converted at 80 °C within 0.8 h ($\text{TOF} = 124\ \text{h}^{-1}$) when the polymerization was conducted with a catalyst loading of 1 mol% (relative to the monomer) at a monomer concentration of $1.5\ \text{mol L}^{-1}$. Even at lower monomer concentration ($0.4\ \text{mol L}^{-1}$), less catalyst loading (0.2 mol%) and lower polymerization temperature (RT), $t\text{Bu-P}_4$ still exhibited high catalytic activity, achieving a monomer conversion up to 74% within 6 h. The ROP of PDL mediated by $t\text{Bu-P}_4$ displayed some "living" polymerization characteristics. For example, the molecular weight of PPDL increased linearly with monomer conversion despite the broad dispersities of PPDL. Both linear and cyclic

PPDL were obtained as supported by the MALDI-TOF-MS spectrum, indicating that significant side reactions occurred during the polymerization process, such as transesterification and back-biting. When tBu-P_4 was used as a single catalyst, only random copolymers of macrolactones and small lactones were obtained due to the intensive transesterification no matter the monomers were added by "sequential feeding" or "one-shot". A "catalyst-switch" strategy was then developed for the preparation of block copolymers of poly(macrolactones-*b*-small lactones) (PMLs-*b*-PSLs) [115]. In this strategy, tBu-P_4 was first employed to catalyze the ROP of macrolactone monomers, followed by neutralizing tBu-P_4 with DPP, and then weaker phosphazene base tBu-P_2 was introduced to catalyze the ROP of second monomer (small lactones). This "catalyst-switch" strategy effectively alleviated the impact of transesterification side reactions on the block ordering of polymers but required a long time to achieve high monomer conversions (>2 days). This strategy has been proven to be applicable to various macrolactones such as PDL, ω-hexadecalactone (HDL), dodecanolactone (DDL), and small lactones including ε-caprolactone (εCL) and δ-valerolactone (δVL).

CTPB has a moderate basicity ($pK_a^{MeCN} = 33.3$) that is comparable with tBu-P_2 ($pK_a^{MeCN} = 33.5$) but exhibits higher catalytic activity toward the ROP of lactones with small and medium-size. Li et al. reported the ROP of PDL in the presence of CTPB as the catalyst and BnOH as the initiator [116]. Almost quantitative monomer conversion was achieved to produce PPDL with a high molecular weight of $M_n = 34\ \text{kg mol}^{-1}$ within 10 min when the ROP of PDL was conducted at a 1 mol% catalyst loading with a monomer concentration of $1.0\ \text{mol L}^{-1}$ at 80 °C. The high catalytic activity of CTPB toward the ROP of PDL was attributed to its bulky and symmetrical spherical molecular structure, which resulted in the formation of a loose ion pair of $[BnO]^-[CTPBH]^+$. Interestingly, CTPB was proved to be a trisbase with three basic sites by acid-base titration experiment, while the other phosphazene base are usually considered as monoacidic bases. Adding a small amount of acid to partially neutralize CTPB can reduce its catalytic activity and suppressed side reactions, such as transesterification and back-biting. Block copolymers of PDL and εCL or δVL (PPDL-*b*-PCL or PPDL-*b*-PVL) were successfully prepared by adjusting the feeding amount of acid to better control the catalytic activity of CTPB. Specifically, after complete conversion of PDL, 0.33 ~ 1 equivalent of benzoic acid was introduced to suppress the transesterification during ROPs of εCL or δVL. Thanks to the high catalytic activity and simple neutralization treatment of CTPB, well-defined block copolyesters can be prepared within 70 min.

4.4 Summary and Perspective

A variety of organocatalysts have been developed for the efficient and controlled ROPs of lactides and lactones. Compared to the organometallic catalysts, organocatalysts provide advantages including low toxicity, easy handle, good stability to oxygen and water as well as facile removal from the resultant polymers. Organocatalysts are especially attractive for the microelectronic and biomedical applications where metal residues are undesirable. In addition, many organocatalysts, such as DBU, TBD, organophosphazene bases and some (thio)ureas, are commercially available or easy to prepare. In contrast, the ligands of organometallic catalysts usually require complicated and tedious synthetic protocols. Some organocatalysts, especially the recently developed (thio)urea anion catalysts, exhibit high catalytic activity and excellent control over the ROP that compare or even exceed those of organometallic catalysts.

Although great successes have been achieved in the past decades as we discussed above, there are still plenty of rooms for the development of organocatalytic ROPs of lactides and lactones. In our opinion, more efforts should be devoted to addressing the following challenges. Firstly, stereoselective ROPs of chiral monomers mediated by organocatalysts remain to be explored. In contrast to the organometallic catalysts, there are few reports on the organocatalytic stereoselective ROPs. Actually, only limit examples address this problem and achieve the stereoselective control at extremely low temperatures. Secondly, the most current studies investigate the ROPs of lactides and lactones at RT in dilute solution. However, melt polymerizations under solvent-free conditions are industrially desirable. The future studies should pay more attentions to achieve the controlled ROPs in melt. As such, high efficient organocatalysts with good thermal stability at elevated temperature are desired to be explored. Finally, although some argue that organocatalysts exhibit low toxicity compared to the organometallic catalysts, the recent studies showed that the residual organocatalysts were considerably cytotoxic [117]. As such, a comprehensive evaluation of toxicity for the organocatalysts is required if the resultant polyesters are used for biomedical applications.

References

1 PlasticsEurope (2021). Plastics – the facts 2021. https://plasticseurope.org/knowledge-hub/plastics-the-facts-2021/

2 Fagnani, D.E., Tami, J.L., Copley, G. et al. (2020). 100th anniversary of macromolecular science viewpoint: redefining sustainable polymers. *ACS Macro Lett.* 10: 41–53.

3 Haque, F.M., Ishibashi, J.S.A., Lidston, C.A.L. et al. (2022). Defining the macromolecules of tomorrow through synergistic sustainable polymer research. *Chem. Rev.* 122: 6322–6373.

4 Millican, J.M. and Agarwal, S. (2021). Plastic pollution: a material problem? *Macromolecules* 54: 4455–4469.

5 Worch, J.C. and Dove, A.P. (2020). 100th anniversary of macromolecular science viewpoint: toward catalytic chemical recycling of waste (and future) plastics. *ACS Macro Lett.* 9: 1494–1506.

6 Haider, T.P., Völker, C., Kramm, J. et al. (2019). Plastics of the future? the impact of biodegradable polymers on the environment and on society. *Angew. Chem. Int. Ed.* 58: 50–62.

7 Tang, X. and Chen, E.Y.X. (2019). Toward infinitely recyclable plastics derived from renewable cyclic esters. *Chem* 5: 284–312.

8 Hong, M. and Chen, E.Y.X. (2017). Chemically recyclable polymers: a circular economy approach to sustainability. *Green Chem.* 19: 3692–3706.

9 Zhu, Y., Romain, C., and Williams, C.K. (2016). Sustainable polymers from renewable resources. *Nature* 540: 354–362.

10 Kamber, N.E., Jeong, W., Waymouth, R.M. et al. (2007). Organocatalytic ring-opening polymerization. *Chem. Rev.* 107: 5813–5840.

11 Kiesewetter, M.K., Shin, E.J., Hedrick, J.L., and Waymouth, R.M. (2010). Organocatalysis: opportunities and challenges for polymer synthesis. *Macromolecules* 43: 2093–2107.

12 Fukushima, K. and Nozaki, K. (2020). Organocatalysis: a paradigm shift in the synthesis of aliphatic polyesters and polycarbonates. *Macromolecules* 53: 5018–5022.

13 Lin, B.H. and Waymouth, R.M. (2018). Organic ring-opening polymerization catalysts: reactivity control by balancing acidity. *Macromolecules* 51: 2932–2938.

14 Lin, B. and Waymouth, R.M. (2017). Urea anions: simple, fast, and selective catalysts for ring-opening polymerizations. *J. Am. Chem. Soc.* 139: 1645–1652.

15 Zhang, X.Y., Jones, G.O., Hedrick, J.L., and Waymouth, R.M. (2016). Fast and selective ring-opening polymerizations by alkoxides and thioureas. *Nat. Chem.* 8: 1047–1053.

16 Pothupitiya, J.U., Hewawasam, R.S., and Kiesewetter, M.K. (2018). Urea and thiourea H-bond donating catalysts for ring-opening polymerization: mechanistic insights via (non)linear free energy relationships. *Macromolecules* 51: 3203–3211.

17 Coderre, D.N., Fastnacht, K.V., Wright, T.J. et al. (2018). H-bonding organocatalysts for ring-opening polymerization at elevated temperatures. *Macromolecules* 51: 10121–10126.

18 Martin, D.P. and Williams, S.F. (2003). Medical applications of poly-4-hydroxybutyrate: a strong flexible absorbable biomaterial. *Biochem. Eng. J.* 16: 97–105.

19 Williams, S.F., Rizk, S., and Martin, D.P. (2013). Poly-4-hydroxybutyrate (P4HB): a new generation of resorbable medical devices for tissue repair and regeneration. *Biomed Tech (Berl)* 58: 439–452.

20 Bozell, J.J. and Petersen, G.R. (2010). Technology development for the production of biobased products from biorefinery carbohydrates—the US department of energy's "top 10" revisited. *Green Chem.* 12: 539–554.

21 Yamashita, K., Yamamoto, K., and Kadokawa, J.-I. (2014). Acid-catalyzed ring-opening polymerization of γ-butyrolactone under high-pressure conditions. *Chem. Lett.* 43: 213–215.

22 Hong, M. and Chen, E.Y. (2016). Completely recyclable biopolymers with linear and cyclic topologies via ring-opening polymerization of gamma-butyrolactone. *Nat. Chem.* 8: 42–49.

23 Yuan, P.J. and Hong, M. (2019). Ring-opening polymerizations of the "non-strained" gamma-butyrolactone and its derivatives: an overview and outlook. *Acta Polym. Sin.* 50: 327–337.

24 Hong, M. and Chen, E.Y.X. (2016). Towards truly sustainable polymers: a metal-free recyclable polyester from biorenewable non-strained γ-butyrolactone. *Angew. Chem. Int. Ed.* 55: 4188–4193.

25 Zhao, N., Ren, C., Li, H. et al. (2017). Selective ring-opening polymerization of non-strained γ-butyrolactone catalyzed by a cyclic trimeric phosphazene base. *Angew. Chem. Int. Ed.* 56: 12987–12990.

26 Shen, Y., Zhang, J., Zhao, N. et al. (2018). Preparation of biorenewable poly(γ-butyrolactone)-*b*-poly(L-lactide) diblock copolyesters via one-pot sequential metal-free ring-opening polymerization. *Polym. Chem.* 9: 2936–2941.

27 Shen, Y., Zhang, J., Zhao, Z. et al. (2019). Preparation of amphiphilic poly(ethylene glycol)-*b*-poly(γ-butyrolactone) diblock copolymer via ring opening polymerization catalyzed by a cyclic trimeric phosphazene base or alkali alkoxide. *Biomacromolecules* 20: 141–148.

28 Walther, P., Frey, W., and Naumann, S. (2018). Polarized olefins as enabling (co)catalysts for the polymerization of γ-butyrolactone. *Polym. Chem.* 9: 3674–3683.

29 Ren, C.L., Zhu, X., Zhao, N. et al. (2019). Polystyrene beads supported phosphazene superbase as recyclable organocatalyst for ring-opening polymerization of delta-valerolactone. *Eur. Polym. J.* 119: 130–135.

30 Yuan, L., Zhou, W., Shen, Y., and Li, Z. (2022). Chemically recyclable polyurethanes based on bio-renewable γ-butyrolactone: from thermoplastics to elastomers. *Polym. Degrad. Stab.* 204: 110116.

31 Zhang, C.-J., Hu, L.-F., Wu, H.-L. et al. (2018). Dual organocatalysts for highly active and selective synthesis of linear poly(γ-butyrolactone)s with high molecular weights. *Macromolecules* 51: 8705–8711.

32 Lin, L., Han, D., Qin, J. et al. (2018). Nonstrained γ-butyrolactone to high-molecular-weight poly(γ-butyrolactone): facile bulk polymerization using economical ureas/alkoxides. *Macromolecules* 51: 9317–9322.

33 Liu, Y., Xu, J., Zhang, Y. et al. (2023). Rapid ring-opening polymerization of γ-butyrolactone toward high-molecular-weight poly(γ-butyrolactone) by an organophosphazene base and bisurea binary catalyst. *Chem. Asian J.* 18: e202201107.

34 Trotta, J.T., Jin, M., Stawiasz, K.J. et al. (2017). Synthesis of methylene butyrolactone polymers from itaconic acid. *J. Polym. Sci., Part A: Polym. Chem.* 55: 2730–2737.

35 Higaki, Y., Okazaki, R., and Takahara, A. (2012). Semirigid biobased polymer brush: poly(α-methylene-γ-butyrolactone) brushes. *ACS Macro Lett.* 1: 1124–1127.

36 Ding, K., John, A., Shin, J. et al. (2015). High-performance pressure-sensitive adhesives from renewable triblock copolymers. *Biomacromolecules* 16: 2537–2539.

37 Hillmyer, M.A. and Tolman, W.B. (2014). Aliphatic polyester block polymers: renewable, degradable, and sustainable. *Acc. Chem. Res.* 47: 2390–2396.

38 Bai, Y., He, J., and Zhang, Y. (2018). Ultra-high-molecular-weight polymers produced by the immortal phosphine-based catalyst system. *Angew. Chem. Int. Ed.* 57: 17230–17234.

39 Zhang, Y., Miyake, G.M., and Chen, E.Y.-X. (2010). Alane-based classical and frustrated lewis pairs in polymer synthesis: rapid polymerization of MMA and naturally renewable methylene butyrolactones into high-molecular-weight polymers. *Angew. Chem. Int. Ed.* 49: 10158–10162.

40 Tang, X., Hong, M., Falivene, L. et al. (2016). The quest for converting biorenewable bifunctional α-methylene-γ-butyrolactone into degradable and recyclable polyester: controlling vinyl-addition/ring-opening/cross-linking pathways. *J. Am. Chem. Soc.* 138: 14326–14337.

41 Shen, Y., Xiong, W., Li, Y. et al. (2021). Chemoselective polymerization of fully biorenewable α-methylene-γ-butyrolactone using organophosphazene/urea binary catalysts toward sustainable polyesters. *CCS Chem.* 3: 620–630.

42 Zhao, Z., Shen, Y., Kou, X. et al. (2020). Organocatalytic ring-opening copolymerization of biorenewable α-methylene-γ-butyrolactone toward functional copolyesters: preparation and composition dependent thermal properties. *Macromolecules* 53: 3380–3389.

43 Zhu, J.B., Watson, E.M., Tang, J., and Chen, E.Y. (2018). A synthetic polymer system with repeatable chemical recyclability. *Science* 360: 398–403.

44 Cywar, R.M., Zhu, J.-B., and Chen, E.Y.X. (2019). Selective or living organopolymerization of a six-five bicyclic lactone to produce fully recyclable polyesters. *Polym. Chem.* 10: 3097–3106.

45 Shi, C., Li, Z.-C., Caporaso, L. et al. (2021). Hybrid monomer design for unifying conflicting polymerizability, recyclability, and performance properties. *Chem* 7: 670–685.

46 Shi, C., Clarke, R.W., McGraw, M.L., and Chen, E.Y. (2022). Closing the "one monomer-two polymers-one monomer" loop via orthogonal (de)polymerization of a lactone/olefin hybrid. *J. Am. Chem. Soc.* 144: 2264–2275.

47 Sanda, F., Sanada, H., Shibasaki, Y., and Endo, T. (2002). Star polymer synthesis from ε-caprolactone utilizing polyol/protonic acid initiator. *Macromolecules* 35: 680–683.

48 Persson, P.V., Schröder, J., Wickholm, K. et al. (2004). Selective organocatalytic ring-opening polymerization: a versatile route to carbohydrate-functionalized poly(ε-caprolactones). *Macromolecules* 37: 5889–5893.

49 Gazeau-Bureau, S., Delcroix, D., Martín-Vaca, B. et al. (2008). Organo-catalyzed ROP of ε-caprolactone: methanesulfonic acid competes with trifluoromethanesulfonic acid. *Macromolecules* 41: 3782–3784.

50 Kakuchi, R., Tsuji, Y., Chiba, K. et al. (2010). Controlled/living ring-opening polymerization of δ-valerolactone using triflylimide as an efficient cationic organocatalyst. *Macromolecules* 43: 7090–7094.

51 Makiguchi, K., Satoh, T., and Kakuchi, T. (2011). Diphenyl phosphate as an efficient cationic organocatalyst for controlled/living ring-opening polymerization of δ-valerolactone and ε-caprolactone. *Macromolecules* 44: 1999–2005.

52 Saito, T., Aizawa, Y., Tajima, K. et al. (2015). Organophosphate-catalyzed bulk ring-opening polymerization as an environmentally benign route leading to block copolyesters, end-functionalized polyesters, and polyester-based polyurethane. *Polym. Chem.* 6: 4374–4384.

53 Connor, E.F., Nyce, G.W., Myers, M. et al. (2002). First example of *N*-heterocyclic carbenes as catalysts for living polymerization: organocatalytic ring-opening polymerization of cyclic esters. *J. Am. Chem. Soc.* 124: 914–915.

54 Nyce, G.W., Glauser, T., Connor, E.F. et al. (2003). In situ generation of carbenes: a general and versatile platform for organocatalytic living polymerization. *J. Am. Chem. Soc.* 125: 3046–3056.

55 Coulembier, O., Lohmeijer, B.G.G., Dove, A.P. et al. (2006). Alcohol adducts of *N*-heterocyclic carbenes: latent catalysts for the thermally-controlled living polymerization of cyclic esters. *Macromolecules* 39: 5617–5628.

56 Dove, A.P., Pratt, R.C., Lohmeijer, B.G.G. et al. (2006). *N*-heterocyclic carbenes: effective organic catalysts for living polymerization. *Polymer* 47: 4018–4025.

57 Kamber, N.E., Jeong, W., Gonzalez, S. et al. (2009). *N*-heterocyclic carbenes for the organocatalytic ring-opening polymerization of ε-caprolactone. *Macromolecules* 42: 1634–1639.

58 Shin, E.J., Brown, H.A., Gonzalez, S. et al. (2011). Zwitterionic copolymerization: synthesis of cyclic gradient copolymers. *Angew. Chem. Int. Ed.* 50: 6388–6391.

59 Jeong, W., Shin, E.J., Culkin, D.A. et al. (2009). Zwitterionic polymerization: a kinetic strategy for the controlled synthesis of cyclic polylactide. *J. Am. Chem. Soc.* 131: 4884–4891.

60 Shin, E.J., Jeong, W., Brown, H.A. et al. (2011). Crystallization of cyclic polymers: synthesis and crystallization behavior of high molecular weight cyclic poly(ε-caprolactone)s. *Macromolecules* 44: 2773–2779.

61 Brown, H.A., Xiong, S., Medvedev, G.A. et al. (2014). Zwitterionic ring-opening polymerization: models for kinetics of cyclic poly(caprolactone) synthesis. *Macromolecules* 47: 2955–2963.

62 Lohmeijer, B.G.G., Pratt, R.C., Leibfarth, F. et al. (2006). Guanidine and amidine organocatalysts for ring-opening polymerization of cyclic esters. *Macromolecules* 39: 8574–8583.

63 Chuma, A., Horn, H.W., Swope, W.C. et al. (2008). The reaction mechanism for the organocatalytic ring-opening polymerization of L-lactide using a guanidine-based catalyst: hydrogen-bonded or covalently bound? *J. Am. Chem. Soc.* 130: 6749–6754.

64 Zhang, L., Nederberg, F., Pratt, R.C. et al. (2007). Phosphazene bases: a new category of organocatalysts for the living ring-opening polymerization of cyclic esters. *Macromolecules* 40: 4154–4158.

65 Alamri, H., Zhao, J., Pahovnik, D., and Hadjichristidis, N. (2014). Phosphazene-catalyzed ring-opening polymerization of ε-caprolactone: influence of solvents and initiators. *Polym. Chem.* 5: 5471–5478.

66 Zhao, J., Pahovnik, D., Gnanou, Y., and Hadjichristidis, N. (2014). A "catalyst switch" strategy for the sequential metal-free polymerization of epoxides and cyclic esters/carbonate. *Macromolecules* 47: 3814–3822.

67 Li, H., Zhao, N., Ren, C. et al. (2017). Synthesis of linear and star poly(ε-caprolactone) with controlled and high molecular weights via cyclic trimeric phosphazene base catalyzed ring-opening polymerization. *Polym. Chem.* 8: 7369–7374.

68 Dharmaratne, N.U., Pothupitiya, J.U., Bannin, T.J. et al. (2017). Triclocarban: commercial antibacterial and highly effective H-bond donating catalyst for ring-opening polymerization. *ACS Macro Lett.* 6: 421–425.

69 Li, Y., Zhao, N., Wei, C. et al. (2019). Binary organocatalytic system for ring-opening polymerization of ε-caprolactone and δ-valerolactone: synergetic effects for enhanced selectivity. *Eur. Polym. J.* 111: 11–19.

70 Zhou, L., Xu, G., Mahmood, Q. et al. (2019). *N*-heterocyclic olefins and thioureas as an efficient cooperative catalyst system for ring-opening polymerization of δ-valerolactone. *Polym. Chem.* 10: 1832–1838.

71 Zhou, L., Wang, Z., Xu, G. et al. (2021). Structure and activity relationship studies of *N*-heterocyclic olefin and thiourea/urea catalytic systems: application in ring-opening polymerization of lactones. *Polym. Chem.* 12: 1806–1815.

72 Xiong, M., Schneiderman, D.K., Bates, F.S. et al. (2014). Scalable production of mechanically tunable block polymers from sugar. *Proc. Natl. Acad. Sci. USA* 111: 8357–8362.

73 Adams, T.B., Greer, D.B., Doull, J. et al. (1998). The FEMA GRAS assessment of lactones used as flavour ingredients. *Food Chem. Toxicol.* 36: 249–278.

74 Prades, A., Dornier, M., Diop, N., and Pain, J.-P. (2012). Coconut water uses, composition and properties: a review. *Fruits* 67: 87–107.

75 Parliment, T.H., Nawar, W.W., and Fagerson, I.S. (1965). Delta-caprolactone in heated milk fat. *J. Dairy Sci.* 48: 615–616.

76 Díaz-Rodríguez, A., Lavandera, I., Kanbak-Aksu, S. et al. (2012). From diols to lactones under aerobic conditions using a laccase/TEMPO catalytic system in aqueous medium. *Adv. Synth. Catal.* 354: 3405–3408.

77 Buntara, T., Noel, S., Phua, P.H. et al. (2011). Caprolactam from renewable resources: catalytic conversion of 5-hydroxymethylfurfural into caprolactone. *Angew. Chem. Int. Ed.* 50: 7083–7087.

78 Hardouin Duparc, V., Shakaroun, R.M., Slawinski, M. et al. (2020). Ring-opening (co)polymerization of six-membered substituted δ-valerolactones with alkali metal alkoxides. *Eur. Polym. J.* 134: 109858.

79 Bandelli, D., Weber, C., and Schubert, U.S. (2019). Strontium isopropoxide: a highly active catalyst for the ring-opening polymerization of lactide and various lactones. *Macromol. Rapid Commun.* 40: e1900306.

80 Schneiderman, D.K. and Hillmyer, M.A. (2016). Aliphatic polyester block polymer design. *Macromolecules* 49: 2419–2428.

81 Zhao, J. and Hadjichristidis, N. (2015). Polymerization of 5-alkyl δ-lactones catalyzed by diphenyl phosphate and their sequential organocatalytic polymerization with monosubstituted epoxides. *Polym. Chem.* 6: 2659–2668.

82 Li, C., Wang, L., Yan, Q. et al. (2022). Rapid and controlled polymerization of bio-sourced δ-caprolactone toward fully recyclable polyesters and thermoplastic elastomers. *Angew. Chem. Int. Ed.* 61: e202201407.

83 Romero-Guido, C., Belo, I., Ta, T.M.N. et al. (2011). Biochemistry of lactone formation in yeast and fungi and its utilisation for the production of flavour and fragrance compounds. *Appl. Microbiol. Biotechnol.* 89: 535–547.

84 Alam, M.I., Khan, T.S., and Haider, M.A. (2019). Alternate biobased route to produce δ-decalactone: elucidating the role of solvent and hydrogen evolution in catalytic transfer hydrogenation. *ACS Sustain. Chem. Eng.* 7: 2894–2898.

85 Cederholm, L., Olsén, P., Hakkarainen, M., and Odelius, K. (2020). Turning natural δ-lactones to thermodynamically stable polymers with triggered recyclability. *Polym. Chem.* 11: 4883–4894.

86 Xu, C., Wang, L., Liu, Y. et al. (2023). Rapid and controlled ring-opening (co)polymerization of bio-sourced alkyl-δ-lactones to produce recyclable (co)polyesters and their application as pressure-sensitive adhesives. *Macromolecules* 56: 6117–6125.

87 Xu, T.-Q., Yu, Z.-Q., and Zhang, X.-M. (2019). Recyclable vinyl-functionalized polyesters via chemoselective organopolymerization of bifunctional α-methylene-δ-valerolactone. *Macromol. Chem. Phys.* 220: 1900150.

88 Li, J., Liu, F., Liu, Y. et al. (2022). Functionalizable and chemically recyclable thermoplastics from chemoselective ring-opening polymerization of bio-renewable bifunctional α-methylene-δ-valerolactone. *Angew. Chem. Int. Ed.* 61: e202207105.

89 Liu, Y., Kou, X., Xu, C. et al. (2023). Chemoselective and controlled ring-opening copolymerization of biorenewable α methylene-δ-valerolactone with ε-caprolactone toward functional copolyesters. *Polym. Chem.* 14: 2326–2332.

90 Liu, S., Ren, C., Zhao, N. et al. (2018). Phosphazene bases as organocatalysts for ring-opening polymerization of cyclic esters. *Macromol. Rapid Commun.* 39: 1800485.

91 Ottou, W.N., Sardon, H., Mecerreyes, D. et al. (2016). Update and challenges in organo-mediated polymerization reactions. *Prog. Polym. Sci.* 56: 64–115.

92 Dove, A.P. (2012). Organic catalysis for ring-opening polymerization. *ACS Macro Lett.* 1: 1409–1412.

93 Xu, J., Wang, X., Liu, J. et al. (2022). Ionic H-bonding organocatalysts for the ring-opening polymerization of cyclic esters and cyclic carbonates. *Prog. Polym. Sci.* 125: 101484.

94 Mezzasalma, L., Dove, A.P., and Coulembier, O. (2017). Organocatalytic ring-opening polymerization of L-lactide in bulk: a long standing challenge. *Eur. Polym. J.* 95: 628–634.

95 Dove, A.P., Pratt, R.C., Lohmeijer, B.G. et al. (2005). Thiourea-based bifunctional organocatalysis: supramolecular recognition for living polymerization. *J. Am. Chem. Soc.* 127: 13798–13799.

96 Pratt, R.C., Lohmeijer, B.G.G., Long, D.A. et al. (2006). Exploration, optimization, and application of supramolecular thiourea–amine catalysts for the synthesis of lactide (co)polymers. *Macromolecules* 39: 7863–7871.

97 Spink, S.S., Kazakov, O.I., Kiesewetter, E.T., and Kiesewetter, M.K. (2015). Rate accelerated organocatalytic ring-opening polymerization of L-lactide via the application of a bis(thiourea) H-bond donating cocatalyst. *Macromolecules* 48: 6127–6131.

98 Kazakov, O.I. and Kiesewetter, M.K. (2015). Cocatalyst binding effects in organocatalytic ring-opening polymerization of L-lactide. *Macromolecules* 48: 6121–6126.

99 Fastnacht, K.V., Spink, S.S., Dharmaratne, N.U. et al. (2016). Bis- and tris-urea H-bond donors for ring-opening polymerization: unprecedented activity and control from an organocatalyst. *ACS Macro Lett.* 5: 982–986.

100 Hewawasam, R.S., Kalana, U.L.D.I., Dharmaratne, N.U. et al. (2019). Bisurea and bisthiourea H-bonding organocatalysts for ring-opening polymerization: cues for the catalyst design. *Macromolecules* 52: 9232–9237.

101 Zhou, W., Xu, C., Liu, Y., and Shen, Y. (2023). Preparation of high-molecular-weight polylactide by ring-opening polymerization of L-lactide using base/bisurea binary organocatalyst. *J. Polym. Res.* 30: 230.

102 Jiang, Z., Zhao, J., and Zhang, G. (2019). Ionic organocatalyst with a urea anion and tetra-*n*-butyl ammonium cation for rapid, selective, and versatile ring-opening polymerization of lactide. *ACS Macro Lett.* 8: 759–765.

103 Mezzasalma, L., De Winter, J., Taton, D., and Coulembier, O. (2018). Benzoic acid-organocatalyzed ring-opening (co)polymerization (ORO(c)P) of L-lactide and ε-caprolactone under solvent-free conditions: from simplicity to recyclability. *Green Chem.* 20: 5385–5396.

104 Mezzasalma, L., Harrisson, S., Saba, S. et al. (2019). Bulk organocatalytic synthetic access to statistical copolyesters from L-lactide and ε-caprolactone using benzoic acid. *Biomacromolecules* 20: 1965–1974.

105 Wang, H., Yao, Z., Li, Z. et al. (2020). Biocompatible and low-cost pyridinium halides catalysts promoted ring-opening polymerizations of cyclic esters in bulk. *Eur. Polym. J.* 127: 109570.

106 Basterretxea, A., Gabirondo, E., Jehanno, C. et al. (2021). Stereoretention in the bulk ROP of L-lactide guided by a thermally stable organocatalyst. *Macromolecules* 54: 6214–6225.

107 Pothupitiya, J.U., Dharmaratne, N.U., Jouaneh, T.M.M. et al. (2017). H-bonding organocatalysts for the living, solvent-free ring-opening polymerization of lactones: toward an all-lactones, all-conditions approach. *Macromolecules* 50: 8948–8954.

108 Stempfle, F., Ortmann, P., and Mecking, S. (2016). Long-chain aliphatic polymers to bridge the gap between semicrystalline polyolefins and traditional polycondensates. *Chem. Rev.* 116: 4597–4641.

109 Todd, R., Tempelaar, S., Lo Re, G. et al. (2015). Poly(ω-pentadecalactone)-*b*-poly(L-lactide) block copolymers via organic-catalyzed ring opening polymerization and potential applications. *ACS Macro Lett.* 4: 408–411.

110 Bouyahyi, M. and Duchateau, R. (2014). Metal-based catalysts for controlled ring-opening polymerization of macrolactones: high molecular weight and well-defined copolymer architectures. *Macromolecules* 47: 517–524.

111 Hodge, P. (2014). Entropically driven ring-opening polymerization of strainless organic macrocycles. *Chem. Rev.* 114: 2278–2312.

112 Bouyahyi, M., Pepels, M.P.F., Heise, A., and Duchateau, R. (2012). ω-pentandecalactone polymerization and ω-pentadecalactone/ε-caprolactone copolymerization reactions using organic catalysts. *Macromolecules* 45: 3356–3366.

113 Naumann, S., Thomas, A.W., and Dove, A.P. (2016). Highly polarized alkenes as organocatalysts for the polymerization of lactones and trimethylene carbonate. *ACS Macro Lett.* 5: 134–138.

114 Ladelta, V., Bilalis, P., Gnanou, Y., and Hadjichristidis, N. (2017). Ring-opening polymerization of ω-pentadecalactone catalyzed by phosphazene superbases. *Polym. Chem.* 8: 511–515.

115 Ladelta, V., Kim, J.D., Bilalis, P. et al. (2018). Block copolymers of macrolactones/small lactones by a "catalyst-switch" organocatalytic strategy. Thermal properties and phase behavior. *Macromolecules* 51: 2428–2436.

116 Zhao, N., Ren, C., Shen, Y. et al. (2019). Facile synthesis of aliphatic ω-pentadecalactone containing diblock copolyesters via sequential ROP with L-lactide, ε-caprolactone, and δ-valerolactone catalyzed by cyclic trimeric phosphazene base with inherent tribasic characteristics. *Macromolecules* 52: 1083–1091.

117 Xia, Y., Shen, J., Alamri, H. et al. (2017). Revealing the cytotoxicity of residues of phosphazene catalysts used for the synthesis of poly(ethylene oxide). *Biomacromolecules* 18: 3233–3237.

5

Organo-Catalyst Catalyzed Ring Opening Polymerization of *N*-Carboxyanhydrides

Chongyi Chen[1] *and Zhibo Li*[2]

[1] *Ningbo University, School of Materials Science and Chemical Engineering, No.818 Fenghua Rd, Ningbo 315211, China*
[2] *Qingdao University of Science and Technology, Key Laboratory of Biobased Polymer Materials, Shandong Provincial Education Department, College of Polymer Science and Engineering, No.53 Zhengzhou Rd, Qingdao 266042, China*

5.1 Introduction

The proteins are polymers composed of different amino acids produced by biological systems. They can exhibit different properties by precisely controlling the sequence of amino acid residues. Numerous efforts have been made to mimic the properties of proteins from the man-made materials. Polypeptides are homopolymers or copolymers of amino acid monomers, which are biocompatible and biodegradable. Moreover, these synthetic polypeptides can self-assemble into hierarchically ordered nanostructures, such as vesicle and condensates. These synthetic polypeptide materials have been explored for more than 100 years. However, their applications are still limited due to the difficulty in synthesis of well-defined polypeptides and copolypeptides. The most economical method for synthesis of long polypeptide chains is the ring-opening polymerization (ROP) of α-amino acid-*N*-carboxyanhydrides (NCAs) [1–5]. Since its discovery, the ROP of NCA has been extensively applied to prepare polypeptide materials for their potential applications in biological and medicine fields, and even energy materials [6–8].

To date, the ROP of NCA has made great progresses from the low polymerization rate, uncontrolled molecular weight and distribution to superfast polymerization rate, well-defined molecular weight and dispersity. Most of progresses in this field are summarized in Table 5.1 according to the timeline. Both initiator and catalyst are required for organo-catalyst catalyzed polymerization of NCA. Initiators for ROP of NCA are always nucleophilic chemicals, such as primary amine and alcohol that can attack the electrophilic C_5 carbonyl group of NCA rings. In this chapter, we will emphasize on the organo-catalysts for ROP of NCA, and three types of catalysts will be discussed. First, some catalysts can activate the initiator, live polymer chains or

Organocatalysts in Polymer Chemistry: Synthesis and Applications, First Edition. Edited by Zhibo Li.

Table 5.1 The timeline of initiators and organo-catalysts for NCA ring-opening polymerization.

Structure	Authors	Time	Refs.
H_2O/R-OH/R-NH_2	Curtius et al. Wessely et al.	1921–1952	[2]
NaOH	Blout et al.	1954	[9]
Tertiary amine	Bamford et al.	1956	[10]
BipyNi(COD) Chiral Ni/Co/Fe/Pt/Ru/Ir complex	Deming et al.	1997–1999	[11] [12–15]
Al complex	Jhurry et al.	2003	[16]
R-NH_2 and R-NH_3^+	Schlaad et al.	2003	[17]
Hexamethyldisilazane (HMDS) *N*-trimethylsilylamine (R-NHTMS) Polynorbornene-NHTMS	Cheng et al.	2007–2009	[18–20]
R-SH	Giani et al.	2010	[21]
Rare earth (RE) catalysts	Shen et al.	2012	[22]
RE tris(borohydride) complexe	Ling et al.	2012	[23]
R-NH_2 BF_4	Vicent et al.	2013	[24]
Alkali-metal borohydrides	Ling et al.	2014	[25]
R-NH_2 + tertiary amine	Schlaad et al.	2015	[26]
Primary + secondary/tertiary amines Thiourea	Hadjichristidis et al.	2015	[27, 28] [29]
Phenyl trimethylsilyl sulfide	Lu et al.	2016	[30]
Amine-borane Lewis pairs	Yang et al.	2017	[31]
Polynorbornene-NHTMS	Cheng et al.	2017	[32]
Methansulfonic acid	Pahovnik et al.	2017	[33]
Trimethylstannyl phenyl sulfide	Lu et al.	2018	[34]
LiHMDS	Liu et al.	2018	[35]
Fluorinated alcohol	Zhao et al.	2019	[36]
Aliphatic diamines	Cheng et al.	2019	[37]
Macroinitiators (biphasic system)	Cheng et al.	2019	[38]
Binaphthol-derived phosphoric acids	Yang et al.	2019	[39]
Dendritic polyamine	Cheng et al.	2020	[40]
Tertiary aminoalcohols	Zhao et al.	2020	[41]
KHMDS/NaHMDS	Liu et al.	2020	[42]
Polyoxometalates	Martín-Rapún et al.	2021	[43]
Crown ether	Cheng et al.	2021	[44]
Fluorinated tertiary alcohol	Tao et al.	2021	[45]
Tetraalkylammonium carboxylate	Liu et al.	2021	[46]
1,1,3,3-tetramethyl guanidine	Li et al.	2022	[47]
Tri-thiourea	Tao et al.	2022	[48]
Crown ether/cosolvents	Song et al.	2023	[49]
MOF + H_2O	Bai et al.	2023	[50]
Amine-modified inorganic nanoparticles	Yin et al.	2023	[51]
Thiourea-quaternary ammonium salt	Li et al.	2023	[52]

monomers. Second, some initiators themselves can also catalyze the polymerization. These two types of catalysts are divided into three sections according to their history and chemical composition. Third, more interestingly, recent research demonstrated the autocatalytic effect of polypeptide chain in the polymerization of NCA, which will be discussed in the last section.

5.2 Organo-Catalysts for NCA Ring Opening Polymerization

5.2.1 Organo-Catalysts Before 2010

Blout et al. firstly employed catalysts in NCA ring opening polymerization (ROP) in 1954 [9]. They catalyzed the polymerization of γ-benzyl-L-glutamate and γ-methyl-L-glutamate NCA using sodium hydroxide. However, they did not propose the detailed catalysis mechanism. According to the recent understanding on the NCA polymerization mechanism, NaOH probably captures the proton of –NH– group on NCA ring and the monomer anion can initiate the polymerization. Two year later, Tertiary amines were firstly used to catalyze ROP of NCA by Ballard and Bamford, and the "activated monomer mechanism" (AMM) was proposed in this work (Figure 5.1) [10]. Since then, there were few researches on the catalysis of

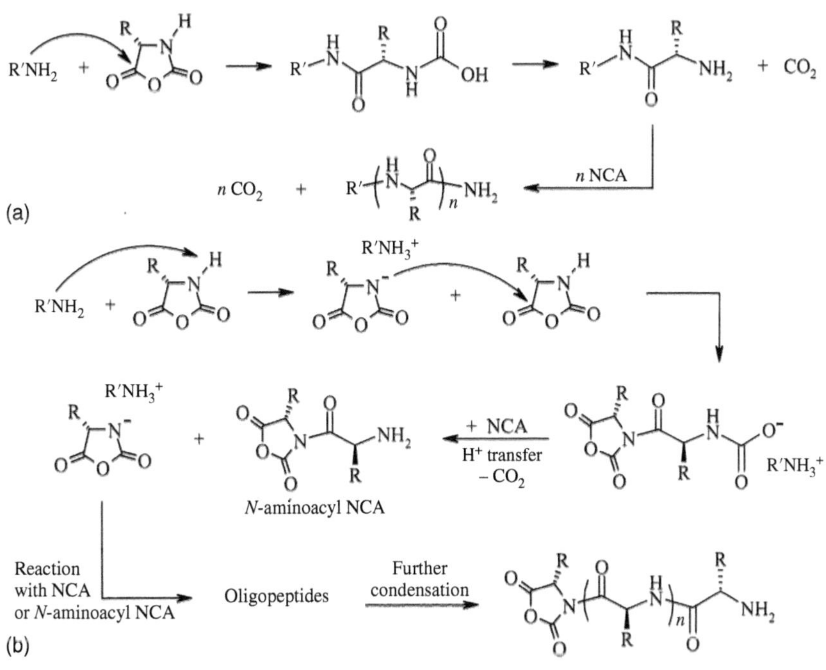

Figure 5.1 Proposed mechanism for NCA polymerization initiated by (a) normal amine (normal amine mechanism NAM) and (b) activated monomer (activated monomer mechanism, AMM). Source: Deming [3]/Springer Nature.

NCA ROP until 1997. A series of groundbreaking transition metal-based catalytic initiators for polymerization of NCA were developed by Deming group from 1997 to 1999, such as bpyNi(COD) and $(PMe_3)_4Co$ [11–15]. The side reactions in NCA polymerizations were significantly suppressed via the usage of transition metal complexes. Later, Cheng et al. discovered that hexamethyldisilazane (HMDS) and substituted TMS exhibited remarkable control of NCA polymerizations in DMF and led to formation of homo/block/brush polypeptides with expected molecular weights and narrow dispersity [18–20]. The above contributions have been systematically discussed in the already published reviews and chapters [1–5].

5.2.2 Amine-Based Organo-Catalysts

Amine has been used for the initiation of NCA polymerization due to its excellent nucleophilicity. However, the living end groups of propagating polypeptide chains are always deactivated by the side reactions during the polymerization. Dimitrov and Schlaad, reported using HCl to partially neutralize amine and realized much better controllable NCA ROP [17]. Vicent and coworkers substituted the hydrochloride by a non-nucleophilic tetrafluoroborate anion (BF_4^-), enabling the preparation of well-defined homo or block copolypeptides in DMF using different initiators, such as neopentyl, *n*-Butyl, PEG amines [24]. Compared to amine hydrochlorides, BF_4^- is more hydrophobic and soluble in in a broad range of organic solvents. The symmetry structure of BF_4^- anion leads to an equal electron distribution with four highly electronegative fluorine atoms. The nucleophilicity of the anion is significantly decreased and can also suppress AMM via the reprotonation of NCA monomer during the polymerization.

Similarly, Yang and coworkers introduced frustrated Lewis pairs (FLP) as organo-catalyst for NCA ROP, including different Lewis pairs [31]. The bulky borane Lewis acid forms intermediate with the amino groups of the initiator and propagation chain in order to control the initiation and propagation. The polymerization of NCA initiated by aniline conducted in DMF, THF and DCM in the absence of Lewis acid, affording polypeptides with narrow dispersity and well-controlled molecular weights. They propose that FLP is formed between aniline and $Fmes_2BF$ because of the steric hindrance, which maintain the nucleophilicity of amino group on aniline for the initiation of polymerization (Figure 5.2). Then, $Fmes_2BF$ can form new FLP with the terminal amino group of the propagation chain. The slow initiation and the fast propagation rate of conventional NCA polymerization always led to the less control of molecular weight and dispersity. In this work, the propagation rate is significantly decreased as well as the initiation rate due to the formation of FLP, which downgraded the controllability of NCA polymerization.

Instead of frustrated Lewis pairs, the combination of organophosphates (binaphthol-derived phosphoric acids) and primary amines was used in the polymerization of NCA [39]. AMM reactions are suppressed due to the addition of binaphthol-derived phosphoric acids, yielding polypeptides with controllable molecular weights and narrow dispersities. They propose that the equilibrium

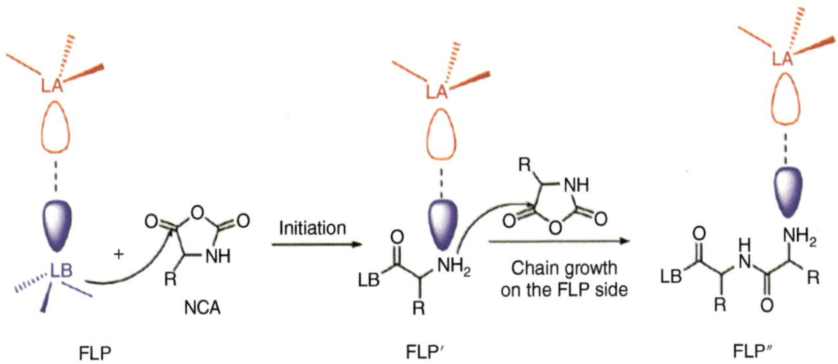

Figure 5.2 Proposed mechanism of NCA polymerization catalyzed by frustrated Lewis pairs. Source: Zhang et al. [31]/Royal Society of Chemistry.

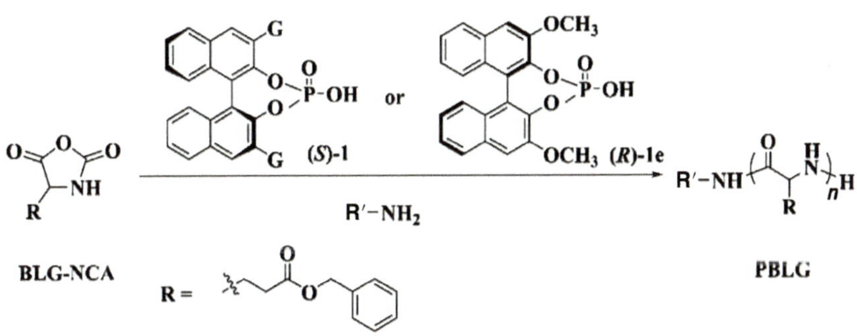

Figure 5.3 The binaphthyl phosphoric acids and primary amines used for the ROP of BLG-NCA. Source: Liang et al. [39]/with permission of ELSEVIER.

between amine and ammonium contributes to the controlled polymerization (Figure 5.3).

Shen and coworkers developed rare earth (RE) complexes (RE = Y and Sc) for the catalysis of NCA polymerization [22]. Among them, the obtained molecular weight of $RE(BH_4)_3(THF)_3$ system can be controlled by the feeding ratio of [NCA]/[RE]. They proposed that NCA monomer was deprotonated by RE complex and then attacked the 5-CO of another NCA (Figure 5.4). The obtained carbamate group was protected by RE for the further propagation. In the same year, RE borohydride complexes $Ln(BH_4)_3(THF)_3$ (Ln = Sc, Y, La, and Dy) were employed for the catalysis of NCA polymerization, obtaining polypeptides and copolypeptides with narrow dispersity [23]. In this system, they found two active species, an yttrium carbamate derivative and a *N*-yttriumlated NCA, which were produced from the attack of RE complexes to 5-CO and 3-NH of NCA, respectively. They initiated the polymerization at the same time, resulting in the hydroxyl and carboxyl terminated polypeptides. When the polymerization temperature was lowered to 0 °C, the deprotonation of NCA at 3-NH was suppressed and hence a pure α-hydroxyl-ω-aminotelechelic polypeptide was obtained.

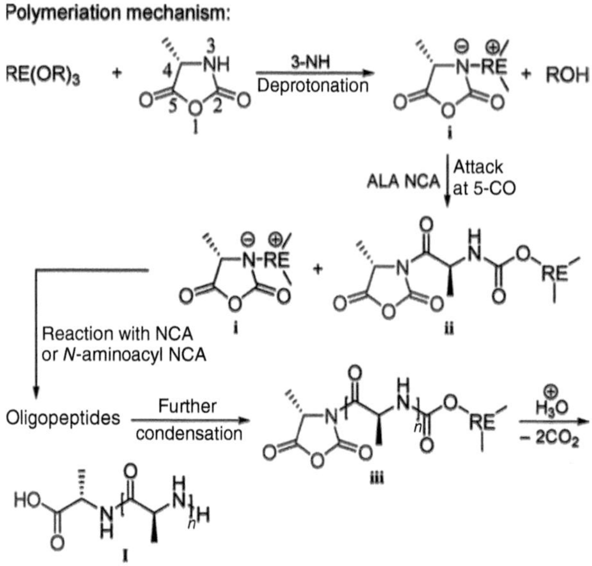

Figure 5.4 Possible polymerization mechanism of Ala NCA catalyzed by RE complexes. Source: Peng et al. [22]/John Wiley & Sons.

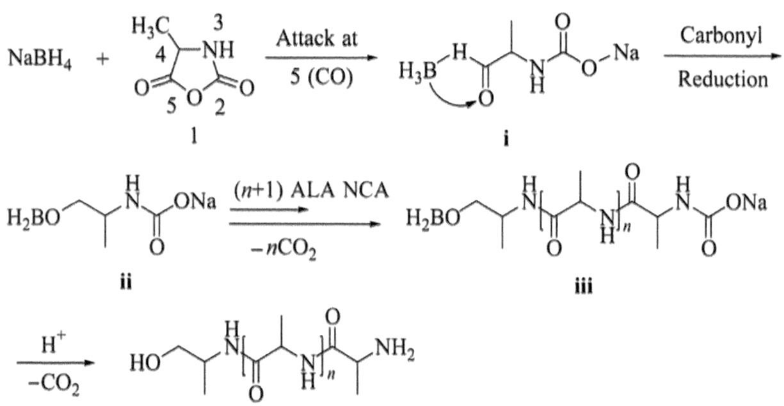

Figure 5.5 Possible polymerization mechanism of Ala NCA catalyzed by $NaBH_4$. Source: Peng et al. [25]/Springer Nature.

Later, they substituted RE by alkali metal, demonstrating the catalysis of NCA polymerization by MBH_4 (M = Li, Na and K) [25]. A similar mechanism was proposed as RE complexes. $NaBH_4$ attacks 5-CO of NCA and the obtained carbamate sodium salt can attack 5-CO of NCA for further propagation of polypeptides (Figure 5.5).

Alkali metal substituted HMDS can also be applied in organo-catalytic NCA polymerization. Liu and coworkers reported a series of HMDS alkali salts (LiHMDS,

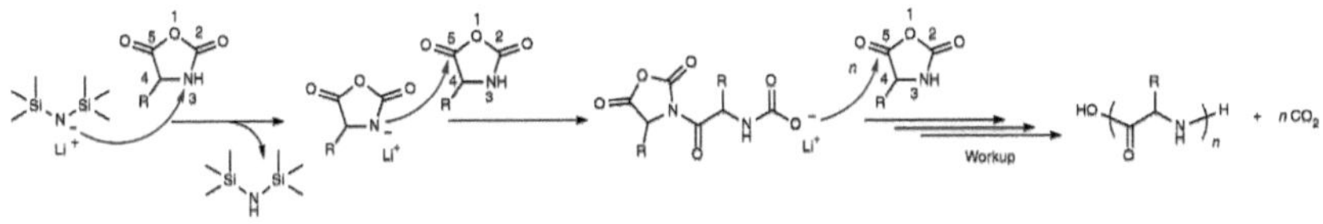

Figure 5.6 Proposed mechanism of LiHMDS-initiated NCA polymerization. Source: Wu et al. [35]/Springer Nature/CC BY 4.0.

KHMDS and NaHMDS) as initiators for the superfast NCA polymerization, producing homo or block copolypeptides in THF with variable lengths (20–1294 mers) and narrow dispersity [35, 42]. This method enables quick library synthesis and large-scale synthesis of polypeptides. Their preliminary studies reveal that THF is better than DMF as the solvent for NCA polymerization. NCA is less stable in DMF because of the possible side reactions caused by the impurities such as water and chloride. Conventional NCA polymerization strategies are highly sensitive to moisture due to the slow polymerization rate. However, HMDS alkali salts initiated NCA polymerization can conducted in an open flask in untreated THF under environment with 60% relative humidity due to its superfast polymerization rate. LiHMDS is a non-nucleophilic strong Bronsted base, therefore, the proposed mechanism is an anionic ring opening NCA polymerization mechanism initiated by LiHMDS. LiHMDS deprotonate the 3-NH in a NCA ring and followed by the nucleophilic addition of 3-N anion to C5-carbonyl of another NCA (Figure 5.6). Then the resulting lithium carbamate can undergo nucleophile addition to a new NCA for chain propagation.

Except for the deprotonation of NCA, the initiator can also be deprotonated by some strong base. 1,1,3,3-tetramethyl guanidine (TMG) was used for the catalysis in ROP of *N*-butyl NCA (Bu-NCA) and sarcosine-derived *N*-thiocarboxyanhydrides (Me-NNTA), producing well-defined polypeptoids with controlled molecular weights and narrow dispersity [53, 54]. Interestingly, compared to Bu-NCA and Me-NNTA, TMG can significantly enhance the polymerization rate of NCA by using amine, hydroxyl and carboxyl-terminated initiators, reported by Li et al. [47]. In the initiation stage, the nucleophilicity of initiating functional groups is significantly enhanced via the formation of hydrogen/ionic bonding between initiator and TMG (Figure 5.7). In the propagation stage, the terminal carbamate of the intermediate is stabilized by the formation of ionic bonding and the nucleophilicity of carbamate is enhanced in the same way. The introduction of TMG gives rise to the straightforward preparation of copolypeptides and PEG-polypeptides using hydroxyl and carboxyl terminated initiators/macroinitiators.

Other organic base, such as triazabicyclo[4.4.0]dec-5-ene (TBD) and diazabicyclo [5.4.0]undec-7-ene (DBU), can also be suitable candidates for the catalysis of NCA polymerization. For example, TBD and DBU have been used for the polymerization of morpholino-2,5-dione derivatives (MDs). Polydepsipeptides (PDPs) are strictly alternating copolymers of α-hydroxy acids and α-amino acids produced via the ROP of MDs. Li et al. reported that PDPs with controlled molecular weights and low dispersity were synthesized using binary catalytic systems of 1-(3,5-bis(trifluoromethyl)-phenyl-3-cyclohexyl-2-thiourea) (TU) with DBU or TBD [55, 56].

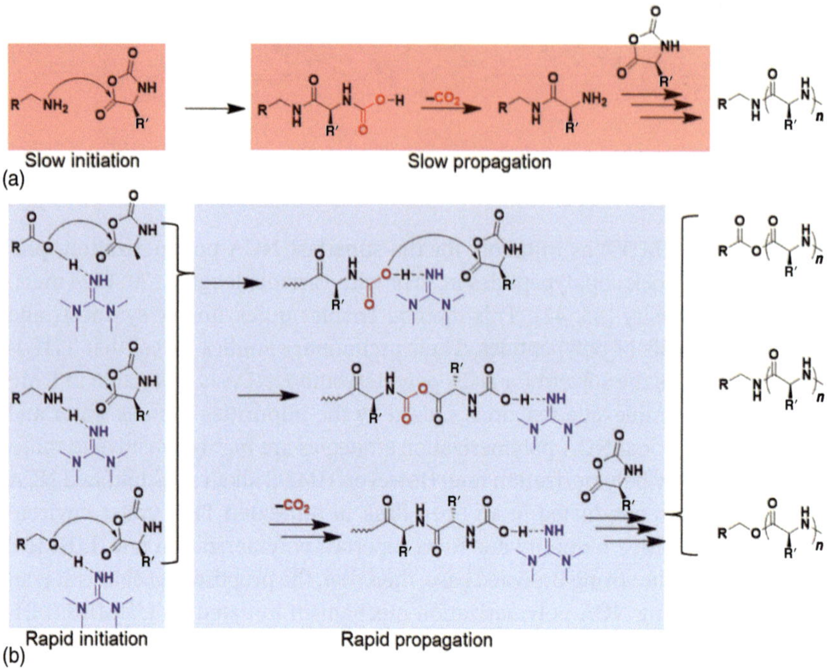

Figure 5.7 (a) The normal amine mechanism (NAM) and (b) the proposed mechanism of ring-opening polymerization of NCA initiated by different initiator/TMG systems. Source: Li et al. [47]/Royal Society of Chemistry.

Polyoxometalates (POMs) are molecular metal-oxides, which have been employed in the fields of catalysis, energy and biology. Bis-amino functionalized POM $[MnMo_6O_{18}((OCH_2)_3CNH_2)_2]^{3-}$ with tetrabutylammonium (TBA) as counter-cations was selected as bifunctional initiator for NCA polymerization in DMF [43]. Hybrid POM-polypeptide materials are prepared straightforward using this approach. Carboxyl anions can undergo nucleophilic addition to the carbonyl group due to the relatively good nucleophilicity. Another tetraalkylammonium-based ion pairs, Tetraalkylammonium carboxylates ion pairs, were developed by Liu and coworkers for the superfast NCA polymerization without the demand of extra catalysts [46]. Tetraalkylammonium carboxylates initiated NCA polymerization in THF can conducted in open vessels at ambient condition due to its superfast polymerization rate, obtaining polypeptides with variable lengths and narrow dispersity. They propose that the NCA polymerization in this study is initiated by nucleophilic addition of carboxylate 1 to C5-carbonyl on the NCA ring (Figure 5.8), resulting in the transition state 1 (TS1). The NCA ring is opened to form compound 2 with a terminal tetraalkylammonium carbamate complex. The new carbamate anion as the propagating species to undergo a further nucleophilic addition to the C5-carbonyl of another NCA ring, resulting in the transition state 2 (TS2). Then chain propagation happens at one concerted step with decarboxylation and nucleophilic attack of N to the carbonyl on a NCA ring to extend the polymer chain

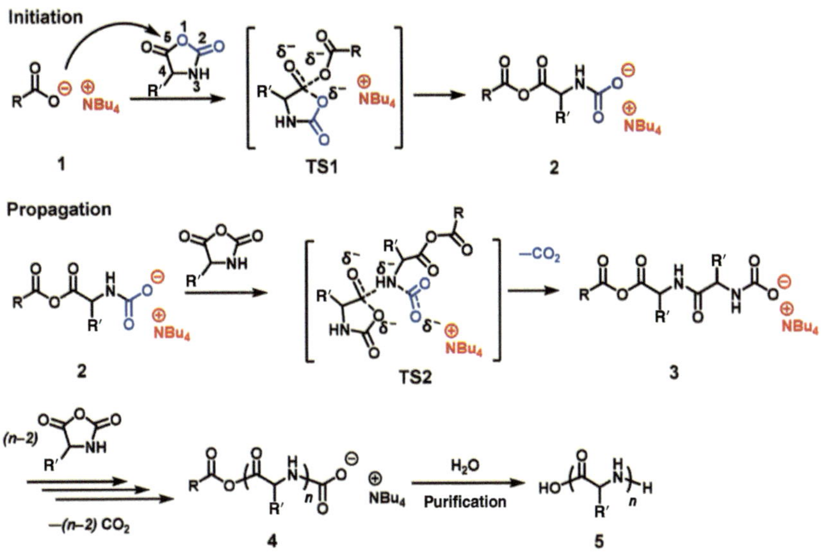

Figure 5.8 Plausible mechanistic pathway of tetraalkylammonium carboxylate-initiated open vessel NCA polymerization. Source: Wu et al. [46]/John Wiley & Sons.

as compound 3. The polymer 4 bearing a C-terminal R group from the initiator and an N-terminal carbamate group undergoes hydrolysis of the anhydride and decarboxylation to afford product 5 during polymer purification. Furthermore, tetraalkylammonium carboxylate-initiated polymerization of NCA can produce block copolypeptides with more than 15 blocks.

The introduction of secondary and tertiary amines is also able to regulate NCA polymerization. Vacogne and Schlaad reported controlled polymerization of NCA in DMF using primary ammonium salt and tertiary amine as initiating system [26]. The initiation and propagation can be activated and inhibited based on the equilibrium between amine and ammonium, which endow polypeptides with well-defined molecular weight and narrow dispersity. The propagation rate can be regulated by altering the ratio of primary ammonium salt to tertiary amine. In the same year, Hadjichristidis and coworkers discovered that primary amines in cooperation with secondary amines/tertiary amines could be used for the fast and living polymerization of NCA [27, 28]. They proposed that the high nucleophilicity of secondary amines/tertiary amines lead to the formation of hydrogen bonding between N: and N–H, which accelerate the addition of primary amine to carbonyl group and ring opening reaction (Figure 5.9).

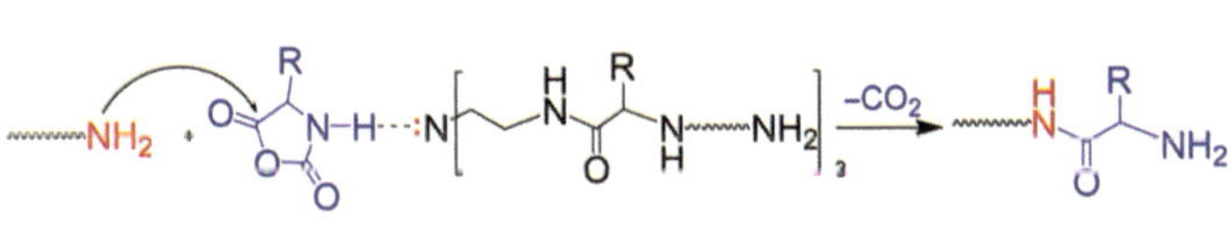

Figure 5.9 The accelerated amine mechanism by monomer activation (AAMMA). Source: Zhao et al. [28]/Royal Society of Chemistry.

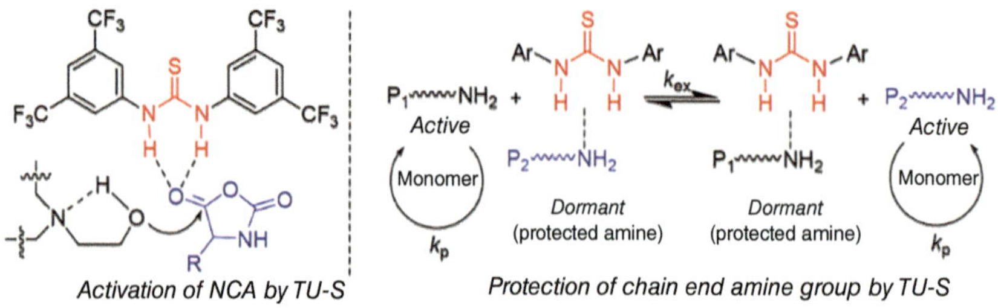

Figure 5.10 The proposed mechanism of TU-S catalyzed NCA ring opening polymerization. Source: Zhao et al. [29]/Royal Society of Chemistry.

5.2.3 Other Organo-Catalysts

Other non-nitrogen catalysts for NCA polymerization are included in this section. Recently, thiourea-based catalysts have been applied in ROP of lactide [57]. Hadjichristidis et al. discovered that the cooperation between aminoalcohols and *N,N′*-bis[3,5-bis(trifluoromethyl)phenyl]thiourea (TU-S) were able to promote the fast and controlled polymerization of NCA in different solvents under mild conditions [29]. The authors propose that TU-S can form hydrogen bonding with both monomer and polypeptide, which can activate the monomer and reversibly deactivate the terminal amino group of polypeptides for the protection of living polymer chain ends (Figure 5.10). Then this group synthesized a series of functionalized aminoalcohols via the Michael addition between acrylates and secondary aminoalcohols [41]. These functionalized aminoalcohols can be used for the initiation and fast and controlled propagation for polymerization of NCA in cooperation with TU-S. Well-defined and site-specific functionalized polypeptides can be synthesized bearing functional groups at the chain end or in the middle.

Compared to catalyst with single thiourea group, Tao and coworkers demonstrated an anion-binding catalytic system via tri-thiourea for polymerization of NCA, producing various polypeptides with narrow molecular weight dispersity [48]. This tri-thiourea catalyst was synthesized via a nucleophilic addition between tris-(2-aminoethyl)amine and isothiocyanate. One of the thiourea group can activate another thiourea group through intramolecular hydrogen bonds cooperatively. In the initiation stage, the initiating sulfide anion bonds to tri-thiourea via anion-binding interaction (Figure 5.11). Then the bound sulfide anion nucleophilic attack the NCA monomer, which has been activated by the hydrogen bonds with tri-thiourea. In the propagating stage, the terminal carbamate anion is stabilized by tri-thiourea via anion-binding interaction and the intramolecular hydrogen bond within the tri-thiourea structure, further minimize the probability of deleterious side reactions.

Li and coworkers combined the thiourea and quaternary ammonium salt in one molecule for the catalysis of NCA polymerization [52]. Propylene oxide (PO) is added to the polymerization system and thiourea-quaternary ammonium salt works cooperatively for the ring opening of PO (Figure 5.12). The resultant alkoxy anion acts as the initiator to nucleophilic attack the carbonyl group of NCA. After initiation, the

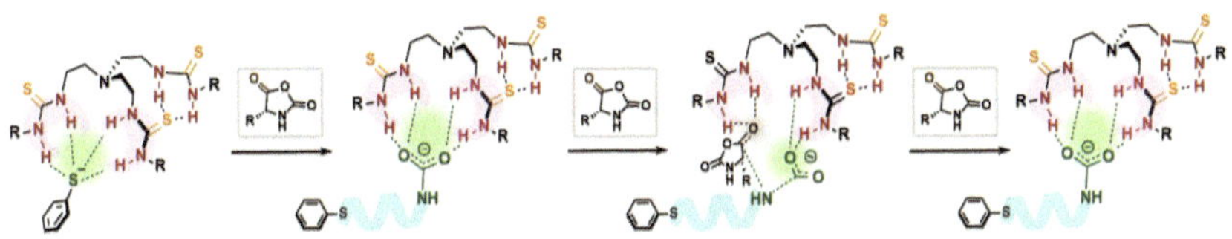

Figure 5.11 Proposed mechanism for living anionic ROP of NCAs enabled by tri-thiourea/PhSNa. Source: Lv et al. [48]/American Chemical Society.

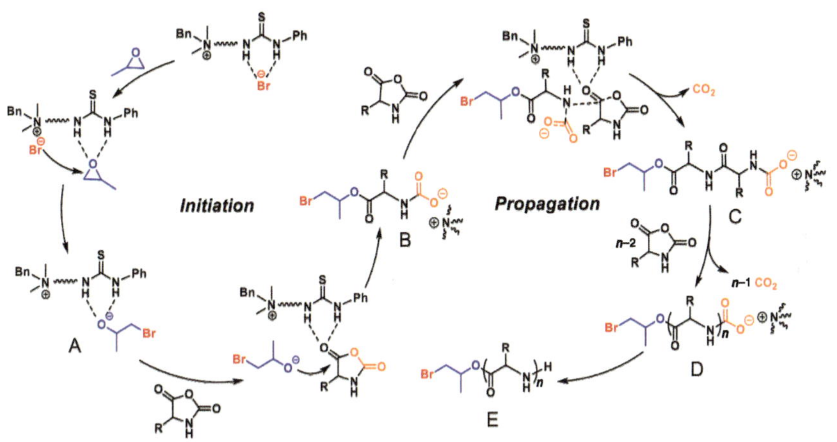

Figure 5.12 Proposed mechanism of NCA polymerization catalyzed by the bifunctional thiourea-quaternary ammonium salt catalyst. Source: Wang et al. [52]/American Chemical Society.

propagating carbamate anion interacts with quaternary ammonium and the NCA monomer is activated by thiourea. The polymerization is controlled by the cooperation between thiourea and quaternary ammonium salt, obtaining polypeptide with well-defined molecular weight and distribution. Different anions, electronic effects on thiourea group, steric hindrance of tertiary amines and different solvents can tune the catalytic performance of thiourea-quaternary ammonium salt.

Except for thiourea, Hadjichristidis and coworkers discovered that a fluorinated alcohol, 1,3-bis(2-hydroxyhexafluoroisopropyl)benzene (1,3-Bis-HFAB), can be used for the catalysis of fast and controlled synthesis of polypeptides, producing polypeptides with narrow dispersity and predicable molecular weights [36]. A plausible mechanism was proposed (Figure 5.13). There exist multiple hydrogen-bonding interactions between the NCA monomer, initiator (*N*,*N*-dimethylethanolamine, DMEA), catalyst (1,3-Bis-HFAB) and the propagating polypeptide chains. At the initiation stage, the initiator is firstly activated by the intramolecular hydrogen bonds between the N atom of amino group and the H atom of hydroxyl group. In the initiation and propagating stage, the NCA monomer was activated by 1,3-Bis-HFAB due to the hydrogen bonding between the H atom of hydroxyl group on 1,3-Bis-HFAB and the 5-carbonyl group of NCA monomer. 1,3-Bis-HFAB can form hydrogen bonding with initiator and the terminal amino group of propagating

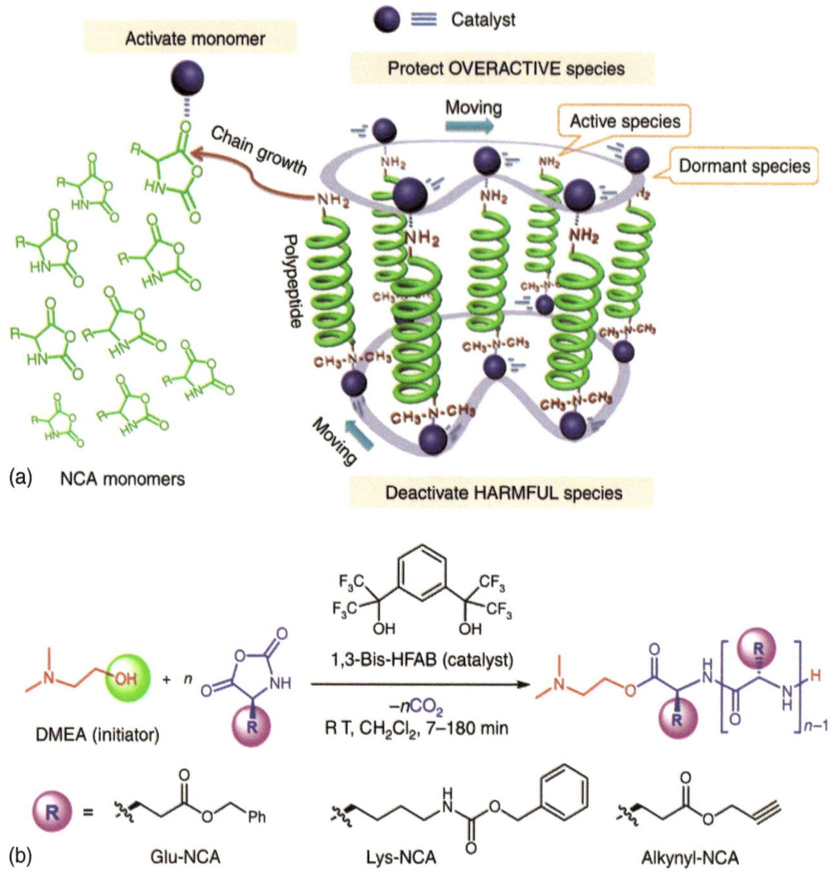

Figure 5.13 Organocatalytic ROP of NCA catalyzed by fluorinated alcohol. (a) Schematic diagram of the roles of fluorinated alcohol catalyst played during polymerization; (b) (co)polymerization of NCA initiated by DMEA in the presence of fluorinated alcohol catalyst 1,3-Bis-HFAB. Source: Zhao et al. [36]/Springer Nature/CC BY 4.0.

polypeptides, thus dynamically protecting the propagating primary amine from side reactions. He and Tao designed a series of bifunctional organo-catalysts based on the combination of fluorinated alcohol and aminoalcohol in one molecule for NCA polymerization [45]. The polymerization catalyzed by the bifunctional catalyst also proceeded in a fast and controlled manner, affording polypeptide with narrow molecular weight dispersity.

In addition to oxygen atom, sulfur atom can also be used for the nucleophilic agent. However, the investigation on the sulfide-based initiator for NCA polymerization remains scarce. Lu and coworkers reported the first sulfide-based controlled polymerization method for NCA polymerization [30]. Phenyl trimethylsilyl sulfide (PhS-TMS) was employed as the initiator, producing different polypeptides with controlled molecular weights and narrow dispersity. The S atom of PhS-TMS proceeds a nucleophilic attack to 5-carbonyl group of NCA monomer, companied

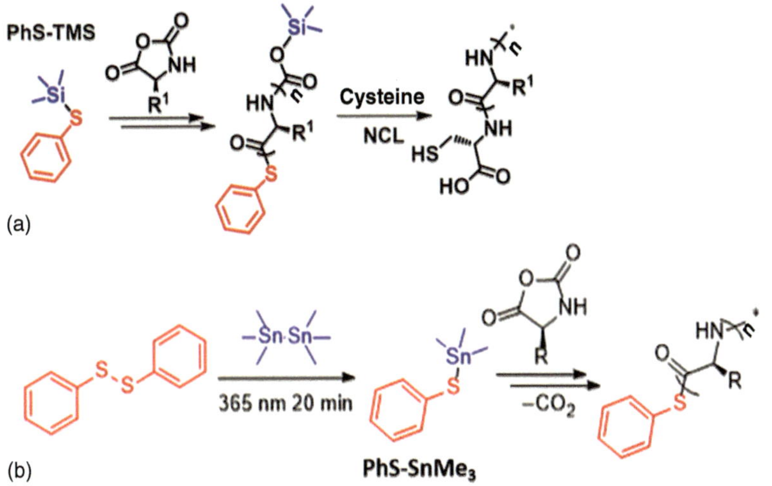

Figure 5.14 The proposed mechanism for (a) PhS-TMS and (b) S-Sn Lewis pair mediated NCA polymerization. Source: Yuan et al. [30] and Yuan et al. [34].

by the formation of intermediate, TMS protected carbamate, as the propagating terminal group (Figure 5.14a). The proposed mechanism is similar to the previous reported HMDS related works [19]. Then, compared to PhS-TMS, the same group reported a S-Sn Lewis pair-based initiation system, trimethylstannyl phenyl sulfide (PhS-$SnMe_3$) for the controlled polymerization of NCA, affording polypeptides with narrow dispersity (Figure 5.14b) [34]. The molecular weight control and polymerization rate of polypeptides is related to the solvent composition (THF/DMF), which can be attributed to the different stabilities of S-Sn bond in different solvents. They propose that the S-Sn bond is more stable in the less polar solvents such as THF, resulting in slow and inefficient initiation. In comparison, the S-Sn bond can be polarized in the highly polar solvents such as DMF, leading to fast and efficient initiation. During the propagation stage, the N-$SnMe_3$ species is proposed as the reactive center, which form looser Lewis pairs in DMF and tighter Lewis pairs in THF.

Pahovnik and coworkers reported a strategy for the preparation of well-defined polypeptides using 3-phenyl-1propanol (PPA) as initiator and methansulfonic acid as catalyst [33]. The nucleophilic attack reaction of hydroxyl group of PPA to NCA ring is slow, and slow initiation and fast propagation will lead to uncontrolled molecular weight of polypeptides. Therefore, they employed the acid catalyst to separate the slow initiation stage and the fast propagation stage. The acid catalyst protonates the formed primary amine group after initiation to inhibit the propagation until all the initiation complete (Figure 5.15). After completion of the initiation, the propagation was further started via the deprotonation of the ammonium group by adding another base.

For a long time, the acids in NCA ROP are always considered as the inactivation agent for the initiators and NCA monomers. Recently, Zhang et al. reported a moisture-tolerant ROP of *N*-methyl-α-amino acids NCA catalyzed by acetic acid, providing well-defined *N*-methyl polypeptides [58]. The acetic acid was proposed

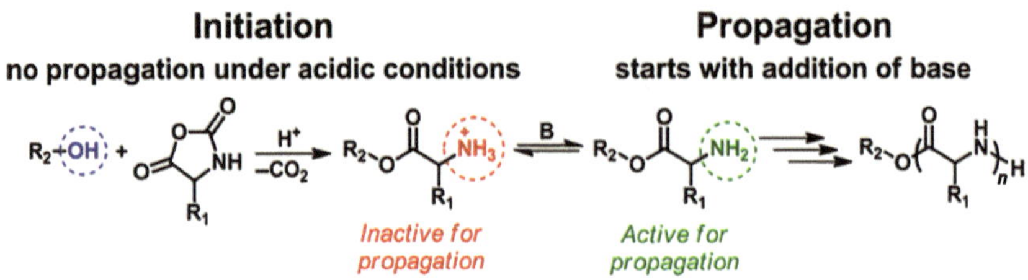

Figure 5.15 Reaction pathway of ROP of NCA using 3-phenyl-1propanol as initiator. Source: Gradišar et al. [33]/American Chemical Society.

to activate the monomer by the hydrogen bonding, promote the release of CO_2 and form an acid-base equilibrium with the secondary amine chain end. Then, Lu and coworkers reported that the organic acid can significantly enhance the polymerization rate of sarcosine NCA (Sar-NCA) [59]. Compared to previous studies, the carboxylic acid-catalyzed Sar-NCA ROP enabled the robust synthesis of polysarcosine with an unprecedented ultrahigh molecular weight up to 586 kDa (DP ~ 8200) and exceptionally narrow dispersity below 1.05. They proposed that the carboxylic acid was able to significantly facilitate proton transfer processes and avoids charge separation, which promote the ring opening of NCA, rather than decarboxylation, as the rate-determining step (Figure 5.16).

5.2.4 Autocatalysis of NCA Ring Opening Polymerization

The organo-catalyzed polymerizations always require the introduction of organo-catalyst. The catalyst can form interactions with the monomer or polymer for the catalysis of polymerization. Is it possible to realize autocatalytic polymerization? For that reason, the formed polymers during the polymerization shall be able to form specific interaction with monomers. For instance, the polymer can absorb monomers to increase the local concentration of monomer; alternatively, the polymer can activate the monomer to decrease the activation energy of polymerization reaction. Interestingly, similar phenomenon was observed in NCA polymerization, namely cooperative covalent polymerization (CCP) [60].

Cheng et al. discovered that the macromolecular initiators (trimethylsilyl amine substituted polynorbornene) were able to significantly enhance the rate of NCA polymerization in low dielectric constant solvents, such as DCM (Figure 5.17a,b) [32]. In fact, they synthesized the same macroinitiator for the synthesis of polypeptide brush about eight year ago [20]. However, at that time, the polymerization was conducted in DMF. DMF is a good solvent for the polypeptide, such as PBLG, because DMF can form hydrogen bonding with the polypeptide for better solvation. The enhancement of polymerization rate can be attributed to the macrodipoles of the α-helices, which formed by PBLG due to the cooperative intramolecular hydrogen bonds. The ordered hydrogen bonds along the polypeptide backbone endow the polypeptide with a large dipole moment. The negative pole of this helical macrodipole located at the C-terminus of polypeptide and the positive pole located at the N-terminus as the active polymerization center (Figure 5.17c). For one single

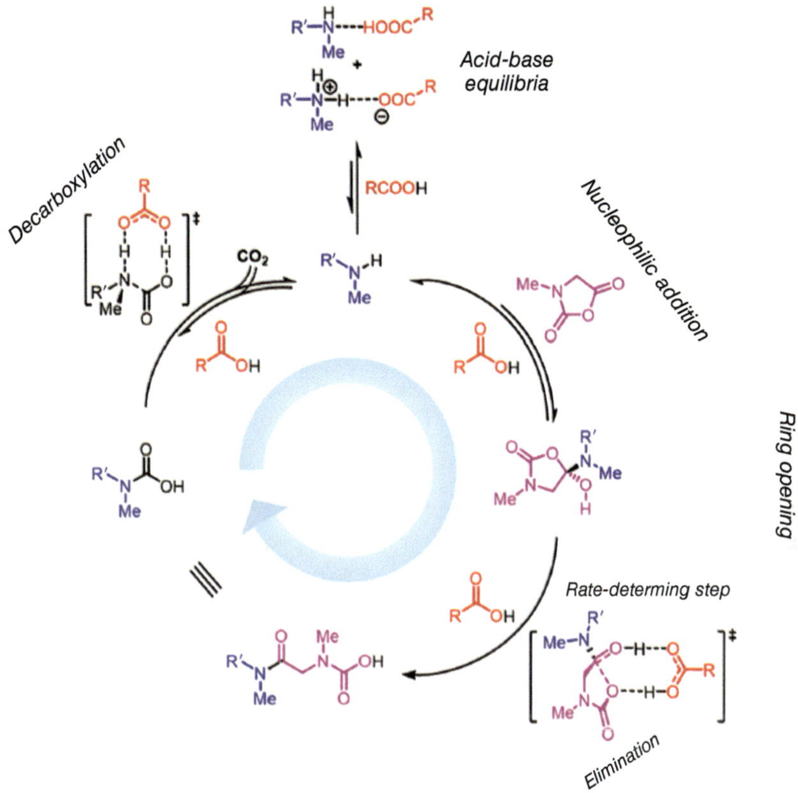

Figure 5.16 Proposed mechanism for ROP of Sar-NCA catalyzed by carboxylic acids. Source: Wang et al. [60]/American Chemical Society.

polypeptide chain, the conformation will transform from coil (nucleation stage) to helix (propagation stage) when the DP higher than 8–12. The formation of macrodipole will significantly change the electrostatic environment of the growing chain end. More interestingly, for the brush system, the macrodipoles are tethered in limited spatial room and the high-density electrostatic environment greatly enhance the polymerization rate. As a result, the polymerization rate of propagation stage is much higher than nucleation stage. The solvent with high dielectric constant, such as DMF, will reduce the macrodipole of polypeptide, resulting in a much slower polymerization rate. Therefore, the polymerization rate can be regulated by changing the solvent. When the macrodipole is diminished by the introduction of racemic DL monomer, compared to the α-helical brush system, the random coil brush system formed by racemic DL monomer results in a slower polymerization rate due to the decreased interaction strength between neighboring polypeptides.

Then, they discovered that CCP was able to proceed in a much simpler system. They reported the NCA polymerization initiated by linear aliphatic diamines [37]. Compared to the NCA polymerization initiated by the hexylamine, the chain propagation rate is increased by more than 600 times when initiated by its diamine analogue (1,6-diaminohexane) (Figure 5.18). The "hinged" polypeptides, two blocks

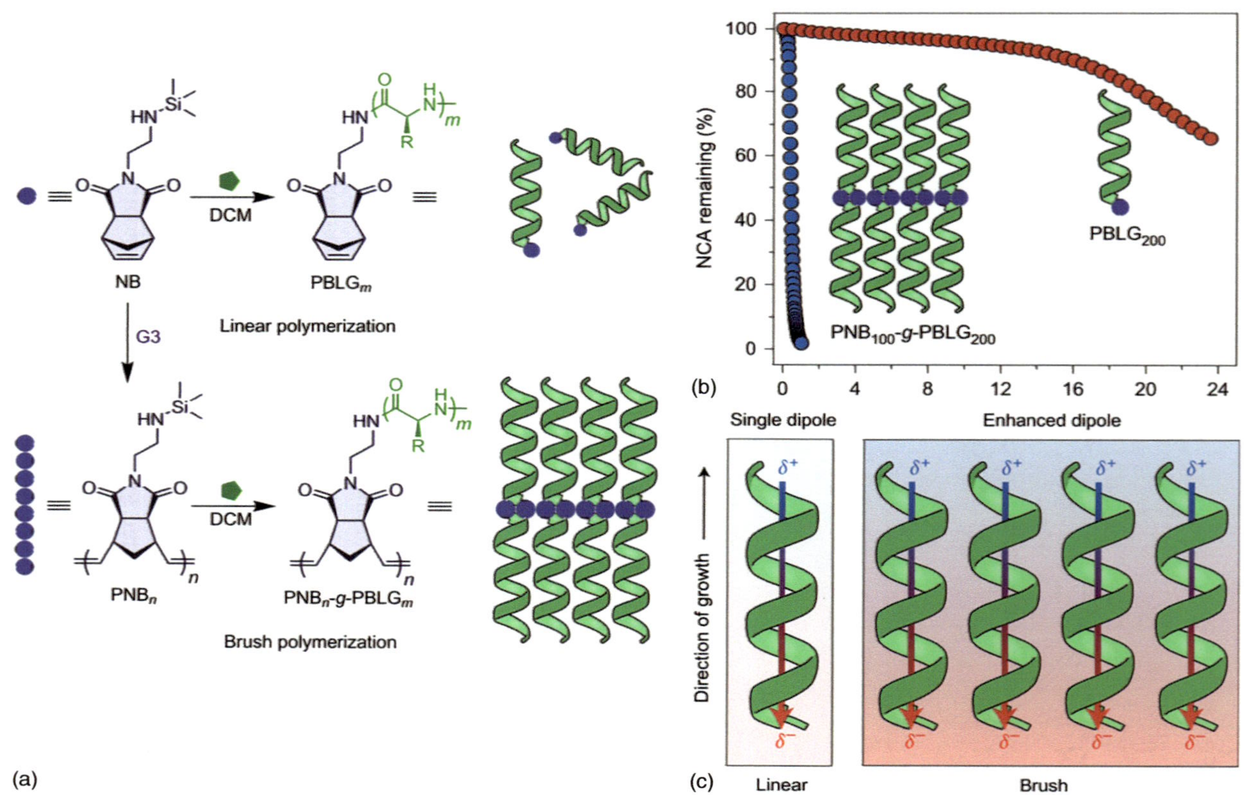

Figure 5.17 (a) NCA polymerization initiated by NB and PNB$_n$. (b) conversion of NCA using NB (red) or PNB$_{100}$ (blue) as initiators. (c) The macrodipole of linear and brush growing α-helix. Source: Baumgartner et al. [32]/Springer Nature.

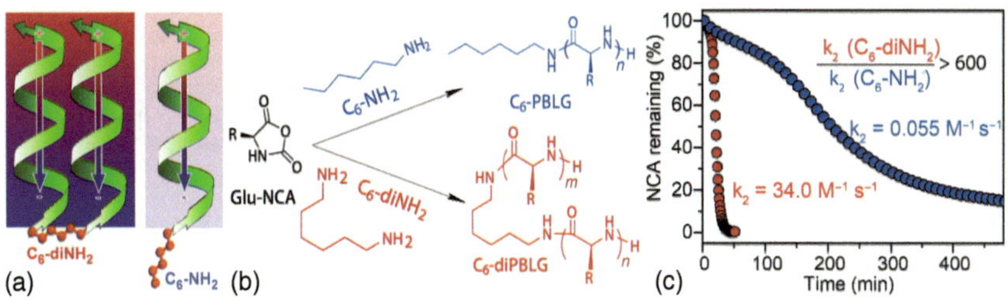

Figure 5.18 Proximity-induced cooperative polymerization in "hinged" helical polypeptides. Source: Chen et al. [37]/American Chemical Society.

of polypeptides connected head-to-head by the diamine molecules, are synthesized accordingly. This polymerization proceeds a two-stage cooperative kinetic as the brush system. They also observed CCP in dendritic polyamine (PAMAM)-initiated NCA polymerization [40]. The polymerization rate is ultrafast, obtaining polypeptides with exceptionally high molecular weights (up to 85 MDa) and low dispersity ($Đ < 1.05$) for unimolecular polypeptide micelles.

They conducted further mechanistic investigations on CCP. They demonstrate the self catalyzed polymerization of NCA induced by the formation of helix in the solvents with low dielectric constants [61]. They observe that there exists binding interaction between the propagating helical polypeptide chains and the incoming monomers (Figure 5.19). The solvents with higher dielectric constants, such as DMF, can form interaction with both polypeptide chains and monomers, which impair the interaction between polypeptide chains and monomers. On the contrary, the solvents with low dielectric constants, such as DCM, are crucial for the enhancement of the interaction between polypeptide chains and monomers.

Based on the systematic mechanistic study, they developed more interesting CCP systems. They reported the application of PEG-PBLG amphiphilic macroinitiators in a water/DCM biphasic system for the polymerization of unpurified NCA monomers [38]. The macroinitiators anchored at the interface of water and dichloromethane phases and the polymerization takes place in the DCM phase, which is accelerated due to the cooperative interactions of macrodipoles between neighboring helices and enhanced binding of monomers (Figure 5.20). More interestingly, the impurities of unpurified NCA monomers can be extracted into the aqueous phase *in situ*. Therefore, this polymerization method allow the synthesis of high molecular weight and low dispersity polypeptides without tedious NCA purification steps and air-free conditions. Then Yin and coworkers discovered that amine-modified inorganic nano-initiators, such as mesoporous silica nanoparticles, were capable of CCP and in situ purification of NCA in the water/DCM biphasic system [51]. Well-defined polypeptide nano-hybrids are prepared via a fast polymerization process.

Except for the molecular design of initiators for CCP, crown ether can be employed as the CCP catalyst for the conventional primary amine initiators. Cheng and coworkers reported the NCA polymerization in DCM catalyzed by crown ether, allowing the accelerated synthesis of well-defined polypeptides (Figure 5.21) [44]. Among the crown ethers 12-crown-4, 15-crown-5, 18-crown-6 and 24-crown-8,

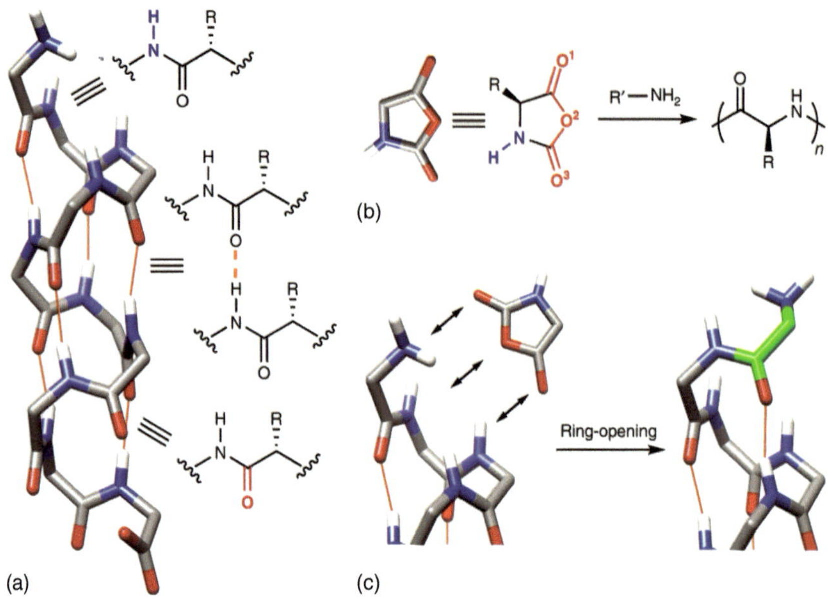

Figure 5.19 Schematic illustration of the catalytic power of an α-helix. Source: Song et al. [61]/Springer Nature/CC BY 4.0.

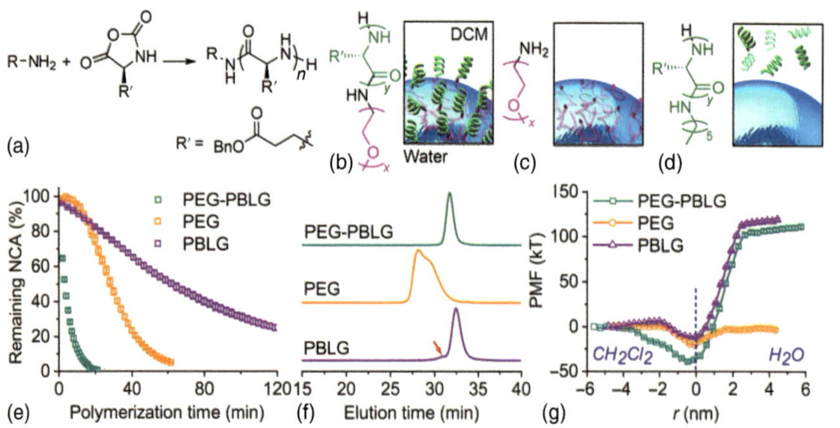

Figure 5.20 Accelerated polymerization rate with localized initiators. Source: Song et al. [38]/National Academy of Sciences.

18-crown-6 enables the more rapid polymerization rate. The size of the cyclic structure played a critical role in the catalysis. The proposed that 18-crown-6 enhances the interactions between propagating polypeptide and NCA monomer, which activate the ring-opening reaction, resulting in the rapid polymerization rate.

Then, nanoscale metalorganic frameworks (MOF) was applied for CCP by Bai and coworkers, which provides fast kinetics and polypeptide with low dispersity [50]. Besides, this heterogeneous catalyst can be recycled after polymerization. The polymerization catalyzed by MOF proceeds via a two-step mechanism. They

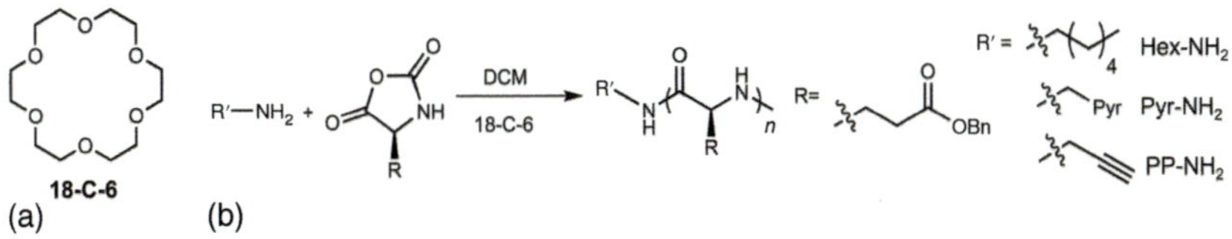

Figure 5.21 CE-catalyzed polymerization of NCA by small-molecular amine initiators. Source: Xia et al. [44]/Springer Nature/CC BY 4.0.

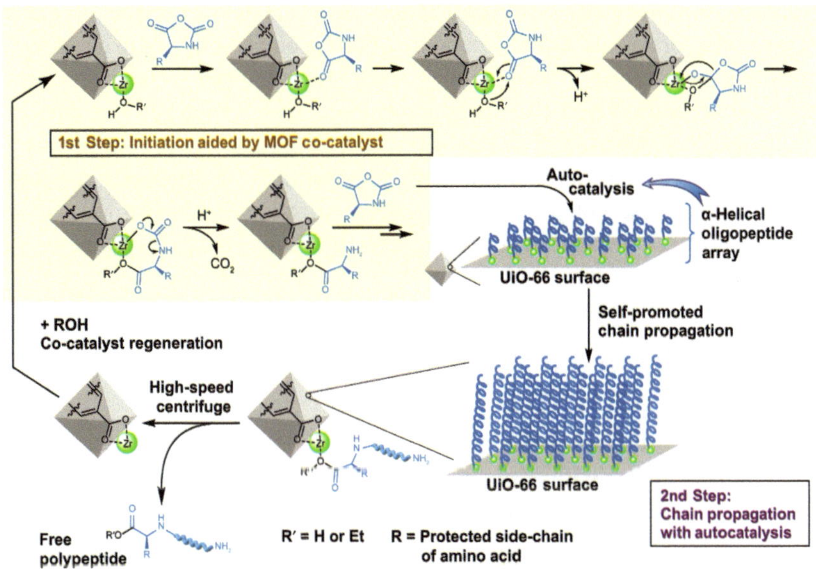

Figure 5.22 The proposed mechanism of water-initiated NCA-ROP on UiO-66 NP. Source: Liu et al. [50]/Springer Nature/CC BY 4.0.

propose that the initiation stage may involve a coordination insertion mechanism (Figure 5.22). The densely packed open Zr centers of MOF nanoparticles activate the carbonyl of NCAs, and the coordinated water molecules in close proximity can attack the carbonyl. Then, the formed carbamates decompose for the creation of a normal amine for the chain propagation. The high density of normal amine initiators can cooperatively enhance the polymerization rate as CCP, producing well-aligned polypeptide arrays anchored on the MOF surfaces. The polypeptides can be separated by high-speed centrifuge and the MOF catalysts are released.

5.3 Conclusions and Future Prospects

The organo-catalytic ROP of NCA proceeded rapidly in the past 30 years. To date, organo-catalysts for NCA polymerization enable the controlled and efficient preparation of versatile polypeptides, ultrafast polymerization rate (in minutes), molecular weight from thousands to millions, well-controlled dispersity, kilogram-scale production and variable end groups. To our knowledge, the mechanism of NCA ROP is very complicated especially due to the diverse structure of polypeptides compared to traditional polymers. Encouragingly, with the advancement of characterization methods and the cooperation of various research groups around the world, the mechanism of organo-catalytic NCA polymerization has been more and more clear, which is conducive to the development of better catalytic systems.

However, there exists some difficulties for the industrial production of polypeptide via NCA ROP. For example, the purification of NCA monomers requires tedious recrystallization or column chromatography. Cheng group and Lu group developed NCA polymerization of *in situ* purified NCA [38] and a moisture-tolerant route to unprotected α/β-amino acid NCA [62], respectively, which both will contributed to the industrial production of polypeptides. In our opinion, there remain many important obstacles to be overcome in the fields of synthetic polypeptides: the poor solubility of NCAs and polypeptides, the copolymerization of different NCAs, the sequence control of polypeptides, the preparation of NCAs without phosgene.

Abbreviations

AMM	activated monomer mechanism
bpy	2,2′-bipyridine
CCP	cooperative covalent polymerization
CD	circular dichroism
COD	1,4-cyclooctadiene
DCM	dichloromethane
DMF	*N,N*-dimethylformamide
DMSO	dimethyl sulfoxide
FTIR	fourier transform infrared spectroscopy
GPC	gel permeation chromatography
HMDS	hexamethyldisilazane
NAM	normal amine mechanism
NCA	α-amino acid-*N*-carboxyanhydride
PAMAM	polyamidoamine
PBLG	poly(γ-benzyl-L-glutamate)
PEG	poly(ethylene glycol)
ROP	ring-opening polymerization
THF	tetrahydrofuran
TMS	trimethylsilyl

References

1 Deming, T.J. (1997). Polypeptide materials: new synthetic methods and applications. *Adv. Mater.* 9: 299–311.

2 Kricheldorf, H.R. (2006). Polypeptides and 100 years of chemistry of α-amino acid *N*-carboxyanhydrides. *Angew. Chem. Int. Ed.* 45: 5752–5784.

3 Deming, T.J. (2006). Polypeptide and polypeptide hybrid copolymer synthesis via NCA polymerization. In: *Peptide Hybrid Polymers* (ed. H.-A. Klok and H. Schlaad), 1–18. Berlin, Heidelberg: Springer.

4 Hadjichristidis, N., Iatrou, H., Pitsikalis, M., and Sakellariou, G. (2009). Synthesis of well-defined polypeptide-based materials via the ring-opening polymerization of α-amino acid *N*-carboxyanhydrides. *Chem. Rev.* 109: 5528–5578.

5 Cheng, J. and Deming, T.J. (2012). Synthesis of polypeptides by ring-opening polymerization of α-amino acid N-carboxyanhydrides. In: *Peptide-Based Materials* (ed. T. Deming), 1–26. Berlin, Heidelberg: Springer Berlin Heidelberg.

6 Song, Z., Fu, H., Wang, R. et al. (2018). Secondary structures in synthetic polypeptides from *N*-carboxyanhydrides: design, modulation, association, and material applications. *Chem. Soc. Rev.* 47: 7401–7425.

7 Song, Z., Han, Z., Lv, S. et al. (2017). Synthetic polypeptides: from polymer design to supramolecular assembly and biomedical application. *Chem. Soc. Rev.* 46: 6570–6599.

8 Nguyen, T.P., Easley, A.D., Kang, N. et al. (2021). Polypeptide organic radical batteries. *Nature* 593: 61–66.

9 Blout, E.R., Karlson, R.H., Doty, P., and Hargitay, B. (1954). Polypeptides. I. The synthesis and the molecular weight of high molecular weight polyglutamic acids and esters. *J. Am. Chem. Soc.* 76: 4492–4493.

10 Ballard, D.G.H. and Bamford, C.H. (1956). 77. Reactions of *N*-carboxy-α-amino-acid anhydrides catalysed by tertiary bases. *J. Chem. Soc.* 381–387.

11 Deming, T.J. (1997). Facile synthesis of block copolypeptides of defined architecture. *Nature* 390: 386–389.

12 Cheng, J. and Deming, T.J. (1999). Screening of optically active nickel initiators for enantioasymmetric polymerization of γ-benzyl glutamate-*N*-carboxyanhydride. *Macromolecules* 32: 4745–4747.

13 Deming, T.J. (1999). Cobalt and iron initiators for the controlled polymerization of α-amino acid-*N*-carboxyanhydrides. *Macromolecules* 32: 4500–4502.

14 Goodwin, A.A., Bu, X., and Deming, T.J. (1999). Reactions of α-amino acid-*N*-carboxyanhydrides (NCAs) with organometallic palladium(0) and platinum(0) compounds: structure of a metallated NCA product and its role in polypeptide synthesis. *J. Organomet. Chem.* 589: 111–114.

15 Seidel, S.W. and Deming, T.J. (2003). Use of chiral ruthenium and iridium amido-sulfonamidate complexes for controlled, enantioselective polypeptide synthesis. *Macromolecules* 36: 969–972.

16 Bhaw-Luximon, A., Jhurry, D., Belleney, J., and Goury, V. (2003). Polymerization of γ-methylglutamate *N*-carboxyanhydride using Al–Schiff's base complexes as initiators. *Macromolecules* 36: 977–982.

17 Dimitrov, I. and Schlaad, H. (2003). Synthesis of nearly monodisperse polystyrene–polypeptide block copolymers via polymerisation of *N*-carboxyanhydrides. *Chem. Commun.* 0: 2944–2945.

18 Lu, H. and Cheng, J. (2007). Hexamethyldisilazane-mediated controlled polymerization of α-amino acid *N*-carboxyanhydrides. *J. Am. Chem. Soc.* 129: 14114–14115.

19 Lu, H. and Cheng, J. (2008). *N*-trimethylsilyl amines for controlled ring-opening polymerization of amino acid *N*-carboxyanhydrides and facile end group functionalization of polypeptides. *J. Am. Chem. Soc.* 130: 12562–12563.

20 Lu, H., Wang, J., Lin, Y., and Cheng, J. (2009). One-pot synthesis of brush-like polymers via integrated ring-opening metathesis polymerization and polymerization of amino acid *N*-carboxyanhydrides. *J. Am. Chem. Soc.* 131: 13582–13583.

21 Zhang, X., Oddon, M., Giani, O. et al. (2010). Novel strategy for ROP of NCAs using thiols as initiators: synthesis of diblock copolymers based on polypeptides. *Macromolecules* 43: 2654–2656.

22 Peng, H., Ling, J., and Shen, Z. (2012). Ring opening polymerization of α-amino acid *N*-carboxyanhydrides catalyzed by rare earth catalysts: polymerization characteristics and mechanism. *J. Polym. Sci., Part A: Polym. Chem.* 50: 1076–1085.

23 Peng, H., Ling, J., Zhu, Y. et al. (2012). Polymerization of α-amino acid *N*-carboxyanhydrides catalyzed by rare earth tris(borohydride) complexes: Mechanism and hydroxy-endcapped polypeptides. *J. Polym. Sci., Part A: Polym. Chem.* 50: 3016–3029.

24 Conejos-Sánchez, I., Duro-Castano, A., Birke, A. et al. (2013). A controlled and versatile NCA polymerization method for the synthesis of polypeptides. *Polym. Chem.* 4: 3182–3186.

25 Peng, H., Chen, W.-L., Kong, J. et al. (2014). Synthesis of α-hydroxy-ω-aminotelechelic polypeptide from α-amino acid *N*-carboxyanhydrides catalyzed by alkali-metal borohydrides. *Chin. J. Polym. Sci.* 32: 743–750.

26 Vacogne, C.D. and Schlaad, H. (2015). Primary ammonium/tertiary amine-mediated controlled ring opening polymerisation of amino acid *N*-carboxyanhydrides. *Chem. Commun.* 51: 15645–15648.

27 Zhao, W., Gnanou, Y., and Hadjichristidis, N. (2015). Fast and living ring-opening polymerization of α-amino acid *N*-carboxyanhydrides triggered by an "alliance" of primary and secondary amines at room temperature. *Biomacromolecules* 16: 1352–1357.

28 Zhao, W., Gnanou, Y., and Hadjichristidis, N. (2015). From competition to cooperation: a highly efficient strategy towards well-defined (co)polypeptides. *Chem. Commun.* 51: 3663–3666.

29 Zhao, W., Gnanou, Y., and Hadjichristidis, N. (2015). Organocatalysis by hydrogen-bonding: a new approach to controlled/living polymerization of [small alpha]-amino acid *N*-carboxyanhydrides. *Polym. Chem.* 6: 6193–6201.

30 Yuan, J., Sun, Y., Wang, J., and Lu, H. (2016). Phenyl trimethylsilyl sulfide-mediated controlled ring-opening polymerization of α-amino acid *N*-carboxyanhydrides. *Biomacromolecules* 17: 891–896.

31 Zhang, H., Nie, Y., Zhi, X. et al. (2017). Controlled ring-opening polymerization of α-amino acid *N*-carboxy-anhydride by frustrated amine/borane Lewis pairs. *Chem. Commun.* 53: 5155–5158.

32 Baumgartner, R., Fu, H., Song, Z. et al. (2017). Cooperative polymerization of α-helices induced by macromolecular architecture. *Nat. Chem.* 9: 614–622.

33 Gradišar, Š., Žagar, E., and Pahovnik, D. (2017). Ring-opening polymerization of *N*-carboxyanhydrides initiated by a hydroxyl group. *ACS Macro Lett.* 6: 637–640.

34 Yuan, J., Zhang, Y., Li, Z. et al. (2018). A S-Sn Lewis pair-mediated ring-opening polymerization of α-amino acid *N*-carboxyanhydrides: fast kinetics, high molecular weight, and facile bioconjugation. *ACS Macro Lett.* 7: 892–897.

35 Wu, Y., Zhang, D., Ma, P. et al. (2018). Lithium hexamethyldisilazide initiated superfast ring opening polymerization of alpha-amino acid *N*-carboxyanhydrides. *Nat. Commun.* 9: 5297.

36 Zhao, W., Lv, Y., Li, J. et al. (2019). Fast and selective organocatalytic ring-opening polymerization by fluorinated alcohol without a cocatalyst. *Nat. Commun.* 10: 3590.

37 Chen, C., Fu, H., Baumgartner, R. et al. (2019). Proximity-induced cooperative polymerization in "hinged" helical polypeptides. *J. Am. Chem. Soc.* 141: 8680–8683.

38 Song, Z., Fu, H., Wang, J. et al. (2019). Synthesis of polypeptides via bioinspired polymerization of in situ purified *N*-carboxyanhydrides. *Proc. Natl. Acad. Sci.* 116: 201901442.

39 Liang, J., Zhi, X., Zhou, Q., and Yang, J. (2019). Binaphthol-derived phosphoric acids as efficient organocatalysts for the controlled ring-opening polymerization of γ-benzyl-L-glutamate *N*-carboxyanhydrides. *Polymer* 165: 83–90.

40 Lv, S., Kim, H., Song, Z. et al. (2020). Unimolecular polypeptide micelles via ultrafast polymerization of *N*-carboxyanhydrides. *J. Am. Chem. Soc.* 142: 8570–8574.

41 Zhao, W., Lv, Y., Li, J. et al. (2020). A synthetic method for site-specific functionalized polypeptides: metal-free, highly active, and selective at room temperature. *Angew. Chem. Int. Ed.* 60: 889–895.

42 Wu, Y.-M., Zhang, W.-W., Zhou, R.-Y. et al. (2020). Facile synthesis of high molecular weight polypeptides via fast and moisture insensitive polymerization of α-amino acid *N*-carboxyanhydrides. *Chin. J. Polym. Sci.* 38: 1131–1140.

43 Soria-Carrera, H., Franco-Castillo, I., Romero, P. et al. (2021). On-POM ring-opening polymerisation of *N*-carboxyanhydrides. *Angew. Chem. Int. Ed.* 60: 3449.

44 Xia, Y., Song, Z., Tan, Z. et al. (2021). Accelerated polymerization of *N*-carboxyanhydrides catalyzed by crown ether. *Nat. Commun.* 12: 732.

45 He, W. and Tao, Y. (2021). Bifunctional fluoroalcohol catalysts enabled sustainable synthesis of poly(amino acid)s. *Chin. J. Chem.* 39: 2119–2124.

46 Wu, Y., Chen, K., Wu, X. et al. (2021). Superfast and water-insensitive polymerization on α-amino acid *N*-carboxyanhydrides to prepare polypeptides using tetraalkylammonium carboxylate as the initiator. *Angew. Chem. Int. Ed.* 60: 26063–26071.

47 Li, K., Li, Z., Shen, Y. et al. (2022). Organobase 1,1,3,3-tetramethyl guanidine catalyzed rapid ring-opening polymerization of α-amino acid

N-carboxyanhydrides adaptive to amine, alcohol and carboxyl acid initiators. *Polym. Chem.* 13: 586–591.

48 Lv, W., Wang, Y., Li, M. et al. (2022). Precision synthesis of polypeptides via living anionic ring-opening polymerization of *N*-carboxyanhydrides by tri-thiourea catalysts. *J. Am. Chem. Soc.* 144: 23622–23632.

49 Li, Q., Lan, Y., Wang, W. et al. (2023). Polymerization of *N*-carboxyanhydride in cosolvents: the balance between the polymerization rate and molecular weight control. *Macromolecules* 56: 7023–7031.

50 Liu, Y., Ren, Z., Zhang, N. et al. (2023). A nanoscale MOF-based heterogeneous catalytic system for the polymerization of *N*-carboxyanhydrides enables direct routes toward both polypeptides and related hybrid materials. *Nat. Commun.* 14: 5598.

51 Zhu, J., He, Y., Ge, C. et al. (2023). Accelerated ring-opening polymerization of α-amino acid *N*-carboxyanhydride via inorganic nano-initiators. *Chin. J. Chem.* 41: 2476–2482.

52 Wang, K.-Y., Li, Z.-Q., Xu, M.-N., and Li, B. (2023). Thiourea-based bifunctional catalysts for epoxide-involved ring-opening polymerization of α-amino acid *N*-carboxyanhydrides. *Macromolecules* 56: 5599–5609.

53 Chan, B.A., Xuan, S., Horton, M., and Zhang, D. (2016). 1,1,3,3-tetramethylguanidine-promoted ring-opening polymerization of *N*-butyl *N*-carboxyanhydride using alcohol initiators. *Macromolecules* 49: 2002–2012.

54 Siefker, D., Chan, B.A., Zhang, M. et al. (2022). 1,1,3,3-tetramethylguanidine-mediated zwitterionic ring-opening polymerization of sarcosine-derived *N*-thiocarboxyanhydride toward well-defined polysarcosine. *Macromolecules* 55: 2509–2516.

55 Guo, Y.-T., Shi, C., Du, T.-Y. et al. (2022). Closed-loop recyclable aliphatic poly(ester-amide)s with tunable mechanical properties. *Macromolecules* 55: 4000–4010.

56 Shi, C.-X., Guo, Y.-T., Wu, Y.-H. et al. (2019). Synthesis and controlled organobase-catalyzed ring-opening polymerization of morpholine-2,5-dione derivatives and monomer recovery by acid-catalyzed degradation of the polymers. *Macromolecules* 52: 4260–4269.

57 Zhang, X., Jones, G.O., Hedrick, J.L., and Waymouth, R.M. (2016). Fast and selective ring-opening polymerizations by alkoxides and thioureas. *Nat. Chem.* 8: 1047–1053.

58 Yu, X., Wang, Y., Dong, Y. et al. (2023). Efficient synthesis of *N*-methyl polypeptides by organic acid promoted controlled ring-opening polymerization of *N*-methyl-α-amino acids *N*-carboxyanhydrides. *Macromolecules* 56: 8899–8911.

59 Wang, S., Lu, M.-Y., Wan, S.-K. et al. (2024). Precision synthesis of polysarcosine via controlled ring-opening polymerization of *N*-carboxyanhydride: fast kinetics, ultrahigh molecular weight, and mechanistic insights. *J. Am. Chem. Soc.* 146: 5678–5692.

60 Wang, W., Fu, H., Lin, Y. et al. (2023). Cooperative covalent polymerization of *N*-carboxyanhydrides: from kinetic studies to efficient synthesis of polypeptide materials. *Acc. Mater. Res.* 4: 604–615.

61 Song, Z., Fu, H., Baumgartner, R. et al. (2019). Enzyme-mimetic self-catalyzed polymerization of polypeptide helices. *Nat. Commun.* 10: 5470.
62 Tian, Z.-Y., Zhang, Z., Wang, S., and Lu, H. (2021). A moisture-tolerant route to unprotected α/β-amino acid *N*-carboxyanhydrides and facile synthesis of hyperbranched polypeptides. *Nat. Commun.* 12: 5810.

6

Organocatalyst Catalyzed Ring Opening Polymerization of O(S)-Carboxyanhydride to Produce High Performance Poly(thio)esters

Yinuo Zhu and Youhua Tao

Changchun Institute of Applied Chemistry, Chinese Academy of Sciences, Key Laboratory of Polymer Ecomaterials, Renmin Street 5625, Changchun 130022, People's Republic of China

6.1 Introduction

The annual production of polymers has reached 500 million tons, making it an indispensable key material in modern society [1]. But it also faces challenges of non-renewability and difficulty in degradation. Substantial efforts have been devoted to develop degradable polymer [2–5]. Poly(hydroxy acid)s including polylactide (PLA), polyglycolide (PGA), are a class of degradable polymer, which find a variety of applications including packaging, agricultural, biomedical and pharmaceutical due to their biocompatibility and (bio)degradability [6–12]. Polythioesters, including poly(thiolactic acid), are another important type of degradable polymers, showcasing enhanced depolymerizability, improved crystallinity and high refractive index [13]. However, the lack of structural diversity limits its mechanical and thermal properties to rival petroleum-based plastics [12, 14]. One common approach to synthesize functional poly(thio)esters is the ring-opening polymerization of functional cyclic (thio)ester monomers. However, preparation of cyclic (thio)ester monomers bearing functional groups in large scale and reasonable yields still faces many challenges [15–19].

Amino acids are naturally available renewable feedstocks, and recent developments in biotechnology have made it possible to produce amino acids in large scale at a cheap price [20, 21]. Nevertheless, the synthesis of functionalized poly(thio)esters using amino acids as building blocks have been rarely reported due to the difficulty of preparing the large-scale monomers and poor yields during the preparation. Since the twenty-first century, O(S)-carboxyanhydrides (O(S)CAs), which can be readily derived from amino acid precursors with inherent rich side chain functionality have emerged as an alternative class of highly active monomers for preparing functional poly(thio)esters [22–35] (Figure 6.1a). However, the functionalization of corresponding lactide monomers often involves multisteps with lower yields [36].

The ring opening polymerization (ROP) of O(S)CAs has been regarded thermodynamically more favorable than that of (thio)lactide due to the liberation

Organocatalysts in Polymer Chemistry: Synthesis and Applications, First Edition. Edited by Zhibo Li.

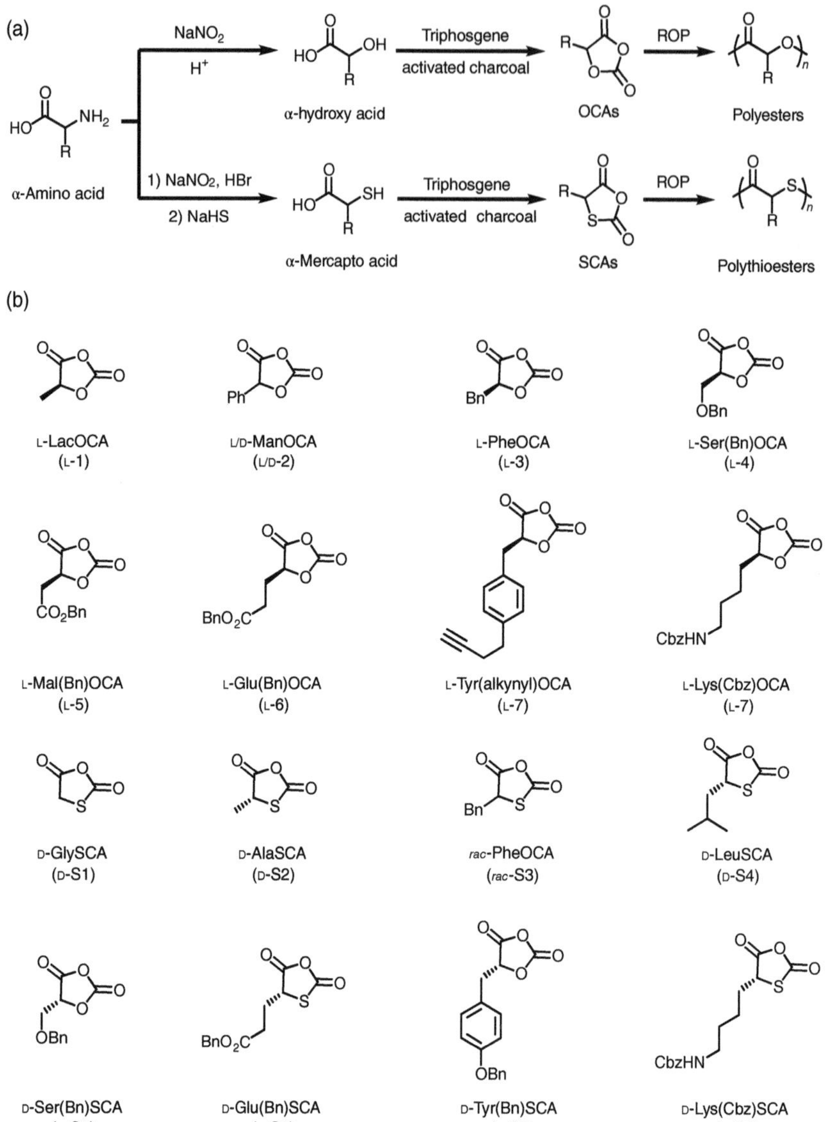

Figure 6.1 (a) Synthesis of sustainable poly(thio)esters via O(S)CAs polymerization. (b) Classes of O(S)CAs discussed herein.

of a carbon dioxide molecule during the polymerization process [37], making it easier to achieve controlled molecular weights and narrow dispersities under mild conditions. So far, a variety of organic and metal-based catalysts have been found capable of catalyzing of the ROP of OCAs [32, 38, 39], but only a few literatures have reported the ROP of SCAs [33–35]. In this regard, this chapter focus on the recent development in the organocatalytic ROP of O(S)CAs and highlight the usage

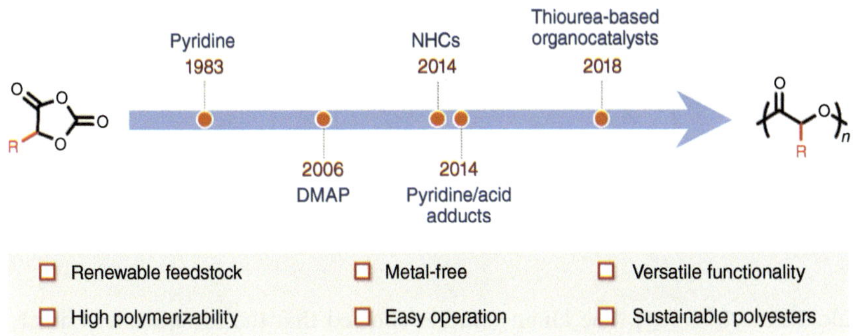

Figure 6.2 Synthesis of sustainable polyesters via organocatalytic ROP of OCAs.

of thiourea-based bifunctional single-molecule organocatalysts for eliminating epimerization and other side reactions in OCAs polymerization that are pertinent to the future development of sustainable isotactic poly(thio)esters (Figure 6.2).

6.2 Synthesis and Purification of O(S)-Carboxyanhydride Monomers

Similar to the synthesis of *N*-carboxyanhydrides (NCAs), Davies first demonstrated the preparation of O(S)CAs by the reaction of α-hydroxy/mercapto acid with phosgene in 1951 [40]. O(S)CAs are typically synthesized by carbonylation of α-hydroxy/mercapto acids using phosgene, diphosgene or triphosgene. Activated charcoal is often used to promote the decomposition of triphosgene to phosgene and sometimes a tertiary amine (e.g. *N*-methylmorpholine) is needed as an acid scavenger [19, 41]. Much effort has been made to prepare new O(S)CAs from amino acids due to the structural similarity between α-amino acids and α-hydroxy/mercapto acids. Up to now, OCAs can be furnished from lactic acid [37, 42], mandelic acid [27, 43], phenylalanine [27, 29, 44], serine [45], malic acid [46], glutamic acid [36], tyrosine [26, 47, 48] and lysine [49, 50]. Although there is limited literature regarding SCA, it can be prepared from glycolic acid, alanine, malic acid, serine, leucine, glutamic acid, tyrosine and lysine [33–35] (Figure 6.1b). In general, recrystallization is sufficient to obtain pure O(S)CA monomers. However, some additional steps are still required to synthesize OCAs (e.g. OCAs from Leucine and Isoleucine) which are hard to recrystallize [45].

6.3 The Racemization of α-Proton

Early work by Kricheldorf and Jonte, realized the ROP of OCAs catalyzed via bases, but accompanied by epimerization [19] (Figure 6.3). The epimerization of 5-arylOCAs was also observed in the alcoholysis catalyzed by a modified cinchona

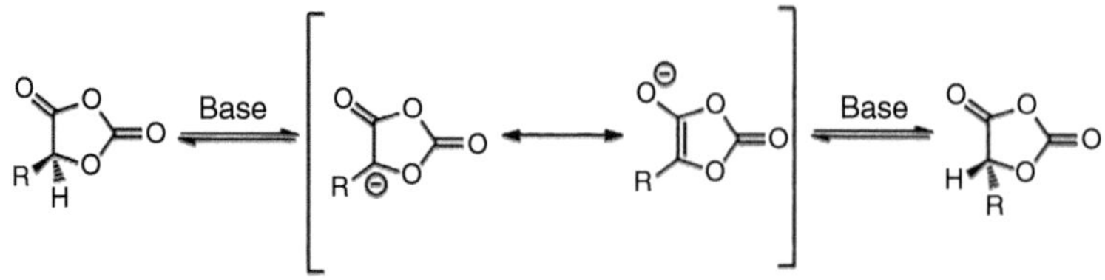

Figure 6.3 Epimerization of α-proton in OCAs ROP.

alkaloid at −78 °C [51]. The kinetic studies showed that the racemize was much faster than the enantioselective alcoholysis. In 2011, Dove and coworkers, also reported the epimerization in the DMAP/neopentanol-mediated ROP of L-MalOCA (L-5), thus leading to a number of undesirable side products [46]. When the aryl groups were substituted with alkyl groups, the acidity of the α-proton decreased, thereby preventing its epimerization by base. This underscores the significance of the substituent group on the 5-position of the OCA monomer.

When the first attempt to ROP of SCAs by K. Bösinger in 1973, they found an issue associated with the use of organocatalysts like DMAP and Et_3N as catalysts would lead to the deprotonation of α-methine hydrogen atoms [35]. This results in the epimerization of α-carbon atom, and the loss of stereoregularity in the polythioesters backbone, leading to changes in the properties of the final polymers. More specifically, these challenges are particularly arduous to solve for SCAs with smallest (TlaSCA) or strong electron-donating (PheSCA) substituent. Recently, Zhu and Tao reported the use of salen-Al catalysts with appropriate ligand suppress the epimerization and thereby producing the high-isotactic ($P_m > 0.99$) crystalline poly(thiolactic acid) with high molecular weight [34].

6.4 Organocatalyst for OCA Polymerization

6.4.1 ROP Catalyzed by DMAP

Although the ROP of OCAs initiated by alcohol under heating, tertiary amines or metal alkoxides was reported in the 1960s–1980s [19, 52–54], the controlled polymerization was first achieved by Bourissou and coworkers until 2006 [37]. They used DMAP as catalyst to synthesize polyesters with predictable molecular weight (MW) and low dispersity (*Đ*) under mild conditions (Figure 6.4). With DMAP/neopentane alcohol, the controlled polymerization of L-LacOCA (L-1) could be realized at room temperatures in DCM within a few minutes to several hours. Remarkably, both primary and secondary alcohols can initiate polymerization, because of the much higher reactivity of L-1 compared to that of L-LA.

Computational studies of the polymerization mechanism suggested that DMAP activation of the initiating/propagating alcohol (Path B) was more favorable compared with nucleophilic activation of the monomer (Path A) involving an acylpyridinium intermediate (Figure 6.5).

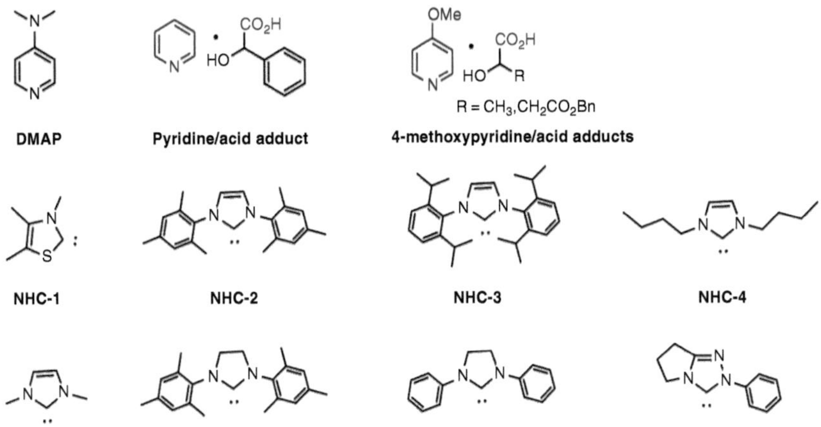

Figure 6.4 Various Organocatalysis for OCAs Polymerization.

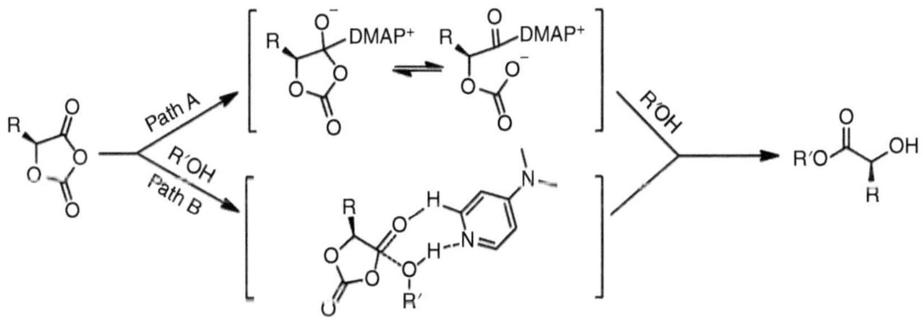

Figure 6.5 The nucleophilic activation monomer (Path A) and basic activation (Path B) for the DMAP-catalyzed ROP of OCAs.

Excitingly, this chemistry was successfully extended to the ROP of other OCAs, including L-Ser(Bn)OCA (L-4), L-Glu(Bn)OCA (L-6), L-Tyr(alkynyl)OCA (L-7) and L-Lys(Cbz) OCA (L-8) [30, 45, 47–49] (Figure 6.1b). For instance, Bourissou and coworkers, reported the homopolyesters as well as block and random copolyesters featuring pendant carboxyl groups could be achieved by DMAP-catalyzed ROP of L-6 in DCM at 25 °C [30]. Cheng and coworkers evaluated the ROP of L-4 using DMAP as the catalyst and isobutanol as the initiator [45]. Thus, water-soluble polyesters with pendant hydroxyl groups were obtained and performed excellent cell compatibility, suggesting their potential being applied as drug delivery materials and as hydrogel scaffolds for tissue engineering applications. Cheng and coworkers, also synthesized poly(L-7) with MWs varying from 5 to 11.6 kg mol^{-1} and low $Đ$ (<1.15) via DMAP/pyrenebutanol [47]. In 2014, Tao and coworkers prepared L-8 from renewable lysine and utilized DMAP to generation polyesters having pendent Cbz-protected amino groups with controlled MW and low $Đ$ in the range of 1.15–1.26 (Figure 6.6) [49]. Well-controlled polyesters bearing amino groups

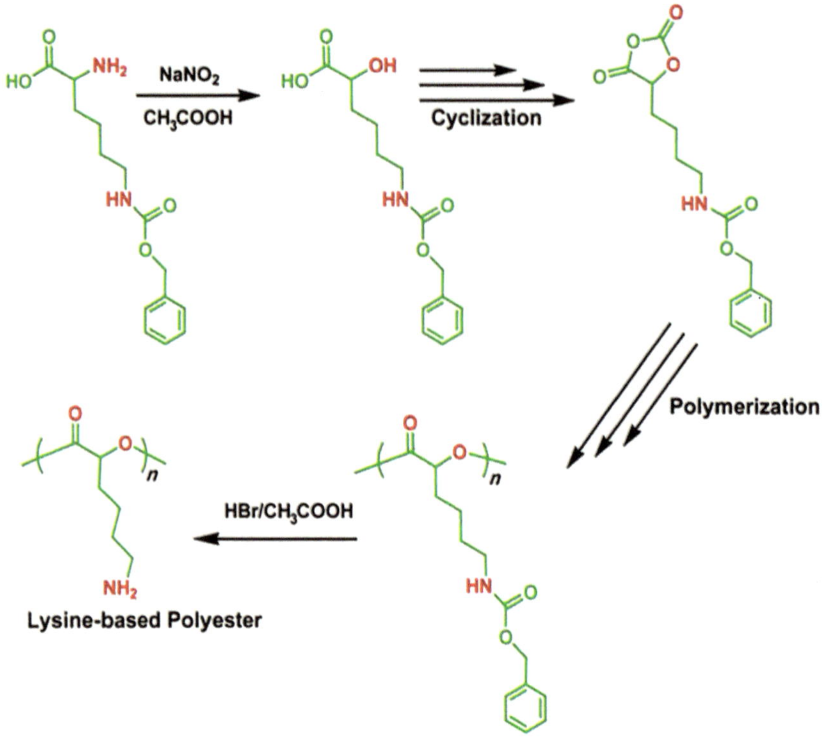

Figure 6.6 Preparation and ROP of L-Lys(Cbz)OCA (L-8).

were recovered by cleavage of the Cbz protecting group. These novel functional polyesters have broad applications in biodegradable plastics and biomedical fields.

6.4.2 ROP Catalyzed by *N*-Heterocyclic Carbenes

N-heterocyclic carbenes (NHCs) is also an important type of organic catalyst, having been widespread development due to its good stability towards air and oxidants, diverse structures and strong nucleophilicity. Li and coworkers utilized NHCs for the polymerization of L-1 in THF at room temperature [42] (Figure 6.4). Controlled MW were obtained via NHC-1, NHC-2 and NHC-3, displaying better control with narrow $Đ$ in the range of 1.10–1.16 for degree of polymerizations (DPs) from 30 to 200. The controlled character of the polymerization was further supported by the plot of the molecular weights vs. monomer to initiator ratio. Compared with NHC-2 and NHC-3, less sterical bulky carbenes (e.g. NHC-4 and NHC-5) were less efficient in MW control.

6.4.3 ROP Catalyzed by Acid/Base Adducts

As mentioned above, some organic-base, such as DMAP, could serve as catalyst to facilitate the ROP of OCAs while also induce epimerization of the α-carbon of OCAs

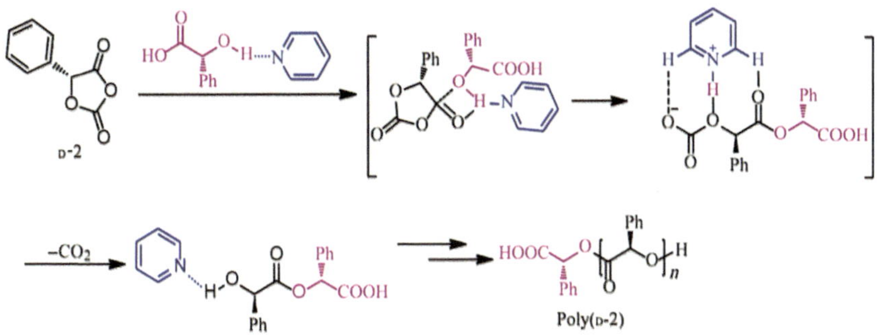

Figure 6.7 The application of acid/base adducts to suppress the epimerization in OCA polymerization.

with activating substituent groups [46]. In 2014, Buchard and coworkers proved that employing acid/base ion pairs (crystalline adducts of mandelic acid and pyridine) as catalysts for the ROP of enantiopure OCA derived from mandelic acid (D-2) could suppress the epimerization and obtain highly isotactic polyesters, which show improved performances compared to atactic polymers [2] (Figure 6.7) [43]. Similarly, the adducts of 4-methoxypyridine with L-lactic acid and β-benzyl α-L-malate could also initiate polymerization of L-5 with minimized racemization; however, those adducts were unable to produce isotactic poly(L-2) [55]. It is worth noting that the acid/base adducts could still result in racemization even with decreased basicity at low [OCA]/[initiator] ratios.

6.4.4 Thiourea-Based Organocatalysts for OCAs Polymerization

Stereocontrol is of great significance because stereoregularity is a prerequisite for crystalline polymers, and its mechanical properties are often superior to those of atactic polymers [7, 10, 56–59]. So far, the stereoregularity polymerization of OCAs catalyzed by organocatalysts still faces challenges. Moreover, these challenges are particularly formidable when dealing with L/D-2, primarily due to the increased C–H acidity. To date, there have no reports on the crystallization behavior of poly(mandelic acid)s (PMAs).

To address this challenge, Tao and coworkers recently developed thiourea-based bifunctional single-molecule organocatalysts for eliminating epimerization in OCAs polymerization (Figure 6.8) [60]. To eliminate epimerization in polymerization, Tao et al., used a bifunctional organocatalysts consisting of a thiourea linked to ortho-substituted chloropyridine in the polymerization (Figure 6.9). The catalyst design was based on the reduced basicity of the ortho-substituted chloropyridine. Polymerization conducted using the single-molecule organocatalyst showed several features of controlled polymerization, and isotactic polyesters with controlled MWs (up to 19.8 kg mol^{-1}) could be furnished.

To emphasize the versatility of the thiourea-based bifunctional single-molecule organocatalysts, Tao et al. applied organocatalyst **3e** for polymerization of other OCAs monomers, including L-1, L-PheOCA(L-3), L-5, L-8 (Figure 6.1b). In all cases,

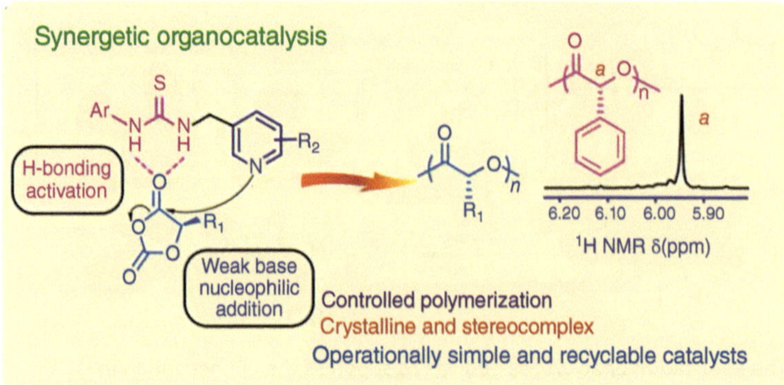

Figure 6.8 Thiourea-based organocatalysis for OCA polymerization.

the ROPs were well-controlled, similar to that for the generation of poly(D-2); and there was no epimerization in the α-methine region. Based on the kinetic experiments and ^{1}H NMR measurements, Tao and coworkers propose the polymerization mechanism catalyzed by **3e**, as shown in Figure 6.10.

All previously obtained PMAs are amorphous, which limits their potential applications. The enantiomers of chiral polymers can be physically blended at stoichiometric ratio to form crystalline stereocomplexed materials, which usually show excellent performances. They observe that a stereocomplex from the 1 : 1 mixture of (D)-PMA and (L)-PMA was semicrystalline, and the DSC data showed a T_m of 150 °C [60], which has not been observed in previous studies.

The degree of polymerization (DP) and molecular weights (MWs) are pivotal parameters governing the physical properties of polymers, exerting significant influence on their performance. Studies on PLA shows that the mechanical properties and crystallization behaviors of PLA are highly dependent on the MW of the polymer [61]. However, until recently, most polymers obtained from the ROP of OCAs have exhibited relatively low MWs (<50 kDa). Only a specific type of metal organic catalytic system (Ni/Ir) has been able to achieve minimal epimerization and efficient chain propagation simultaneously [62], but the metal residues impede the applications of these polyesters in optoelectronic and biomedical materials.

In this regard, Tao and coworkers utilized a bifunctional catalyst, **3b**, which incorporates a thiourea anion-binding unit and an anchored DMAP group for ROP of OCAs [63]. While other catalytic approaches for the ROP of carboxyanhydrides still suffer from low chain propagation or stereocontrol selectivity, this approach facilitates the synthesis of highly isotactic polyesters with high MW up to 150.0 kDa. As expected, the mechanical properties of high MW Ph-PLA (Mn = 150.0 kDa) were quite different from those of low MW congener (Mn = 41.1 kDa). The measured tensile strength and the elongation at break of high MW Ph-PLA was respectively 4.3 times and 30 times higher than its low MW counterpart.

The anion binding of the thiourea unit to the carboxylate of the growing end of the polymer chain shows several advantageous effects such as an enhanced of the moisture tolerance and stability of the active polymerization species, as well as selective elongation process. Hence, after the initial nucleophilic opening of the cyclic

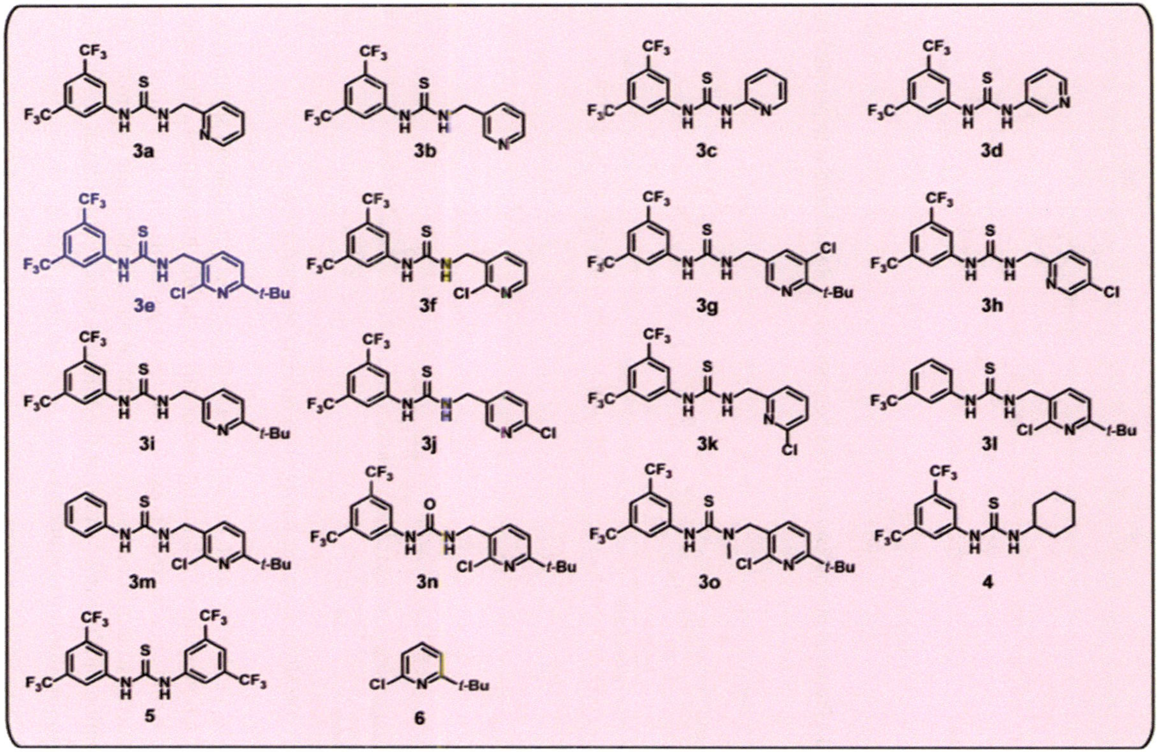

Figure 6.9 Bifunctional organocatalysts for OCAs polymerization.

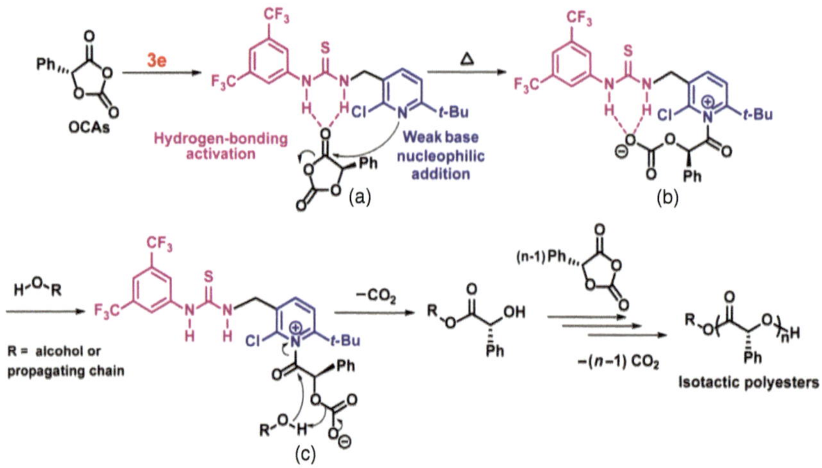

Figure 6.10 Proposed mechanism for OCAs polymerization. Reproduced with permission.

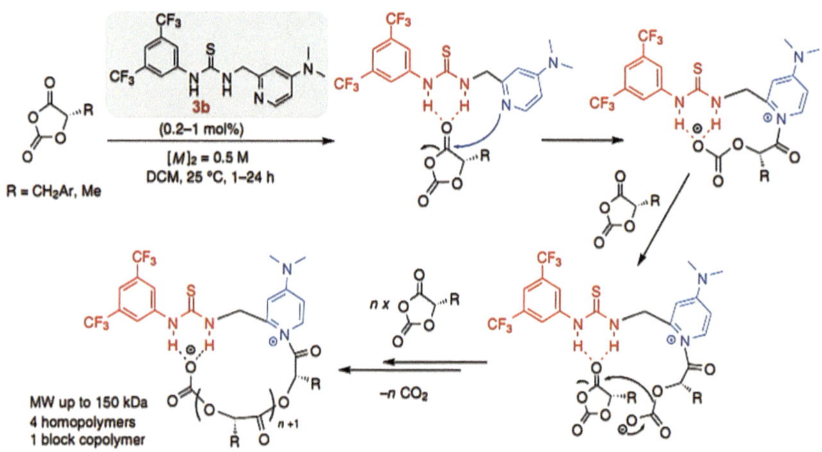

Figure 6.11 DMAP-thiourea cocatalyzed ring-opening polymerization.

carbonate by the DMAP unit, the thiourea facilitates the propagation step through a dynamic anion–carbonate binding and decarboxylation that ensures a high selectivity for the ring opening over possible side reactions. Moreover, DFT calculations were conducted to reinforce the experimental observations on the key role of the dynamic anion-binding interaction of the thiourea moiety. The detail mechanism catalyzed by **3b** shown in Figure 6.11.

6.5 Organocatalyst for SCA Polymerization

The initial investigation on ROP of SCAs was conducted by Bösinger with tertiary bases as catalysts, leading to uncontrolled polymerizations and the prevalence of

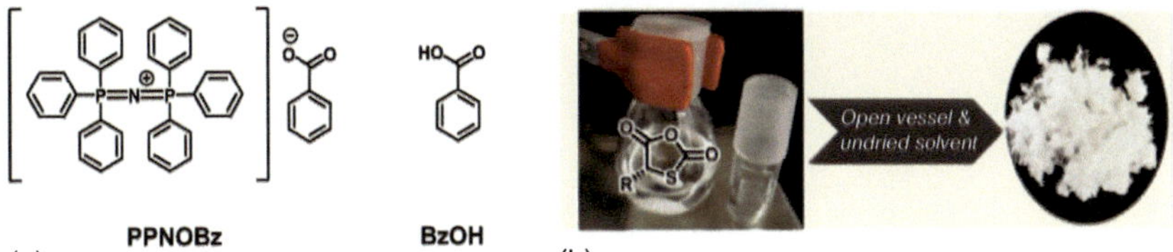

(a) (b)

Figure 6.12 (a) Structures of initiators and chain transfer agent. (b) Open-vessel polymerization of SCA at 5 g scale.

side reactions (e.g. transthioesterification and epimerization) [35]. The obtained polymers were insoluble in common organic solvents, thus preventing accurate characterization of polythioesters. Moreover, they failed to realize that the SCA monomer possessed hyperactivity, which resulted in almost no report on the ROP of SCA appearing in the subsequent decades.

Until recently, Tao and coworkers developed ultrafast and selective approach for ROP of SCAs, using PPNOBz ([PPN] = bis(triphenylphosphine)-iminium) as initiator along with the chain transfer agent benzoic acid. The polymerizations can be operated in open vessels and reach complete conversion within just 1–2 minutes at room temperature, yielding polythioesters with predictable molecular weight and low dispersities [33] (Figure 6.12). Most importantly, the stereoregularity of the isolated PTE prepared with the [PPN]OBz/BzOH system was retained, indicating that this system does not induce racemization of SCAs or of the resulting polythioesters.

A fascinating feature of this method is the functional groups can easily incorporating along the polymer backbone due to the diverse structures of starting α-amino acids. Hence, this approach was successfully extended to the ROP of D-Ser(Bn)SCA, D-Glu(Bn)SCA, D-Lys(Cbz)SCA, D-Tyr(Bn)SCA, and *rac*-PheSCA (Figure 6.1b). All polymerizations were rapid and controlled, and the random or sequential copolymerization of D-SerSCA and D-GluSCA gave well defined copolymer as well.

DFT computations provided support for the selectivity towards monomer propagation over transthioesterification. This selectivity arises from the high polymerizability of SCAs, with the calculated energy barrier for chain back-biting involving nucleophilic attack onto the terminal anhydride being higher than that of chain propagation (Figure 6.13). Based on these results, a plausible mechanism for the ROP of SCAs mediated by PPNOBz/BzOH was proposed in Figure 6.14. The reaction starts with the nucleophilic attack of [PPN]OBz onto the highly reactive monomer, with concomitant release of CO_2. Then, the formed thiolate is protonated by the carboxylic acid used as CTA, generating a thiol and another carboxylate. This initiation step proceeds iteratively until all carboxylic acid is consumed. After that, proton transfer between dormant (thiol) and active chain-ends (thiolate) facilitates propagation. Due to the significant acidity discrepancy between the initiating and propagating species, chain growth shows an induction period, as evidenced by experimental kinetic studies.

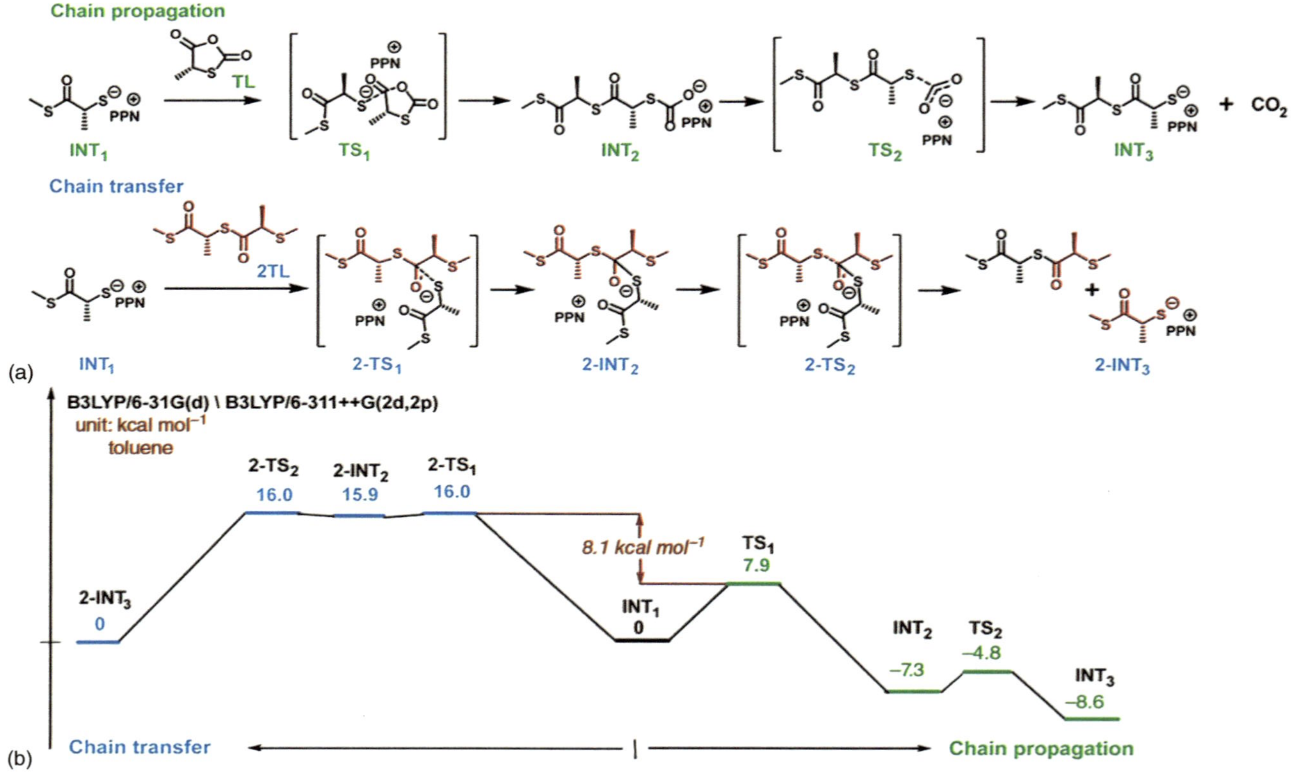

Figure 6.13 (a) Postulated propagation and transfer intermediates (INT) and transition states (TS) in the ROP of SCA in toluene. (b) Calculated free energy of each INT and TS.

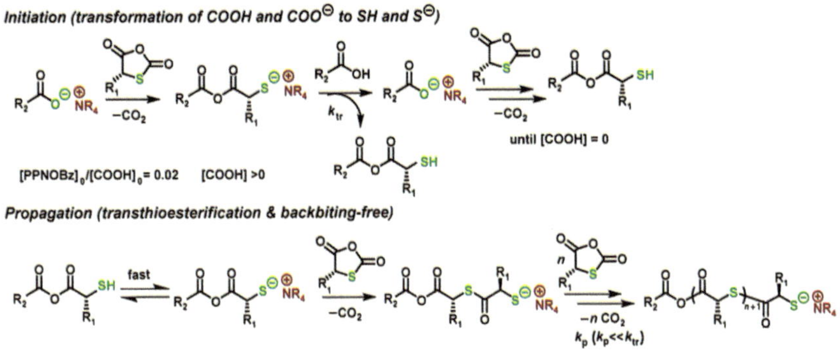

Figure 6.14 Plausible mechanistic pathway for the [PPN]OBz/BzOH-mediated ROP of SCAs based on DFT calculations.

6.6 Conclusion and Perspective

As illustrated and summarized in this review, the development of organocatalyst for O(S)CAs polymerization has steadily progressed over the past few decades considering the challenges associated with metal residues and operational complexities of metal-based catalysts,. However, achieving stereoregular polymerization from O(S)CAs remains challenging, and generating high molecular weight poly(thio)esters concurrently presents even greater difficulty. A noteworthy breakthrough in this field is the development of thiourea-based bifunctional single-molecule organocatalysts, which demonstrated minimal epimerization for ROP of various OCAs, leading to highly isotactic sustainable polyesters with high MWs up to 150.0 kDa.

Nevertheless, the study of O(S)CAs remains in its nascent stages despite ongoing research efforts, with few catalyst, whether organic or metal-based, showing siginificant promise for industrial production or commercialization. The detailed chain-propagation mechanism, including non-regioselective ring-opening of O(S)CAs, and the existence of active-monomer mechanism [31], have not been thoroughly investigated. The standardized and systemic studies, together with the use of visualized experimental procedures, can be helpful for understanding the propagation process. On the other hand, while a few metal catalysts have demonstrated the ability to prepare polyesters with various microstructures [26, 64], achieving stereoselective organocatalytic ROP of O(S)CAs with diverse functional groups remains a formidable challenging. Therefore, there is growing anticipation for the development of different approaches aimed at designing and refining organocatalysts capable of catalyzing the stereoselective ROP of racemic OCAs. This will provide access to functional polyesters with distinct microstructures, exhibiting varied mechanical performances, and thereby expanding their potential applications.

References

1 MacLeo, M., Arp, H.P.H., Tekman, M.B., and Jahnke, A. (2021). The global threat from plastic pollution. *Science* 373 (6550): 61–65.

2 Shi, C., Reilly, L.T., Phani Kumar, V.S. et al. (2021). Design principles for intrinsically circular polymers with tunable properties. *Chem* 7 (11): 2896–2912.

3 Coates, G.W. and Getzler, Y.D.Y.L. (2020). Chemical recycling to monomer for an ideal, circular polymer economy. *Nat. Rev. Mater.* 5 (7): 501–516.

4 Tang, X. and Chen, E.Y.X. (2019). Toward infinitely recyclable plastics derived from renewable cyclic esters. *Chem* 5 (2): 284–312.

5 Hong, M. and Chen, E.Y.X. (2017). Chemically recyclable polymers: a circular economy approach to sustainability. *Green Chem.* 19 (16): 3692–3706.

6 Ottou, W.N., Sardon, H., Mecerreyes, D. et al. (2016). Update and challenges in organo-mediated polymerization reactions. *Prog. Polym. Sci.* 56: 64–115.

7 Stanford, M.J. and Dove, A.P. (2010). Stereocontrolled ring-opening polymerisation of lactide. *Chem. Soc. Rev.* 39 (2): 486–494.

8 Bourissou, D., Moebs-Sanchez, S., and Martín-Vaca, B. (2007). Recent advances in the controlled preparation of poly(α-hydroxy acids): metal-free catalysts and new monomers. *C. R. Chim.* 10 (9): 775–794.

9 Dechy-Cabaret, O., Martin-Vaca, B., and Bourissou, D. (2004). Controlled ring-opening polymerization of lactide and glycolide. *Chem. Rev.* 104 (12): 6147–6176.

10 Chamberlain, B.M., Cheng, M., Moore, D.R. et al. (2001). Polymerization of lactide with zinc and magnesium β-diiminate complexes: stereocontrol and mechanism. *J. Am. Chem. Soc.* 123 (14): 3229–3238.

11 Deng, X.M., Yuan, M.L., Li, X.H., and Xiong, C.D. (2000). Polymerization of lactides and lactones VII. Ring-opening polymerization of lactide by rare earth phenyl compounds. *Eur. Polym. J.* 36 (6): 1151–1156.

12 Jacobsen, S., Degée, P.H., Fritz, H.G. et al. (1999). Polylactide (PLA) – a new way of production. *Polym. Eng. Sci.* 39 (7): 1311–1319.

13 Xiong, W. and Lu, H. (2023). Recyclable polythioesters and polydisulfides with near-equilibrium thermodynamics and dynamic covalent bonds. *Science China Chem.* 66: 725–738.

14 Farah, S., Anderson, D.G., and Langer, R. (2016). Physical and mechanical properties of PLA, and their functions in widespread applications – A comprehensive review. *Adv. Drug Deliv. Rev.* 107: 367–392.

15 Jing, F. and Hillmyer, M.A. (2008). A bifunctional monomer derived from lactide for toughening polylactide. *J. Am. Chem. Soc.* 130 (42): 13826–13827.

16 Becker, G. and Wurm, F.R. (2018). Functional biodegradable polymers via ring-opening polymerization of monomers without protective groups. *Chem. Soc. Rev.* 47 (20): 7739–7782.

17 Pounder, R.J. and Dove, A.P. (2010). Towards poly(ester) nanoparticles: recent advances in the synthesis of functional poly(ester)s by ring-opening polymerization. *Polym. Chem.* 1 (3): 260–271.

18 Gerhardt, W.W., Noga, D.E., Hardcastle, K.I. et al. (2006). Functional lactide monomers: methodology and polymerization. *Biomacromolecules* 7 (6): 1735–1742.

19 Kricheldorf, H.R. and Jonte, J.M. (1983). New polymer syntheses. 8. Synthesis and polymerization of L-lactic acid *O*-carboxyanhydride (5-methyl-dioxolan-2,4-dione). *Polym. Bull.* 9 (6-7): 276–283.

20 Deng, W.P., Wang, Y.Z., Zhang, S. et al. (2018). Catalytic amino acid production from biomass-derived intermediates. *Proc. Natl. Acad. Sci. USA* 115 (20): 5093–5098.

21 D'Este, M., Alvarado-Morales, M., and Angelidaki, I. (2018). Amino acids production focusing on fermentation technologies – a review. *Biotechnol. Adv.* 36 (1): 14–25.

22 Zhong, Y.L., Feng, Q.Y., Wang, X.Q. et al. (2020). Functionalized polyesters via stereoselective electrochemical ring-opening polymerization of *O*-carboxyanhydrides. *ACS Macro Lett.* 9 (8): 1114–1118.

23 Jiang, J.X., Cui, Y.Q., Lu, Y.G. et al. (2020). Weak Lewis pairs as catalysts for highly isoselective ring-opening polymerization of epimerically labile *rac*-*O*-carboxyanhydride of mandelic acid. *Macromolecules* 53 (3): 946–955.

24 Jia, Z.W., Chen, S.C., Jiang, J.X. et al. (2020). Ring-opening alternating copolymerization of *O*-carboxyanhydrides of lactic acid and malic acid using hafnium alkoxide initiators with different stereo selectivities and activities. *Inorg. Chem.* 59 (14): 10353–10360.

25 Cui, Y.Q., Jiang, J.X., Pan, X.B., and Wu, J.C. (2019). Highly isoselective ring-opening polymerization of *rac*-*O*-carboxyanhydrides using a zinc alkoxide initiator. *Chem. Commun.* 55 (86): 12948–12951.

26 Sun, Y.Y., Jia, Z.W., Chen, C.J. et al. (2017). Alternating sequence controlled copolymer synthesis of α-hydroxy acids via syndioselective ring-opening polymerization of *O*-carboxyanhydrides using zirconium/hafnium alkoxide initiators. *J. Am. Chem. Soc.* 139 (31): 10723–10732.

27 Wang, R.B., Zhang, J.W., Yin, Q. et al. (2016). Controlled ring-opening polymerization of *O*-carboxyanhydrides using a β-diiminate zinc catalyst. *Angew. Chem. Int. Edit.* 55 (42): 13010–13014.

28 Jia, F., Chen, X.Y., Zheng, Y. et al. (2015). One-pot atom-efficient synthesis of bio-renewable polyesters and cyclic carbonates through tandem catalysis. *Chem. Commun.* 51 (40): 8504–8507.

29 Yin, Q., Tong, R., Xu, Y.X. et al. (2013). Drug-initiated ring-opening polymerization of *O*-carboxyanhydrides for the preparation of anticancer drug-poly (*O*-carboxyanhydride) nanoconjugates. *Biomacromolecules* 14 (3): 920–929.

30 du Boullay, O.T., Bonduelle, C., Martin-Vaca, B., and Bourissou, D. (2008). Functionalized polyesters from organocatalyzed ROP of gluOCA, the *O*-carboxyanhydride derived from glutamic acid. *Chem. Commun.* 15: 1786–1788.

31 Bonduelle, C., Martín-Vaca, B., Cossío, F.P., and Bourissou, D. (2008). Monomer versus alcohol activation in the 4-dimethylaminopyridine-catalyzed ring-opening polymerization of lactide and lactic *O*-carboxylic anhydride. *Chem-Eur J* 14 (17): 5304–5312.

32 Feng, Q.Y., Zhong, Y.L., Xie, L.H., and Tong, R. (2017). Recent advances in ring-opening polymerization of *O*-carboxyanhydrides. *Synlett* 28 (15): 1857–1866.

33 Wang, Y., Li, M., Wang, S. et al. (2021). *S*-carboxyanhydrides: ultrafast and selective ring-opening polymerizations towards well-defined functionalized polythioesters. *Angew. Chem. Int. Ed. Engl.* 60 (19): 10798–10805.

34 Zhu, Y. and Tao, Y. (2024). Stereoselective ring-opening polymerization of *S*-carboxyanhydrides using salen aluminum catalysts: a route to high-isotactic functionalized polythioesters. *Angew. Chem. Int. Ed.* 63 (9).

35 Kricheldorf, H.R. and Bösinger, K. (1973). Über die polymerisation von 2,5-dioxo-1,3-oxathiolan und 2,4-dioxo-1,3-dithiolan. *Die Makromolekulare Chemie* 173 (1): 67–80.

36 Yu, Y., Zou, J., and Cheng, C. (2014). Synthesis and biomedical applications of functional poly(α-hydroxyl acid)s. *Polym. Chem.* 5 (20): 5854–5872.

37 du Boullay, O.T., Marchal, E., Martin-Vaca, B. et al. (2006). An activated equivalent of lactide toward organocatalytic ring-opening polymerization. *J. Am. Chem. Soc.* 128 (51): 16442–16443.

38 Vaca, B.M. and Bourissou, D. (2015). *O*-carboxyanhydrides: useful tools for the preparation of well-defined functionalized polyesters. *ACS Macro Lett.* 4 (7): 792–798.

39 Yin, Q., Yin, L.C., Wang, H., and Cheng, J.J. (2015). Synthesis and biomedical applications of functional poly(α-hydroxy acids) via ring-opening polymerization of *O*-carboxyanhydrides. *Acc. Chem. Res.* 48 (7): 1777–1787.

40 Davies, W.H. (1951). Anhydrocarboxy-derivatives of hydroxy-acids and mercapto-acids. *J. Chem. Soc.* 1357–1359.

41 Vandenbossche, C.P., de Croos, P., Singh, S.P. et al. (2010). Formation of (*S*)-5-cyclohexyl-5-phenyl-1,3-dioxolane-2,4-dione: a key intermediate in the synthesis of (*S*)-oxybutynin hydrochloride. *Org. Process. Res. Dev.* 14 (4): 921–925.

42 Xia, H.D., Kan, S.L., Li, Z.J. et al. (2014). *N*-heterocyclic carbenes as organocatalysts in controlled/living ring-opening polymerization of *O*-carboxyanhydrides derived from L-lactic acid and L-mandelic acid. *J. Polym. Sci., Part A: Polym. Chem.* 52 (16): 2306–2315.

43 Buchard, A., Carbery, D.R., Davidson, M.G. et al. (2014). Preparation of stereoregular isotactic poly(mandelic acid) through organocatalytic ring-opening polymerization of a cyclic *O*-carboxyanhydride. *Angew. Chem. Int. Edit.* 53 (50): 13858–13861.

44 Liang, J., Yin, T., Han, S., and Yang, J. (2020). Synthesis of macrocyclic poly(α-hydroxyl acids) via DABCO-mediated ROP of *O*-carboxylanhydrides derived from L-phenylalanine even in the presence of an alcohol. *Polym. Chem.* 11 (43): 6944–6952.

45 Lu, Y.B., Yin, L.C., Zhang, Y.F. et al. (2012). Synthesis of water-soluble poly(α-hydroxy acids) from living ring-opening polymerization of *O*-benzyl-L-serine carboxyanhydrides. *ACS Macro Lett.* 1 (4): 441–444.

46 Pounder, R.J., Fox, D.J., Barker, I.A. et al. (2011). Ring-opening polymerization of an *O*-carboxyanhydride monomer derived from L-malic acid. *Polym. Chem.* 2 (10): 2204–2212.

47 Zhang, Z.H., Yin, L.C., Xu, Y.X. et al. (2012). Facile functionalization of polyesters through thiol-yne chemistry for the design of degradable, cell-penetrating and gene delivery dual-functional agents. *Biomacromolecules* 13 (11): 3456–3462.

48 Zhang, Z.H., Yin, L.C., Tu, C.L. et al. (2013). Redox-responsive, core cross-linked polyester micelles. *ACS Macro Lett.* 2 (1): 40–44.

49 Chen, X.Y., Lai, H.W., Xiao, C.S. et al. (2014). New bio-renewable polyester with rich side amino groups from L-lysine via controlled ring-opening polymerization. *Polym. Chem.* 5 (22): 6495–6502.

50 Liu, S.S., Zhang, X.J., Li, M.S. et al. (2017). Precision synthesis of sustainable thermoplastic elastomers from lysine-derived monomers. *J. Polym. Sci., Part A: Polym. Chem.* 55 (2): 349–355.

51 Tang, L. and Deng, L. (2002). Dynamic kinetic resolution via dual-function catalysis of modified cinchona alkaloids: asymmetric synthesis of α-hydroxy carboxylic acids. *J. Am. Chem. Soc.* 124 (12): 2870–2871.

52 Smith, I.J. and Tighe, B.J. (1976). Studies in ring-opening polymerization. 4. Thermal polymerization of phenyl-substituted 1,3-dioxolan-2,4-diones. *J. Polym. Sci., Part A: Polym. Chem.* 14 (4): 949–960.

53 Smith, I.J. and Tighe, B.J. (1973). Phenyl and pentafluorophenyl substituted 1,3-dioxolan-2,4-diones – novel group of heterocyclic-monomers. *Chem. Ind. - London* 14: 695–696.

54 Tighe, B.J. (1969). Ring opening reactions of 5,5-dimethyl-1,3-dioxolan-2,4-dione. *Chem. Ind. - London* 51: 1837.

55 Bexis, P., De Winter, J., Coulembier, O., and Dove, A.P. (2017). Isotactic degradable polyesters derived from *O*-carboxyanhydrides of L-lactic and L-malic acid using a single organocatalyst/initiator system. *Eur. Polym. J.* 95: 660–670.

56 Worch, J.C., Prydderch, H., Jimaja, S. et al. (2019). Stereochemical enhancement of polymer properties. *Nat. Rev. Chem.* 3 (9): 514–535.

57 Thomas, C.M. (2010). Stereocontrolled ring-opening polymerization of cyclic esters: synthesis of new polyester microstructures. *Chem. Soc. Rev.* 39 (1): 165–173.

58 Ligny, R., Hänninen, M.M., Guillaume, S.M., and Carpentier, J.F. (2018). Steric vs. electronic stereocontrol in syndio- or iso-selective ROP of functional chiral β-lactones mediated by achiral yttrium-bisphenolate complexes. *Chem. Commun.* 54 (58): 8024–8031.

59 Li, H., Shakaroun, R.M., Guillaume, S.M., and Carpentier, J.F. (2020). Recent advances in metal-mediated stereoselective ring-opening polymerization of functional cyclic esters towards well-defined poly(hydroxy acid)s: from stereoselectivity to sequence-control. *Chem. Eur. J.* 26 (1): 128–138.

60 Li, M.S., Tao, Y., Tang, J.D. et al. (2019). Synergetic organocatalysis for eliminating epimerization in ring opening polymerizations enables synthesis of stereoregular isotactic polyester. *J. Am. Chem. Soc.* 141 (1): 281–289.

61 Garlotta, D. (2001). A literature review of poly(lactic acid). *J. Polym. Environ.* 9 (2): 63–84.

62 Feng, Q.Y. and Tong, R. (2017). Controlled photoredox ring-opening polymerization of *O*-carboxyanhydrides. *J. Am. Chem. Soc.* 139 (17): 6177–6182.

63 Li, M., Zhang, S., Zhang, X. et al. (2021). Unimolecular anion-binding catalysts for selective ring-opening polymerization of *O*-carboxyanhydrides. *Angew. Chem. Int. Ed.* 60 (11): 6003–6012.

64 Feng, Q.Y., Yang, L., Zhong, Y.L. et al. (2018). Stereoselective photoredox ring-opening polymerization of *O*-carboxyanhydrides. *Nat. Commun.* 9: 1559.

7

Organocatalyzed Stereoselective Ring-Opening (Co)Polymerization

Shaofeng Liu[1] *and Daniel Taton*[2]

[1]*Qingdao University of Science and Technology, Key Laboratory of Biobased Polymer Materials, College of Polymer Science and Engineering, No. 53 Zhengzhou Road, Qingdao, 266042, China*
[2]*Univ. Bordeaux, CNRS, Bordeaux INP, LCPO, UMR 5629, F-33600, Pessac, France*

7.1 Introduction

Controlling the stereochemistry during polymer synthesis is a powerful means to tune the thermomechanical properties of polymeric materials [1]. Stereoselective polymerization of propylene, a prochiral monomer, that produces isotactic polypropylene from Ziegler-Natta or metallocene catalysts is an emblematic success story of the last century of polymer science [2]. An alternative method to control polymer tacticity is by polymerizing chiral monomers, a prototypal example of which is the ring-opening polymerization (ROP) of lactide (LA) [3–6]. As LA possesses two chiral centers, the ROP of the three different stereoisomers and racemic monomer can produce a range of microstructures, including atactic, syndiotactic, isotactic and heterotactic PLAs with different physico-chemical properties (Scheme 7.1) [7]. For example, atactic PLA is an amorphous material, whereas isotactic PLLA and PDLA are semicrystalline materials with a melting temperature (T_m) around 160–180 °C and high mechanical strength. Moreover, physical blending of PLLA and PDLA in a molar ratio of 1 : 1 can afford a stereocomplex by co-crystallization of both polymeric enantiomers, increasing the T_m value up to 230 °C [8, 9].

Alternatively, instead of blending the two homopolymers of opposite configuration, such PLA stereocomplexes can be directly achieved via stereoselective ROP of racemic lactide (*rac*-LA), potentially representing a less costly synthetic route and increasing the value of the racemic monomer [4, 5]. Since the first report in 1996 by Spassky et al. regarding the stereoselective ROP of *rac*-LA using chiral Salen-Al complexes [10], a tremendous amount of work aimed at developing metal-based catalysts, providing both high catalytic activity and high stereoselectivity, has been reported. However, metallic residues can contaminate the final PLA products, impeding their applications in food packing, biomedical, and microelectronic fields, owing to their often-proven cytotoxicity. In addition, organometallic catalysts often

Organocatalysts in Polymer Chemistry: Synthesis and Applications, First Edition. Edited by Zhibo Li.

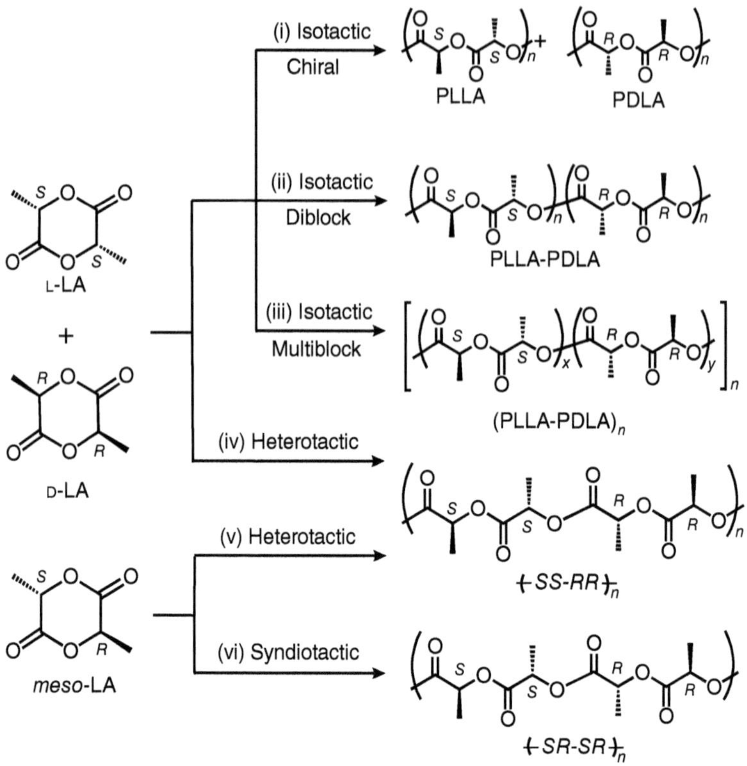

Scheme 7.1 Lactide stereochemistry and PLA microstructures from stereo-controlled ROP of L-lactide (L-LA), D-lactide (D-LA), racemic lactide (*rac*-LA) or meso lactide (*meso*-LA).

require demanding or tedious design and storage. In contrast, organic catalysts have emerged as promising alternatives to access metal-free polymers, in addition of being easily accessible, if not commercially available. Since pioneer work in 2001 by Hedrick and coworkers reporting the organocatalyzed ROP of lactide using 4-dimethylaminopyridine (DMAP) [11], this field has witnessed considerable booming. Organic catalysis of polymerization, in its general sense, has become today a powerful strategy in polymer chemistry for achieving a wide array of metal-free polymer structures. The topic has been the focus of several reviews and book chapters to which the reader can refer [12–18]. This chapter discusses the recent advances pertaining to the organocatalyzed stereoselective ROP, with a special emphasis on the structures and catalytic performances of the various organocatalysts reported, and the underlying mechanisms.

7.2 Mechanisms in the Organocatalyzed Stereoselective ROP

From a mechanistic standpoint, stereoselective ROP of *rac*-LA can be mediated following either the enantiomorphic site control (ESC) mechanism or the chain end control (CEC) mechanism [19]. The ESC mechanism (Scheme 7.2i) usually operates with a chiral catalyst, which determines the chirality of the next monomer to be inserted and involves differentiation between L-LA and D-LA by kinetic resolution. Effective ROP of *rac*-LA by the ESC mechanism thus produces isotactic PLA chains with almost all monomer units having the same configuration, e.g. PLLA, leaving the other enantiomer (D-LA) unreacted (Scheme 7.2i). In ROP reactions involving the CEC mechanism, control of chirality is due to the propagating PLA chain end that, in the transition state of the next monomer incorporation step, drives the insertion of the next monomer unit. In the latter case, the organic catalyst is usually nonchiral but sterically hindered to reinforce a chiral environment at the chain end. Therefore, the ROP reaction occurs without enantiomeric differentiation of *rac*-LA (Scheme 7.2ii, <A>). The monomer with the same chirality as that of the last inserted monomer is then preferentially enchained by the CEC mechanism (Scheme 7.2ii, <B>). Occasionally, a stereoerror can be created during ROP in a PLA growing chain, i.e. the latter incidentally incorporates a monomer unit of opposite configuration to the previous one (Scheme 7.2ii, <C>). If growing via the CEC mechanism, stereoblocks $(PLLA–PDLA)_n$ or multi-stereoblocks (Scheme 7.2ii) can be generated, owing to the occurrence of successive inversions of configuration from one stereo-sequence to another. Conversely, PLA chains grown exclusively by an ESC mechanism will incorporate single and isolated stereoerrors, e.g. –*RRRRSSRRR*– or –*SSSSRRSSSS*–. Depending on the organocatalytic system employed, both ESC and CEC mechanisms may concomitantly and concurrently intervene, which eventually compromises stereocontrol of the ROP process [20, 21]. The following sections highlight the various classes of organic catalysts enabling stereoselective ROP and also discuss the related mechanisms.

7.3 Organic Acids for Stereoselective ROP of *rac*-LA

ROP reactions catalyzed by organic acids generally use alcohols as initiators and proceed via an electrophilic activation of the monomer [17, 22]. In molecular chemistry, chiral Brønsted acids have been widely used for a variety of enantioselective organic transformations [23, 24]. Among them, binaphthol (BINOL)-derived monophosphoric acids (**1**) represent an important family of such organocatalysts. In the context of ROP, two essential features of these compounds, namely, the presence of both a hydrogen bond donor group and a chiral motif, have made them potent organocatalysts for the stereoselective ROP of *rac*-LA. Binaphthol-containing compounds thus proved efficient for the enantioselective ROP of *rac*-LA, providing high stereoselectivity in toluene in presence of 3-phenyl-1-propanol (PPA) as the initiator (Scheme 7.3a) [25]. The electronic properties of the substituent group (G) was found to have a more profound impact on the stereoselectivity. With an electronic-withdrawing group such as C_6F_5–, (*R*)-**1a** led to a selectivity ratio, $s = k_D/k_L$, of 28.3 and an enantiomeric excess (*ee*) of unreacted LA of 80.6%, for a conversion of *rac*-LA = 49% (M_n = 3730 g mol^{-1}; $Đ$ = 1.13). As expected, the enantioselectivity decreased upon increasing the reaction temperature, from 75 to 90 °C. What was unexpected was that decrease in stereocontrol was noted by

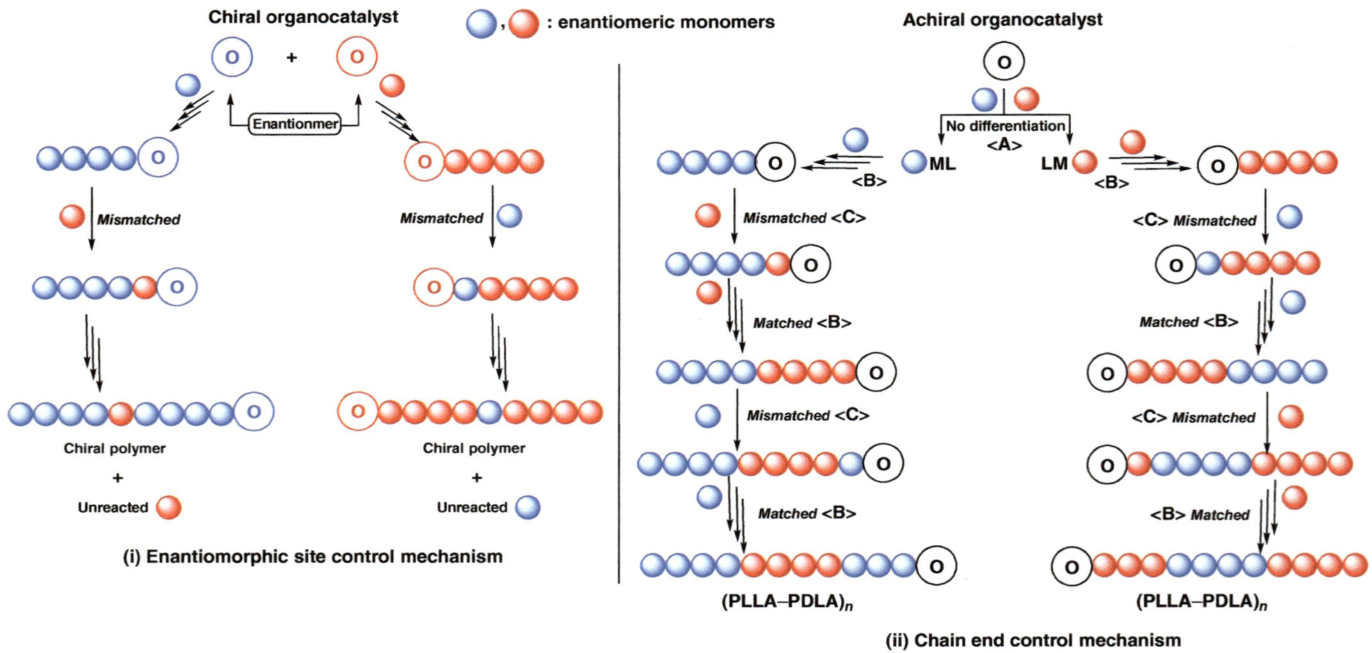

Scheme 7.2 Stereoselective ROP of *rac*-LA via the (i) enantiomorphic site control (ESC) mechanism and the (ii) chain end control (CEC) mechanism.

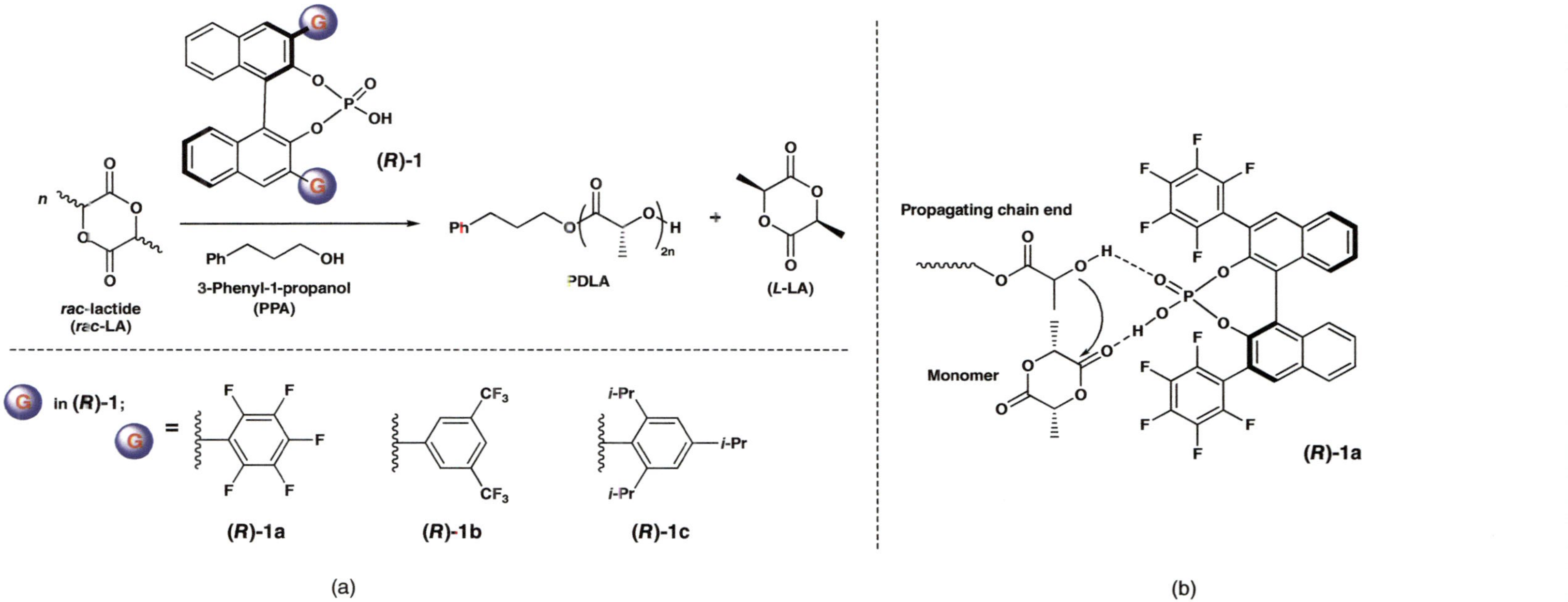

Scheme 7.3 Enantioselective ROP of *rac*-LA catalyzed by (a) binaphthol (BINOL)-derived monophosphoric acids (**1**) and (b) proposed reaction mechanism of dual activation. Source: Makiguchi et al. [25]/Royal Society of Chemistry.

decreasing the temperature from 75 to 60 °C. This was likely due to the fact that the ROP reaction turned heterogeneous at 60 °C, also leading to a loss of control over the PLA molar masses. The high enantioselectivity of binaphthol (BINOL)-derived monophosphoric acids (**1**) was attributed to a dual activation mechanism of both the monomer and the propagating chain-end. As illustrated in Scheme 7.3b, the phosphoric acid site would activate the monomer, while the phosphonate moiety (conjugate base of the acid) would serve to activate the growing PLA chain in its alcohol form.

7.4 Organic Bases for Stereoselective ROP of *rac*-LA

7.4.1 *N*-Heterocyclic Carbenes (NHCs)

N-heterocyclic carbenes (NHCs) are neutral cyclic molecules featuring a divalent carbon atom surrounded by six-electron valency. NHCs mainly behave as strong nucleophiles due to their strong σ-donating properties. As such, NHCs have been extensively used as ligands of transition metals, and as organic catalysts as well for molecular and polymer synthesis [26], obviously including for ROP of LA. Variation of the substitution pattern of NHCs allows modulating their steric and electronic properties, hence their catalytic activity and selectivity for ROP. In a preliminary work, Hillmyer and coworkers found that NHC **2**, referred to IMes (Scheme 7.4), led to slightly isotactic-enriched PLA (P_m = 0.59) in CH_2Cl_2 at 25 °C [27]. By lowering the temperature to −20 °C, an increase in isotacticity was noted (P_m = 0.75). The resulting PLA (M_n = 15.9 kg mol^{-1}; $Đ$ = 1.26), produced in 71% yield from *rac*-LA, showed a T_m of 153 °C by differential scanning calorimetry (DSC), consistent with the presence of a stereo-sequences and formation of a semicrystalline material.

Hedrick and coworkers later studied a number of unsaturated chiral and achiral NHCs with relatively bulky substituents for the ROP of *rac*-LA at different temperatures (Scheme 7.4, **3**, (*R*,*R*)-**4**, and **5**) [28]. The authors attempted to correlate the structure of these NHCs to their catalytic activity. The use of **3** as catalyst provided PLAs with desired degrees of polymerization (DP), narrow distribution of molar masses, and high catalytic activities (TOF = 95 min^{-1}) at room temperature. For instance, **3** gave a PLA in just 1 minute at 95% conversion. By carrying out the reaction at lower temperatures, a significant increase in stereocontrol was noted (Scheme 7.4a), with P_m value increasing from 0.59 at room temperature to 0.90 at −70 °C. ROP reactions involving *meso*-LA proved less active than those with *rac*-LA, the *meso* isomer being also less soluble at low temperature than the *racemic* mixture. Yet, NHC **3** was found particularly efficient for the stereoselective ROP of *meso*-LA, affording an heterotactic PLA with P_m of 0.83 at −40 °C (Scheme 7.4b). Unexpectedly, the sterically hindered and saturated chiral NHC **5** showed no activity toward the ROP of *rac*-LA carried out in presence of an initiating alcohol, even at elevated temperatures. The stereoselective ROP of *rac*-LA and *meso*-LA from both achiral (**3**) and chiral ((*R*,*R*)-**4**) NHCs was a strong indication of the

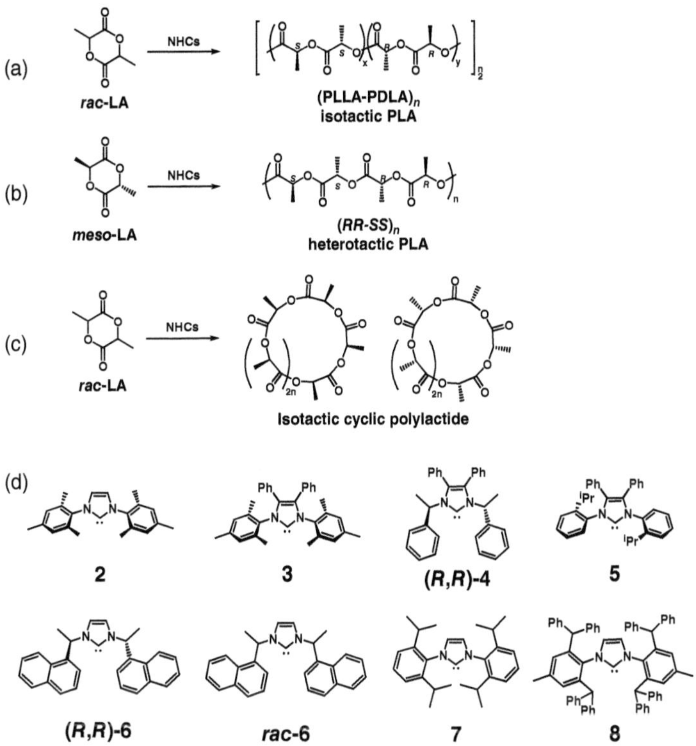

Scheme 7.4 Stereoselective ROP of *rac*-LA and *meso*-LA by various chiral and achiral NHC catalysts. Source: Jensen et al. [27], Dove et al. [28], Li et al. [29], Si et al. [30].

occurrence of chain growth by the CEC mechanism in this case. The role of chiral motifs indeed appeared to be inconsequential and seemed to have no effect on the stereoselectivity, and only steric hindrance appeared to influence the polymer tacticity.

Bulky chiral NHCs incorporating naphthyl groups on both nitrogen atoms (Scheme 7.4, **6**) were designed as catalysts for the ROP of *rac*-LA [29]. Investigations into the reaction conditions showed that low temperature (−78 °C) and selection of a suitable solvent (toluene or DCM) could favor some stereocontrol. An isotactic-rich PLA, very likely arranging into stereoblocks (P_m = 0.81), obtained with high catalytic activity (conv. = 98% in 30 min) and narrow dispersity ($Đ$ = 1.05) was indeed achieved from the chiral organic catalyst (*R*,*R*)-**6** at −78 °C in DCM. A comparative study between (*R*,*R*)-**6** and *rac*-**6** revealed that stereoselectivity originated from chain-growth taking place by the CEC mechanism. The steric hindrance provided by the substituents on the nitrogen atoms was presumably the key factor favoring the stereoselective ROP of *rac*-LA.

NHCs **2**, **7** and **8** (Scheme 7.4) were also used for the stereoselective zwitterionic ROP -referred as to ZROP- of *rac*-LA, proceeding in absence of an alcohol initiator and in this case producing isotactic cyclic PLA at low temperatures in tetrahydrofuran (Scheme 7.4c) [30]. By no surprise, the stereoselectivity of the ZROP process was found to dramatically increase with a decrease in the reaction temperature. At room temperature, the commercially available NHC **7** catalyzed the ZROP of *rac*-LA, forming basically atactic PLA (P_m = 0.57), indicative of a poor stereocontrol. At −30 °C, the P_m value reached 0.78 (T_m = 180 °C), and the greatest isoselectivity was achieved at −70 °C, giving a P_m value of 0.91 and enabling the corresponding PLA (M_n = 5000 g mol^{-1}; Đ = 1.11) to crystallize with a T_m as high as 193 °C. This excellent stereocontrol could be rationalized by PLA chains growing by the CEC mechanism during ZROP at low temperature. The cyclic topology of the as-obtained PLA was supported by ^{1}H NMR and MALDI-TOF MS analyses.

7.4.2 Amidines and Guanidines

Amidines and guanidines are organobases generally behaving as powerful deprotonating agents of the initiator, then of the chain-end, during the ROP of cyclic monomers. A representative example is the 1,8-diazabicyclo[5.4.0]-undec-7-ene DBU (Scheme 7.5, **9**) amidine, which activates the initiating alcohol. In contrast, the 1,5,7-triazabicyclo[4.4.0]dec-5-ene TBD (Scheme 7.5, **10**) guanidine acts as a dual catalyst, i.e. activating both the monomer and the alcohol in a cooperative manner [31, 32]. Interestingly, the ROP of *rac*-lactide by TBD showed a slight isotactic enhancement with a P_m value of 0.58 at room temperature. No significant difference in stereoselectivity was noted between TBD, DBU, and *N*-methyl-TBD MTBD (Scheme 7.5, **11**; P_m value of 0.58–0.60) [32]. Decreasing the temperature of the reaction enabled to achieve much better stereocontrol for the ROP of *rac*-LA and *meso*-LA. Thus, TBD was found to polymerize *rac*-LA leading to a highly isotactic PLA, with a P_m value of 0.88, while *meso*-LA yielded a heterotactic PLA (P_m ~ 0.8) at −75 °C [33]. A chiral guanidine catalyst, namely 4,8-diphenyl-1,5,7-triazabicyclo[4.4.0]dec-5-ene (**12**, Scheme 7.5), was also designed for the ROP of *rac*-LA [34], which provided a high activity and produced PLAs with controlled molecular weight, low dispersity, and high end group fidelity. A relatively modest isoselectivity (P_m = 0.68) was however achieved when polymerizing 400 equiv. of *rac*-LA at room temperature, following a CEC mechanism, according to the authors.

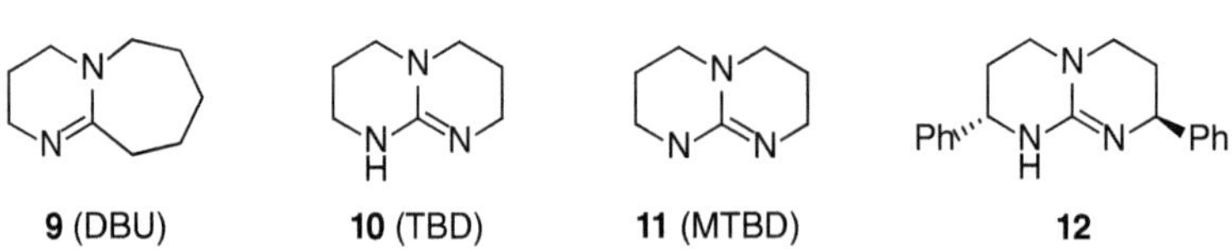

Scheme 7.5 Typical amidines and guanidines used for the stereoselective ROP of *rac*-LA. Source: Pratt et al. [31], Lohmeijer et al. [32], Moins et al. [33].

7.4.3 Phosphazenes

Phosphazenes are super strong organic Brønsted bases showing, in most cases, a poor nucleophilicity [14]. Owing to their highly basic and non-nucleophilic nature, phosphazene bases (PBs) can easily convert protic reagents into nucleophilic initiating species through deprotonation, which could further initiate fast and controlled ROP reactions. The steric hindrance of many PBs is of particular interest, as it can restrict the mobility of reactive species, thus exerting a form of stereocontrol. The absence of chiral motifs in the PB structure suggests a CEC-type mechanism. PBs, such as 2-*tert*-butylimino-2-diethylamino-1,3-dimethylperhydro-1,3,2-diazaphosphorine BEMP (Scheme 7.6, **13**) or *tert*-butylimino-tris(dimethylamino)phosphorane P_1-*t*-Bu (Scheme 7.6, **14**), were found active organocatalysts for the ROP of *rac*-LA or L-LA, in the presence of PyBuOH or BnOH as the initiator [35]. Polymerization carried out in toluene at room temperature for 66 h achieved a conversion of 97% of *rac*-LA, affording an isotactic-enriched PLA (P_m = 0.70; M_n = 15,000 g mol^{-1}; Đ = 1.05). Results obtained by NMR confirmed that stereocontrol was due to the occurrence of the CEC mechanism. 1-*tert*-Butyl-2,2,4,4,4-pentakis (dimethylamino)-2λ^5, 4λ^5-catenadi (phosphazene) P_2-*t*-Bu (Scheme 7.6, **15**) with a higher basicity exhibited an even higher catalytic activity, the ROP of *rac*-LA with 1-pyrenebutanol as initiator at room temperature producing a PLA with M_n of 13,300 g mol^{-1} and Đ of 1.06 at 85% conversion [36]. Use of P_2-*t*-Bu enabled the ROP of *rac*-LA to be conducted at −75 °C with the resulting PLA (M_n = 27,200 g mol^{-1}; Đ = 1.11) being obtained with a conversion of 99% within 3 h. According to the Markovian statistics and analyses by NMR, the P_m values of these PLAs increased, from 0.72 at room temperature, up to 0.95 at −75 °C. Thermal analysis further revealed that the latter PLA exhibited a high degree of crystallinity, with a glass transition temperature (T_g) of 62 °C and a T_m value as high as 201 °C, which was considerably higher than that of a PLLA (T_m = 163 °C), strongly suggesting the formation of stereocomplexes, as discussed above.

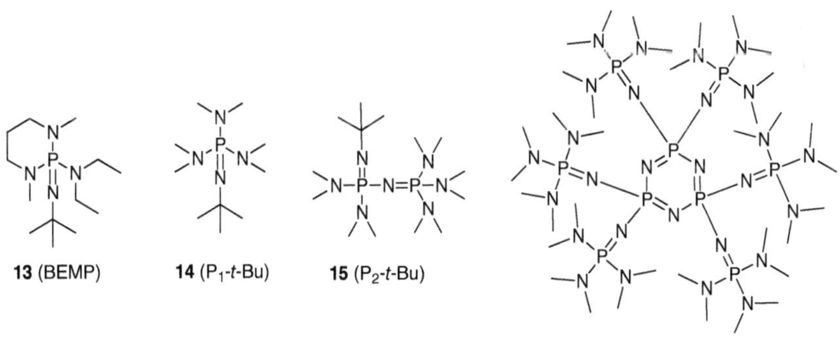

13 (BEMP) **14** (P_1-*t*-Bu) **15** (P_2-*t*-Bu) **16** (CTPB)

Scheme 7.6 Phosphazene bases used for the stereoselective ROP of *rac*-LA. Source: Zhang et al. [35, 36], Liu et al. [37].

The cyclic trimeric PB CTPB (**16**, Scheme 7.6) was reported to efficiently catalyze the stereoselective ROP of *rac*-LA, leading to isotactic PLAs arranged in stereoblocks [37]. The CTPB base has a spherical structure with a larger diameter (approx. 1.4 nm) than P_2-*t*-Bu. It is therefore reasonable to postulate that CTPB can induce a stereoselective ROP process *via* a CEC mechanism. Investigations into the ROP parameters showed that the tacticity of the as-obtained PLA did depend on the temperatures and solvents. Using BnOH as the initiator, *rac*-LA was quantitatively converted into an atactic PLA (P_m = 0.52) within 10 min, at room temperature (M_n = 12,500 g mol^{-1}; $Đ$ = 1.42). In contrast, an isotactic stereoblock-type PLA (P_m = 0.93, T_m = 184 °C) with M_n of 11,100 g mol^{-1} and $Đ$ of 1.09 was achieved with 92% conversion within 3 h, at −75 °C. The kinetic studies revealed a faster rate for ROP of L-LA, as compared to *rac*-LA under otherwise the same conditions, supporting the occurrence of the CEC mechanism during chain-growth.

7.5 Dual Organocatalysts

Dual organocatalysts consist of either a single (mono-component) molecule or bicomponent molecular co-catalysts. Typical functional groups used for the purpose of organic catalysis of polymerization include thioureas, ureas, squaramides or catalytic pairs combining a protic acid and a base. As already mentioned, dual organocatalysts enable a cooperative mechanism to take place during the ROP process, and incorporate both a hydrogen bond donor (HBD) motif for monomer activation, and a hydrogen bond acceptor (HBA) motif for initiator (initiation step) or chain-end (propagation steps) activation [12, 13, 15]. Dual organocatalysis in the context of ROP was pioneered by Hedrick and coworkers [38], and has become a powerful tool for engineering a wide variety of polymeric materials.

7.5.1 Mono-Component Bifunctional Organocatalysts

The mono-component Takemoto's catalyst incorporating both a thiourea and a tertiary amino group typically operates by a bifunctional cooperative mechanism [39]. In this case, the amino moiety activates the alcohol initiator, while the thiourea motif activates the heterocyclic monomer. Various tethered bifunctional thiourea-amines featuring different alkyl linkages were initially reported for the ROP *rac*-LA (**17**, Scheme 7.7). Despite excellent control over the molecular weight (M_n in the range 16,700–21,100 g mol^{-1}) and dispersity ($Đ$ < 1.10) was noted, relatively modest stereoselectivities could be achieved at room temperature [40]. Use of the enantiomerically pure organic catalyst (*R*,*R*)-**17a** did not significantly change the PLA stereochemistry, when compared with the racemic version, *rac*-**17a** (P_m = 0.76 vs. P_m = 0.67), which was in accordance with the prevalence of CEC mechanism. Later on, however, both enantiomeric versions of the Takemoto's catalyst, namely, (*S*,*S*)-**17a** and (*R*,*R*)-**17a**, were re-investigated for the ROP of *rac*-LA either in CH_2Cl_2 or in toluene. Benzyl alcohol (BnOH) served as initiator in a temperature range of 25–85 °C [20]. Intriguingly, DSC analysis of the (*S*,*S*)-**17a**-derived PLA sample revealed a melting transition at T_m = 152 °C, thus contrasting what was expected from a PLA with an anticipated P_m = 0.76 based on previous report [40]. However, using an ESC statistical model, a P_m = 0.87 was calculated, much more in line with the observation of crystallinity of the resultant PLA. In the case of (*R*,*R*)-**17a**, a P_m value as high as 0.90 could be achieved at room temperature in toluene, that is, a less polar than DCM and which very likely increased the probability for forming

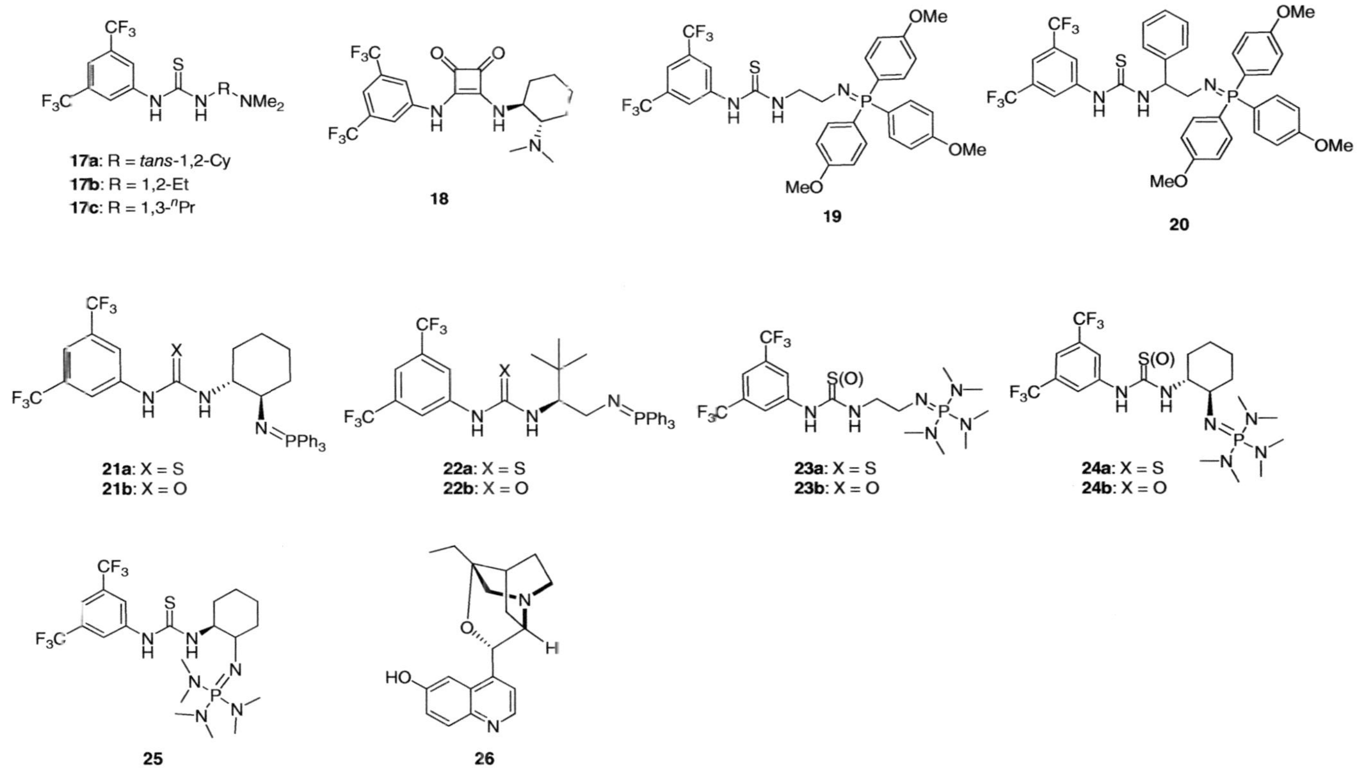

Scheme 7.7 Mono-component bifunctional organocatalysts for the stereoselective ROP of *rac*-LA [20, 40–45].

H-bonds. Combined kinetic investigations and stereochemical analysis of PLAs by homonuclear decoupled ^{1}H NMR spectroscopy suggested that both the ESC and the CEC mechanisms eventually occurred during the ROP of *rac*-LA by these chiral Takemoto's organic catalysts.

Similar observations were reported for the bifunctional H-bonding organic catalyst possessing both a basic tertiary amino group and a squaramide motif (**18**, Scheme 7.7) [41]. This compound (**18**) was found to catalyze the ROP of *rac*-LA in DCM, with 51% monomer conversion reached in 24 h, affording a structurally well-controlled PLA (M_n = 5900 g mol^{-1}, $Đ$ = 1.08), the DSC analysis of which revealed the formation of a semi-crystalline material with a T_m value of 153 °C. The corresponding P_m values were found equal to 0.76 and 0.88, as determined according to the CEC and the ESC mechanism, respectively. The rather good stereocontrol was ascribed to a synergistic effect due to the close vicinity of the squaramide and the amine motifs in **18**.

As a means to increase the catalytic activity in ROP, bifunctional organic catalysts **19** and **20** (Scheme 7.7), associating both a thiourea moiety and a triaryl iminophosphorane group, were conceived [42]. The ROP of *rac*-LA using **19** at room temperature gave a PLA with a P_m value of 0.64, indicating a slight isotactic enrichment. Conducting the polymerization at −78 °C slightly increased the P_m value to 0.74, but a conversion of only 36% was reached within 4 h. Use of catalyst **20** containing a bulky phenyl-substituent did not significantly affect the tacticity.

Chiral and achiral urea-iminophosphoranes **21b** and **22b** (Scheme 7.7) were designed as mono-component bifunctional catalysts for the ROP of *rac*-LA, which was carried out at various temperatures. A comparison of these organic catalysts with thiourea-based analogues was also made (Scheme 7.7, **21a** and **22a**) [43]. ROP reactions proved efficient, affording PLAs with controlled molecular weights, narrow distribution and well defined end groups. Under similar conditions, urea-derivatives **21b** and **22b** were found to be superior to the thiourea-derivatives **21a** and **22a**, in terms of catalytic activity and stereoselectivity. The highest P_m value (P_m = 0.80) was obtained using **21b** at 25 °C in toluene. The mechanistic study led the authors to conclude for a stereoselective ROP process occurring by the CEC mechanism.

The strength of the basic moiety in such mono-component bifunctional catalysts can be modulated by introduction of phosphazene motifs. A series of bifunctional phosphazene-thiourea/urea organic catalysts (**23–25**, Scheme 7.7) was thus designed and applied for the ROP of various cyclic monomers [44]. The stereoselective ROP of *rac*-LA could be achieved using **24b** possessing a rigid cyclohexyl linker, reaching 83% conversion of monomer within 14 h, at room temperature in CH_2Cl_2, yielding a PLA sample with a P_m value of 0.79 (M_n = 11,500 g mol^{-1}, $Đ$ = 1.11).

β-Isocupreidine (**26**, Scheme 7.7) consisting of both a chiral nucleophilic amino group and an electrophilic hydroxy moiety was reported as the first organocatalyst to perform the kinetic resolution ROP of *rac*-LA [45]. In presence of BnOH as the initiator, **26** proved highly efficient, achieving conversions without noticeable epimerization or transesterification side reactions, thereby producing a PLA with a controlled M_n and low dispersity. In CH_2Cl_2 at 25 °C, **26** thus afforded a crystalline isotactic-rich, stereogradient PLA with a P_m value of 0.75. The sample exhibited multiple melting-transition temperatures, as observed by DSC, which resulted from a partial kinetic resolution polymerization. In this case, L-LA was found to be preferentially incorporated, with a selectivity factor ($s = k_L/k_R$) of 4.4 at 48.4% monomer conversion.

A Lewis pair combining 1,4-diazabicyclo[2.2.2]octane (DABCO) (**27**, Scheme 7.8) and $B(C_6F_5)_3$ (**28**) was reported to catalyze rapid and quantitative epimerization of *meso*-LA into *rac*-LA. The resulting *rac*-LA was next kinetically polymerized into isotactic PLLA and optically resolved D-LA, using a series of bifunctional chiral catalysts (**29–36**, Scheme 7.8) [46]. Among the organocatalysts screened, compound **32** combining three key elements into a single organic molecule, namely, a β-isocupreidine core, a thiourea functionality and a chiral BINAM, showed the highest stereoselectivity. A selectivity factor ($s = k_L/k_R$) as high as 53 was thus determined at 50.6% monomer conversion in *o*-diflurobenzene at 25 °C. Thus, epimerization followed by enantioselective ROP can be combined into a one-pot process for directly transforming *meso*-LA into PLLA and D-LA (Scheme 7.8).

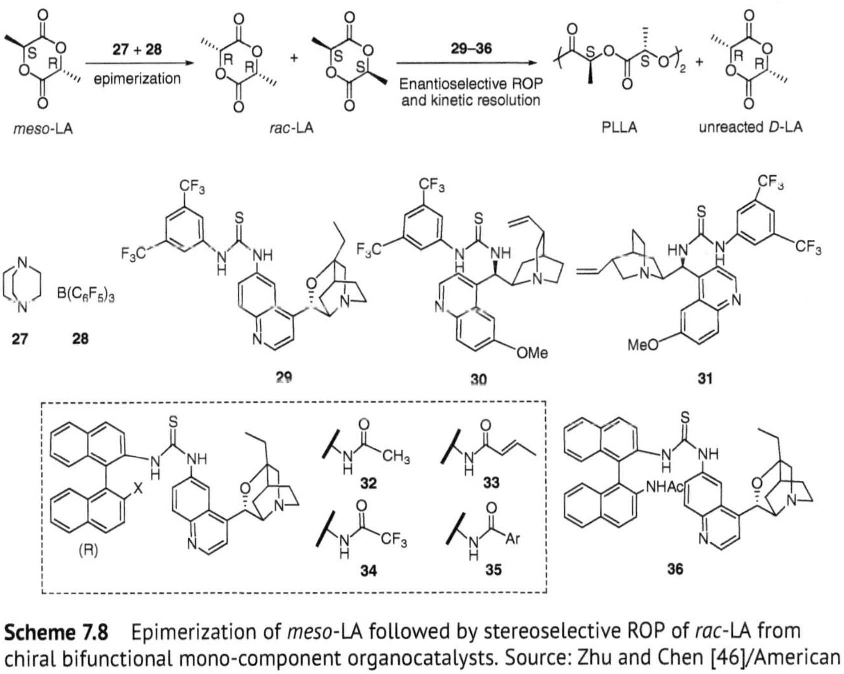

Scheme 7.8 Epimerization of *meso*-LA followed by stereoselective ROP of *rac*-LA from chiral bifunctional mono-component organocatalysts. Source: Zhu and Chen [46]/American Chemical Society.

7.5.2 Bi-Component Organocatalysts

Resorting to bi-component organocatalysts enable to independently tune each component, affording a wide structural diversity to vary both the catalytic performances and the stereocontrol. A typical example of such a strategy is the use of a binary catalytic system based on a thiourea (TU) motif, providing the HBD capability, and an amine as a base. In this case, both the acidity of the TU and the basicity of the amine have a profound impact on the outcomes of the ROP of *rac*-LA [40]. For instance, TU incorporating electron withdrawing substituents on the phenyl ring showed improved activity, due to an increase in acidity of the –NH protons. Among all amines tested, (−)-sparteine (**37**, Scheme 7.9) used in conjunction with TU **39** exhibited both the highest activity and stereoselectivity toward the ROP of *rac*-LA. Complete conversion was thus achieved within 2 h, producing a PLA (M_n = 20,700 g mol^{-1}, $Đ$ = 1.05) with a P_m value of 0.77. Such a two-component organocatalytic system proved to be about 25 times more reactive than the Takemoto thiourea alone. The superior catalytic properties observed with (−)-sparteine was attributed to the rigidity of the backbone with the lone pairs of electrons of the tertiary amines favoring hydrogen bonding. However, as (−)-sparteine has become difficult to acquire through commercial sources, benzyl bispidine (**38**, Scheme 7.9) was proposed as a structural analogue. Compound **38** was thus used in combination with **39** for the ROP of *rac*-LA in $CDCl_3$ at room temperature, producing a predominantly isotactic PLA with a P_m value of 0.74 [22]. DBU (Scheme 7.9, **9**) was also used with a variety of chiral ureas (**40–42**, Scheme 7.9) to catalyze the

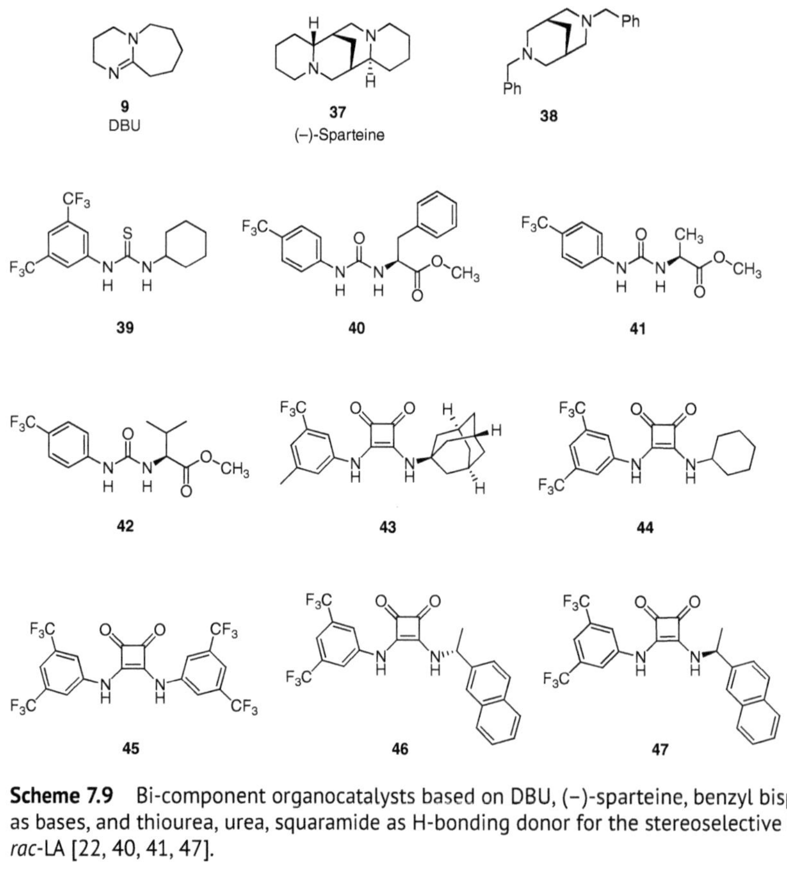

Scheme 7.9 Bi-component organocatalysts based on DBU, (−)-sparteine, benzyl bispidine as bases, and thiourea, urea, squaramide as H-bonding donor for the stereoselective ROP of *rac*-LA [22, 40, 41, 47].

stereoselective ROP of *rac*-LA [47]. A PLA very likely arranged into stereoblocks (P_m up to 0.87) was obtained from DBU/**40** at room temperature. The P_m could be further increased up to 0.90 upon decreasing the temperature to −20 °C, fostering the CEC mechanism.

In addition, squaramides instead of (thio)ureas (**43–47**, Scheme 7.9) were employed in duos with DBU to efficiently promote the ROP of *rac*-LA at room temperature [41]. Owing to the reinforced chiral environment resulting from the sterically hindered squaramide, the peculiar dual catalyst DBU/**43** exhibited the highest stereoselectivity (P_m = 0.88), without transesterification side reactions. The resulting PLAs proved to have controlled M_n values as high as 22,200 g mol^{-1} and narrow distributions ($Đ$ = 1.10–1.24).

A photo-responsive azobenzene-based TU **49** (Scheme 7.10) was designed for triggering a switchable ROP of *rac*-LA [48]. Compound **49** indeed consists of both the *Z*- and the *E*-isomer, which can mutually and easily be transformed by appropriate light irradiation. The *E*-isomer (*E*-**49**, Scheme 7.10) exhibits an open active site of the TU motif, while the thiourea group in its *Z*-isomer form (Scheme 7.10, *Z*-**49**) is blocked by the formation of intramolecular hydrogen bonds with the nitro group. Consequently, the catalytic activity of **49** could be switched between slow and fast states by alternating exposure to UV and ambient light. When used in combination with PMDETA (**48**, Scheme 7.10), the *E*-**49** isomer showed a higher activity during the ROP of *rac*-LA, compared to the *Z*-**49** isomer. However, a PLA with a relatively modest isoselectivity (P_m = 0.73–0.75) was obtained.

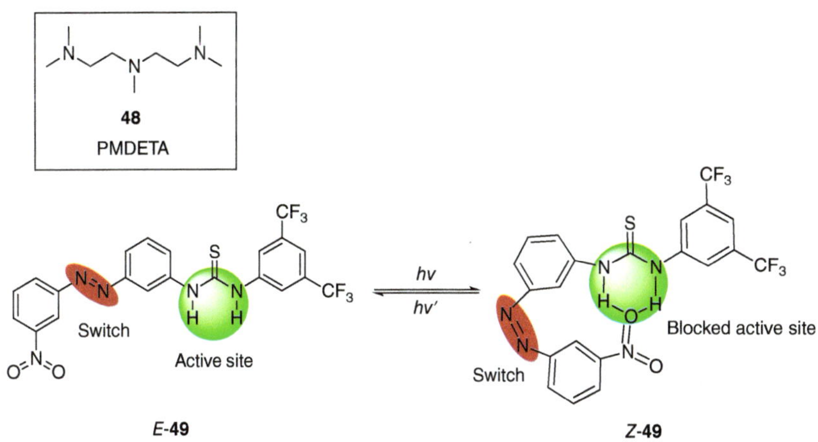

Scheme 7.10 PMDETA/azobenzene-based thiourea bi-component organocatalyst for the photo-switchable stereoselective ROP of *rac*-LA. Source: Dai et al. [48]/Royal Society of Chemistry.

NHCs were also employed as strong bases in conjunction with the commercially available chiral TU **17a**, (Scheme 7.11) or its urea homologue, **53**, to trigger the stereoselective ROP of *rac*-LA. ROP reactions were conducted at room temperature in toluene in presence of NHCs with varied basicity and steric hindrance [49]. These catalytic systems showed a synergistic effect, providing both a high catalytic activity (>90% monomer conversion in 5–9 h) and a high stereoselectivity (P_m of 0.82–0.93 and T_m of 132–189 °C). It was found that compound **50** with the least steric hindrance and higher basicity (pK_a = 25.2 in H_2O) showed the highest stereoselectivity, achieving 90% monomer conversion within 3 h, when associated with the chiral **17a**. The resulting isotactic PLA possessed predictable M_n of 28,000 g mol^{-1}, narrow distribution ($Đ$ = 1.06), high P_m of 0.93 and high T_m of 189 °C. This T_m value being significantly higher than that of PLLA, formation of stereoblock-type PLA materials was suggested.

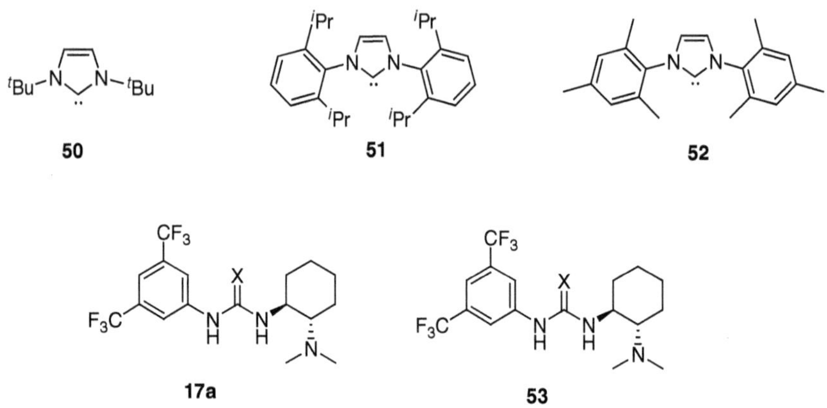

Scheme 7.11 NHC and (thio)urea bi-component organocatalysts (X = S, **17a**; X = O, **53**) for stereoselective ROP of *rac*-LA. Source: Zaky et al. [49]/American Chemical Society.

Besides NHCs, a *N*-heterocyclic olefin (NHO; **57**, Scheme 7.12) was also evaluated for the stereoselective ROP of *rac*-LA, when used in duo with various mono(thio)ureas (**39** and **63**, Scheme 7.12) or bis(thio)ureas (**69–72**, Scheme 7.12) [50]. ROP reactions were investigated in THF both at room temperature and at −70 °C. The catalytic activity was found to increase in the following order: bisthioureas **69** and **70** < monothiourea **39** < bisureas **71** and **72** < urea **63**. In the presence of BnOH as the initiator, the polymerization utilizing **57**/**39** proceeded rapidly, reaching 96% conversion in 5 min at room temperature and affording a PLA with a controlled molecular weight (M_n = 5900 g mol^{-1}, $Đ$ = 1.19), and a high P_m value of 0.81. Attempts to increase the stereocontrol was successfully achieved by conducting the ROP at −70 °C with **57**/**63**, which yielded an isotactic PLA with M_n of 5940 g mol^{-1}, $Đ$ of 1.16, and a P_m value up to 0.91. Analysis by chiral HPLC of the unreacted monomer demonstrated that the high stereoselectivity could be explained by the occurrence of the CEC mechanism. A structure-catalytic activity could be established for the different NHOs (**54–58**, Scheme 7.12) associated to a series of thioureas (TUs) and ureas (Us) (**39** and **59–72**, Scheme 7.12) for the ROP of various lactones, including δ-valerolactone, ε-caprolactone and *rac*-LA [51]. Thus, for a given NHO, the catalytic activity proved to increase first, then to decrease, as the acidity of the TU or the U decreased in the pK_a range of 13.8–22.8. On the other hand, for a given TU or U with different NHOs, the catalytic activity increased gradually as the basicity of the NHOs increase. However, moderate stereoselectivities (P_m in the range 0.56–0.77) could only be achieved at room temperature. Decreasing the temperature to −78 °C allowed the authors to significantly improve the stereocontrol, P_m values in the range 0.89–0.93 being determined. In particular, the organocatalytic pair combining **56** and **63** led to a P_m value as high as 0.93.

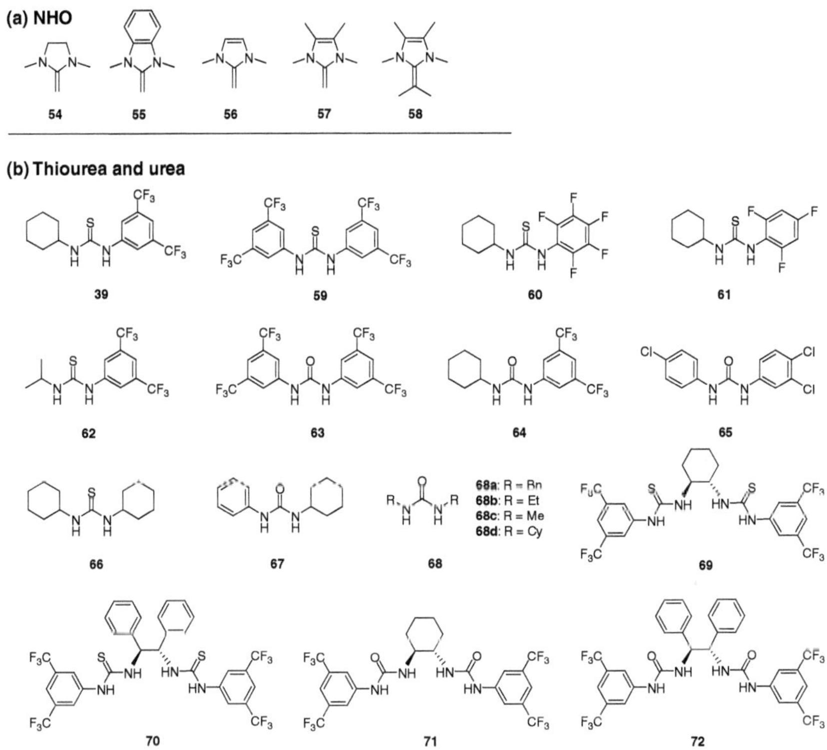

Scheme 7.12 Bi-component organocatalysts based on (a) NHO and (b) (thio)urea for the stereoselective ROP of *rac*-LA. Source: Wang et al. [50], Zhou et al. [51].

Phosphazene bases exhibit high catalytic activity for the for ROP of *rac*-LA, but cannot provide any stereoselectivity, unless if operating under cryogenic conditions (very low temperature) [36, 37]. On the other hand, Takemoto's organocatalysts provide some stereocontrol at room temperature, but they exhibit a poor catalytic activity, requiring several days to achieve high conversion [20]. Combining such organic catalysts in duos afforded excellent dual catalytic systems, in terms of both stereocontrol and catalysis activity. Thus, catalytic pairs based on phosphazenes **14** (P_1-*t*-Bu), **15** (P_2-*t*-Bu) and **73** (P_4-*t*-Bu; Scheme 7.13) and the chiral Takemoto's organocatalyst ((*S*,*S*)-**17a** and (*R*,*R*)-**17a**, Scheme 7.13) or some (thio)urea analogues (**78–82**, Scheme 7.13) proved particularly efficient to achieve relatively high polymerization rates and high stereoselectivity for the ROP of *rac*-LA at ambient temperature.

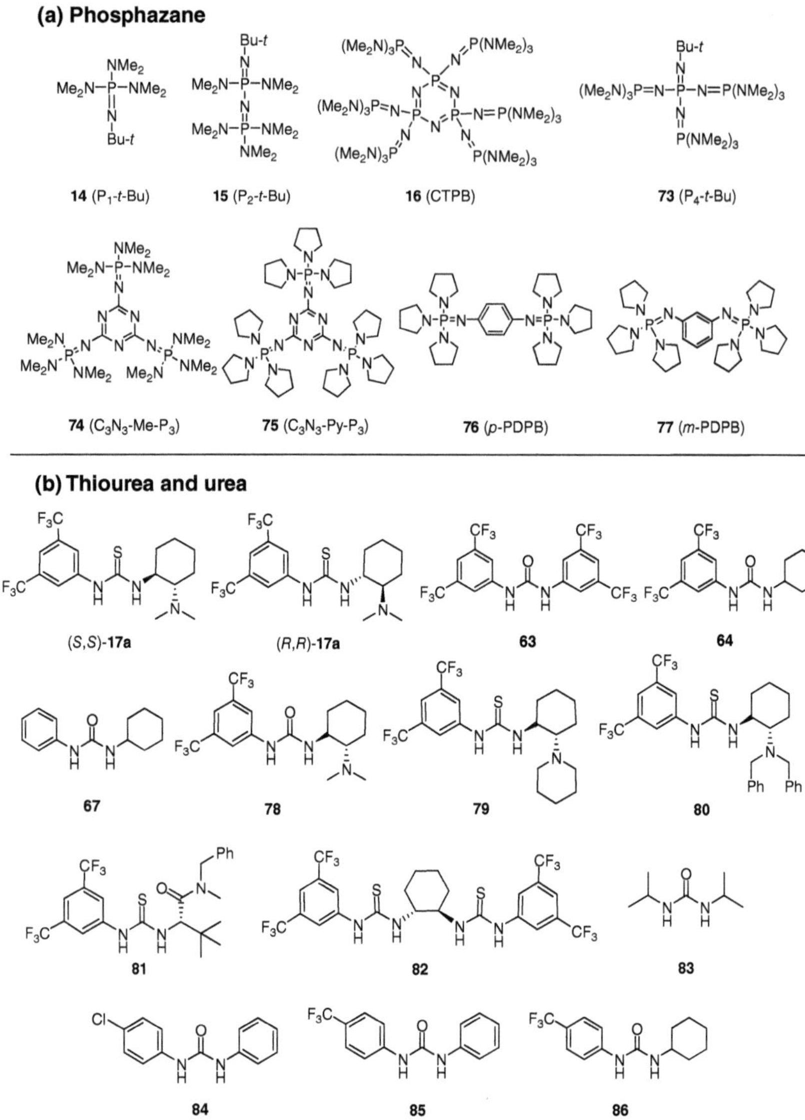

Scheme 7.13 Bi-component organocatalysts based on (a) phosphazene and (b) (thio)urea for the stereoselective ROP of *rac*-LA. Source: Zaky et al. [52, 53], Li et al. [54], Liu et al. [55].

Metal-free PLAs with T_m and P_m values in the range 138–187 °C and 0.72–0.96, respectively, could be produced in this way, following the ESC mechanism [52, 53]. The most basic P_4-*t*-Bu base (**73**, Scheme 7.13) was found the most active, but showed the lesser stereoselectivity. In this case, an anionic-like mechanism, involving a phosphazenium thioimidate ion pair, was postulated. In contrast, the weakest P_1-*t*-Bu base (**14**, Scheme 7.13) promoted an improved stereoselectivity via an associated-type mechanism involving H-bonding. A good tradeoff to provide both high reactivity and high selectivity could eventually be achieved from pairs associating P_2-*t*-Bu (**15**, Scheme 7.13) and some chiral amino(thio)ureas. While the Takemoto's catalyst/P_2-*t*-Bu pair operated by an ionic-like mechanism involving an ion pair during ROP of *rac*-LA, an associated-like mechanism was proposed upon using P_2-*t*-Bu and the urea analogue of the Takemoto's catalyst, highlighting that a subtle change in the catalytic pair had a significant effect on the reaction outcomes.

In a similar context, a library of achiral ureas (**63**, **64**, **67**, and **83–86**, Scheme 7.13), were employed to construct binary organocatalysts with different phosphazenes (Scheme 7.13), including **15** (P_2-*t*-Bu), **16** (CTPB), **73** (P_4-*t*-Bu), **74** (C_3N_3-Me-P_3), **75** (C_3N_3-Py-P_3), **76** (*p*-PDPB), **77** (*m*-PDPB). These catalytic pairs were then evaluated for the ROP of *rac*-LA under mild conditions [54, 55]. Interestingly, the duo constituted of **75** with **67** proved to be of compatible basicity/acidity, as it led to both a high activity (92% conversion within 10 min) and excellent stereocontrol (P_m = 0.92) at 20 °C in toluene. The resultant stereoblock-like PLA exhibited a predictable molecular weight (M_n = 12,200 g mol^{-1}), a narrow distribution ($Đ$ = 1.07) and proved to be semi-crystalline with a T_m value as high as 190 °C [55]. As these catalytic systems were nonchiral, the ROP process was proposed to proceed by the CEC mechanism, as verified by kinetic studies involving ROP of D-LA, L-LA, and *rac*-LA at 20 °C.

The diastereomeric densely substituted amino acids (2*S*,3*R*,4*S*,5*S*)-1-methyl-4-nitro-3,5-diphenylpyrrolidine-2-carboxylic acid (*endo*-**87**, Scheme 7.14) and (2*S*, 3*S*,4*R*,5*S*)-1-methyl-4-nitro-3,5-diphenylpyrrolidine-2-carboxylic acid (*exo*-**87**, Scheme 7.14) were designed as chiral oranocatalysts for the ROP of *rac*-LA [56]. In

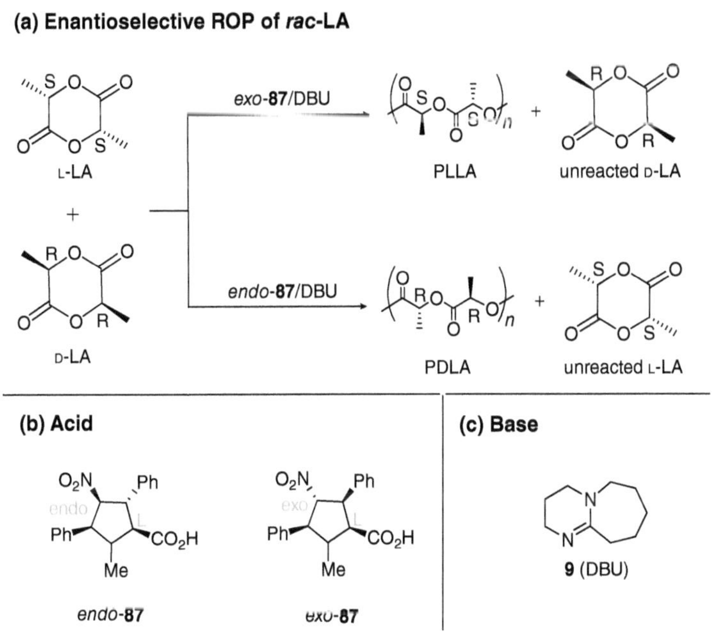

Scheme 7.14 (a) Enantioselective ROP of *rac*-LA by (b) diastereomeric densely substituted amino acids and (c) DBU bi-component organocatalysts. Source: Sanchez-Sanchez et al. [56]/American Chemical Society.

combination with DBU (**9**, Scheme 7.14) as a co-catalyst, the *exo*-**87** preferentially polymerized L-LA, while the *endo*-**87** showed a preference for D-LA as substrate. Thus, a highly efficient transfer of chirality from (*exo*-**87**) or (*endo*-**87**) to the as-produced PLA (M_n = 900–23 000 g mol^{-1}; $Đ$ = 1.1–1.5; P_m = 0.87–0.96) took place during the ROP process that was carried out in CH_2Cl_2 at room temperature (Scheme 7.14). The origin of this unique stereoselectivity and the underlying ROP mechanism were further investigated by density functional theory (DFT) calculations, which accounted that the chirality of the catalyst did define the stereochemistry of the monomer insertion.

The dynamic achiral rotaxanes consisting of a macrocyclic crown ether and an axis containing an ammonium and a TU or a triazole group (Scheme 7.15, **88–91**) were specifically designed. In the context of the ROP of *rac*-LA, these compounds used in the presence of $KN(SiMe_3)_2$ as the base and BnOH as the initiator, showed a good stereoselectivity [57]. The crown ether moiety was found to reside over the protonated ammonium group via intramolecular hydrogen bonding in the neutral compound, so that in this form, the polymerization was totally inhibited. In contrast, deprotonation of the amine allowed the free movement of the macrocycle along the axis, hence the resulting deprotonated rotaxane proved active for ROP reactions. In addition, the macrocycle-axis translocation was also crucial to achieve a high stereoselectivity. Replacing the TU motif by a triazole (**91**, Scheme 7.15), which provided a weaker hydrogen bond donating capacity in the rotaxane, led to ROP showing a

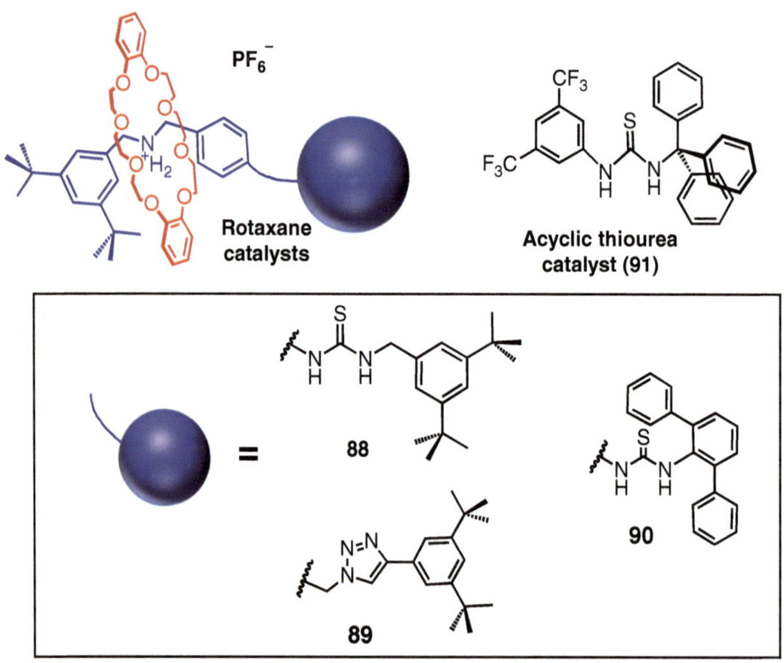

Scheme 7.15 Dynamic achiral rotaxanes as organocatalysts for the enantioselective ROP of *rac*-LA in the presence of $KN(SiMe_3)_2$. Source: Lim et al. [57]/John Wiley & Sons/CC BY 4.0.

lower stereoselectivity. The sterically hindered compound **90** (Scheme 7.15) also led to a decreased stereoselectivity, because the monomer coordination site for stereocontrol was hindered. Although P_m values were not that high (P_m = 0.54–0.81), the dynamic exchange associated with the rotaxane motif represents a new means to control the tacticity during the ROP of *rac*-LA.

7.6 Organocatalysts for Stereoselective ROP of Other Cyclic Monomers

Besides the of ROP of LA utilizing organic catalysts, which has been the topic of numerous studies in the past few years, a very few other cyclic monomers has been evaluated in the same context [58, 59]. In particular, synthesis of the eight-membered cyclic diolide (DL) was recently developed to further access isotactic poly(3-hydroxybutyrate) (P3HB). Interestingly, the latter polymer exhibits similar properties to those of the massively produced and engineered isotactic polypropylene (*i*-PP) [60, 61]. As DL possesses two chiral centers, like LA, the ROP of the racemic monomeric version, *rac*-DL, can give rise to stereoregular polymers, ranging from isotactic to atactic to heterotactic P3HB, depending on the preference of the employed catalyst to select a specific monomer. Using the chiral (*R*)-BINOL-derived CPAs (**92–98**, Scheme 7.16) the stereoselective ROP of *rac*-DL was conducted in toluene at 60 °C, affording well-defined gradient-type isotactic multiblock P3HB-based materials, with a P_m value of 0.71 (M_n = 3500–15,900 g mol^{-1}; $Đ$ = 1.01–1.07) [62]. The stereoselectivity and activity

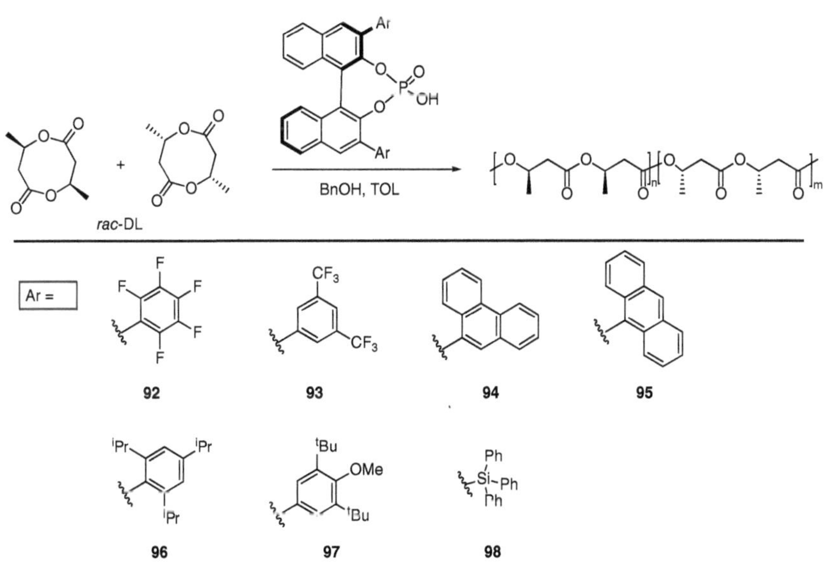

Scheme 7.16 Chiral phosphoric acid organocatalysts for the stereoselective ROP of *rac*-DL. Source: Liu et al. [62]/Royal Society of Chemistry.

of CPAs can be accommodated by varying the substituents as a means to balance both the steric and electronic effects. In particular, CPA **96** with a bulky 2,4,6-triisopropylphenyl substituent at the 3,3′-position achieved 42% conversion of *rac*-DL within 48 h, providing a P3HB with controlled M_n of 3700 g mol^{-1}and a proportion of isotactic [*mm*] triad of 71%. Experimental and computational analysis suggested a combination of ESC and CEC mechanisms to account for the stereoselectivity.

The chiral BINOL-derived CPAs ((*R*)-**99**, (*S*)-**99**, and (*S*)-**100**, Scheme 7.17) were also employed for the ROP of racemic seven-membered cyclic substituted caprolactones (CLs). A moderate selectivity factor (*s*), i.e. in the range of 2.0–7.3 was calculated, depending on the monomers, including 2-Bn-CL, 6-Bn-CL, 6-aryl-CL and 6-Me-CL (Scheme 7.17) [63–65]. Analysis of the unreacted monomer by chiral HPLC revealed that BINOL-derived CPAs preferentially polymerized the monomer with the same configuration to that of the catalyst. In the end, gradient stereoblock-type copolymers were produced at high conversions.

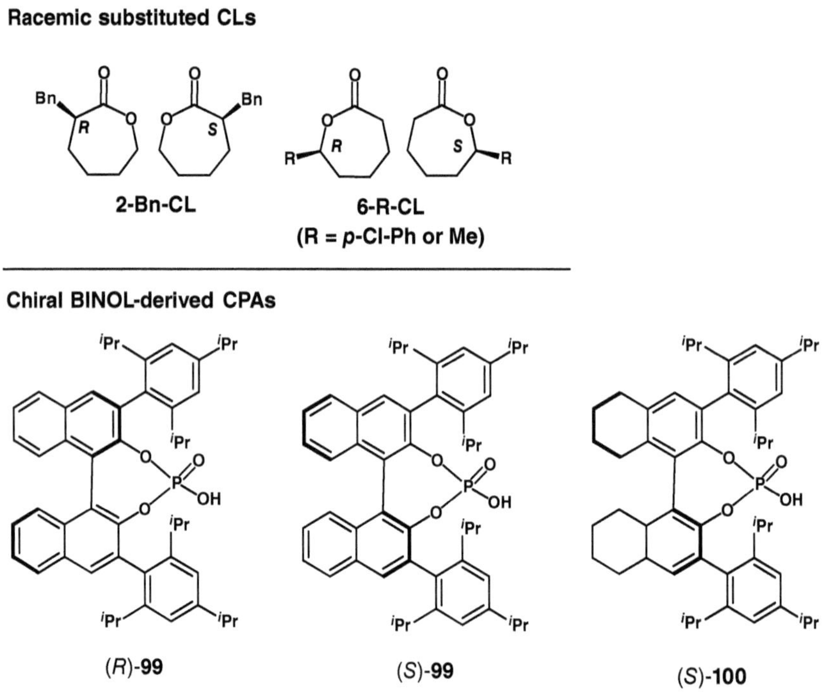

Scheme 7.17 Chiral phosphoric acid organocatalysts for the stereoselective ROP of substituted *rac*-CL. Source: Lv et al. [63–65].

7.7 Conclusion

Stereoregular polymeric materials usually show significantly different properties than their stereorandom counterparts. Control over the tacticity of polymeric chain thus stands for a powerful tool to access functional polymers with tunable physical and thermomechanical properties. The most popular example refers to the stereoselective ROP of *rac*-LA, which has been extensively investigated from a wide range of metal-based catalysts. However, organocatalyzed polymerization has emerged as an attractive and practical alternative to access polymers of controlled tacticity under metal-free conditions. Once again, *rac*-LA has been the focus of most of the studies published over the past decade, whereas the stereoselective metal-free ROP of other cyclic monomers featuring chiral centers has been underexplored. A series of organic acids (binaphthol (BINOL)-derived monophosphoric acids), organic bases (*N*-heterocyclic carbenes (NHCs), amidines and guanidines, phosphazenes), and dual organocatalysts have successfully achieved the stereoselective ROP of *rac*-LA. Particularly, the latter consisting of a hydrogen bond donor (HBD) for monomer activation and a hydrogen bond acceptor (HBA) for initiator activation or chain-end (initiation step and propagation step, respectively) has shown great potential to exquisitely control the activity and stereoselectivity for the ROP of *rac*-LA, the field having drawn a great deal of attention in recent years. Despite significant advances have been made, organocatalyzed stereoselective ROP is still in the infancy. It is anticipated that the search for organocatalysts combining both high activity and stereoselectivity for ROP that can be performed under mild conditions or under industrially relevant conditions (bulk polymerization at high temperature) will continue to develop in the near future.

Acknowledgements

Financial support by Ministry of Education of the People's Republic of China, and the 111 Project (no. D17004) is gratefully acknowledged.

References

1. Coates, G.W. (2000). Precise control of polyolefin stereochemistry using single-site metal catalysts. *Chem. Rev.* 100: 1223–1252.
2. Resconi, L., Cavallo, L., Fait, A., and Piemontesi, F. (2000). Selectivity in propene polymerization with metallocene catalysts. *Chem. Rev.* 100: 1253–1345.
3. Dijkstra, P.J., Du, H., and Feijen, J. (2011). Single site catalysts for stereoselective ring-opening polymerization of lactides. *Polym. Chem.* 2: 520–527.
4. Thomas, C.M. (2010). Stereocontrolled ring-opening polymerization of cyclic esters: synthesis of new polyester microstructures. *Chem. Soc. Rev.* 39: 165–173.
5. Stanford, M.J. and Dove, A.P. (2010). Stereocontrolled ring-opening polymerisation of polylactide. *Chem. Soc. Rev.* 39: 486–494.
6. Dechy-Cabaret, O., Martin-Vaca, B., and Bourissou, D. (2004). Controlled ring-opening polymerization of lactide and glycolide. *Chem. Rev.* 104: 6147–6176.

7 Chamberlain, B.M., Cheng, M., Moore, D.R. et al. (2001). Polymerization of lactide with zinc and magnesium β-diiminate complexes: stereocontrol and mechanism. *J. Am. Chem. Soc.* 123: 3229–3238.

8 Tsuji, H., Horii, F., Hyon, S.H., and Ikada, Y. (1991). Stereocomplex formation between enantiomeric poly(lactic acid)s. 2. Stereocomplex formation in concentrated solutions. *Macromolecules* 24: 2719–2724.

9 Ikada, Y., Jamshidi, K., Tsuji, H., and Hyon, S.H. (1987). Stereocomplex formation between enantiomeric poly(lactides). *Macromolecules* 20: 904–906.

10 Spassky, N., Wisniewski, M., Pluta, C., and Le Borgne, A. (1996). Highly stereoelective polymerization of *rac*-(D,L)-lactide with a chiral schiff's base/aluminum alkoxide initiator. *Macromol. Chem. Phys.* 197: 2627–2637.

11 Nederberg, F., Connor, E.F., Möller, M. et al. (2001). New paradigms for organic catalysts: the first organocatalytic living polymerization. *Angew. Chem., Int. Ed.* 40: 2712–2715.

12 Jain, I. and Malik, P. (2020). Advances in urea and thiourea catalyzed ring opening polymerization: a brief overview. *Eur. Polym. J.* 133: 109791.

13 Dharmaratne, N.U., Pothupitiya, J.U., and Kiesewetter, M.K. (2019). The mechanistic duality of (thio)urea organocatalysts for ring-opening polymerization. *Org. Biomol. Chem.* 17: 3305–3313.

14 Liu, S., Ren, C., Zhao, N. et al. (2018). Phosphazene bases as organocatalysts for ring-opening polymerization of cyclic esters. *Macromol. Rapid Commun.* 39: 1800485.

15 Thomas, C. and Bibal, B. (2014). Hydrogen-bonding organocatalysts for ring-opening polymerization. *Green Chem.* 16: 1687–1699.

16 Dove, A.P. (2012). Organic catalysis for ring-opening polymerization. *ACS Macro Lett.* 1: 1409–1412.

17 Kiesewetter, M.K., Shin, E.J., Hedrick, J.L., and Waymouth, R.M. (2010). Organocatalysis: opportunities and challenges for polymer synthesis. *Macromolecules* 43: 2093–2107.

18 Dove, A., Sardon, H., and Naumann, S. (2018). *Organic Catalysis for Polymerisation*. London: The Royal Society of Chemistry.

19 Nomura, N., Ishii, R., Yamamoto, Y., and Kondo, T. (2007). Stereoselective ring-opening polymerization of a racemic lactide by using achiral salen- and homosalen-aluminum complexes. *Chem. - Eur. J.* 13: 4433–4451.

20 Orhan, B., Tschan, M.J.L., Wirotius, A.-L. et al. (2018). Isoselective ring-opening polymerization of *rac*-lactide from chiral takemoto's organocatalysts: elucidation of stereocontrol. *ACS Macro Lett.* 7: 1413–1419.

21 Chisholm, M.H., Patmore, N.J., and Zhou, Z. (2005). Concerning the relative importance of enantiomorphic site vs. chain end control in the stereoselective polymerization of lactides: reactions of (*R*,*R*-salen)- and (*S*,*S*-salen)–aluminium alkoxides $LAlOCH_2R$ complexes (R = CH_3 and *S*-CHMeCl). *Chem. Commun.* 127–129.

22 Todd, R., Rubio, G., Hall, D.J. et al. (2013). Benzyl bispidine as an efficient replacement for (−)-sparteine in ring opening polymerisation. *Chem. Sci.* 4: 1092–1097.

23 Akiyama, T., Itoh, J., Yokota, K., and Fuchibe, K. (2004). Enantioselective mannich-type reaction catalyzed by a chiral brønsted acid. *Angew. Chem., Int. Ed.* 43: 1566–1568.

24 Uraguchi, D. and Terada, M. (2004). Chiral brønsted acid-catalyzed direct mannich reactions via electrophilic activation. *J. Am. Chem. Soc.* 126: 5356–5357.

25 Makiguchi, K., Yamanaka, T., Kakuchi, T. et al. (2014). Binaphthol-derived phosphoric acids as efficient chiral organocatalysts for the enantiomer-selective polymerization of *rac*-lactide. *Chem. Commun.* 50: 2883–2885.

26 Fèvre, M., Pinaud, J., Gnanou, Y. et al. (2013). *N*-heterocyclic carbenes (NHCs) as organocatalysts and structural components in metal-free polymer synthesis. *Chem. Soc. Rev.* 42: 2142–2172.

27 Jensen, T.R., Breyfogle, L.E., Hillmyer, M.A., and Tolman, W.B. (2004). Stereoelective polymerization of D,L-lactide using *N*-heterocyclic carbene based compounds. *Chem. Commun.* 2504–2505.

28 Dove, A.P., Li, H., Pratt, R.C. et al. (2006). Stereoselective polymerization of *rac*- and *meso*-lactide catalyzed by sterically encumbered *N*-heterocyclic carbenes. *Chem. Commun.* 2881–2883.

29 Li, H., Ai, B.-R., and Hong, M. (2018). Stereoselective ring-opening polymerization of *rac*-lactide by bulky chiral and achiral *N*-heterocyclic carbenes. *Chin. J. Polym. Sci.* 36: 231–236.

30 Si, G., Zhang, S., Pang, W. et al. (2018). Stereoselective zwitterionic ring-opening polymerization of *rac*-lactide. *Polymer* 154: 148–152.

31 Pratt, R.C., Lohmeijer, B.G.G., Long, D.A. et al. (2006). Triazabicyclodecene: a simple bifunctional organocatalyst for acyl transfer and ring opening polymerization of cyclic esters. *J. Am. Chem. Soc.* 128: 4556–4557.

32 Lohmeijer, B.G.G., Pratt, R.C., Leibfarth, F. et al. (2006). Guanidine and amidine organocatalysts for ring-opening polymerization of cyclic esters. *Macromolecules* 39: 8574–8583.

33 Moins, S., Hoyas, S., Lemaur, V. et al. (2020). Stereoselective ROP of *rac*- and *meso*-lactides using achiral TBD as catalyst. *Catalysts* 10: 620.

34 Mahmood, Q., Xu, G., Zhou, L. et al. (2020). Chiral 1,5,7-triazabicyclo[4.4.0]dec-5-ene (TBD)-catalyzed stereoselective ring-opening polymerization of *rac*-lactide: high reactivity for isotactic enriched polylactides (PLAs). *Polymers* 12: 2365.

35 Zhang, L., Nederberg, F., Pratt, R.C. et al. (2007). Phosphazene bases: a new category of organocatalysts for the living ring-opening polymerization of cyclic esters. *Macromolecules* 40: 4154–4158.

36 Zhang, L., Nederberg, F., Messman, J.M. et al. (2007). Organocatalytic stereoselective ring-opening polymerization of lactide with dimeric phosphazene bases. *J. Am. Chem. Soc.* 129: 12610–12611.

37 Liu, S., Li, H., Zhao, N., and Li, Z. (2018). Stereoselective ring-opening polymerization of *rac*-lactide using organocatalytic cyclic trimeric phosphazene base. *ACS Macro Lett.* 7: 624–628.

38 Dove, A.P., Pratt, R.C., Lohmeijer, B.G.G. et al. (2005). Thiourea-based bifunctional organocatalysis: supramolecular recognition for living polymerization. *J. Am. Chem. Soc.* 127: 13798–13799.

39 Okino, T., Hoashi, Y., and Takemoto, Y. (2003). Enantioselective Michael reaction of malonates to nitroolefins catalyzed by bifunctional organocatalysts. *J. Am. Chem. Soc.* 125: 12672–12673.

40 Pratt, R.C., Lohmeijer, B.G.G., Long, D.A. et al. (2006). Exploration, optimization, and application of supramolecular thiourea–amine catalysts for the synthesis of lactide (co)polymers. *Macromolecules* 39: 7863–7871.

41 Jiang, X., Zhao, N., and Li, Z. (2021). Stereoselective ring-opening polymerization of *rac*-lactide catalyzed by squaramide derived organocatalysts at room temperature. *Chin. J. Chem.* 39: 2403–2409.

42 Goldys, A.M. and Dixon, D.J. (2014). Organocatalytic ring-opening polymerization of cyclic esters mediated by highly active bifunctional iminophosphorane catalysts. *Macromolecules* 47: 1277–1284.

43 Lv, C., Zhou, L., Yuan, R. et al. (2020). Isoselective ring-opening polymerization and asymmetric kinetic resolution polymerization of *rac*-lactide catalyzed by bifunctional iminophosphorane–thiourea/urea catalysts. *New J. Chem.* 44: 1648–1655.

44 Yuan, R., Xu, G., Lv, C. et al. (2020). Bifunctional phosphazene-thiourea/urea catalyzed ring-opening polymerization of cyclic esters. *Mater. Today Commun.* 22: 100747.

45 Miyake, G.M. and Chen, E.Y.X. (2011). Cinchona alkaloids as stereoselective organocatalysts for the partial kinetic resolution polymerization of *rac*-lactide. *Macromolecules* 44: 4116–4124.

46 Zhu, J.-B. and Chen, E.Y.X. (2015). From *meso*-lactide to isotactic polylactide: epimerization by B/N Lewis pairs and kinetic resolution by organic catalysts. *J. Am. Chem. Soc.* 137: 12506–12509.

47 Kou, X., Shen, Y., and Li, Z. (2020). Stereoselective ring-opening polymerization of *rac*-lactide using chiral urea/strong organobase binary catalyst system. *Acta Polym. Sin.* 51: 1121–1129.

48 Dai, Z., Cui, Y., Chen, C., and Wu, J. (2016). Photoswitchable ring-opening polymerization of lactide catalyzed by azobenzene-based thiourea. *Chem. Commun.* 52: 8826–8829.

49 Zaky, M.S., Wirotius, A.-L., Coulembier, O. et al. (2022). Reaching high stereoselectivity and activity in organocatalyzed ring-opening polymerization of racemic lactide by the combined use of a chiral (thio)urea and a *N*-heterocyclic carbene. *ACS Macro Lett.* 11: 1148–1155.

50 Wang, Z.-Y., Xu, G.-Q., Zhou, L. et al. (2021). Isoselective ring-opening polymerization of racemic lactide catalyzed by *N*-heterocyclic olefin/(thio)urea organocatalysts. *Chin. J. Polym. Sci.* 39: 709–715.

51 Zhou, L., Wang, Z., Xu, G. et al. (2021). Structure and activity relationship studies of *N*-heterocyclic olefin and thiourea/urea catalytic systems: application in ring-opening polymerization of lactones. *Polym. Chem.* 12: 1806–1815.

52 Zaky, M.S., Wirotius, A.L., Coulembier, O. et al. (2021). A chiral thiourea and a phosphazene for fast and stereoselective organocatalytic ring-opening-polymerization of racemic lactide. *Chem. Commun.* 57: 3777–3780.

53 Zaky, M.S., Guichard, G., and Taton, D. (2023). Structural effect of organic catalytic pairs based on chiral amino(thio)ureas and phosphazene bases for the isoselective ring-opening polymerization of racemic lactide. *Macromolecules* 56: 3607–3616.

54 Li, Z., Duan, S., Zhao, N., and Li, Z. (2023). Stereoselective ring-opening polymerization of *rac*-lactide catalyzed by phenyl diphosphazene/urea binary catalytic system. *J. Polym. Sci.* 61: 1688–1696.

55 Liu, Y., Zhang, J., Kou, X. et al. (2022). Highly active organocatalysts for stereoselective ring-opening polymerization of racemic lactide at room temperature. *ACS Macro Lett.* 11: 1183–1189.

56 Sanchez-Sanchez, A., Rivilla, I., Agirre, M. et al. (2017). Enantioselective ring-opening polymerization of *rac*-lactide dictated by densely substituted amino acids. *J. Am. Chem. Soc.* 139: 4805–4814.

57 Lim, J.Y.C., Yuntawattana, N., Beer, P.D., and Williams, C.K. (2019). Isoselective lactide ring opening polymerisation using [2]rotaxane catalysts. *Angew. Chem., Int. Ed.* 58: 6007–6011.

58 Xie, X., Huo, Z., Jang, E., and Tong, R. (2023). Recent advances in enantioselective ring-opening polymerization and copolymerization. *Commun. Chem.* 6: 202.

59 Tschan, M.J.L., Gauvin, R.M., and Thomas, C.M. (2021). Controlling polymer stereochemistry in ring-opening polymerization: a decade of advances shaping the future of biodegradable polyesters. *Chem. Soc. Rev.* 50: 13587–13608.

60 Tang, X., Westlie, A.H., Watson, E.M., and Chen, E.Y.X. (2019). Stereosequenced crystalline polyhydroxyalkanoates from diastereomeric monomer mixtures. *Science* 366: 754–758.

61 Tang, X. and Chen, E.Y.X. (2018). Chemical synthesis of perfectly isotactic and high melting bacterial poly(3-hydroxybutyrate) from bio-sourced racemic cyclic diolide. *Nat. Commun.* 9: 2345.

62 Liu, J., Wang, J., Li, M., and Tao, Y. (2022). Gradient isoselective ring-opening polymerization of racemic cyclic diolide driven by chiral phosphoric acid catalysis. *Polym. Chem.* 13: 6328–6334.

63 Lv, C., Xu, G., Yang, R. et al. (2021). Stereogradient polycaprolactones formed by asymmetric kinetic resolution polymerization of 6-methyl-ε-caprolactone. *Polym. Chem.* 12: 4856–4863.

64 Lv, C., Yang, R., Xu, G. et al. (2020). Phosphoric acids catalyzed asymmetric kinetic resolution polymerization of benzyl substituted ε-caprolactones: efficient protocol for stereogradient polycaprolactones. *Eur. Polym. J.* 133: 109792.

65 Lv, C., Xu, G., Yang, R. et al. (2020). Chiral phosphoric acid catalyzed asymmetric kinetic resolution polymerization of 6-aryl-ε-caprolactones. *Polym. Chem.* 11: 4203–4207.

8

Ring-Opening Polymerization of Epoxides Through Organocatalysis

Lijun Liu, Yuxuan Zhou, Xingyu Tang, and Junpeng Zhao

Faculty of Materials Science and Engineering, South China University of Technology, Guangzhou 510640, China

8.1 Introduction

Epoxides, featured by easy accessibility, a rich catalog of substituents, high polarity, and reactivity, are fundamental raw materials for the synthesis of main-chain oxygenated polymers [1–3]. Typically, aliphatic polyethers and derivative materials obtained through ring-opening polymerization (ROP) of epoxides have reached an annual production of millions of tons for the widespread applications in biomedicine, pharmaceutics, cosmetics, lubricants, surfactants, polyurethanes (PUs), etc [4–7]. The development of controlled and selective homo- and copolymerization of ethylene oxide (EO) and monosubstituted epoxides, especially propylene oxide (PO) and glycidyl ethers, has delivered polyethers with enormously abundant structures and compositions [8–11], which is vital to meet the rising demands for tailored and diversified polymer properties [12–15]. The great value of epoxides also lies in its statistical, alternating, and block copolymerization with a variety of non-epoxide compounds, such as cyclic esters, cyclic anhydrides, carbon dioxide (CO_2), vinyl monomers, etc., which have been forged into a powerful toolbox for sophisticated macromolecular design and construction [16–19]. Since most of these copolymerization reactions are elaborated in other chapters, we will make our discussion on only a small portion of the synthesis of polyether-based block copolymers.

Facilitated by the high ring-strain of the three-membered oxacycles, epoxides are highly polymerizable by cationic, coordination-insertion, and anionic mechanistic pathways [2, 20]. Although strong acids, including several organic ones, readily trigger cationic ROP of epoxides, poor control over macromolecular characteristics is usually inflicted by the extensive chain transfer reactions [21, 22], which is therefore not discussed in depth in this chapter. The coordination-insertion mechanism may allow excellent control and, particularly, chemo-, regio-, and stereoselectivity, but almost exclusively clings to metal-based catalysts such as alkylmagnesium [23, 24], alkylzinc [25], double metal cyanide [26, 27], and (salen)Co/Cr [3, 28–30], which is therefore not included here unless aligned with organocatalysts.

Organocatalysts in Polymer Chemistry: Synthesis and Applications, First Edition. Edited by Zhibo Li.

Living/controlled anionic polymerization has been regarded as the simple and reliable synthetic pathway towards well-defined macromolecular structures [31–33]. Hydroxides and alkoxides of alkali metals (Na, K, Cs) are commonly employed initiators for the anionic ROP of epoxides in laboratories and industries [34–39], but limited by the relatively low activity and poor solubility in organic media due to the strong ion pairing and association [2]. Also, the need for elevated reaction temperature and/or polar solvent enhances the chance of irreversible chain transfer (to monomer or solvent) [40]. In the case of lithium alkoxide initiators, ROP barely occurs due to the overly strong ionic interaction between oxyanion and lithium cation (Li^+) [31]. A number of organic ligands, including crown ethers [41, 42], cryptates [43], phosphazene bases (PBs) [32, 33], etc., were used to complex with the alkali metal cations and elevate the nucleophilicity of the initiating/propagating oxyanions. The increasing demands for metal-free polymers also encouraged a series of studies using organic salts, such as *tetra-n*-butylammonium fluoride (TBAF) and *tetra-n*-butylammonium acetate (TBAA), as the initiators [44, 45] These efforts were made before the commencement of boosted development of organocatalytic polymerization at the beginning of the twenty-first century [46–49]. In the last two decades, anionic ROP of epoxides has been one of the most appealing subjects of research in this area, with a variety of small-molecule organocatalysts discovered and/or created enabling high efficiency, high chemoselectivity, and high structural control [50–52]. The anionic pathways will thus be the main focus of this chapter. Catalysts and mechanisms of different types will be discussed first. Special focus will be placed in the next two sections on the recent progress in the synthetic methods for functionalized polyethers and polyether-based block copolymers harnessing the unique activity and chemoselectivity disclosed for the organocatalysts.

8.2 Catalyst and Mechanism

In general, organobases complex with the cationic counterions (e.g. H^+, Li^+, K^+) to separate the ion pair and improve the nucleophilicity of the anionic initiators or the propagating chain ends. Organic acids, including Brønsted acids and Lewis acids, activate the monomers by interacting with the *O* atoms of the epoxy rings, which substantially reduce the energy barrier of the nucleophilic ring-opening reactions. The strengths and limits of these single-component catalysts have encouraged combination of basic and acidic catalysts into two-component catalytic systems or one bifunctional catalytic molecules, aiming to enhance the catalytic activity and selectivity through concurrent activation of chain ends and monomers in a compact domain [52]. The above-mentioned quaternary ammonium salts (TBAF and TBAA) are also frequently referred to as (co)catalysts [44, 45]. In fact, these substances play identical roles as the alkali metal hydroxides/alkoxides, i.e. with the anionic moieties attacking the epoxides to start the ROP and covalently connecting to the polyether products, and therefore are considered as initiators here for the clarity of discussion. Some merged uses of organobases and metal-based catalysts/initiators have greatly inspired the exploration into analogous organocatalytic systems, which are therefore briefed here as well.

8.2.1 Single-Component Catalyst

8.2.1.1 Base

Organobases, typically cyclic amidines, guanidines, and PBs, have been widely used in synthetic polymer chemistry [46, 47]. In an oxyanionic ROP reaction, the Lewis base generally takes effect by binding/complexing with the cationic counterion, which loosens the ion pair and enhances the nucleophilicity of the active species. In the 1980s, crown ethers and cryptates, termed additives then, were used to promote the anionic ROP of heterocyclic monomers initiated by alkali metal alkoxides by forming stable cation inclusion complexes and preventing the formation of inactive ionic aggregates (Figure 8.1) [41–43]. For instance, the ROP rate of PO was increased nearly eightfold at 110 °C by the addition of 18-crown-6 [53].

The neutral PBs (^{t}BuP$_4$, ^{t}BuP$_2$; Figure 8.1) have received a large part of attention with regard to the anionic ROP of epoxides [54, 55]. The complexation of bulky and strongly basic ^{t}BuP$_4$ with Li$^+$ effectively separates the contact *O-Li* ion pair. ROP of EO was well-performed from either small-molecule or macromolecular lithium alkoxide initiators (Figure 8.1) [56–60]. But it still needed a few days to ensure complete conversion of EO even at 50 °C. For protonic initiators, typically alcohols, the hydroxyl group is deprotonated by ^{t}BuP$_4$ to form the oxyanionic initiating and propagating species which then nucleophilically attacks EO (Figure 8.1) [61, 62]. The large and soft protonated counterion, [^{t}BuP$_4$H]$^+$, greatly improves the activity and solubility of the initiators in organic media, such as tetrahydrofuran (THF). The non-nucleophilic nature of PB is also highly beneficial for ensuring that EO only reacts with the initiator or the propagating chain end which is key for achieving poly(ethylene oxide) (PEO) with controlled molar mass and complete end-group fidelity.

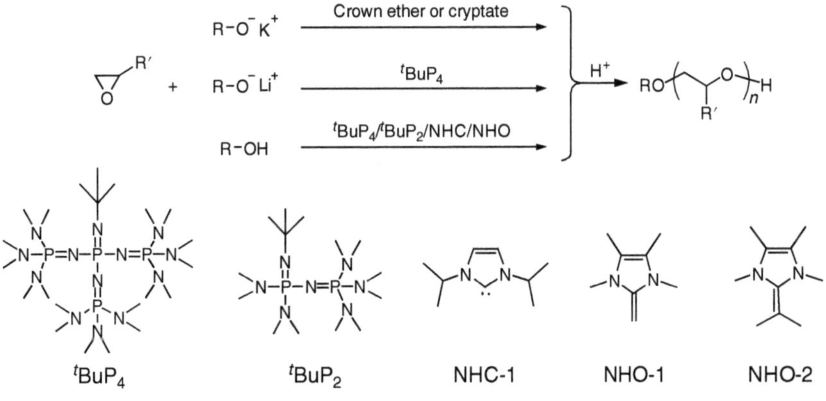

Figure 8.1 General schematic illustration of the organobase-catalyzed ROP of epoxides with initiators bearing different counterions.

A variety of non-alcohol protonic compounds, including phenols, amides, and carboxylic acids, were also shown effective initiators for the ROP of epoxides catalyzed by ^{t}BuP$_4$. Polystyrene(PS)-*b*-poly(*p*-hydroxystyrene) diblock copolymers were prepared through the Li^+-based anionic polymerization of the vinyl monomers, and then the pendant phenolic hydroxyl groups served as initiating sites for the ^{t}BuP$_4$-catalyzed ROP of EO at 45 °C for 48 h, yielding amphiphilic block-graft copolymers with high molar mass and low molar mass distribution ($Đ_M$) [63–65]. Amide groups, either in small molecules or the pendant groups of polyacrylamides, with acidity between alcoholic and phenolic hydroxyl groups could also initiate the ROP of EO, PO, and 1,2-butylene oxide (BO) when activated by ^{t}BuP$_4$ [66–68]. Especially, the direct "grafting from" route exhibited its simplicity and effectiveness for synthesizing graft copolymers with polyolefin/polyacrylamide backbones and polyether side chains. Compared with the base-stable connections formed after chain initiation in these cases, i.e. ether and amide bonds, ROP of EO initiated by carboxylic acids was accompanied by fast end-group transesterification, resulting in a large number of diesterified and dihydroxy PEO byproducts in addition to the targeted α-ester-ω-hydroxyl PEO [69].

The catalytic activity of ^{t}BuP$_4$ was also confirmed in the ROP of monosubstituted epoxides such as PO [64–66], BO [70–72], styrene oxide (SO) [73, 74], allyl glycidyl ether (AGE) [75], ethoxyethyl glycidyl ether (EEGE) [76], and some other glycidyl ethers [77–83]. Although living/controlled ROP was shown for most of these monomers, the case of PO suffered eminently from irreversible chain transfer to monomer when catalyzed by ^{t}BuP$_4$ alone [64–66]. This side reaction caused ill-defined poly(propylene oxide) (PPO) with limited molar mass, loss of chain-end fidelity, and unsaturated end groups. PB with a lower basicity, ^{t}BuP$_2$, could also complete the ROP of EO at 50 °C in THF in a few days and SO at 80 °C in bulk in 12 h [84, 85]. With further reduced basicity, ^{t}BuP$_1$ could not trigger the ROP of epoxides, including EO, even under high temperature and bulk conditions [85, 86].

N-heterocyclic carbenes (NHC) or *N*-heterocyclic olefin (NHO) were used as nucleophilic catalyst/initiator for the ROP of epoxide (Figure 8.1). The strong nucleophilicity of NHC facilitated a direct attack at the methylene group of EO, generating an imidazolium alkoxide which was further propagated by a zwitterionic mechanism [87]. NHC could be released from the polyether end and replaced by nucleophilic terminating agents, which was used as a synthetic strategy for end-functionalized PEO. NHC-initiated ROP of PO was carried out in bulk at 50 °C for three days, obtaining only low monomer conversion ($\leq$40%) and low-molar-mass PPO (4.5 kg mol^{-1}) with unsaturated end groups [88]. On the other hand, NHO delivered high yield of PPO with molar mass up to 12.0 kg mol^{-1} and $Đ_M < 1.09$ [89]. The existence of a C=C double bond in the *N*-heterocyclic ring and two methyl groups on the exocyclic carbon was proven essential to suppress side reactions. In addition, PPO-*b*-PEO-*b*-PPO triblock copolymers (Pluronic R) were achieved by use of PEO diols as macroinitiators [90, 91].

Although organobase-catalyzed ROP of epoxides follows simple mechanistic pathways, the reactions only rely on activation of initiators and chain ends. So, catalytic efficiency is limited and high basicity, high catalyst loading, high temperature, and long reaction time are usually required which tend to cause low chemoselectivity and induce disruptive side reactions such as irreversible chain transfer to monomers, solvents, and base-labile end or pendant groups. It was also disclosed that the residues of $^{t}BuP_4$ and $^{t}BuP_2$ were considerably cytotoxicity, which may devalue the polyethers for bio-related applications [92].

8.2.1.2 Acid

Several organic Brønsted acids were used as the initiators or catalysts for the cationic ROP of epoxides [93, 94]. The ROP conducted without an externally added nucleophile followed conventional "activated chain end" mechanism with tertiary oxonium ion as the cationic active species [95]. However, inter- and intramolecular chain transfer reactions were extensively accompanied and caused uncontrolled molar mass of polyethers together with the formation of large amounts of cyclic oligomers. It was found that addition of hydroxy compounds could evidently suppress the side reactions and the cationic active species changed from the tertiary oxonium ion to the protonated monomer [96]. In such an "activated monomer" (AM) pathway, the protonated monomer was adequately electrophilic to be attacked by the non-activated hydroxy species, which was the key to the enhanced controllability of the ROP [97]. However, molar mass of the polyether was still limited to $\leq$3.0 kg mol^{-1}, and the formation of cyclic oligomers was not fully inhibited. To date, these drawbacks are still not well-addressed and cationic ROP of epoxides has thus been receiving much less attention [98].

Though boron is technically a metalloid (non-metal) element, catalytic organoboron compounds are also widely regarded as organocatalysts, and therefore are termed here likewise. The highly Lewis acidic tris(pentafluorophenyl)borane ($B[C_6F_5]_3$) was used as an organocatalyst for the ROP of PO initiated by alcohols [99]. PPOs with low molar mass and relatively high $Đ_M$ were obtained at room temperature (RT). In another report, ROP of glycidyl phenyl ether was initiated by $B(C_6F_5)_3$ in bulk or dichloromethane in the absence of a protonic compound, generating cyclic polyethers in a zwitterionic pathway [100]. Instead, hydroxyl-terminated linear polyethers were formed when a small amount of water was introduced in the reaction. So far, the mechanistic pathway of $B(C_6F_5)_3$-catalyzed/initiated ROP has not been clearly defined to be anionic or cationic.

Onium salts could initiate the anionic ROP of epoxides without heating when used in cooperation with Lewis acids [101–105]. In the case of triisobutylaluminum ($^{i}Bu_3Al$), PPO with low $Đ_M$ (<1.2) and controlled molar mass (up to 150 kg mol^{-1}) were obtained in minutes to hours, while cooling at 0 °C was favorable due to the instability of $^{i}Bu_3Al$ (e.g. chain transfer to catalyst) [106, 107]. Organoboranes, typically triethylborane (Et_3B), featuring higher stability than $^{i}Bu_3Al$, were also combined with onium salts for the ROP of EO at RT and glycidyl azide (GA) at 0 °C [108–110]. Later, triaryl and trimethyl boroxines were employed with bis-(triphenylphosphine) iminium chloride as the initiator [111]. The reaction also needed to be conducted at 0 °C to avoid decomposition of the oxygen-rich boron-based catalysts. The adduct of poly(4-vinylphenol) and 9-borabicycl[3.3.1]nonane (9-BBN) was shown to be moisture and air stable, and

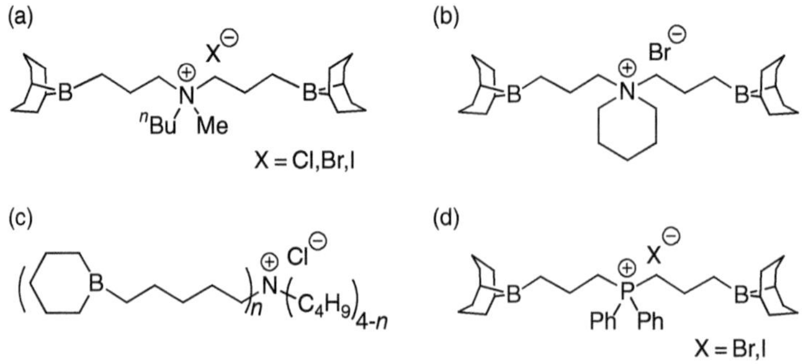

Figure 8.2 Representative boron-based bifunctional catalysts used for ROP of epoxides.

used as multifunctional macromolecular Lewis acid to deliver PPOs with molar mass up to 85 kg mol^{-1} and $Đ_M \leq 1.6$ at RT in combination with onium salt initiators [112].

The efficacy of combined onium salt initiators and boron-based Lewis acid catalysts was elevated to a unprecedentedly high level through integration of the latter into the cationic part of the former. For example, bifunctional catalysts composed of a halide anion as the initiating species and two catalytic boron centers derived from 9-BBN on the ammonium cation allowed ROP of epoxides to proceed at −20 or 0 °C in an exceedingly efficient and highly controlled manner with turnover frequency (TOF) reaching 3.0×10^4 h^{-1}, molar mass up to 1000 kg mol^{-1}, and $Đ_M \leq 1.3$ (Figure 8.2(a)) [113, 114]. Controlled ROP at RT was realized when a hydroxy chain transfer agent (CTA) was added in 10–40 equiv. of the catalyst. An intramolecular Lewis acid–base cooperative catalytic approach was indicated by control experiments using separated initiator and catalyst, and further corroborated by DFT calculations. A deeper investigation into the structure–activity relationship of the boron-based bifunctional catalysts was performed to disclose the impacts of steric hindrance at the ammonium cationic center (N^+), the number of boron centers and their distance to N^+ [115]. The dinuclear catalyst with a rigid ring-like pentamethylene substituent on N^+ showed the highest catalytic efficiency for the ROP of EO with TOF reaching 6.1×10^4 h^{-1} at 0 °C (Figure 8.2(b)). Triblock copolyethers including PEO-*b*-PPO-*b*-PEO (Pluronic) and Pluronic R were obtained through one-pot sequential monomer addition in the presence of dihydroxy CTAs.

Later, a series of bifunctional catalysts comprising 2, 3, or 4 less bulky pentamethylene-substituted borane moieties were prepared (Figure 8.2(c)) and showed remarkably enhanced activity with TOF reaching 1.8×10^5 h^{-1} and molar mass up to 1500 kg mol^{-1} ($Đ_M < 1.3$) for the ROP of PO. The loaded amount of catalyst could be minimized to <50 ppb (molar fraction) for the ROP of EO [116]. With the *N* atom replaced by a *P* atom, the phosphonium halide type bifunctional catalysts containing two 9-BBN-derived boron centers exhibited similarly high efficiency for the synthesis of polyethers (Figure 8.2(d)) [117, 118]. It is clear that anchoring Lewis acidic organoborane moieties to the cationic part of onium salts has been an ingenious catalyst design and brought exciting results for the ROP of epoxides. Some remaining issues may await to be explained and/or addressed, such as the necessity of low reaction temperature (−20 to 0 °C) in the absence of large quantities of CTAs.

8.2.2 Two-Component Catalyst

In 2012, a two-component catalyst comprising ^{t}BuP$_4$ and excess ^{i}Bu$_3$Al was found effective for the ROP of PO from alcohols (Table 8.1) [119]. Despite the weakened nucleophilicity of the ^{t}BuP$_4$-activated hydroxy species resulting from the formation of "ate" complex with ^{i}Bu$_3$Al, the epoxide activation by excess Lewis acid ensured high polymerization rate at RT. The alkoxide-^{i}Bu$_3$Al complexation also played a crucial role in the control of ROP by minimizing the basicity and thus suppressing irreversible chain transfer to monomer. In 2018, organic Lewis pair comprising PB with lower basicity (^{t}BuP$_1$, ^{t}BuP$_2$) and Et$_3$B was developed fulfilling efficient and controlled ROP of epoxides at RT (Table 8.1) [120, 121]. The catalyst loading for the ROP of EO in bulk could be lowered to <50 ppm (weight fraction) and TOF could reach >6000 h^{-1}. MMT assay of the crude PEO products synthesized with ^{t}BuP$_1$-Et$_3$B excluded the cytotoxicity of the catalyst residues on the basis that the cell viability remained around 100% as the polymer concentration was raised to 10 mg ml^{-1}. ROP of PO catalyzed by ^{t}BuP$_2$-Et$_3$B was also highly controlled and, importantly, free of chain transfer to monomer, so that well-defined PPO with molar mass up to >200 kg mol^{-1} and $Đ_M \leq 1.2$ were achieved in bulk at RT. When the catalyst loading was substantially increased, ROP of PO could be completed within two minutes with TOF up to 7500 h^{-1} reached and high chemoselectivity maintained (Table 8.1) [121]. Stereoregular PPO was obtained by use of stereospecific PO monomer.

The catalytic method was easily extended to larger-sized monosubstituted epoxides, including BO, SO, hexene oxide, octene oxide, AGE, and *tert*-butyl glycidyl ether (^{t}BGE) [120, 122]. It was also revealed that ROP of PO and step-growth polymerization (SGP) of the resultant PPO diol and diisocyanate could be performed in a one-pot one-catalyst route to obtain polyether-based PU [120]. In addition to PB, 1,8-diazabicyclo [5.4.0]undec-7-ene (DBU) and 7-methyl-1,5,7 triazabicyclo[4.4.0]dec-5-ene (MTBD) could also be combined with Et$_3$B to catalyze the ROP of EO and PO at RT, while showing much lower catalytic activity. Here, we would like to note that it might be difficult to make quantitative comparison between different catalysts regarding their activity by use of TOF values because of the significantly varied reaction conditions in these studies, including solvent, temperature, concentration, initiator, etc.

Chemoselective ROP of epoxides from carboxylic acids was accomplished by PB-Et$_3$B with transesterification fully suppressed to yield well-defined α-carboxylic ester-ω-hydroxyl polyethers [123]. An α,ω-dihydroxy PPO with two ester groups embedded in the center was generated from a dicarboxylic acid initiator and transformed in situ to PPO-based PU which exhibited distinct enzymatic degradability. An induction period of ca. 1–2 h followed by first-order kinetics was detected, indicating this living/controlled ROP features a slow-initiation mode. This unique

Table 8.1 Selected ROP systems using two-component organocatalysts.

Initiator	Base	Acid	Monomer	Condition	Reference
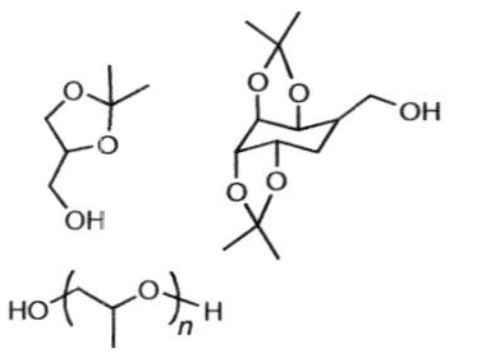	${}^{t}BuP_4$	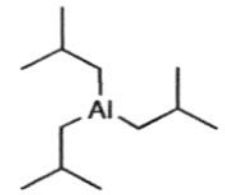 ($^{i}Bu_3Al$)	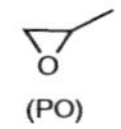	20 °C, in toluene	[119]
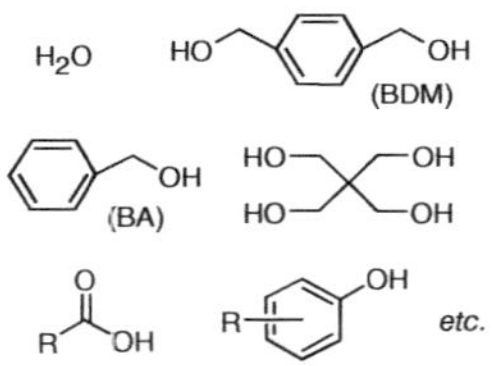	$^{t}BuP_2$ 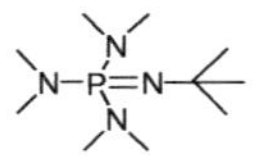 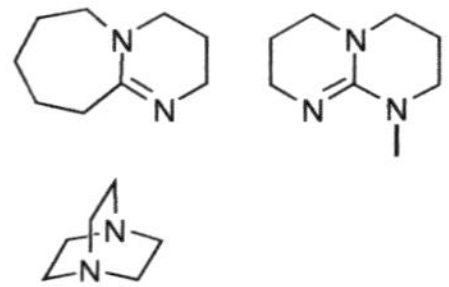	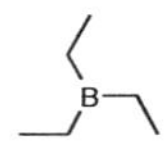 (Et_3B)	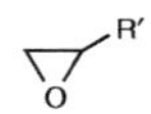 R′ = H, methyl, ethyl, 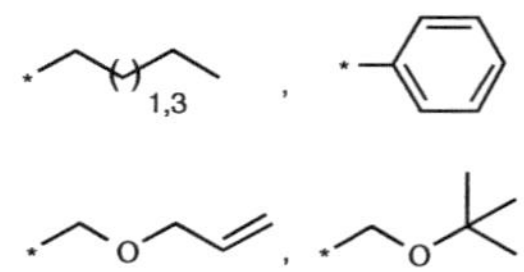	25 °C, in THF or bulk	[120–124]
BA	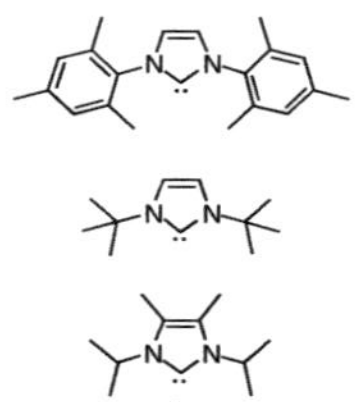 ($I^{i}Pr$)	$^{i}Bu_3Al$	PO	25 °C, in Me-THF	[125]
BA, BDM, H_2O	$I^{i}Pr$	Et_3B	PO	25 °C, in THF or bulk	[126]
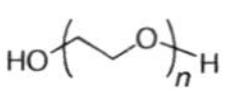	NHO-1, NHO-2	Et_3B	PO	50–80 °C, in bulk	[127, 128]

character was attributed to self-inhibited propagation during the initiation step. Namely, with a low catalyst-to-initiator ratio, a small portion of the carboxyl groups was deprotonated to trigger the chain initiation and form oxyanions (propagating species) which were immediately neutralized by the nonactivated carboxyl groups thus losing the ability to react with epoxide (propagation step). The newly formed carboxylate continued with the initiation until (almost) all the carboxyl groups were esterified. Then the hydroxyl groups could have the chance to be activated and proceed to the propagation step. The slow-initiation living/controlled ROP was also shown by phenolic initiators following the same mechanistic pathway [124]. Phenols with higher acidity resulted in phenolates with lower nucleophilicity and consequently longer initiation steps, but without affecting the control of ROP.

NHCs and Lewis acids were also combined (Table 8.1). With smaller steric hindrance and higher basicity, 1,3-bis(isopropyl)4,5(dimethyl)imidazol-2-ylidene (I^iPr), together with Et_3B showed higher catalytic efficiency for the ROP of PO in THF or bulk at RT with chain transfer to monomer prevented [126]. Bimodal molar mass distribution was observed and attributed to the initiation from adventitious water generating PEO diols with doubled molar mass. When $^iBu_3Al$ was used as the Lewis acid, unimodal distribution was achieved owing to its reaction with water forming inactive aluminum hydroxide [125]. Both NHC-Et_3B and NHC-$^iBu_3Al$ were effective for sequential polymerization of PO and BO in a living/controlled manner, and believed to follow the same chain-end-complexation and monomer-activation synergetic mechanistic pathway as PB-Et_3B. NHO and Et_3B also constituted effective catalyst for ROP of PO at 50–80 °C in bulk with PEO diols as the initiators (Table 8.1) [127, 128]. The presence of Et_3B significantly accelerated PO enchainment and eliminated side reactions.

8.3 Synthesis of Functionalized Polyether

Incorporation of functional groups, which are structurally distinct from the main body, at chain ends or as pendant groups is crucial for tailoring and diversification of polyether properties [129, 130]. Great challenges are often posed by chemically reactive functional groups, especially the electrophilic ones, due to their instability in the anionic ROP system. Conventionally, the issue has been circumvented by incorporation of functional groups that are compatible with the ROP, followed by transformation into the target ones by, in most cases, coupling reactions, though additional atoms or groups are likely to be introduced as linkers [131, 132]. The continuous development of the chemoselective catalytic/initiating systems for the anionic ROP has been accompanied by improvements in functional group tolerance. Especially, the two-component and bifunctional catalysts have allowed direct embedding of reactive functional groups or their easily deprotected precursors by use of functional initiators or monomers in the anionic ROP of epoxides, which have greatly simplified the synthesis of functionalized polyethers. Such endeavors will be the focus of this section.

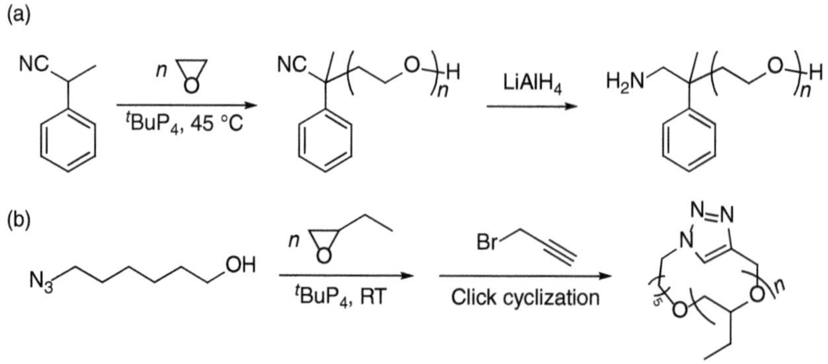

Figure 8.3 Examples of synthesis and post-transformation of end-functionalized polyethers through ^{t}BuP$_4$-catalyzed ROP using functional initiators.

8.3.1 End-Group Functionalization

Bearing functional end groups with designated reactivity is indispensable for PEO/PPO to undergo structural/compositional evolution and coupling with other synthetic polymers, bio(macro)molecules, drugs, solid surfaces, nano- and micro-sized particles, etc [133–136]. In 2001, following a series of successes in ^{t}BuP$_4$-catalyzed ROP of epoxides, the first attempt at metal-free synthesis of α,ω-heterobifunctional PEO was made with α-methylbenzyl cyanide as the initiator (Figure 8.3(a)) [137]. The cyano group was stable during the ROP and reduced afterwards by $LiAlH_4$ to afford an α-amino-ω-hydroxyl PEO with controlled molar mass of 2.5 kg mol^{-1}.

Stability of azido group in ^{t}BuP$_4$-catalyzed ROP system was also demonstrated and used for the synthesis of macrocyclic (co)polyethers from BO, 2-(2-[2-methoxyethoxy]ethoxy)ethyl glycidyl ether (EGE), decyl glycidyl ether (DGE), etc. For example, ROP of BO was performed from an azido-functionalized alcohol, and the hydroxyl end group was transformed into ethynyl group which then "clicked" with the azido group to form a monocyclic polyether (Figure 8.3(b)) [70]. Block copolymers of DGE and EGE were synthesized similarly with varied number and location of the azido and alkynyl groups which were used to form more complicated amphiphilic block copolyethers with linear, cyclic, figure-eight, tadpole, trefoil, quatrefoil, and cage-shaped architectures [71, 77–79]. It was found that the cyclic topology enabled more stable micellization in aqueous solutions as compared with the linear and tadpole analogs.

In addition to water used for α,ω-dihydroxy PPO, a variety of functionalized alcohols were used as CTAs in the ROP of PO catalyzed/initiated by a phosphonium bromide containing two 9-BBN-derived borane moieties (PBB-Br) for the synthesis of α,ω-heterobifunctional PPOs (Figure 8.4) [117, 138]. CTAs containing compatible functional groups could be directly employed to deliver the end-functionalized PPOs. For example, α-(meth)acrylate-ω-hydroxyl PPOs were obtained by use of 2-hydroxyethyl acrylate and 2-hydroxyethyl methacrylate as the CTAs. Alternatively, protection-deprotection strategy was used for incompatible functional groups. For

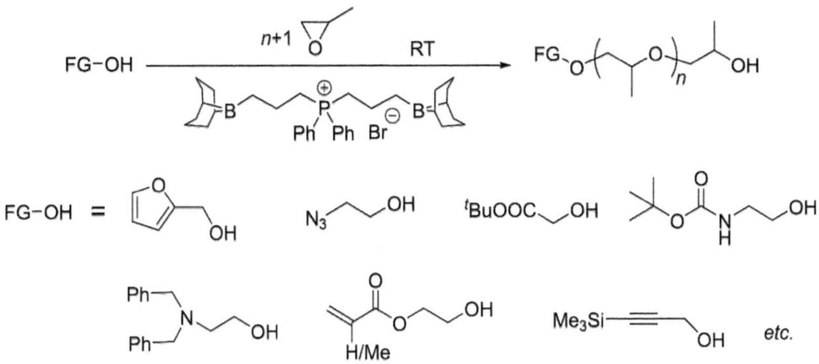

Figure 8.4 ROP of PO catalyzed/initiated by PBB-Br with representative CTAs listed.

example, ROP of PO was performed first in the presence of 2-*tert*-butyl glycolate or 2-(Boc-amino)ethanol and followed by acid-catalyzed hydrolysis to obtain the target carboxyl- or amino-end-functionalized PPOs.

As mentioned above, ROP of epoxide occurs chemoselectively in the presence of carboxylic esters with transesterification fully inhibited when using PB-Et_3B two-component organocatalysts. The potency of this method was further exploited by use of hydroxy(di)carboxylic esters as functional initiators to yield well-defined α-(di)carboxylic ester-ω-hydroxyl PEO/PPO in one step with complete end-group fidelity (Figure 8.5(a)) [139]. The end-group ester groups could undergo in situ organobase-catalyzed hydrolysis to afford α-(di)carboxyl-ω-hydroxyl PEO/PPO. In addition, well-defined α-amide-ω-hydroxyl PEO was delivered by in situ aminolysis.

Chemoselectivity of PB-Et_3B was also exhibited in the ROP of EO/PO using thiocarboxylic acids as the initiators (Figure 8.5(b)) [141]. Avoidance of end-group transesterification was clearly confirmed, despite the high electrophilicity of the thioester group formed in the initiation step. The resultant α-thioester-ω-hydroxyl PEO/PPO with controlled molar mass (2–14.2 kg mol^{-1}) and low $Đ_M$ (<1.1) readily underwent in situ aminolysis to facilitate one-pot synthesis of well-defined α-thiol-ω-hydroxyl PEO/PPO with high end-group fidelity. The replacement of the *O* atom with a *S* atom significantly enhanced the nucleophilicity of the initiating species, and therefore linear kinetic plots were shown, indicating a fast-initiation mode which was distinct from the case of carboxylic acid initiators [123]. The stability of thioester group in the ROP system was further harnessed for convenient and controlled synthesis of α-thiol-ω-dihydroxyl PEO starting from a thioester-containing dihydroxy initiator. Compared with the linear counterpart, the Y-shaped structure was shown equally effective for protein resistance with a considerably lower grafted amount on a solid surface.

Amino group is also a vital end-functionality for the applications of PEO/PPO, but cannot be directly introduced by amino-bearing initiators due to the prompt and uncatalyzed reaction with the monomer. Conventional synthetic routes appealed either to multistep post-polymerization modification of the hydroxyl end groups, or to covalent protection and post-polymerization deprotection of the amino-initiators [137, 142]. Based on the two-component organocatalyst, a

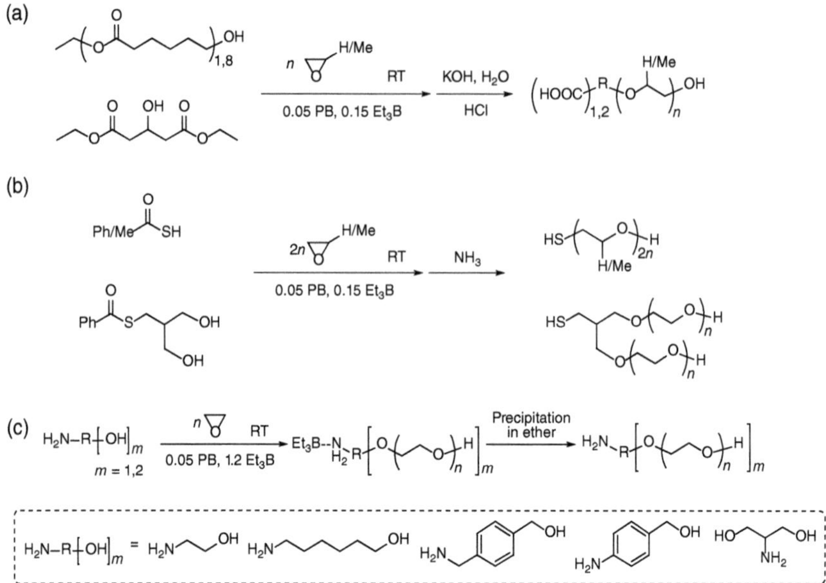

Figure 8.5 Chemoselective ROP catalyzed by organic Lewis pair using functional initiators for facile synthesis of end-functionalized PEO/PPO, including (a) α-(di)carboxyl-ω-hydroxyl PEO/PPO from hydroxycarboxylic esters, (b) α-thiol-ω-(di)hydroxyl PEO/PPO from thiocarboxylic acids, and (c) α-amino-ω-(di)hydroxyl PEO from amino alcohols. Source: Adapted from [140].

noncovalent protection strategy was recently devised and succeeded in one-step synthesis of α-amino-ω-(di)hydroxyl PEOs with molar mass up to 10 kg mol^{-1} and $Đ_M < 1.1$ [140]. The key to the inertness of amino group during the ROP is its complexation with Et_3B which ensured that EO selectively reacts with the hydroxyl group through the catalysis of PB and the rest of Et_3B. After the ROP, Et_3B was easily removable by precipitation of PEO in ether. It was also disclosed that enlarging the size of spacer between the amino group and PEO main body could effectively suppress amino oxidation and enhance storage stability.

8.3.2 Pendant-Group Functionalization

The large and easily expandable scope of functional substituents has made epoxides powerful building blocks for pendant-group functionalized polymers [2, 8]. ROP of epoxides bearing functional groups or the protected precursor groups is the main path to functionalized polyethers. Polymers carrying multiple amino groups have received widespread attention for the pH responsiveness, positive chargeability, and related bio-applications [143]. The ^{t}BuP$_4$-alcohol system was used for the ROP of glycidylamine derivatives such as *N,N*-dibenzylglycidylamine (DBGA) (Figure 8.6) [82]. Compatibility of the organocatalyst allowed controlled synthesis of (co)polyethers with pendant tertiary amino groups which could be transformed into primary or secondary amino groups by debenzylation. Block copolymerization

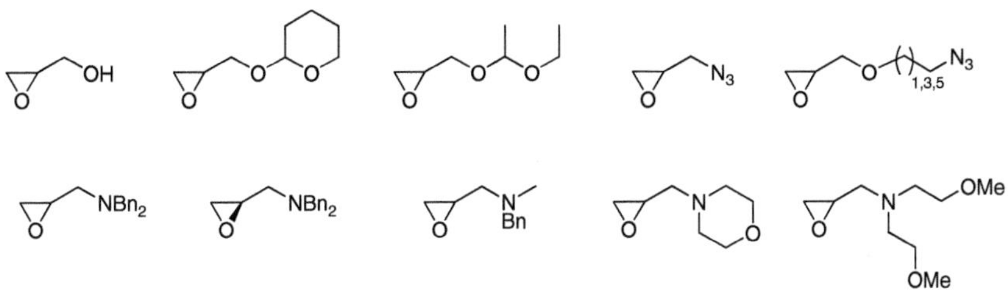

Figure 8.6 Chemical structures of functional monosubstituted epoxides used for organocatalytic synthesis of pendant-group functionalized (co)polyethers.

of DBGA and benzyl glycidyl ether was performed and followed by deprotection to yield copolyethers with pendant hydroxyl and amino groups.

AGE and EEGE, with base-stable and post-modifiable substituents, were sequentially polymerized by ^{t}BuP$_4$. The resultant block copolyethers underwent orthogonal deprotection and/or coupling reactions to deliver a variety of functionalized copolyethers (Figure 8.6) [75]. EEGE has been frequently employed as a hydroxyl-protected monomer to afford linear polyglycerol (PGC) after the ROP by acidic hydrolysis of the acetal groups. In addition, tetrahydropyranyl glycidyl ether (TGE), the cyclic analogue of EEGE (Figure 8.6), could also undergo controlled ROP in the ^{t}BuP$_4$-alcohol system and afford linear PGC [144]. Micelles were formed in water by the amphiphilic block copolymers of EO and TGE, and the cyclic pendant groups allowed slower acidic deprotection and better controlled release of encapsulants.

Although ^{t}BuP$_4$-catalyzed ROP of GA suffered intensively from β-proton abstraction and irreversible chain transfer, azidoalkyl glycidyl ethers (Figure 8.6), with prolonged spacers between the epoxy and azido groups reducing the acidity of β-proton, could polymerize in much more controlled manners in toluene at RT to deliver well-defined azide-functionalized polyethers with molar mass up to 11 kg mol^{-1} and $Đ_M < 1.2$ [83]. The azido groups appended to the backbone were readily transformed into other functionalities through azide-alkyne cycloaddition or Staudinger reduction. Controlled ROP of GA was realized by the combination of onium salt initiators and excess Et$_3$B at 0 °C, reaching molar mass up to 11 kg mol^{-1} and $Đ_M < 1.2$ [110]. The bulky Oct$_4$N$^+$ cation, together with the alkoxide-Et$_3$B complexation, effectively prevented undesired chain transfer (nucleophilic substitution of the azido group).

Anionic ROP of unprotected glycidol was also conducted with onium salt as the initiator and excess (5 equiv.) of Et$_3$B as the catalyst. Quasilinear PGC with a relatively low degree of branching (DB < 0.1) and molar mass up to 43.1 kg mol^{-1} was achieved at elevated temperature (80–90 °C) and with slow monomer addition (0.1 ml h^{-1}) [145]. The large excess of Et$_3$B was essential for monomer activation, on top of forming "ate" complex with the propagating oxyanionic species thus restraining chain transfer among hydroxy species. In comparison, molar mass of PGC synthesized from glycidol by other (cationic and coordination-insertion) methods was limited to <2.0 kg mol^{-1} and the lowest DB was 0.2 [146, 147].

Linear PGC has been shown to be a multipurpose hydrophilic material and a promising PEO substitute [10, 130]. ROP of acetal-bearing epoxides, such as

EEGE and TGE, followed by acidic hydrolysis has been the most frequently followed pathway to well-defined linear PGC [37–39]. An alternative method was established by virtue of the chemoselectivity of PB-Et_3B two-component organocatalyst. Commercially available glycidyl butyrates (GBs) were polymerized in a controlled and transesterification-free fashion. Carboxylic ester pendant groups of the resultant poly(glycidyl ester) were hydrolyzed under basic conditions to afford well-defined linear PGC (Figure 8.7) [148]. Random copolymerization of GB and PO was performed to afford copolyethers with tunable numbers of hydroxyl groups and aqueous thermosensitivity. Moreover, PGC-based amphiphilic and double-hydrophilic copolyethers were achieved through block copolymerization of GB with ^{t}BGE and EO, respectively.

8.4 Synthesis of Polyether-Based Block Copolymer

Amphiphilic block copolymers, especially those with PEO as the hydrophilic component, are extensively investigated in terms of microphase separation, self-assembly, drug delivery, etc [149–153]. The intrinsically occurring hydroxyl end groups of PEO synthesized by anionic ROP are widely employed as initiators or CTAs for other monomers either directly or after simple chemical modification [154]. The epoxide-based block copolyethers, typically, Pluronic and Pluronic R, are usually achievable simply by sequential monomer addition without intermediate alteration of reaction conditions, and therefore are not included in this section. Block copolymers prepared in two pots and multiple steps based on pre-synthesis of macroinitiators, including the commercialized ones, are excluded as well. Here, we focus on the synthesis of polyether-based block copolymers by one-pot two-step and one-step block copolymerization of epoxides and non-epoxide monomers in which the polyether segments are formed through organocatalytic ROP.

8.4.1 One-Pot Two-Step Block Copolymerization

One-pot block copolymer synthesis can be easily performed when the polymerizations of the two monomers share the same active species and catalyst. For example, alternating copolymerization of *o*-phthalaldehyde and PO was first performed in the presence of an onium salt and Et_3B at RT (Figure 8.8) [155]. After that, another batch of PO was added and polymerized from the end group of the first block to afford a polyacetal-*b*-polyether block copolymer with controlled composition and low $Đ_M$.

One-pot continuous construction of block copolymers is more often challenged when the monomers are suited to different and incompatible polymerization mechanisms or catalysts. An additional intermediate step after formation of the first block might be effective to optimize the reaction condition for the second monomer. Classic alkyllithium-initiated living anionic vinyl-addition polymerization of e.g. styrene, isoprene, or 1,4-butadiene is a well-established method for controlled synthesis of polyolefins [31, 156, 157]. But the direct crossover to anionic ROP of EO is obstructed by the strong *O-Li* ionic interaction [57]. Addition of $^t BuP_4$ could effectively separate

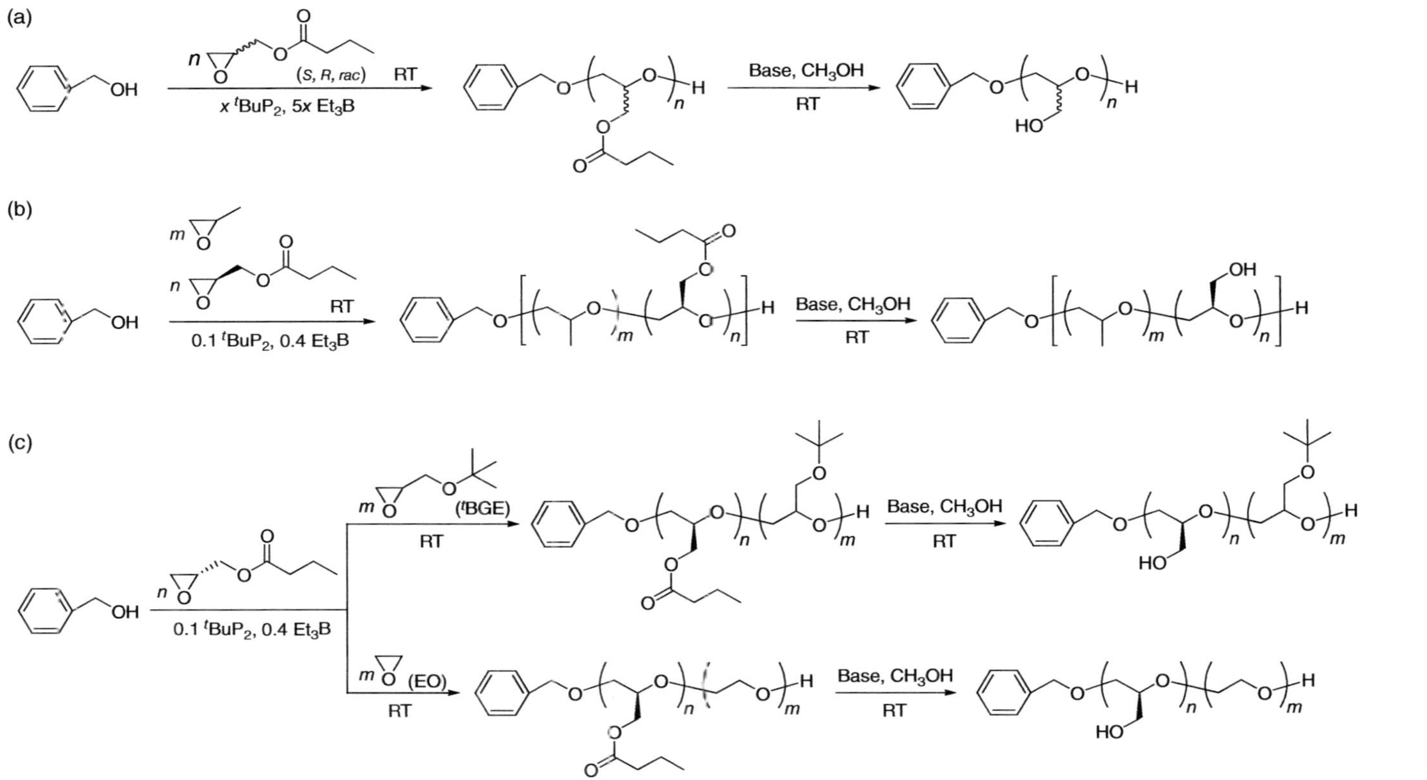

Figure 8.7 Chemoselective ROP of GB and copolymerization of GB with other epoxides catalyzed by ${}^{t}BuP_2$-Et_3B and followed by methanolysis to afford (a) linear PGC (with or without stereoregularity), (b) P(PO-*co*-*R*GC) random copolymer, and (c) amphiphilic P*S*GC-*b*-P^{t}BGE and double-hydrophilic P*S*GC-*b*-PEO block copolymers.

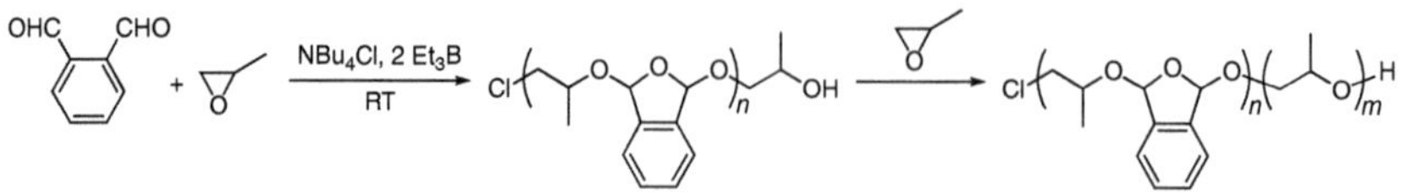

Figure 8.8 One-pot two-step synthesis of polyacetal-*b*-polyether block copolymer by copolymerization of *o*-phthalaldehyde and PO aided by a large excess of Et_3B.

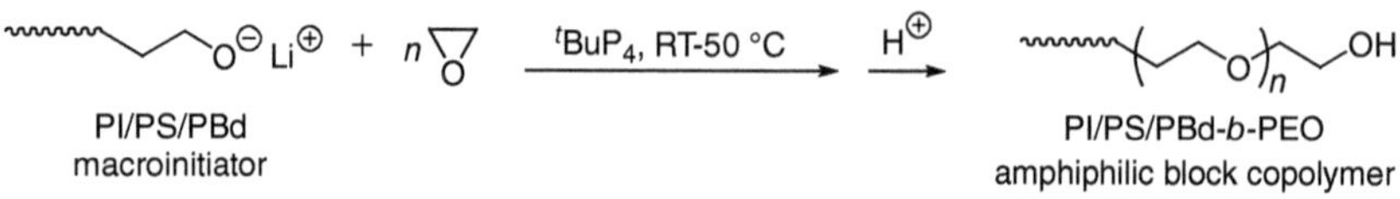

Figure 8.9 Synthetic approach to polyolefin-*b*-PEO hybrid amphiphilic block copolymers by welding anionic vinyl-addition polymerization and ^{t}BuP$_4$-catalyzed ROP of EO.

the contact *O-Li* ion pair thus significantly enhancing the activity of the oxyanionic chain end. In this way, one-pot synthesis of di-/triblock co/terpolymers, including PS-*b*-PEO [56], polyisoprene(PI)-*b*-PEO [58, 59], polybutadiene(PBd)-*b*-PEO [59, 158], PS-*b*-PI-*b*-PEO [159], PBd-*b*-PI-*b*-PEO [160], and PEO-*b*-PI-*b*-PEO [60], were implemented successfully (Figure 8.9). In the presence of ^{t}BuP$_4$, the oxyanion at the end of PS-*b*-PEO could further trigger the ROP of EEGE, and the as-afforded block terpolymer could be in situ transformed into PS-*b*-PEO-*b*-PGC with multiple pendant hydroxyl groups [76, 161]. These materials have been receiving profound attention and exhibiting interesting phase behavior, as documented earlier and very recently [162, 163].

Alcohol-initiated ROP of cyclic esters and cyclic carbonates is also among the main research subjects in the field of organocatalytic polymerization [164]. In general, milder basicity is favorable for the organobase catalysts due to the higher base susceptibility of the polyesters and polycarbonates compared with polyethers. For instance, ^{t}BuP$_4$ can trigger highly efficient ROP of cyclic esters, but sacrifices control over polyester structures due to violent macromolecular transesterification, thus being an unsuitable catalyst for sequential block copolymerization with EO [165]. With milder basicity, ^{t}BuP$_2$ enabled much more controlled sequential ROP of EO, ε-caprolactone (CL), and L-Lactide (LLA), yielding PEO-*b*-poly(ε-caprolactone)-*b*-poly(L-lactide) (PEO-*b*-PCL-*b*-PLLA) triblock terpolymers and PLLA-*b*-PCL-*b*-PEO-*b*-PCL-*b*-PLLA pentablock terpolymers (Figure 8.10(a)) [84]. However, the activity of ^{t}BuP$_2$ is insufficient for involving substituted epoxides.

Indeed, acidic organocatalysts showed better control and chemoselectivity for the ROP of cyclic esters [18, 167]. The successive use of ^{t}BuP$_4$ and diphenyl phosphate (DPP) via a base-to-acid catalyst switch strategy realized one-pot synthesis of a diversity of polyether-*b*-polyester/polycarbonate block copolymers (Figure 8.10(b)) [165]. Specifically, ROP of EO and/or BO was first performed from an alcohol with a small amount of ^{t}BuP$_4$. Then DPP was added in excess to convert to an acidic catalytic condition that suits the ROP of subsequently added CL, δ-valerolactone

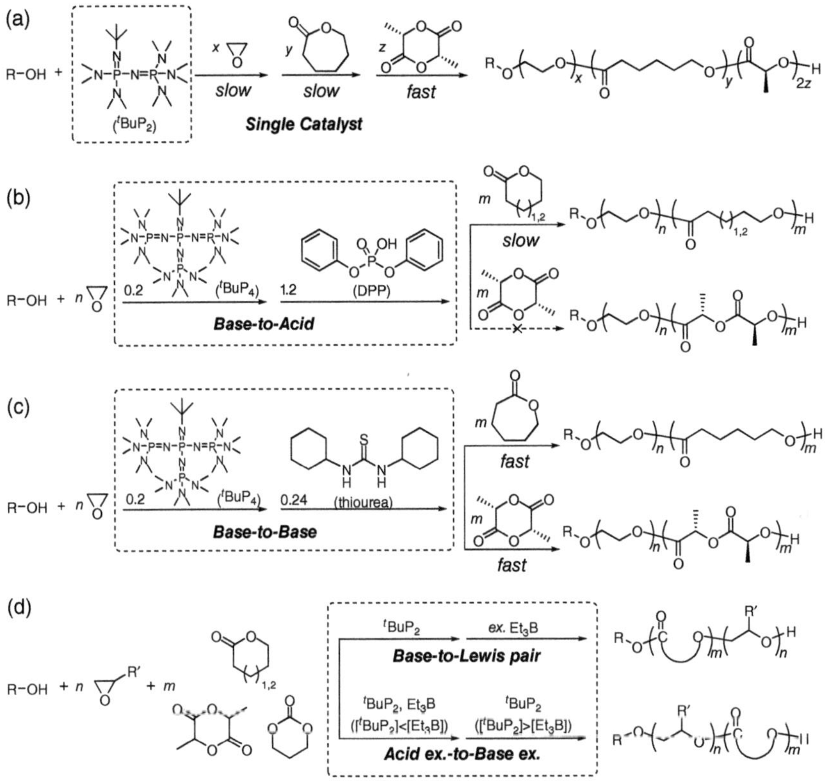

Figure 8.10 One-pot continuous synthesis of polyether-*b*-polyester block copolymers using (a) a single-component PB catalyst, (b) a base-to-acid catalyst switch strategy, (c) a base-to-base catalyst switch strategy, and (d) a biased Lewis pair catalyst which also achieved the sequence-reversed polyester-*b*-polyether block copolymers. Source: Adapted from [166].

and alkylated derivatives, or trimethylene carbonate [168, 169]. This strategy also enabled tandem implementation of ROP of epoxides and SGP of the polyether diol with a diisocyanate. Crosslinking that would have been caused by the strongly basic catalyst was effectively inhibited due to the catalyst switch [170]. On the other hand, the phosphazenium phosphate salt generated in the switching step is catalytically inactive and exerted a retardation effect on the DPP-catalyzed ROP of cyclic esters. So, a relatively large amount of DPP was needed to guarantee a reasonable reaction time scale. Besides, DPP is ineffective for lactides (LAs) and easily deactivated by traces of unreacted epoxides [165].

The catalyst switch strategy was then updated into a base-to-base version with DPP replaced by thiourea (Figure 8.10(c)) [166]. The much lower acidity of thiourea ensured that its deprotonated form still had adequate basicity for catalyzing the ROP of cyclic esters, including CL and LLA, thus avoiding the use of excess thiourea even in the presence of unreacted epoxide. Also, the hydrogen-donor-acceptor bifunctional nature of the thiourea anion enabled highly efficient and controlled formation of the polyester segments in hours or even minutes with macromolecular transesterification evidently suppressed. Onium salt was used as the initiator along with the thiourea-induced activity switch to obtain well-defined poly(glycidyl phenyl ether)-*b*-polylactide block copolymers in one pot [171]. A combinational base-to-acid-to-base strategy through successive addition of ^{t}BuP$_4$, DPP, and ^{t}BuP$_2$ was formulated for the one-pot synthesis of polyether-*b*-PCL-*b*-PLLA triblock terpolymers [74, 172].

Synthesis of polyether-*b*-polyester block copolymers with reverse sequence structures was a major challenge because anionic ROP of epoxide under conventional strongly basic conditions would induce severe transesterification on the polyester blocks [173, 174]. A biased Lewis pair catalytic strategy was developed and shown effective to solve this problem. In the mixture of two monomers and initiator, ROP of cyclic ester was performed first in the presence of ^{t}BuP$_2$ (or base-excess Lewis pair) while the epoxide stayed intact. Then ROP of the epoxide was launched from the polyester chain end after the addition of Et_3B to outnumber ^{t}BuP$_2$ (Figure 8.10(d)) [175]. The easily switchable monomer selectivity based on catalyst composition (acid-to-base ratio) allowed the one-pot block copolymerization to occur in both ester-first and ether-first orders, or to undergo several successive switches and deliver a multiblock copolymer. The same method was successfully applied to the alternating copolymerization of isothiocyanate and epoxide, coupled with the homopolymerization of epoxide, to afford sequence-reversible polythioimidocarbonate-*b*-polyether block copolymers [176]. The enriched sequence structures will expand fundamental understanding about the structure–property relationship of block copolymers and assist related materials design.

8.4.2 One-Step Block Copolymerization

Spontaneous formation of chemically distinct and covalently connected polymer segments in one synthetic step from mixed monomers, initiator, and catalyst is the most ideal way of block copolymer synthesis. However, successive polymerization of monomers from one initiating species in a strictly sequence-defined manner is usually challenged by stringent kinetic requirement [177]. Typically, the second monomer does polymerize from the chain end of the first block but only after complete conversion of the first monomer is reached. It is also required that there should be no disruptive and irreversible chain transfer or termination in the whole process. Plenty of successes have been made in recent decades on one-step sequence-selective ring-opening copolymerization, also widely referred to as (self-)switchable (co)polymerization, taking advantage of the large variety of heterocyclic monomers (with distinct polymerizing groups) and their substantially large reactivity difference [17]. While most of the achievements are made in (quasi)three-component mixed monomers, especially those involving a cyclic anhydride and an epoxide [178], only a few successes have been found in the literature for conventional A_nB_m type block copolymers from binary mixed monomers. In 2018, *N*-sulfonyl aziridine (Az) and EO were reported to block copolymerize in an

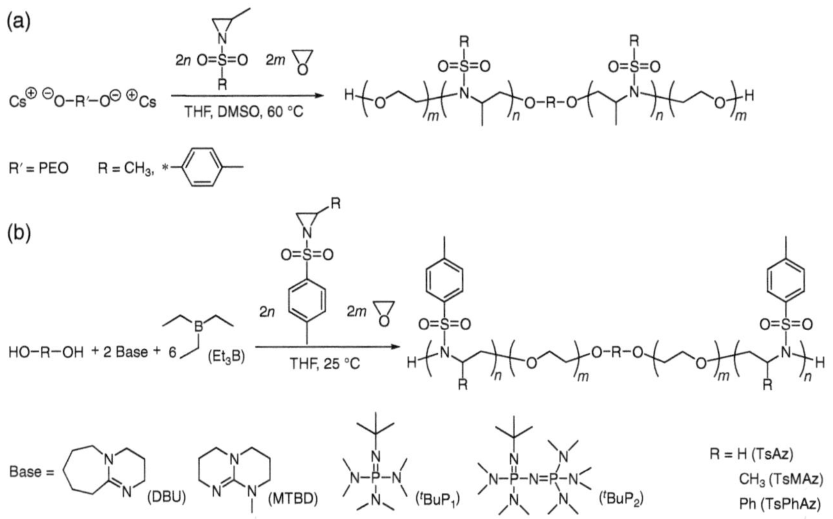

Figure 8.11 One-step block copolymerization of EO and Az by (a) cesium alkoxide-initiated anionic copolymerization in an Az-first order and (b) alcohol-initiated organocatalytic copolymerization in a reverse (EO-first) order. Source: Adapted from [180].

Az-first order using cesium alkoxide as the initiator and fulfilled one-step synthesis of amphiphilic block copolymers (Figure 8.11(a)) [179]. The efficient crossover from sulfonamide-anionic to oxyanionic ROP was the key to the continuous block copolymerization and the distinct activity difference ensured that ROP of EO only began when Az was almost depleted.

Recently, the PB-Et_3B two-component organocatalyst was employed for the copolymerization of EO and Az (Figure 8.11(b)) [180]. Optimal Lewis acid-to-base ratio allowed the two monomers to strictly block-copolymerize in a reverse (EO-first) order. Livingness of the copolymerization facilitated one-pot synthesis of multiblock copolymers by addition of mixed monomers in batches. Calculation results revealed that Et_3B accelerated the enchainment of EO and decelerated that of Az because of its much stronger complexation with the oxyanion than the sulfonamide anion. This unique Janus catalytic effect is key to enlarging the activity difference and fulfilling strictly reversed monomers enchainment order. The sequence structure was shown to significantly influence the aqueous solubility and potential applications of the amphiphilic triblock copolymers.

A similar principle was also well-applied to the binary monomer mixture of phthalic anhydride (PA) and EO/PO (Figure 8.12(a)) [181, 182]. The difference was that the compositionally optimized catalyst allowed strict alternating copolymerization to occur first until PA was fully consumed, and then the excess epoxide acted as "the third monomer" actualizing one-step synthesis of $(AB)_nB_m$ type block copolymers with controlled molar mass up to 60.0 kg mol^{-1} and low $Đ_M$ (<1.2) [181]. The distinct physicochemical properties of the polyether and polyester segments facilitated microphase-separation in bulk and self-assembly in aqueous or alcoholic solutions to form monodisperse nanoscale micelle aggregates. The Janus

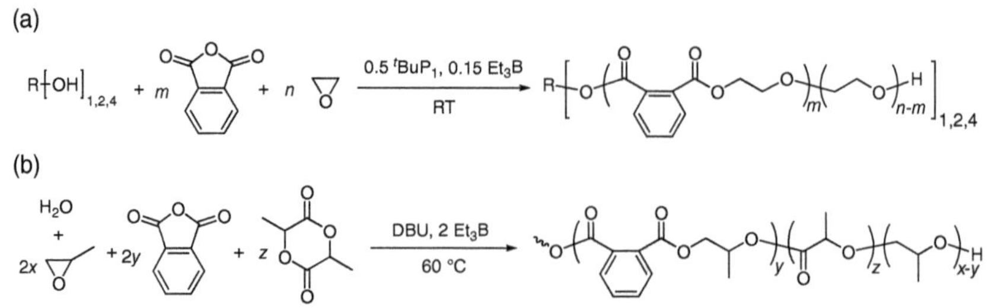

Figure 8.12 Kinetically controlled one-step approaches to polyester-*b*-polyether block copolymers in a cyclic anhydride and epoxide binary system and b cyclic anhydride, epoxide, and lactide ternary system.

catalytic effect was clearly exhibited in this system, as well as the previous one, for regulating the relative activity of the comonomers and the microstructure of the copolymers, i.e. from gradient to block with the same molar mass and composition, simply by changing the acid-to-base ratio [180, 181, 183].

Similar block copolymers were also synthesized from *exo*-norbornene anhydride, tetrahydrophthalic anhydride, etc., and epoxides such as BO, AGE, etc [182]. With the loaded amounts of DBU and Et_3B being 1 and 2 equiv. of the initiator (water), alternating copolymerization of PA and PO, ROP of LA, and ROP of PO consecutively occurred at 60 °C allowing one-step synthesis of PPO-*b*-PLLA-*b*-P(PA-*alt*-PO)-*b*-PLLA-*b*-PPO pentablock terpolymers (Figure 8.12(b)) [184]. The large excess of Et_3B might be key to the formation of polyether blocks, because it was found in a later study that when the amount of Et_3B was lower than that of initiating species, the carboxyl end groups existing during the copolymerization of cyclic anhydride and epoxide at elevated temperature would transform Et_3B by protonolysis into oxygenated boron species which lost the catalytic activity for the ROP of epoxide [185].

As an indispensable alternative strategy, bifunctional initiator that would launch the polymerizations of two chemically distinct monomers, either simultaneously or separately, has also received a large amount of attention [187]. Theoretically, block efficiency and sequence selectivity were independent of the kinetic characters of either polymerization, but on the premise that two mechanistically different polymerizations are compatible with each other. The development of organocatalytic ROP has been accompanied by its conjunction with other polymerization techniques, especially controlled radical polymerization of vinyl monomers, for the synthesis of hybrid block copolymers [17, 50, 188]. Due to the instability of radical initiators and/or CTAs towards strong nucleophiles, organobase-catalyzed anionic ROP of epoxides may be hardly compatible with these techniques. However, the $^tBuP_2$-Et_3B two-component organocatalyst has fulfilled such a purpose. With hydroxyl-functionalized trithiocarbonate photoiniferter as a dual initiator, light-induced reversible addition-fragmentation chain transfer (RAFT) polymerization of acrylamides such as *N, N*-dimethylacrylamide and ROP of PO/BO could be performed at RT in arbitrary order or simultaneously (Figure 8.13) [186]. The unique chemoselectivity of the organic Lewis pair catalyst

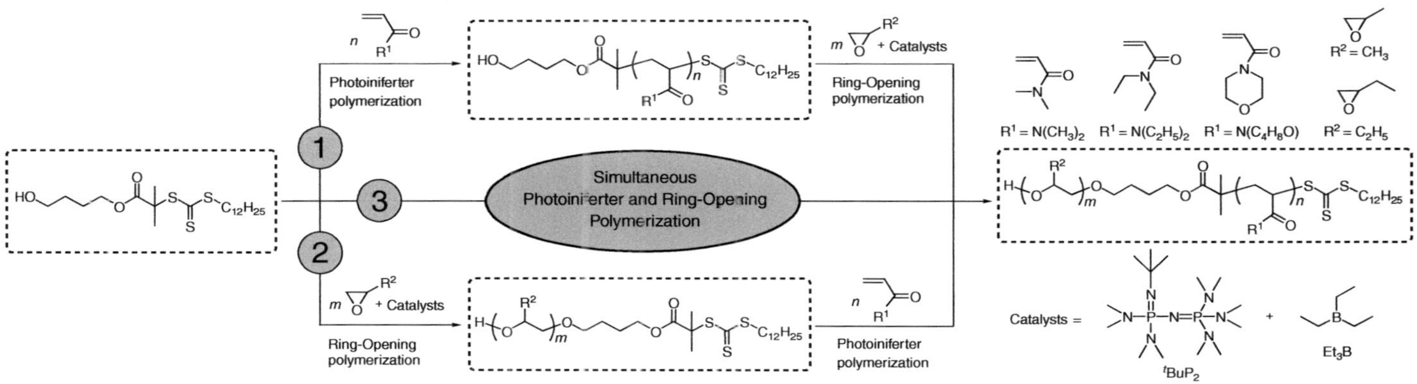

Figure 8.13 Three routes to polyacrylamide-*b*-polyether block copolymers by merged photoiniferter (RAFT) polymerization of acrylamides and ROP of epoxides catalyzed by $^{t}BuP_2$-Et_3B. Route 1: photoiniferter polymerization followed by ROP; route 2: ROP followed by photoiniferter polymerization; route 3: simultaneous photoiniferter polymerization and ROP. Source: Adapted from [186].

against transesterification, together with a mild reaction condition, ensured that the oxyanionic species reacted only with the epoxide, avoiding alcoholysis of the photoiniferter.

8.5 Conclusion and Outlook

Among the most popular research subjects in the realm of organocatalytic polymerization, anionic ROP of epoxides has been through rapid progress from single-component to two-component or bifunctional catalysts. The latter shows enormously enhanced catalytic activity especially when the Lewis acidic components are tethered to the cationic counterion. Also, the monomer-activation and chain-end-complexation synergetic effect has significantly improved the chemoselectivity against irreversible chain transfer to monomers and polymers containing base-labile moieties. The structural diversity of functionalized polyethers and block copolymer derivatives are therefore being noticeably expanded with simplified synthetic routes. It is foreseeable that organocatalytic ROP of epoxides is marching towards multidimensional control of polyether-based materials. A few challenges are inevitably to be addressed in this way, including but not limited to stereoselective ROP from racemic monomers, site-specific ROP involving multifunctional (bio)substances, tolerance for highly reactive functional groups, compatibility and merging with organometallic catalysts, large-scale and industrialized production of metal-free and biosafety polyethers.

References

1 Klein, R. and Wurm, F.R. (2015). Aliphatic polyethers: classical polymers for the 21st century. *Macromol. Rapid Commun.* 36: 1147.

2 Herzberger, J., Niederer, K., Pohlit, H. et al. (2016). Polymerization of ethylene oxide, propylene oxide, and other alkylene oxides: synthesis, novel polymer architectures, and bioconjugation. *Chem. Rev.* 116: 2170.

3 Childers, M.I., Longo, J.M., Van Zee, N.J. et al. (2014). Stereoselective epoxide polymerization and copolymerization. *Chem. Rev.* 114: 8129.

4 Taton, D. (2018). Metal-Free Polyether Synthesis by Organocatalyzed Ring-Opening Polymerization. In: *Organic Catalysis for Polymerisation*. The Royal Society of Chemistry.

5 von Wandruszka, R. (1998). *Nonionic Surfactants: Organic Chemistry*. ACS Publications.

6 Kabanov, A.V., Batrakova, E.V., and Alakhov, V.Y. (2002). Pluronic® block copolymers as novel polymer therapeutics for drug and gene delivery. *J. Control. Release* 82: 189.

7 Fusco, S., Borzacchiello, A., and Netti, P.A. (2006). Perspectives on: PEO-PPO-PEO triblock copolymers and their biomedical applications. *J. Bioact. Compat. Polym.* 21: 149.

8 Mangold, C., Wurm, F., and Frey, H. (2012). Functional PEG-based polymers with reactive groups via anionic ROP of tailor-made epoxide. *Polym. Chem.* 3: 1714.

9 Brocas, A.-L., Mantzaridis, C., Tunc, D., and Carlotti, S. (2013). Polyether synthesis: from activated or metal-free anionic ring-opening polymerization of epoxides to functionalization. *Prog. Polym. Sci.* 38: 845.

10 Thomas, A., Müller, S.S., and Frey, H. (1935). Beyond poly(ethylene glycol): linear polyglycerol as a multifunctional polyether for biomedical and pharmaceutical applications. *Biomacromolecules* 2014: 15.

11 Xia, Y. and Zhao, J. (2018). Macromolecular architectures based on organocatalytic ring-opening (co)polymerization of epoxides. *Polymer* 143: 343.

12 Zalipsky, S. (1995). Functionalized poly(ethylene glycols) for preparation of biologically relevant conjugates. *Bioconjugate Chem.* 6: 150.

13 Chen, Y., Liu, S., and Zhao, J. (2019). Organocatalytic/metal-free ring-opening polymerization and copolymerization of epoxides. *Acta Polym. Sin.* 10: 1067.

14 Herman, S., Hooftman, G., and Schacht, E. (1995). Poly(ethylene glycol) with reactive endgroups: I. Modification of proteins. *J. Bioact. Compat. Polym.* 10: 145.

15 Zalipsky, S. (1995). Chemistry of polyethylene glycol conjugates with biologically active molecules. *Adv. Drug Delivery. Rev.* 16: 157.

16 Zhang, C., Geng, X., Zhang, X. et al. (2023). Alkyl borane-mediated metal-free ring-opening (co)polymerizations of oxygenated monomers. *Prog. Polym. Sci.* 136: 101644.

17 He, G., Gao, S., Li, H., and Zhao, J. (2023). One-step synthesis of hybrid block copolymers through metal-free catalysis. *Materials China* 42: 442.

18 Song, Q., Hu, S., Zhao, J., and Zhang, G. (2017). Organocatalytic copolymerization of mixed type monomers. *Chin. J. Polym. Sci.* 35: 581.

19 Hadjichristidis, N., Iatrou, H., Pitsikalis, M., and Mays, J. (2006). Macromolecular architectures by living and controlled/living polymerizations. *Prog. Polym. Sci.* 31: 1068.

20 McGraw, M.L. and Chen, E.Y.X. (2020). Lewis pair polymerization: perspective on a ten-year journey. *Macromolecules* 53: 6102.

21 Kubisa, P. and Penczek, S. (1999). Cationic activated monomer polymerization of heterocyclic monomers. *Prog. Polym. Sci.* 24: 1409.

22 Penczek, S., Cypryk, M., Duda, A. et al. (2007). Living ring-opening polymerizations of heterocyclic monomers. *Prog. Polym. Sci.* 32: 247.

23 Ghosh, S., Lund, H., Jiao, H., and Mejía, E. (2017). Rediscovering the isospecific ring-opening polymerization of racemic propylene oxide with dibutylmagnesium. *Macromolecules* 50: 1245.

24 Tsuruta, T., Inoue, S., and Yokota, Y. (1967). Synthesis of optically active polymers by asymmetric catalysts. VII. Stereoselective polymerization of propylene oxide catalyzed by organomagnesium compounds. *Die Makromolekulare Chemie* 103: 164.

25 Furukawa, J., Tsuruta, T., Sakata, R. et al. (1959). Polymerization of propylene oxide by diethylzine in the presence of cocatalysts. *Die Makromolekulare Chemie* 32: 90.

26 Chen, S., Xu, N., and Shi, J. (2004). Structure and properties of polyether polyols catalyzed by Fe/Zn double metal cyanide complex catalyst. *Prog. Org. Coat.* 49: 125.

27 Ionescu, M. (2005). *Chemistry and Technology of Polyols for Polyurethanes*. Rapra Technology.

28 Lipinski, B.M., Morris, L.S., Silberstein, M.N., and Coates, G.W. (2020). Isotactic poly (propylene oxide): a photodegradable polymer with strain hardening properties. *J. Am. Chem. Soc.* 142: 6800.

29 Peretti, K.L., Ajiro, H., Cohen, C.T. et al. (2005). A highly active, isospecific cobalt catalyst for propylene oxide polymerization. *J. Am. Chem. Soc.* 127: 11566.

30 Hirahata, W., Thomas, R.M., Lobkovsky, E.B., and Coates, G.W. (2008). Enantioselective polymerization of epoxides: a highly active and selective catalyst for the preparation of stereoregular polyethers and enantiopure epoxides. *J. Am. Chem. Soc.* 130: 17658.

31 Hadjichristidis, N. and Hirao, A. (2015). *Anionic Polymerization: Principles, Practice, Strength, Consequences and Applications*. Springer.

32 Boileau, S. and Illy, N. (2011). Activation in anionic polymerization: why phosphazene bases are very exciting promoters. *Prog. Polym. Sci.* 36: 1132.

33 Zhao, J., Hadjichristidis, N., and Gnanou, Y. (2014). Phosphazene-promoted anionic polymerization. *Polimery* 59: 49.

34 Ezra, G. and Zilkha, A. (1970). Anionic graft polymerization of propylene oxide on starch. *Eur. Polym. J.* 6: 1305.

35 Jannasch, P. (2000). Preparation and characterisation of aggregating comblike poly(propylene oxide). *Polymer* 41: 6701.

36 Price, C.C. and Carmelite, D.D. (1966). Reactions of epoxides in dimethyl sulfoxide catalyzed by potassium t-butoxide. *J. Am. Chem. Soc.* 88: 4039.

37 Mangold, C., Wurm, F., Obermeier, B., and Frey, H. (2010). Hetero-multifunctional poly(ethylene glycol) copolymers with multiple hydroxyl groups and a single terminal functionality. *Macromol. Rapid Commun.* 31: 258.

38 Thomas, A., Niederer, K., Wurm, F., and Frey, H. (2014). Combining oxyanionic polymerization and click-chemistry: a general strategy for the synthesis of polyether polyol macromonomers. *Polym. Chem.* 5: 899.

39 Schömer, M. and Frey, H. (2012). Water-soluble "poly(propylene oxide)" by random copolymerization of propylene oxide with a protected glycidol monomer. *Macromolecules* 45: 3039.

40 Deffieux, A., Carlotti, S., and Barrère, A. (2012). Anionic ring-opening polymerization of epoxides and related nucleophilic polymerization processes. In: *Polymer Science: A Comprehensive Reference* (ed. K. Matyjaszewski and M. Möller), 117. Amsterdam: Elsevier.

41 Allgaier, J., Willbold, S., and Chang, T. (2007). Synthesis of hydrophobic poly(alkylene oxide)s and amphiphilic poly(alkylene oxide) block copolymers. *Macromolecules* 40: 518.

42 Ding, J., Heatley, F., Price, C., and Booth, C. (1991). Use of crown ether in the anionic polymerization of propylene oxide—2. Molecular weight and molecular weight distribution. *Eur. Polym. J.* 27: 895.

43 Deffieux, A., Graf, E., and Boileau, S. (1981). Anionic polymerization of ethylene oxide with cryptates as counterions: 2. *Polymer* 22: 549.

44 Morinaga, H., Ochiai, B., and Endo, T. (2007). Metal-free ring-opening polymerization of glycidyl phenyl ether by tetrabutylammonium fluoride. *Macromolecules* 40: 6014.

45 Morinaga, H., Ujihara, Y., Yamanaka, T. et al. (2011). Metal-free ring-opening polymerization of glycidyl phenyl ether initiated by tetra-n-butylammonium acetate and its application to the hydroxyl-terminated telechelic polymer. *J. Polym. Sci., Part A: Polym. Chem.* 49: 4092.

46 Kiesewetter, M.K., Shin, E.J., Hedrick, J.L., and Waymouth, R.M. (2010). Organocatalysis: opportunities and challenges for polymer synthesis. *Macromolecules* 43: 2093.

47 Ottou, W.N., Sardon, H., Mecerreyes, D. et al. (2016). Update and challenges in organo-mediated polymerization reactions. *Prog. Polym. Sci.* 56: 64.

48 Zhang, X., Fevre, M., Jones, G.O., and Waymouth, R.M. (2018). Catalysis as an enabling science for sustainable polymers. *Chem. Rev.* 118: 839.

49 Dove, A., Sardon, H., and Naumann, S. (2018). *Organic Catalysis for Polymerisation*. The Royal Society of Chemistry.

50 Hu, S., Zhao, J., Zhang, G., and Schlaad, H. (2017). Macromolecular architectures through organocatalysis. *Prog. Polym. Sci.* 74: 34.

51 Naumann, S. (1834). Borane catalysis for epoxide (co)polymerization. *Polym. Chem.* 2023: 14.

52 Zhang, Y.-Y., Yang, G.-W., and Wu, G.-P. (2022). Recent progress in synthesizing polyethers by use of organocatalysts. *Synlett* 33: 8.

53 Ding, J., Price, C., and Booth, C. (1991). Use of crown ether in the anionic polymerization of propylene oxide—1. Rate of polymerization. *Eur. Polym. J.* 27: 891.

54 Schwesinger, R. and Schlemper, H. (1987). Peralkylated Polyaminophosphazenes—Extremely Strong, Neutral Nitrogen Bases. *Angew. Chem. Int. Ed.* 26: 1167.

55 Schwesinger, R., Schlemper, H., Hasenfratz, C. et al. (1996). Extremely strong, uncharged auxiliary bases; monomeric and polymer-supported polyaminophosphazenes (P2–P5). *Liebigs Ann.* 1996: 1055.

56 Esswein, B. and Möller, M. (1996). Polymerization of ethylene oxide with alkyllithium compounds and the phosphazene base “tBu-P4”. *Angew. Chem. Int. Ed.* 35: 623.

57 Schmalz, H., Lanzendörfer, M.G., Abetz, V., and Müller, A.H.E. (2003). Anionic polymerization of ethylene oxide in the presence of the phosphazene base ButP4–kinetic investigations using in-situ FT-NIR spectroscopy and MALDI-ToF MS. *Macromol. Chem. Phys.* 204: 1056.

58 Floudas, G., Vazaiou, B., Schipper, F. et al. (2001). Poly(ethylene oxide-*b*-isoprene) diblock copolymer phase diagram. *Macromolecules* 34: 2947.

59 Förster, S. and Krämer, E. (1999). Synthesis of PB-PEO and PI-PEO block copolymers with alkyllithium initiators and the phosphazene base t-BuP4. *Macromolecules* 32: 2783.

60 Pispas, S. (2006). Double hydrophilic block copolymers of sodium(2-sulfamate-3-carboxylate)isoprene and ethylene oxide. *J. Polym. Sci. A Polym. Chem.* 44: 606.

61 Eßwein, B., Steidl, N.M., and Möller, M. (1996). Anionic polymerization of oxirane in the presence of the polyiminophosphazene base t-Bu-P4. *Macromol. Rapid Commun.* 17: 143.

62 Zhao, J., Schlaad, H., Weidner, S., and Antonietti, M. (2012). Synthesis of terpene–poly(ethylene oxide)s by t-BuP4-promoted anionic ring-opening polymerization. *Polym. Chem.* 3: 1763.

63 Zhao, J., Mountrichas, G., Zhang, G., and Pispas, S. (2009). Amphiphilic polystyrene-*b*-poly (p-hydroxystyrene-g-ethylene oxide) block–graft copolymers via a combination of conventional and metal-free anionic polymerization. *Macromolecules* 42: 8661.

64 Zhao, J., Mountrichas, G., Zhang, G., and Pispas, S. (2010). Thermoresponsive core–shell brush copolymers with poly (propylene oxide)-*block*-poly(ethylene oxide) side chains via a "grafting from" technique. *Macromolecules* 43: 1771.

65 Zhao, J., Zhang, G., and Pispas, S. (2010). Thermoresponsive brush copolymers with poly (propylene oxide-ran-ethylene oxide) side chains via metal-free anionic polymerization "grafting from" technique. *J. Polym. Sci., Part A: Polym. Chem.* 48: 2320.

66 Zhao, J., Alamri, H., and Hadjichristidis, N. (2013). A facile metal-free "grafting-from" route from acrylamide-based substrate toward complex macromolecular combs. *Chem. Commun.* 49: 7079.

67 Zhao, J. and Schlaad, H. (2011). Controlled anionic graft polymerization of ethylene oxide directly from poly(*N*-isopropylacrylamide). *Macromolecules* 44: 5861.

68 Dentzer, L., Bray, C., Noinville, S. et al. (2015). Phosphazene-promoted metal-free ring-opening polymerization of 1, 2-epoxybutane initiated by secondary amides. *Macromolecules* 48: 7755.

69 Zhao, J., Pahovnik, D., Gnanou, Y., and Hadjichristidis, N. (2014). Phosphazene-promoted metal-free ring-opening polymerization of ethylene oxide initiated by carboxylic acid. *Macromolecules* 47: 1693.

70 Misaka, H., Tamura, E., Makiguchi, K. et al. (1941). Synthesis of end-functionalized polyethers by phosphazene base-catalyzed ring-opening polymerization of 1, 2-butylene oxide and glycidyl ether. *J. Polym. Sci., Part A: Polym. Chem.* (2012): 50.

71 Satoh, Y., Matsuno, H., Yamamoto, T. et al. (2017). Synthesis of well-defined three-and four-armed cage-shaped polymers via "topological conversion" from trefoil-and quatrefoil-shaped polymers. *Macromolecules* 50: 97.

72 Isono, T., Kamoshida, K., Satoh, Y. et al. (2013). Synthesis of star- and figure-eight-shaped polyethers by t-Bu-P4-catalyzed ring-opening polymerization of butylene oxide. *Macromolecules* 46: 3841.

73 Misaka, H., Sakai, R., Satoh, T., and Kakuchi, T. (2011). Synthesis of high molecular weight and end-functionalized poly(styrene oxide) by living ring-opening polymerization of styrene oxide using the alcohol/phosphazene base initiating system. *Macromolecules* 44: 9099.

74 Alamri, H. and Hadjichristidis, N. (2016). One-pot synthesis of well-defined polyether/polyester block copolymers and terpolymers by a highly efficient catalyst switch approach. *Polym. Chem.* 7: 3225.

75 Kwon, W., Rho, Y., Kamoshida, K. et al. (2012). Well-defined functional linear aliphatic diblock copolyethers: a versatile linear aliphatic polyether platform for selective functionalizations and various nanostructures. *Adv. Funct. Mater.* 22: 5194.

76 Toy, A.A., Reinicke, S., Müller, A.H.E., and Schmalz, H. (2007). One-pot synthesis of polyglycidol-containing block copolymers with alkyllithium initiators using the phosphazene base t-BuP4. *Macromolecules* 40: 5241.

77 Satoh, Y., Miyachi, K., Matsuno, H. et al. (2016). Synthesis of well-defined amphiphilic star-block and miktoarm star copolyethers via t-Bu-P4-catalyzed ring-opening polymerization of glycidyl ethers. *Macromolecules* 49: 499.

78 Isono, T. (2021). Synthesis of functional and architectural polyethers via the anionic ring-opening polymerization of epoxide monomers using a phosphazene base catalyst. *Polym. J.* 53: 753.

79 T. Isono, Y. Satoh, K. Miyachi, Y. Chen, S.-i. Sato, K. Tajima, T. Satoh, T. Kakuchi, Synthesis of linear, cyclic, figure-eight-shaped, and tadpole-shaped amphiphilic block copolyethers via t-Bu-P4-catalyzed ring-opening polymerization of hydrophilic and hydrophobic glycidyl ethers *Macromolecules* 2014, **47**, 2853.

80 Isono, T., Miyachi, K., Satoh, Y. et al. (2017). Design and synthesis of thermoresponsive aliphatic polyethers with a tunable phase transition temperature. *Polym. Chem.* 8: 5698.

81 Park, J., Kim, A., and Kim, B.-S. (2023). Anionic ring-opening polymerization of functional epoxide monomers in the solid state. *Nat. Commun.* 14: 5855.

82 Isono, T., Asai, S., Satoh, Y. et al. (2015). Controlled/living ring-opening polymerization of glycidylamine derivatives using t-Bu-P4/alcohol initiating system leading to polyethers with pendant primary, secondary, and tertiary amino groups. *Macromolecules* 48: 3217.

83 Lee, J., Han, S., Kim, M., and Kim, B.-S. (2019). Anionic polymerization of azidoalkyl glycidyl ethers and post-polymerization modification. *Macromolecules* 53: 355.

84 Zhao, J., Pahovnik, D., Gnanou, Y., and Hadjichristidis, N. (2014). Sequential polymerization of ethylene oxide, ε-caprolactone and l-lactide: a one-pot metal-free route to tri-and pentablock terpolymers. *Polym. Chem.* 5: 3750.

85 Zhang, H., Hu, S., Zhao, J., and Zhang, G. (2017). Expanding the scope of organocatalysis for alternating copolymerization of dihydrocoumarin and styrene oxide. *Eur. Polym. J.* 95: 693.

86 Zhang, H., Hu, S., Zhao, J., and Zhang, G. (2017). Phosphazene catalyzed alternating copolymerization of dihydrocoumarin and ethylene oxide: weaker is better. *Macromolecules* 50: 4198.

87 Raynaud, J., Absalon, C., Gnanou, Y., and Taton, D. (2009). *N*-heterocyclic carbene-induced zwitterionic ring-opening polymerization of ethylene oxide and direct synthesis of α, ω-difunctionalized poly(ethylene oxide)s and poly(ethylene oxide)-*b*-poly(ε-caprolactone) block copolymers. *J. Am. Chem. Soc.* 131: 3201.

88 Raynaud, J., Ottou, W.N., Gnanou, Y., and Taton, D. (2010). Metal-free and solvent-free access to α, ω-heterodifunctionalized poly(propylene oxide)s by N-heterocyclic carbene-induced ring opening polymerization. *Chem. Commun.* 46: 3203.

89 Naumann, S., Thomas, A.W., and Dove, A.P. (2015). N-Heterocyclic olefins as organocatalysts for polymerization: preparation of well-defined poly(propylene oxide). *Angew. Chem. Int. Ed.* 54: 9550.

90 Balint, A., Papendick, M., Clauss, M. et al. (2018). Controlled preparation of amphiphilic triblock-copolyether in a metal-and solvent-free approach for tailored structure-directing agents. *Chem. Commun.* 54: 2220.

91 Markus, F., Bruckner, J.R., and Naumann, S. (2020). Controlled synthesis of "reverse pluronic"-type block copolyethers with high molar masses for the preparation of hydrogels with improved mechanical properties. *Macromol. Chem. Phys.* 221: 1900437.

92 Xia, Y., Shen, J., Alamri, H. et al. (2017). Revealing the cytotoxicity of residues of phosphazene catalysts used for the synthesis of poly(ethylene oxide). *Biomacromolecules* 18: 3233.

93 Biedron, T., Kubisa, P., and Penczek, S. (1991). Polyepichlorohydrin diols free of cyclics: synthesis and characterization. *J. Polym. Sci., Part A: Polym. Chem.* 29: 619.

94 Bednarek, M., Kubisa, P., and Penczek, S. (1999). Coexistence of activated monomer and active chain end mechanisms in cationic copolymerization of tetrahydrofuran with ethylene oxide. *Macromolecules* 32: 5257.

95 Kubisa, P. (1988). Activated monomer mechanism in the cationic polymerization of cyclic ethers. *Makromol Chem., Macromol. Symp* 13: 203.

96 Wojtania, M., Kubisa, P., and Penczek, S. (1986). Polymerization of propylene oxide by activated monomer mechanism. Suppression of macrocyclics formation. *Makromol. Chem., Macromol. Symp.* 6: 201.

97 Brzezińska, K., Szymański, R., Kubisa, P., and Penczek, S. (1986). Activated monomer mechanism in cationic polymerization, 1. Ethylene oxide, formulation of mechanism. *Makromol. Chem., Rapid Commun.* 7: 1.

98 Hino, T. and Endo, T. (2004). Cationic ring-opening polymerization of an epoxide by tropylium salts as thermal-and photolatent initiators. *J. Polym. Sci., Part A: Polym. Chem.* 42: 2166.

99 Chakraborty, D., Rodriguez, A., and Chen, E.Y.X. (2003). Catalytic ring-opening polymerization of propylene oxide by organoborane and aluminum Lewis acids. *Macromolecules* 36: 5470.

100 Asenjo-Sanz, I., Veloso, A., Miranda, J.I. et al. (2014). Zwitterionic polymerization of glycidyl monomers to cyclic polyethers with $B(C_6F_5)_3$. *Polym. Chem.* 5: 6905.

101 Carlotti, S., Labbé, A., Rejsek, V. et al. (2008). Living/controlled anionic polymerization and copolymerization of epichlorohydrin with tetraoctylammonium bromide–triisobutylaluminum initiating systems. *Macromolecules* 41: 7058.

102 Gervais, M., Labbé, A., Carlotti, S., and Deffieux, A. (2009). Direct synthesis of α-azido, ω-hydroxypolyethers by monomer-activated anionic polymerization. *Macromolecules* 42: 2395.

103 Meyer, J., Keul, H., and Möller, M. (2011). Poly(glycidyl amine) and copolymers with glycidol and glycidyl amine repeating units: synthesis and characterization. *Macromolecules* 44: 4082.

104 Herzberger, J. and Frey, H. (2015). Epicyanohydrin: polymerization by monomer activation gives access to nitrile-, amino-, and carboxyl-functional poly(ethylene glycol). *Macromolecules* 48: 8144.

105 Gervais, M., Forens, A., Ibarboure, E., and Carlotti, S. (2018). Anionic polymerization of activated oxetane and its copolymerization with ethylene oxide for the synthesis of amphiphilic block copolymers. *Polym. Chem.* 9: 2660.

106 Labbé, A., Carlotti, S., Billouard, C. et al. (2007). Controlled high-speed anionic polymerization of propylene oxide initiated by onium salts in the presence of triisobutylaluminum. *Macromolecules* 40: 7842.

107 Rejsek, V., Sauvanier, D., Billouard, C. et al. (2007). Controlled anionic homo-and copolymerization of ethylene oxide and propylene oxide by monomer activation. *Macromolecules* 40: 6510.

108 Chidara, V.K., Gnanou, Y., and Feng, X. (2022). Using triethylborane to manipulate reactivity ratios in epoxide-anhydride copolymerization: application to the synthesis of polyethers with degradable ester functions. *Molecules* 27: 466.

109 Varghese, J.K., Hadjichristidis, N., Gnanou, Y., and Feng, X. (2019). Degradable poly (ethylene oxide) through metal-free copolymerization of ethylene oxide with l-lactide. *Polym. Chem.* 10: 3764.

110 Boopathi, S.K., Hadjichristidis, N., Gnanou, Y., and Feng, X. (2019). Direct access to poly (glycidyl azide) and its copolymers through anionic (co-)polymerization of glycidyl azide. *Nat. Commun.* 10: 293.

111 Gu, G.-G., Wang, L. Y., Zhang, R. et al. (2021). Synthesis of polyethers from epoxides via a binary organocatalyst system. *Polym. Chem.* 12: 6436.

112 Gu, Y., Kou, X., Wang, X., and Li, Z. (2023). Creating remarkably moisture- and air-stable macromolecular Lewis acid by integrating borane within the polymer chain: a highly active catalyst for homo(co)polymerization of epoxides. *Angew. Chem. Int. Ed.* e202318645.

113 Yang, G.W., Zhang, Y.Y., Xie, R., and Wu, G.P. (2020). High-activity organocatalysts for polyether synthesis via intramolecular ammonium cation assisted SN2 ring-opening polymerization. *Angew. Chem. Int. Ed.* 59: 16910.

114 Yang, G.-W., Zhang, Y.-Y., and Wu, G.-P. (2021). Modular organoboron catalysts enable transformations with unprecedented reactivity. *Acc. Chem. Res.* 54: 4434.

115 Qi, H., Xie, R., Yang, G.-W. et al. (2022). Rational optimization of bifunctional organoboron catalysts for versatile polyethers via ring-opening polymerization of epoxides. *Macromolecules*.

116 Chen, C., Gnanou, Y., and Feng, X. (2022). Borinane boosted bifunctional organocatalysts for ultrafast ring-opening polymerization of cyclic ethers. *Macromolecules* 55: 10662.

117 Wang, X., Hui, J., Shi, M. et al. (2022). Exploration of the synergistic effect in a one-component Lewis pair system: serving as a dual initiator and catalyst in the ring-opening polymerization of epoxides. *ACS Catalysis* 12: 8434.

118 Zhang, Y.-Y., Lu, C., Yang, G.-W. et al. (2022). Mechanism-inspired upgradation of phosphonium-containing organoboron catalysts for epoxide-involved copolymerization and homopolymerization. *Macromolecules* 55: 6443.

119 Brocas, A.-L., Deffieux, A., Le Malicot, N., and Carlotti, S. (2012). Combination of phosphazene base and triisobutylaluminum for the rapid synthesis of polyhydroxy telechelic poly(propylene oxide). *Polym. Chem.* 3: 1189.

120 Chen, Y., Shen, J., Liu, S. et al. (2018). High efficiency organic Lewis pair catalyst for ring-opening polymerization of epoxides with chemoselectivity. *Macromolecules* 51: 8286.

121 Zhang, C.-J., Duan, H.-Y., Hu, L.-F. et al. (2018). Metal-Free route to precise synthesis of poly(propylene oxide) and its blocks with high activity. *ChemSusChem* 11: 4209.

122 Hong, F.T., Ladelta, V., Gautam, R. et al. (2021). Polyether-based block co(ter) polymers as multifunctional lubricant additives. *ACS Appl. Polym. Mater.* 3: 3811.

123 Chen, Y., Liu, S., Zhao, J. et al. (2019). Chemoselective polymerization of epoxides from carboxylic acids: direct access to esterified polyethers and biodegradable polyurethanes. *ACS Macro Lett.* 8: 1582.

124 Zhang, Q., Zhou, Y., Chen, Y., and Zhao, J. (2021). Ethoxylation of phenols catalyzed by metal-free Lewis pairs: living/Controlled polymerization in a slow-initiation mode. *Chin. J. Chem.* 39: 2579.

125 Song, Q., Zhao, J., Zhang, G. et al. (2020). *N*-Heterocyclic carbene/Lewis acid-mediated ring-opening polymerization of propylene oxide. Part 1: triisobutylaluminum as an efficient controlling agent. *Eur. Polym. J.* 134: 109819.

126 Song, Q., Zhao, J., Zhang, G. et al. (2020). *N*-Heterocyclic carbene/Lewis acid-mediated ring-opening polymerization of propylene oxide. Part 2: toward dihydroxytelechelic polyethers using triethylborane. *Eur. Polym. J.* 134: 109839.

127 Vogler, C. and Naumann, S. (2020). A simplified approach for the metal-free polymerization of propylene oxide. *RSC Advances* 10: 43389.

128 Markus, F., Vogler, C., Bruckner, J.R., and Naumann, S. (2021). Ordered mesoporous carbons via self-assembly of tailored block copolyethers for pore size-dependent applications. *ACS Appl. Nano Mater* 4: 3486.

129 Matyjaszewski, K. (2011). Architecturally complex polymers with controlled heterogeneity. *Science* 333: 1104.

130 Obermeier, B., Wurm, F., Mangold, C., and Frey, H. (2011). Multifunctional poly(ethylene glycol)s. *Angew. Chem. Int. Ed.* 50: 7988.

131 Pohlit, H., Worm, M., Langhanki, J. et al. (2017). Silver oxide mediated monotosylation of poly(ethylene glycol)(PEG): heterobifunctional PEG via polymer desymmetrization. *Macromolecules* 50: 9196.

132 Hiki, S. and Kataoka, K. (2010). Versatile and selective synthesis of "click chemistry" compatible heterobifunctional poly(ethylene glycol)s possessing azide and alkyne functionalities. *Bioconjugate Chem.* 21: 248.

133 Yilmaz, F., Cianga, I., Ito, K. et al. (2003). Synthesis and characterization of α,ω-heterofunctional poly(ethylene oxide) macromonomers. *Macromol. Rapid Commun.* 24: 316.

134 Thompson, M.S., Vadala, T.P., Vadala, M.L. et al. (2008). Synthesis and applications of heterobifunctional poly(ethylene oxide) oligomers. *Polymer* 49: 345.

135 Harris, J.M. and Chess, R.B. (2003). Effect of pegylation on pharmaceuticals. *Nat. Rev. Drug Discov.* 2: 214.

136 Veronese, F.M. and Pasut, G. (2005). PEGylation, successful approach to drug delivery. *Drug Discov. Today* 10: 1451.

137 Schlaad, H., Kukula, H., Rudloff, J., and Below, I. (2001). Synthesis of α, ω-heterobifunctional poly(ethylene glycol)s by metal-free anionic ring-opening polymerization. *Macromolecules* 34: 4302.

138 Wang, X., Shi, M., Hui, J. et al. (2023). Controllable synthesis of hetero-functionalized polyethers via intramolecular phosphonium bisborane Lewis pair-catalyzed immortal ring-opening polymerization. *Macromolecules* 56: 4030.

139 Liu, S., Liu, L., Chen, Y., and Zhao, J. (2020). Selective polymerization of epoxides from hydroxycarboxylic esters: expediting controlled synthesis of α-carboxyl-ω-hydroxyl polyethers. *Chem. Commun.* 56: 12186.

140 Chen, Y., Zhang, P., Liu, S. et al. (2021). Noncovalent protection for direct synthesis of α-amino-ω-hydroxyl poly(ethylene oxide). *ACS Macro Lett.* 10: 737.

141 Liu, L., Zhao, J., and Zhang, G. (2023). Chemoselectivity streamlines the approach to linear and Y-shaped thiol-polyethers starting from thiocarboxylic acids. *ACS Macro Lett.* 12: 1185.

142 Wurm, F., Klos, J., Räder, H.J., and Frey, H. (2009). Synthesis and non-covalent protein conjugation of linear-hyperbranched PEG-poly(glycerol) α,ω_n-telechelics. *J. Am. Chem. Soc.* 131: 7954.

143 Pang, B., Yu, Y., and Zhang, W. (2021). Thermoresponsive polymers based on tertiary amine moieties. *Macromol. Rapid Commun.* 42: 2100504.

144 Song, J., Palanikumar, L., Choi, Y. et al. (2017). The power of the ring: a pH-responsive hydrophobic epoxide monomer for superior micelle stability. *Polym. Chem.* 8: 7119.

145 Alagi, P., Gnanou, Y., and Feng, X. (2023). Quasilinear polyglycidols by triethyborane-controlled anionic polymerization of unprotected glycidol. *Chem. Commun.* 59: 2958.

146 Tran, C.H., Lee, M.W., Kim, S.A. et al. (2020). Kinetic and mechanistic study of heterogeneous double metal cyanide-catalyzed ring-opening multibranching polymerization of glycidol. *Macromolecules* 53: 2051.

147 Spears, B.R., Marin, M.A., Montenegro-Burke, J.R. et al. (2022). Aqueous epoxide ring-opening polymerization (AEROP): green synthesis of polyglycidol with ultralow branching. *Macromolecules* 2016: 49.

148 Liu, S., Liu, L., Zhou, Y. et al. (2022). Selective ring-opening polymerization of glycidyl esters: a versatile synthetic platform for glycerol-based (co)polyethers. *Polym. Chem.* 13: 3650.

149 Hadjichristidis, N., Pispas, S., and Floudas, G. (2003). *Block Copolymers: Synthetic Strategies, Physical Properties, and Applications*. John Wiley & Sons.

150 Rösler, A., Vandermeulen, G.W.M., and Klok, H.-A. (2001). Advanced drug delivery devices via self-assembly of amphiphilic block copolymers. *Adv. Drug Delivery. Rev.* 53: 95.

151 Bates, C.M. and Bates, F.S. (2017). 50th anniversary perspective: block polymers—pure potential. *Macromolecules* 50: 3.

152 Wei, X., Gong, C., Gou, M. et al. (2009). Biodegradable poly (ϵ-caprolactone)–poly (ethylene glycol) copolymers as drug delivery system. *Int. J. Pharm.* 381: 1.

153 Castillo, R.V. and Müller, A.J. (2009). Crystallization and morphology of biodegradable or biostable single and double crystalline block copolymers. *Prog. Polym. Sci.* 34: 516.

154 Boyer, C., Bulmus, V., Davis, T.P. et al. (2009). Bioapplications of RAFT polymerization. *Chem. Rev.* 109: 5402.

155 Patil, N., Gnanou, Y., and Feng, X. (2022). Anionic copolymerization of *o*-phthalaldehyde with epoxides: facile access to degradable polyacetals and their copolymers under ambient conditions. *Macromolecules* 55: 7817.

156 Lee, P.-C., Wang, C.-C., and Chen, C.-Y. (2020). Synthesis of high-vinyl isoprene and styrene triblock copolymers via anionic polymerization with difunctional t-BuLi initiator. *Eur. Polym. J.* 124: 109476.

157 Vazaios, A. and Hadjichristidis, N. (2005). Anionic polymerization of styrenic macromonomers of polyisoprene, polybutadiene, and polystyrene. *J. Polym. Sci., Part A: Polym. Chem* 43: 1038.

158 Pispas, S. and Hadjichristidis, N. (2003). Aggregation behavior of poly (butadiene-*b*-ethylene oxide) block copolymers in dilute aqueous solutions: effect of concentration, temperature, ionic strength, and type of surfactant. *Langmuir* 19: 48.

159 Bailey, T.S., Pham, H.D., and Bates, F.S. (2001). Morphological behavior bridging the symmetric AB and ABC states in the poly(styrene-*b*-isoprene-*b*-ethylene oxide) triblock copolymer system. *Macromolecules* 34: 6994.

160 Schmalz, H., Knoll, A., Müller, A.J., and Abetz, V. (2002). Synthesis and characterization of ABC triblock copolymers with two different crystalline end blocks: influence of confinement on crystallization behavior and morphology. *Macromolecules* 35: 10004.

161 Siebert, M., Keul, H., and Möller, M. (2012). Synthesis of well-defined polystyrene-*block*-polyglycidol (PS-b-PG) block co-polymers by anionic polymerization. *Des. Monomers Polym.* 13: 547.

162 Lee, H., Kwon, S., Min, J. et al. (2024). Thermodynamically stable plumber's nightmare structures in block copolymers. *Science* 383: 70.

163 Yuan, J., Chiu, P.-T., Liu, X. et al. (2023). Cross-domain chirality transfer in self-assembly of chiral block copolymers. *Angew. Chem. Int. Ed.* e202317102.

164 Song, Q., Pascouau, C., Zhao, J. et al. (2020). Ring-opening polymerization of γ-lactones and copolymerization with other cyclic monomers. *Prog. Polym. Sci.* 110.

165 Zhao, J., Pahovnik, D., Gnanou, Y., and Hadjichristidis, N. (2014). A "catalyst switch" strategy for the sequential metal-free polymerization of epoxides and cyclic esters/carbonate. *Macromolecules* 47: 3814.

166 Xia, Y., Chen, Y., Song, Q. et al. (2016). Base-to-base organocatalytic approach for one-pot construction of poly (ethylene oxide)-based macromolecular structures. *Macromolecules* 49: 6817.

167 Shibasaki, Y., Sanada, H., Yokoi, M. et al. (2000). Activated monomer cationic polymerization of lactones and the application to well-defined block copolymer synthesis with seven-membered cyclic carbonate. *Macromolecules* 33: 4316.

168 Zhao, J. and Hadjichristidis, N. (2015). Polymerization of 5-alkyl δ-lactones catalyzed by diphenyl phosphate and their sequential organocatalytic polymerization with monosubstituted epoxides. *Polym. Chem.* 6: 2659.

169 Zhao, J., Pahovnik, D., Gnanou, Y., and Hadjichristidis, N. (2015). One-pot synthesis of linear-and three-arm star-tetrablock quarterpolymers via sequential metal-free ring-opening polymerization using a "catalyst switch" strategy. *J. Polym. Sci., Part A: Polym. Chem.* 53: 304.

170 Chen, Y., Song, Q., Zhao, J. et al. (2018). Betulin-constituted multiblock amphiphiles for broad-spectrum protein resistance. *ACS Appl. Mater. Interfaces* 10: 6593.

171 Liu, Y., Wang, X., Li, Z. et al. (2018). A switch from anionic to bifunctional H-bonding catalyzed ring-opening polymerizations towards polyether–polyester diblock copolymers. *Polym. Chem.* 9: 154.

172 Ladelta, V., Zapsas, G., Abou-hamad, E. et al. (2019). Tetracrystalline tetrablock quarterpolymers: four different crystallites under the same roof. *Angew. Chem. Int. Ed.* 58: 16267.

173 Hu, S., Zhao, J., and Zhang, G. (2016). Noncopolymerization approach to copolymers via concurrent transesterification and ring-opening reactions. *ACS Macro Lett.* 5: 40.

174 Hu, S., Liu, L., Li, H. et al. (2023). Tuning the properties of ester-based degradable polymers by inserting epoxides into poly(ε-caprolactone). *Chem-Asian J.* 18: e202201097.

175 Liu, S., Bai, T., Ni, K. et al. (2019). Biased Lewis pairs: a general catalytic approach to ether-ester block copolymers with unlimited ordering of sequences. *Angew. Chem. Int. Ed.* 58: 15478.

176 Lai, T., Zhang, P., Zhao, J., and Zhang, G. (2021). Simple and precision approach to polythioimidocarbonates and hybrid block copolymer derivatives. *Macromolecules* 54: 11113.

177 von Tiedemann, P., Blankenburg, J., Maciol, K. et al. (2019). Copolymerization of isoprene with *p*-alkylstyrene monomers: disparate reactivity ratios and the shape of the gradient. *Macromolecules* 52: 796.

178 Zhu, Y., Radlauer, M.R., Schneiderman, D.K. et al. (2018). Multiblock polyesters demonstrating high elasticity and shape memory effects. *Macromolecules* 51: 2466.

179 Gleede, T., Rieger, E., Blankenburg, J. et al. (2018). Fast access to amphiphilic multiblock architectures by the anionic copolymerization of aziridines and ethylene oxide. *J. Am. Chem. Soc.* 140: 13407.

180 Song, Q., Qiu, H., Liu, L. et al. (2023). Janus effect of Lewis acid enables one-step block copolymerization of ethylene oxide and *N*-sulfonyl aziridine. *Angew. Chem. Int. Ed.* 62: e202300187.

181 Li, H., He, G., Chen, Y. et al. (2019). One-step approach to polyester–polyether block copolymers using highly tunable bicomponent catalyst. *ACS Macro Lett.* 8: 973.

182 Ji, H.-Y., Song, D.-P., Wang, B. et al. (2019). Organic Lewis pairs for selective copolymerization of epoxides with anhydrides to access sequence-controlled block copolymers. *Green Chem.* 21: 6123.

183 Zdovc, B., Li, H., Zhao, J. et al. (2022). Influence of microstructure on the elution behavior of gradient copolymers in different modes of liquid interaction chromatography. *Anal. Chem.* 94: 7844.

184 Zhu, S., Wang, Y., Ding, W. et al. (2020). Lewis pair catalyzed highly selective polymerization for the one-step synthesis of $A_zC_y(AB)_xC_yA_z$ pentablock terpolymers. *Polym. Chem.* 11: 1691.

185 Luo, H., Zhou, Y., Li, Q. et al. (1907). Oxygenated boron species generated in situ by protonolysis enables precision synthesis of alternating polyesters. *Macromolecules* 2023: 56.

186 Xia, Y., Scheutz, G., Easterling, C. et al. (2021). Hybrid block copolymer synthesis by merging photoiniferter and organocatalytic ring-opening polymerizations. *Angew. Chem. Int. Ed.* 60: 18537.

187 Yildirim, I., Weber, C., and Schubert, U.S. (2018). Old meets new: combination of PLA and RDRP to obtain sophisticated macromolecular architectures. *Prog. Polym. Sci.* 76: 111.

188 Bernaerts, K.V. and Du Prez, F.E. (2006). Dual/heterofunctional initiators for the combination of mechanistically distinct polymerization techniques. *Prog. Polym. Sci.* 31: 671.

9

Organo-Borane Catalysis for Ring Opening Polymerization of Epoxides

Xiaowu Wang[1] *and Zhibo Li*[1,2]

[1] *College of Chemical Engineering, Qingdao University of Science and Technology, Qingdao, 266042, China*
[2] *Key Laboratory of Biobased Polymer Materials, Shandong Provincial Education Department; College of Polymer Science and Engineering, Qingdao University of Science and Technology, No. 53, Zhengzhou Road, Qingdao, 260042, China*

9.1 Introduction

Borane catalysis has emerged as a powerful and versatile tool in the realm of polymer chemistry, particularly in the context of epoxide polymerization. The unique Lewis acidic properties of borane offer opportunities for precise control over the molecular weight (M_w), molecular weight dispersity (*Đ*), polymer chain structure, stereochemistry. By harnessing the chemical structure of borane and the reactivity of borane towards epoxides, researchers have been able to explore innovative strategies for the synthesis of functionalized polyethers, block copolymers and other topological polymers with various applications.

In this chapter, we delve into the fundamental principles, catalyst design, mechanistic insights, and applications of borane based catalyst for the ring opening polymerization (ROP) of epoxides. Through a comprehensive exploration of this topic, we aim to show the pivotal role of borane atom in organocatalyst in shaping the landscape of polyether synthesis and its potential impact on practical industrial applications.

9.2 Fundamentals of Borane Compounds

9.2.1 General Characteristics of Boron and Boranes

Boron (B), a metalloid element belonging to group 13 (IIIA), exhibits chemical properties that are more akin to those of carbon and silicon than to other elements in its group. However, boron is more electron-deficient compared to both carbon and silicon, as discussed further below. With a relative abundance of about 10 ppm in the Earth's crust, boron is the second most common element in group 13 after aluminum

Organocatalysts in Polymer Chemistry: Synthesis and Applications, First Edition. Edited by Zhibo Li.

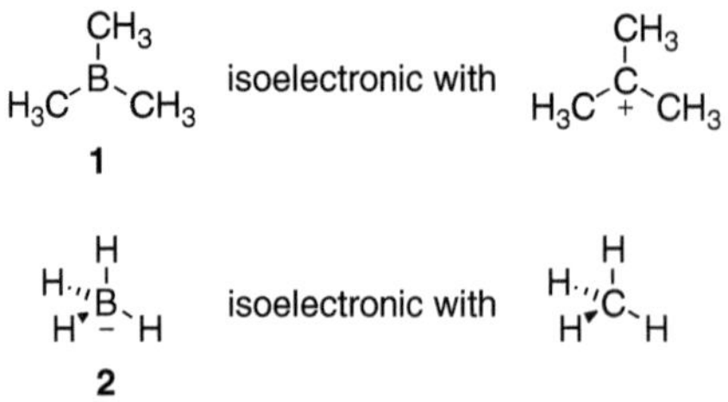

Figure 9.1 Chemical structures of trimethylborane and borohydride ion.

and ranks as the 38th most abundant element overall [1]. Boron can be considered a proton-deficient version of a carbon atom, with the difference between the nuclei of ^{11}B and ^{12}C being precisely one proton. As a result, neutral tricoordinate boron compounds like trimethylborane (**1**) are isoelectronic with carbocations, while negatively charged tetracoordinate borates such as the borohydride ion (**2**) are isoelectronic with neutral carbon compounds (Figure 9.1).

Boron has three valence electrons, with a ground state electron configuration of $1s^2 2s^2 2p^1$. It commonly forms trivalent neutral compounds, such as boron trifluoride (BF_3), where the boron atom possesses six valence electrons. In these compounds, boron is usually sp^2 hybridized, featuring an empty p-orbital and adopting a trigonal planar geometry. Additionally, boron can form negatively charged tetravalent compounds that take on a tetrahedral geometry. Organoboron compounds are straightforward to synthesize and handle, and they generally pose minimal risks in terms of reactivity, toxicity, or environmental contamination.

9.2.2 Bond Length and Strengths

In tri-coordinated boron compounds, such as boranes, the carbon-boron bond length ranges from 1.55 to 1.59 Å, which is longer than a typical carbon–carbon bond [2]. Conversely, the boron-oxygen bond in trivalent organoboron compounds is relatively short, measuring between 1.31 and 1.38 Å, compared to the carbon–oxygen bond in ethers, which is 1.43 Å. This shortening is due to the partial double bond character of the B-O bond, resulting from the interaction between the oxygen's lone pairs and the empty p-orbital on boron. The weak electron donation observed in boron functional groups can be attributed to the difference in electronegativity between boron (2.05) and carbon (2.55). B-C π-conjugation, due to the empty p-orbital on boron, has been observed in aryl- and alkenylboron compounds. A key aspect of organoborane chemistry is the ease of oxidative cleavage of boron-carbon bonds, although these B-C bonds are not inherently weak. The average bond energy of a B-C bond is approximately 322.9 kJ mol^{-1}, compared to the carbon–carbon bond energy of 357.5 kJ mol^{-1}. A more significant difference is seen with boron-oxygen bonds, which have a bond energy of 519.2 kJ mol^{-1}, compared to carbon–oxygen bonds at 383.5 kJ mol^{-1}. This substantial difference in bond energies explains the thermodynamic basis for the ease of oxidation in organoboranes, though further details are necessary to apply these values to specific chemical problems [3].

9.2.3 Characterization

Boron has two stable isotopes, ^{10}B and ^{11}B, along with 14 other known radioactive isotopes [4]. Both stable isotopes are detectable by nuclear magnetic resonance (NMR) spectroscopy, but ^{11}B NMR spectroscopy is commonly used for characterizing boron-containing compounds due to several advantages. ^{10}B has a natural abundance of 19.6% and a nuclear spin of 3, while ^{11}B has a higher natural abundance of 80.4% and a nuclear spin of 3/2. ^{11}B also exhibits greater magnetic receptivity and a wider chemical shift range. However, measuring small coupling constants (less than 10 Hz) in ^{11}B NMR spectra can be challenging. The boron signal in boron trifluoride diethyl etherate, $BF_3{\cdot}O(C_2H_5)_2$, is typically used as an external reference. In ^{11}B NMR spectra, trivalent boron compounds show chemical shift values ranging from approximately −10 to +100 ppm, whereas tetravalent boron compounds typically have shifts between +10 and − 130 ppm. For instance, boronic acids and boronic esters resonate between 25 and 35 ppm, while trialkylboranes resonate from 75 to 95 ppm. The quadrupolar relaxation mechanism of ^{11}B nuclei often causes broadening of the ^{13}C NMR signal of carbon atoms directly bonded to boron, making quaternary carbons with attached boryl groups frequently undetectable in ^{13}C NMR spectra [5].

9.2.4 Lewis Acidity of (Organo)Borane

Acidity and basicity are fundamental concepts in chemistry. According to Lewis theory, acids are defined as electron pair acceptors, while bases are defined as electron pair donors [6]. Borane compounds are examples of Lewis acids due to the electron deficiency of the central boron atom, which possesses a vacant p-orbital (Figure 9.2).

The Lewis acidity of borane compounds plays a vital role in catalysis and polymerization reactions. Significant efforts have been made to fine-tune the Lewis acidity of boranes to enhance their performance and expand the scope of borane-mediated chemical reactions. A major challenge in designing new borane Lewis acids is accurately quantifying their strength. Unlike Brønsted acidity, which can be precisely measured using the *p*Ka scale, quantifying Lewis acid strength is more complex, and no universal criterion exists. Since Lewis introduced his acid–base theory in 1923, various approaches have been proposed to quantify Lewis acidity, including correlations with thermodynamic data (such as the enthalpy of Lewis acid-Lewis base adduct formation), chemical reactivity, and spectroscopic data [7, 8]. Although establishing a universal scale for Lewis acidity is impossible, specific Lewis acidity scales can still be used with a particular reference Lewis base. Today, three primary scales are widely used to estimate the Lewis acidity of boron compounds. The

Figure 9.2 Chemical structure of borane Lewis acid.

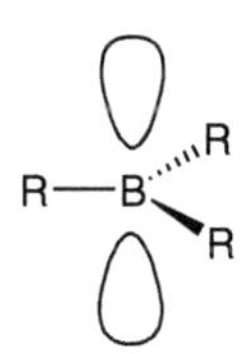

first scale relies on the fluoride ion affinity (FIA) of Lewis acids as a quantitative measure of their acidity, [9] which can be determined experimentally or through quantum chemical calculations. The other two popular scales are based on NMR spectroscopy data. The first method, known as the Gutmann-Beckett method, measures the change in the ^{31}P NMR chemical shift of Et_3PO when it complexes with a Lewis acid [10, 11]. The second method, developed by Childs et al. [12], measures the change in ^{1}H NMR chemical shift of the β-olefinic proton in crotonaldehyde when a Lewis acid binds to the carbonyl oxygen atom. Later, Marks used this method to place various perfluoroaryl boranes on a scale alongside other main group Lewis acids (relative to BBr_3 at 1.00) [13]. Currently, the Lewis acidity of many boron-based Lewis acids has been experimentally determined using these two methods.

9.3 Et_3B-Mediated ROP of Epoxides

9.3.1 Introduction of ROP of Epoxides

Aliphatic polyethers, such as poly(ethylene oxide) (PEO) and poly(propylene oxide) (PPO), are a significant class of polymers that have found extensive commercial use, particularly in the polyurethane (PU) industry [14]. These materials exhibit unique properties due to their flexible backbones, which result in low glass transition temperatures below −60 °C, and their hydrophilicity, stemming from the presence of C−O−C bonds [15]. Since the first report of oxyanionic polymerization of propylene oxide (PO) in 1927 [16], considerable effort has been dedicated to discovering new initiators or catalysts that preserve the living polymerization characteristics while achieving higher molecular weights and diverse polyether architectures. The ROP of epoxides is driven by their high ring strain, which is approximately 110–115 kJ mol^{-1} for ethylene oxide [17]. Due to the chemical structure of epoxides, as the smallest cyclic ether, a wide variety of catalysts can be employed. The Lewis basic nature of the endocyclic oxygen atom [18] allows for monomer activation, either through coordination with a Lewis acid or protonation. As a result, epoxides can undergo polymerization via cationic, anionic, or "coordination-insertion" mechanisms [19]. Since a detailed discussion of each epoxide would lead too far, readers are encouraged to refer to excellent review publications for epoxide homo- and copolymerization [14, 15, 20–25] However, to assure the readers be familiar with epoxide homopolymerization, we take the ROP of PO for example. During PO polymerization, chain growth can proceed in two ways (Figure 9.3): (i) through attack at the methylene group, producing a secondary alkoxide active species, or (ii) through attack at the methine group, resulting in inversion and the formation of a primary alkoxide active species.

Alkali metal derivatives have traditionally been effective initiators for the anionic polymerization of PO. However, these initiators faced several limitations, including restricted molecular weights of the resulting polyethers, low polymerization efficiency, demanding reaction conditions, loss of the living polymerization characteristic, and chain transfer to monomer side reactions [15]. To address

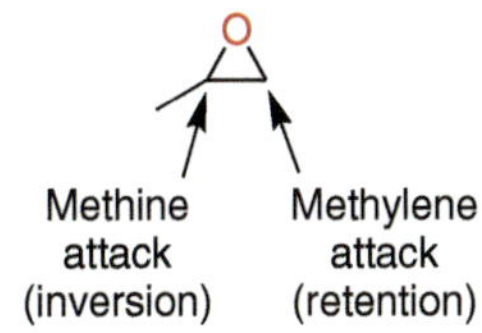

Figure 9.3 Regio-chemistry of S_N2-type epoxide ring opening.

these challenges, various strategies were employed, such as the addition of crown ethers or the use of aluminum metal complexes, which enabled the production of poly(propylene oxide) (PPO) with controlled molecular weights ranging from 10.0 to 15.0 kg mol^{-1} [26, 27]. A notable system called the double metal cyanide (DMC) catalytic system, specifically $Zn_3[Co(CN)_6]_2$, has been utilized for industrial applications [28–31]. Additionally, leveraging the activated monomer strategy and dual catalytic cooperative effects, Carlotti et al. [29] and Coates et al. [32, 33] succeeded in producing high molecular-weight PPO with number-average molecular weights (M_n) reaching up to 150.0 kg mol^{-1}. In terms of heterogeneous catalysts, zeolites have also shown potential as catalytic systems for PPO production [34]. Although DMC and metal-based catalysts have significantly advanced the ROP of epoxides [35, 36], DMC catalysts are associated with issues such as polyether degradation and metal residue contamination, limiting their applicability in fields like food and medical industries [37]. Consequently, the development of efficient, environmentally friendly catalysts that can produce polyethers with controlled molecular weights and the desired main-chain sequence is crucial. Over the past few decades, significant attention has been directed towards the development of metal-free catalysts [25, 38, 39]. In this context, strong phosphazene-type organobases have shown the ability to polymerize ethylene oxide (EO) [40]. However, the overly high basicity cause the chain transfer to PO and other substituted epoxides, thus generating additional unsaturated initiators and limiting the effectiveness in synthesizing high molecular weight polyethers [41]. Strong nucleophiles, such as *N*-heterocyclic carbenes (NHCs) [42] and *N*-heterocyclic olefins (NHOs) [43, 44] have been successful in producing PPO with well-defined end groups and controlled molecular weights. However, these systems tend to be inefficient (yielding 30–40%) and require extended reaction times (up to three days) at 50 °C. Additionally, NHC carboxylates (NHC–CO_2) have been used to generate PPO oligomers, but this approach requires high catalyst loadings and elevated temperatures [45]. These metal-free systems generally proceed through oxoanionic ROP of PO.

Besides anionic polymerization, PO can also undergo cationic polymerization using organic acids, borane trifluoride (BF_3), [11] or trispentafluorophenylborane ($B[C_6F_5]_3$) [46, 47]. However, cationic polymerization often results in poorly controlled molecular weights and complex mixtures of linear and cyclic structures. Since the discovery in 2016 of the remarkable catalytic properties of Lewis acid triethylborane (Et_3B) in copolymerizing epoxides and carbon dioxide [48], research interest in Et_3B mediated ROP and ROCOP has significantly increased [49–52]. Further study revealed that the precise control of borane relative to the active species is crucial for controlling the copolymerization with CO_2. In general, higher Et_3B loading relative to carboxylate favors homopolymerization of PO (polyether linkages), while lower concentrations promote alternating copolymerization with CO_2 (resulting in higher carbonate content) [53, 54]. Therefore, Et_3B-mediated copolymerization of epoxides and CO_2 inspire research efforts in the ROP of epoxides and opens the door to the organoborane catalyzed ROP of epoxides. The following contents will demonstrate the borane catalysis contributing to the ROP of epoxides.

9.3.2 Et_3B-Based Binary Lewis Pair Systems in the ROP of Epoxides

Inspired by Al/Mg based Lewis pair mediated polymerization of polar conjugated monomer [55] and ROP of epoxides [43, 56]. In 2018, the research group led by Zhao [57] and Zhang [53] independently reported the Et_3B-catalyzed homopolymerization of epoxides using organobases and alcohol-type initiators. These studies not only underscored the remarkable catalytic efficiency but also provided important insights into the underlying polymerization mechanism. Zhao's detailed work demonstrated that using ^tBu-P_1/Et_3B system in a 1 : 3 molar ratio under optimized conditions for EO homopolymerization achieved an impressive turnover frequency (TOF) of $6000\,h^{-1}$, with low catalyst loadings (144 ppm of combined base and borane). This approach produced PEO with well-defined properties, including a dispersity ($Đ < 1.3$). High molecular weights exceeding $100\,kg\,mol^{-1}$ were obtained within a relatively short reaction time (a few hours). This performance is not only significantly better than KOH-mediated polymerization but also surpasses earlier organocatalysts such as *N*-heterocyclic carbenes (NHCs) [42] and *N*-heterocyclic olefins (NHOs) [43]. Other epoxides such as 1,2-butylene oxide (BO), phenyl glycidyl ether (PGE), allyl glycidyl ether (AGE), 2-ethylhexyl glycidyl ether (EHGE) or tert-butyl glycidyl ether (TBGE) were also successfully converted into their corresponding homopolymers, though at a slower polymerization rate, as observed by Zhao [57] and Zhang [53]. However, styrene oxide (SO), cyclohexene oxide (CHO), and epichlorohydrin (ECH) could not be converted by this method.

Zhao and Zhang found that milder amidine or guanidine compounds like 1,8-diazabicyclo[5,4,0]undec-7-ene (DBU) or 7-methyl-1,5,7-triazabicyclo[4.4.0]dec-5-ene (MTBD) could also be used effectively to polymerize EO and PO [53, 57]. Subsequently, Naumann and coworkers demonstrated that even weaker nitrogen bases like 1,4-diazabicyclo[2.2.2]octane (DABCO) or 4-dimethylaminopyridine (DMAP) could be suitable for the polymerization process [58]. The polymerization rates were observed to correlate with the basicity of the organobase, leading to rapid conversion with ^tBu-P_2 and more gradual monomer incorporation with DABCO or DMAP. Additionally, *N*-heterocyclic carbenes (NHCs) have been employed in conjunction with Et_3B for the synthesis of polyethers [59]. Controlling the molar mass of the resultant polyether with a chain transfer agent (CTA) is a critical consideration prior to initiating the batch production of downstream products. In Zhao's subsequent studies, elevating the hydroxyl concentration in the ratio of [^tBu-P_2]/[Et_3B]/[hydroxyl]/[PO] from 1 : 3 : 2 : 1600 to 1 : 3 : 40 : 1600 resulted in a reduction of the polymerization rate by half [57]. Moreover, an increase in the ratio of Et_3B to organobase led to an anticipated acceleration of the polymerization rate.

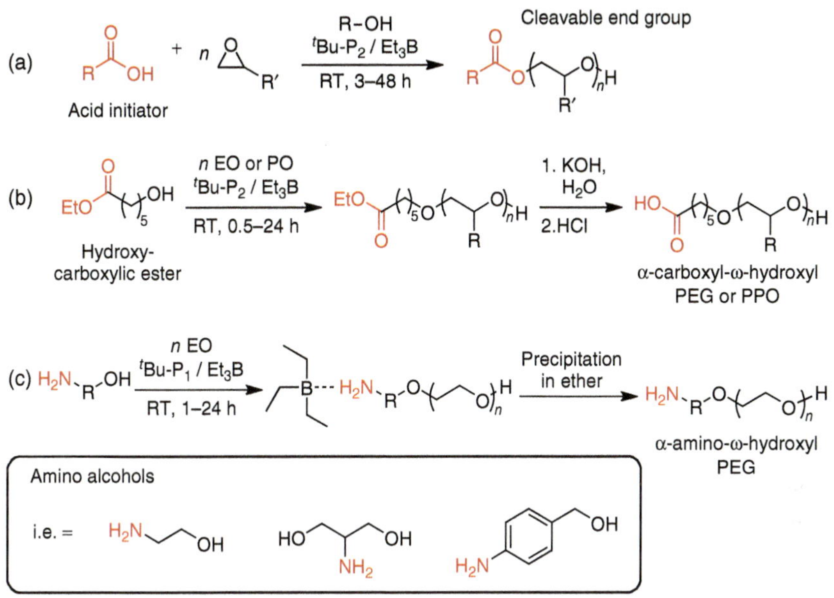

Figure 9.4 Et_3B-mediated polymerization catalysis featured a distinctive functional group tolerance. (a) Synthesis of α-ester/ω-OH polyether. (b) Synthesis of α-carboxylic acid/ω-OH polyether. (c) Synthesis of α-amino/ω-OH polyether.

Conversely, when the borane content was reduced to less than two equivalents, the rate of polymerization decreased significantly, and the reaction often failed to initiate with less than one equivalent of borane. In their subsequent study, Zhao's group demonstrated the capability for selective and functional group-tolerant polymerization, especially in creating end-functionalized polyethers like polyethylene glycol (PEG). Carboxylic (di)acids were applied as initiators for the polymerization of EO, PO, BO, or AGE, using ^{t}Bu-P_2/Et_3B binary catalyst [60]. The well defined structures and narrow molecular distributions ($Đ < 1.12$) are both achieved owing to the inhibition of transesterification. The ability to use non-alcoholic initiators paves the way for selective cleavage or modification of the polymer, providing greater flexibility in post-polymerization processing and functionalization. An example involved constructing an enzymatically degradable polyurethane from a diacid initiator to generate PPO macrodiol was provided (Figure 9.4(a)). The polymerization process was noted for its initial induction period of one to two hours, followed by rapid epoxide conversion, with carboxylate and PO reaction occurring during this induction period, as observed by ^{1}H NMR. The acid species, being a weaker nucleophile compared to activated hydroxyl, allows for controlled polymer growth. The anionic charge primarily resides on the unreacted acid initiator, keeping newly formed hydroxyls dormant until most of the acid is consumed. This slow initiation doesn't lead to a broader molar mass distribution.

Zhao's group also highlighted the use of Et_3B-catalyzed epoxide polymerization for creating α-carboxyl-ω-hydroxyl PEG, a structure commonly used in bioconjugation, through a more straightforward method involving hydroxyl-carboxylic esters as initiators (Figure 9.4(b)). NMR and MALDI-TOF-MS confirmed the successful polymerization and the existence of hydrolysis of ester moiety. Nevertheless, achieving higher molecular weights of polyether with carboxylic acid end group becomes challenging and requires longer reaction times. Alternatively, reacting the polymer with primary amines yields α-amide-ω-hydroxyl PEG, which is highly valued for PEGylation processes due to the amino group's superior reactivity in conjugation reactions. Therefore, the unique aspects of Et_3B catalysis reduces the complexity of functionalized PEG synthesis. Moreover, amino alcohols are directly used as initiators with ^{t}Bu-P_1/Et_3B binary system (Figure 9.4(c)), where polymer growth preferentially occurs at the –OH end. This preference is unexpected as the primary amine is a stronger nucleophile. In this case, Et_3B acting as a noncovalent protector and effectively shielding the amine, prevents amine from participating in the polymerization. Control experiments confirm that without Et_3B, amines react with EO, whereas no reaction occurs with an equivalent amount of Et_3B, due to its complexation with amine. Applying an excess of Et_3B and a minimal phosphazene results in a successful polymerization, yielding amino-terminated polyethers with defined molecular weights (up to 10 kg mol^{-1}) and narrow dispersity ($Đ < 1.09$).

9.3.3 Mechanism of Et_3B Mediated ROP of Epoxides

Detailed spectroscopic studies by using both ^{1}H and ^{11}B NMR spectroscopy provide evidence for the polymerization mechanism, as outlined in Figure 9.5. The polymerization is primarily based on two pathways: the activated monomer (AM) and the activated hydroxyl (AH). Both are formed through equilibria driven by Et_3B, designated as (1) and (2). The efficiency of AH in promoting polymer growth is strongly influenced by the basicity of the organobase. The stronger basicity of organobase accelerates the propagation process in accordance with the experimental observations.

The borane group also plays a pivotal role by stabilizing the negatively charged growing polymer chain end, which, in turn, lowers the hydroxyl's pKa, making even weaker organobases sufficient for initiating polymerization. This stabilization also diminishes the electron density at the alkoxide oxygen, reducing its likelihood of undergoing base-induced side reactions, thereby suppressing transfer-to-monomer reactions and facilitating the production of polymers with high molecular weights. However, it's critical to recognize that AH alone is insufficient for effective polymer propagation. Uncomplexed Et_3B is essential for AM formation, and any variable that alters AM concentration directly affects the rate of polymerization. One such variable is the competition for Et_3B complexation, particularly in the presence of higher concentrations of hydroxyl groups. Since alcohols show stronger basicity than epoxides, they are more effective at competing for complexation with Et_3B in bulk polymerizations, such as that of PO. This leads to a reduced availability of Et_3B for AM formation and consequently resulting in a lower polymerization rate. This decrease in activity can be partially mitigated by adding extra borane. Furthermore, this competition explains the challenging polymerization of functionalized monomers like

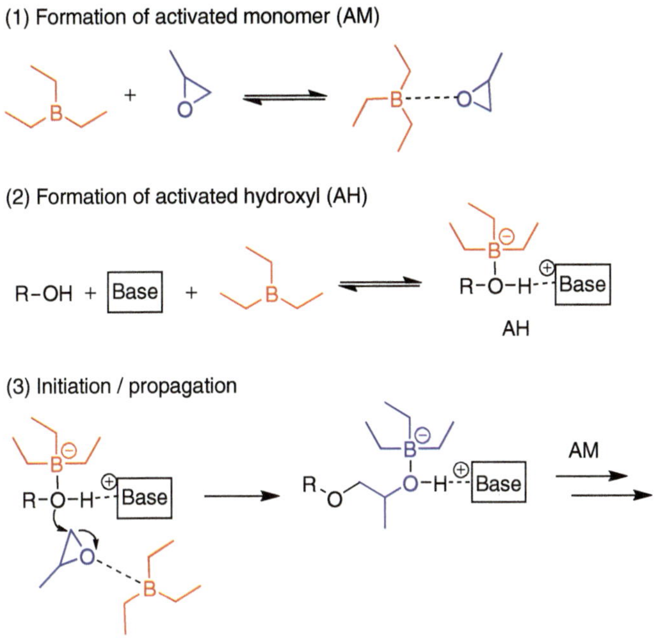

Figure 9.5 Proposed polymerization mechanism for Et_3B-mediated epoxide homopolymerization. Source: [57]/American Chemical Society.

AGE, as their exocyclic oxygen and double bonds weakly interact with Et_3B. Factors such as temperature, solvent polarity, and steric hindrance also influence these equilibria, though further investigation is needed to elucidate their specific effects. It's noteworthy that using a weak organobase might not result in the consumption of an exact equivalent of borane due to the less effective complexation of Et_3B by the less activated hydroxyl. This further leads to slow polymerization even when the concentration of Et_3B is less than that of the organobase. Lastly, the exchange between dormant hydroxyls and AH occurs much more rapidly than the chain propagation, a characteristic behavior in anionic epoxide polymerizations, ensuring uniform chain growth and narrow molecular weight distributions.

9.3.4 Poly(alkylene oxide) Structures: Novel and Innovative Polyether Architectures

The capacity for generating precise polyether structures has been rapidly adopted for enhanced material applications. Specifically, Hadjichristidis, Sarathy, and their colleagues have developed various diblock and triblock all-polyether architectures for utilization as lubricant additives, highlighting their relevance in terms of energy conservation. Polyethers are inherently advantageous in this context due to their low pour points, high viscosity indices, and significant flash points. However, optimal performance in terms of thermal stability, rheological characteristics, and anti-friction capabilities necessitates precise adjustments. This optimization

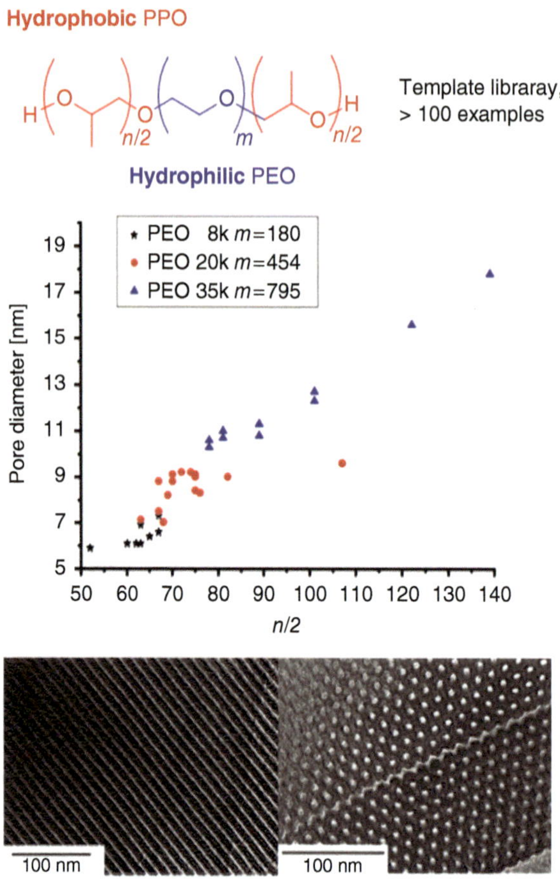

Figure 9.6 Amphiphilic block copolyether series developed using PEO macroinitiators with varying molecular weights for OMC/OMS preparation. Middle: correlation of *n*/2 and pore diameter (OMC). Bottom: TEM of OMC based on PO_{82}-EO_{454}-PO_{82} (9 nm pore diameter). Source: Reproduced with permission from [61]/American Chemical Society.

was efficiently conducted through the use of ^{t}Bu-P_2/Et_3B, leading to the synthesis of polyethers like PHO-*b*-PPO or POO-*b*-PPO-*b*-PSO through the sequential addition of monomers. The resulting block copolymers were highly well-defined ($Đ$ = 1.04–1.12), allowing for the identification of specimens with either superior lubrication efficacy or enhanced thermal stability. Furthermore, borane-mediated homopolymerization of epoxides has been leveraged to generate customized variants of "Pluronic" or "Reverse Pluronic" triblock copolyethers (Figure 9.6). [53, 57, 58, 61, 62] The precise control over polymerization degree, hydrophilic to lipophilic balance (HLB), and the overall molar mass enabled the creation of polymer libraries with stepwise and systematic variations of properties. This precise control is especially advantageous for utilizing these amphiphilic polymers in (soft-) templating processes aimed at generating mesoporous materials. Both the mesopore diameter and pore arrangement symmetry are highly sensitive to these parameters and exhibit significant variability [63, 64]. The standard grades available on the market, such as F127 and P123, have limitations in this regard, especially regarding long PPO blocks. Thus, through the use of NHO/Et_3B, Naumann and Bruckner engineered a series of polyethers with the structure $PO_{n/2}$-EO_m-$PO_{n/2}$

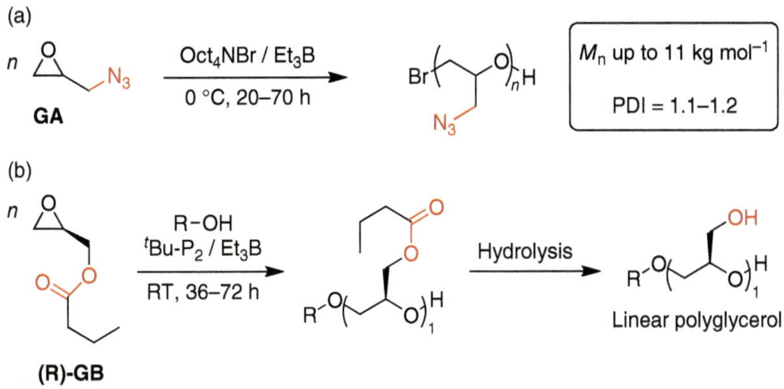

Figure 9.7 Homopolymerization of GA and (R)-GB catalyzed by Et_3B binary system. (a) Synthesis of pendant azide functionalized polyether. (b) Synthesis of linear polyglycerol.

(where m = 90–795 and n/2 = 10–280) and utilized these block copolyethers as templates for producing ordered mesoporous carbon (OMC) or silica (OMS) materials [62]. The dimensions of the resultant mesopores were finely adjustable between approximately 5–20 nm, displaying a linear correlation between n/2 and the diameter of the mesopores (Figure 9.6). It is important to highlight that the effectiveness of these applications relies on the precise architectural control of polyethers and no limitations in configuring the m or n/2 parameters. This highlights why borane-mediated polymerization catalysis has gained widespread use in this scientific field.

The polymerization process facilitated by Et_3B-catalyzed epoxide homopolymerization is usually under mild conditions (0–50 °C). Notably, the propagating polymer chain end is effectively masked, resulting in decreased basicity. This leads to a remarkable tolerance of various functional groups during polymerizations. In 2019, glycidyl azide (GA) was successfully homopolymerized using Oct_4NBr/Et_3B at 0 °C (Figure 9.7(a)), as demonstrated by Gnanou and Feng group [50]. NMR studies indicated weak interactions of Et_3B with both the oxygen atom in the monomer and the azide group. Although transfer reactions were not completely eliminated, a well-defined PGA with M_n up to 11 kg mol^{-1} and $Đ$ values ranging from 1.1 to 1.2 was still achieved. This achievement is particularly notable considering the azide group's sensitivity to nucleophiles and acids, marking it as the first example of this monomer being polymerized. Furthermore, copolymerization with carbon dioxide was successful and post-functionalization of PGA using "click chemistry" worked out.

Zhao and his colleagues successfully polymerized enantiopure glycidyl butyrates (GBs), compounds that contain both an epoxide moiety and an ester group [65]. Using the binary catalyst $Et_3B/^tBu-P_2$ (5,1, with a diol initiator), they achieved selective ROP through the epoxide moiety. DFT calculations revealed a preferred interaction between borane and epoxide, while coordination with the carbonyl oxygen is disfavored, probably due to steric hindrance. This explains why a high concentration of Et_3B does not lead to transesterification. The resulting polymers

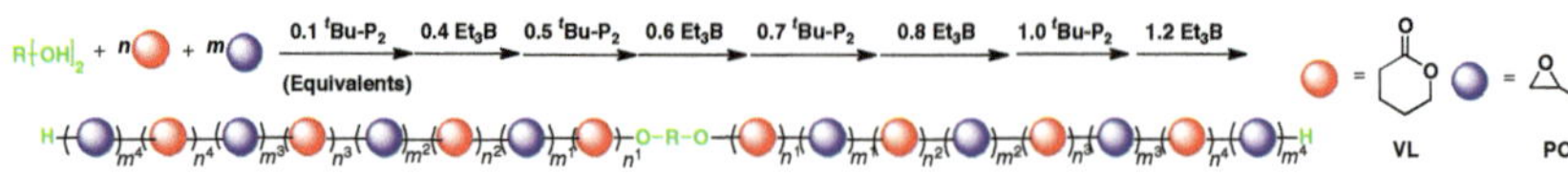

Figure 9.8 Synthesis of well-defined poly(ester-ether) multiblock copolymers using the Et_3B/organobase stoichiometry. Source: [66]/John Wiley & Sons.

are well-defined ($Đ < 1.15$) with molecular weights (M_n) up to 40 kg mol^{-1}. Importantly, the ester side chains could be removed through methanolysis in a subsequent step, allowing for the synthesis of isotactic, linear polyglycerol (Figure 9.7(b)). This synthetic strategy is highly advantageous, as direct polymerization of glycidol (GD) is not feasible due to its inimer nature, and traditional anionic polymerization of GB would compromise the ester groups. Thus, borane catalysis serves as a viable method for synthesizing these hydrophilic, linear polyglycerols. In 2019, Zhao et al. extended these findings by demonstrating that the polymerization mechanism can endure (poly)esters and selectively polymerize cyclic esters and epoxides from one-pot mixtures. This technique resulted in the production of highly defined poly(ester-ether) multiblock copolymers (up to pentadecablocks, see Figure 9.8) [66]. This achievement relies on a careful balance of organobase and Et_3B stoichiometry. In the absence of the borane Lewis acid or when the organobase is used in excess, cyclic esters such as ε-caprolactone (CL), δ-valerolactone (VL), or lactide (LA) are polymerized, leaving epoxides intact. Conversely, an excess of Et_3B selectively polymerizes epoxides while transesterification does not happen. This slight imbalance effectively switches the catalyst system's preference between the ester and epoxide monomers on and off, facilitating the creation of block copolymers suitable for an array of potential applications. This strategy allows access to several polymer architectures typically challenging or inaccessible through standard methods, potentially benefiting bio-related fields and the development of polymeric additives.

9.4 Modified Organoborane Mediated ROP of Epoxides

The binary organo-system based on Et_3B has been successfully applied in the homopolymerization and copolymerization of epoxide monomers, showing good tolerance to functional groups and controllability over the polymerization process, thereby exhibiting increased practicality. For example, it can achieve rapid consumption of monomers, no metal residues, and very low catalyst loading, etc. Moreover, it enables precise control over polyethers, composite structures containing polyethers, and copolymers of various epoxides, further expanding the potential applications of these polymers. However, Et_3B has inherent characteristics and some volatility (boiling point 95 °C), requiring extra caution when used. Additionally, in order to produce low molecular weight polyethers, the presence of a large number of hydroxyl species significantly reduces the Et_3B catalytic efficiency or even deactivates polymerization activity. This is not beneficial for the industrial production of low molecular weight polyethers. Lastly, the conditions for using Et_3B

are typically in solvents, like tetrahydrofuran or hexane, where the introduction of solvent has a "dilution effect" on the polymerization reaction, significantly reducing the rate of polymerization [67].

Inspired by the concepts of intramolecular Lewis pairs [68] and bifunctional catalysts, [67] covalent linking a Lewis acid and a Lewis base within a single molecule is an effective and promising strategy to enhance the catalytic performance of organoborane in the ROP of epoxides. Such catalysts can effectively reduce the entropic effects of the catalysts; fully utilize the acid–base sites and the covalent framework of bridged acid–base sites to synergistically catalyze the ROP of epoxide monomers; and achieve an efficient homopolymerization/copolymerization of epoxide monomers under conditions of low catalyst loading. Compared to the Et_3B containing binary system, synthesizing one-component functionalized organoborane by combining a Lewis acid and a Lewis base within one molecule requires additional synthesis. The hydroboration reaction is an efficient synthetic method for producing one-component functionalized organoborane, and it also provides theoretical and practical assurance for the feasibility of synthesizing diverse one-component functionalized organoborane. 9-Borabicyclo[3.3.1]nonane (9-BBN) is a cost-effective and region-selective hydroboration reagent with an acceptor number (AN = 21.7) close to that of Et_3B (AN = 24.1), indicating they possess similar Lewis acidity. Additionally, the rigid structure of the BBN bicyclic substituent ensures the boron center has certain spatial steric hindrance, making its B—C bond more resistant to hydrolysis and more thermally stable than Et_3B.

9.4.1 One-Component Ammonium Borane-Based Lewis Pair Systems Mediated ROP of Epoxides

In 2020, Wu and co-workers published an array of studies presenting novel bi-or multifunctional borane initiator/catalysts (Figure 9.9) [69]. We called them initiator and catalyst due to the dual functions of the designed system serving both as initiator and catalyst. The defining characteristic of these borane catalysts is that the two borane centers are connected by an ammonium group in the backbone, along with a nucleophilic counter ion (such as a halide). This nucleophile is stabilized through both Coulombic interactions with the positively charged nitrogen atom and weak dative coordination to the boron atom. This dynamic stabilization becomes even more pronounced when the catalyst possesses more than one boron functionality, resulting in a distinctive pattern in ^{11}B NMR analyses, characterized by broadened signals. Such catalyst architecture is especially beneficial for the polymerization mechanisms previously discussed, where two equivalents of borane are often needed. Consequently, the bifunctional, multinuclear borane catalysts have been adopted as a powerful method for polyether synthesis. Wu and his team employed a catalyst consisting of two borane units derived from 9-BBN (**3**) (Figure 9.10) to achieve an organocatalytic system with unparalleled efficiency in the homopolymerization of epoxides. For instance, the system was capable of converting 3000 equivalents of propylene oxide (PO) completely within 6 h, maintaining a narrow molecular weight distribution. Using 30,000 equivalents of PO, the

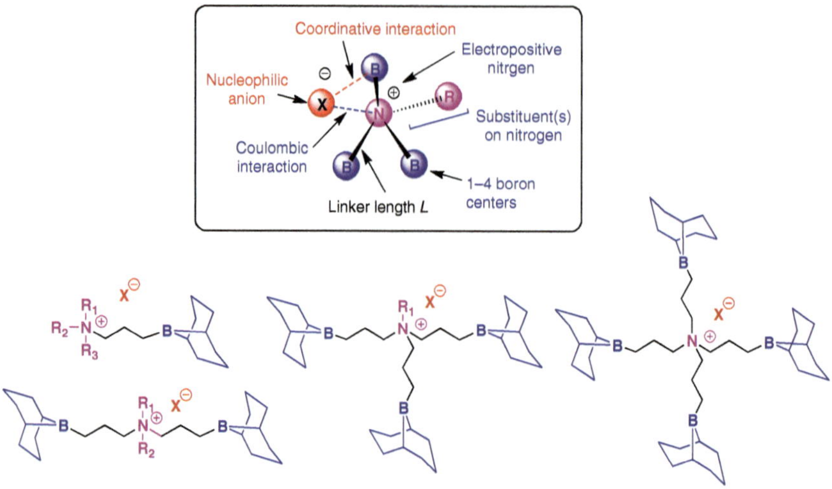

Figure 9.9 Chemical structures of bifunctional mono-, di-, tri and tetranuclear organoboron initiators/catalysts as published by Wu, along with a summary of key tuning sites. Source: [69]/American Chemical Society.

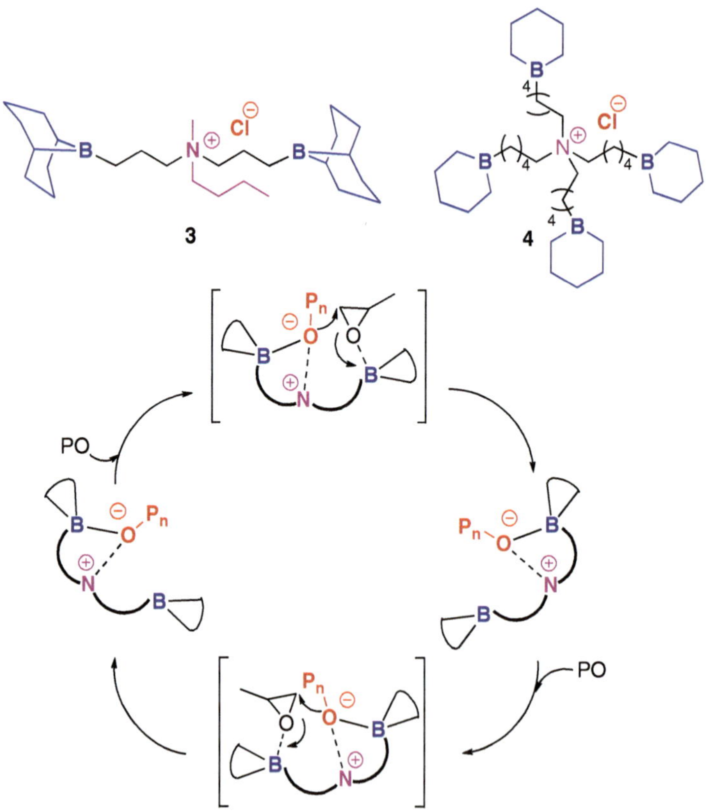

Figure 9.10 Bifunctional borane initiators/catalysts for epoxide homopolymerization and proposed mechanism. Source: [69]/American Chemical Society.

reaction achieved full monomer conversion after 60 h, producing poly(propylene oxide) (PPO) with a molecular weight M_n exceeding 1000 kg mol^{-1}. Ethylene oxide (EO), being more reactive, allowed for the conversion of 10,000 equivalents in 20 minutes (TOF = 30,000 h^{-1}), making the reaction rate of EO over 25 times faster than PO. Various experiments, including chain extension, kinetic studies, and MALDI-TOF-MS, confirmed the living nature of the polymerization. Importantly, the polymerization showed zero-order dependence on the concentration of PO. However, a limitation of this catalyst system was the need to carry out reactions at temperatures between −20 °C and 0 °C to maintain control. At higher temperatures, the exothermic nature of the ring-opening and the rapid reaction rates caused a quick loss of control. Mechanistically, the effectiveness of catalyst **3** was attributed to the spatial arrangement of the two borane groups, linked by ammonium flexible linker, allowing the propagating chain end on one boron atom to readily attack the activated epoxide on the adjacent boron. The quaternary nitrogen atom facilitates the propagation step by stabilizing the alkoxide chain end and facilitating its transfer to the next borane site. Control experiments with a similar diborane linked by a neutral (pure methylene) chain showed significantly lower efficiency, highlighting the presence of a synergistic effect in the original catalyst design. DFT calculations showed that the nitrogen atom distributes a significant portion of its positive charge onto the hydrogens of the methylene groups in the α and β positions.

Previous work indicated that steric hindrance around the boron centers plays a significant role in influencing the polymerization of epoxides. Owing to their experience in alkylborane mediated polymerization, in 2022, Feng and coworkers replaced 9-BBN with borinane to reduce the steric hindrance around the boron centers. Interestingly, reducing steric hindrance around the Lewis acid by incorporating the borinane motif, along with the use of a four-armed catalyst molecule (**4**), significantly improved performance, resulting in outstanding outcomes in the bulk polymerization of PO [70]. In the bulk polymerization of PO, 30000 equivalents of PO were fully polymerized in less than 10 minutes at 0 °C. The efficiency of polymerization decreased with an analogous catalyst having less borane functionalities. Those similar to **4** but with three borane groups achieved 68% conversion, and those with two reached only 21%. The number of Lewis acid sites in the catalyst directly influences its effectiveness. Notably, using catalyst **4** for PO polymerization resulted in an extraordinarily high turnover frequency (TOF) of 180 000 h^{-1}, a significant improvement over previous results with catalyst **3**. This setup not only facilitated the creation of polymers with very high molecular weights (M_n > 1000 kg mol^{-1}, $Đ$ = 1.2) but also allowed for the precise synthesis of lower molecular weight polymers and oligomers. This adaptability is crucial for many polyether applications and typically necessitates a high catalyst-to-monomer ratio or, more practically, polymerization in the presence of an excess of alcohol-type chain transfer agents (CTAs). While Et_3B-based systems suffer from slow monomer conversion under such conditions. The use of catalyst **4** eliminates this issue, enabling the use of a variety of alcohols (and water) to accurately control the degree of polymerization. Remarkably, for the highly reactive ethylene oxide (EO), catalyst loadings can be reduced in the parts per billion range (with 250 000 equivalents of CTA relative to the catalyst), which led to 90% EO conversion within 48 hours, equating to an average TOF of 187 500 h^{-1} and a turnover number (TON) of 9×10^6. Therefore, a gram of catalyst **4** can produce up to 600 kg of well-defined polyethylene glycol (PEG) ($Đ$ = 1.03), easily surpassing existing commercial processes. The scope of monomers was extended beyond EO and PO to include butylene oxide (BO), octylene oxide (OO), allyl glycidyl ether (AGE), and phenyl glycidyl ether (PGE), with slightly lower TONs and TOFs for the glycidyl ethers but still enabling rapid polymerization, low catalyst loadings, and affording well-defined polyethers.

Wu and coworkers carried out a systematic study on examining how the structures of catalysts influence their performance in epoxide homopolymerization [71]. Their research concentrated on four key areas: (i) the electronic and steric characteristics of the borane group, (ii) the steric hindrance surrounding the ammonium nitrogen atom, (iii) the distance between boron and nitrogen (B–N distance), and (iv) the numbers of borane groups within the catalyst structure. Initially, they explored a series of monoboron catalysts, varying the Lewis acidity of the boron centers and guiding their selection by acceptor numbers (AN). Comparisons among groups with diminishing Lewis acidity, specifically from borinane (AN = 23.4) to BBN (21.7), BCy_2 (18.8), and Bpin (15.5), showed minimal epoxide (PO) conversion regardless of the chosen borane. This is in accordance with earlier findings for Et_3B, highlighting the necessity of excess borane for effective monomer activation, with one equivalent borane needed for end-capping the growing polymer chain. Since monoboron bifunctional catalysts introduce only one boron per halide initiator, the absence of significant polymerization was in line with expectations. Subsequent investigation into diboron catalysts with a fixed linker length (L = 3) (Figure 9.11) under the same conditions (10 000 eq. of PO, 6 hours, −20 °C) revealed interesting outcomes. The BBN-based catalyst achieved 35% conversion with a TOF of 585 h^{-1}, while the less Lewis acidic BCy_2 or Bpin derivatives showed negligible conversion. Strikingly, the borinane motif reached a 63% conversion in just two hours, suggesting a preference for moderately strong Lewis acidic boron centers. The current state of research indicates that the optimal range for boron Lewis acidity is suggested to be AN = 20–40.

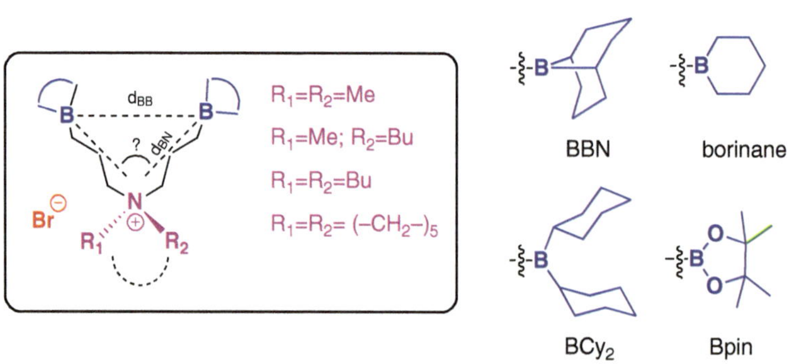

Figure 9.11 Dinuclear organoboron catalysts evaluated by Yang and Wu for epoxide homopolymerization.

Further exploration of varying linker lengths (L = 3, 4, 5, and 6) demonstrated TOFs decreased in the following order: TOF of 585 h^{-1}, 380 h^{-1}, 325 h^{-1} and 113 h^{-1}. A very comparable trend was also found for EO polymerization (TOF = 24,600 h^{-1}, 16,150 h^{-1}, 12,900 h^{-1} and 7750 h^{-1}, respectively). This gradual reduction in the dinuclear catalysts' activity as the linker length increased highlights the importance of spatial arrangement of boron centers for effective propagation. Their study also emphasized the impact of steric hindrance around the nitrogen atom. With the same borane (BBN) and linker length (L = 3), different substitutions at the nitrogen atom significantly influenced the TOF (Me_2, 258 h^{-1}; Bu_2, 690 h^{-1}, pentamethylene, 1205 h^{-1}), indicating that steric hindrance at nitrogen can favorably limit rotational conformations, facilitating the optimal spatial positioning of the boron centers. Crystal structure analysis and Newman projections further supported these findings, showing that less active catalysts had borane moieties in anti-positions, while larger substituents promoted a gauche orientation, enhancing catalyst activity. Computational studies confirmed that greater steric demand around the nitrogen atom reduces the energy barrier for achieving the ideal conformation. As previously observed, catalyst activity increased with the number of boron groups, with tri- and tetranuclear structures outperforming dinuclear ones.

9.4.2 One-Component Phosphonium Borane-Based Lewis Pair Systems Mediated ROP of Epoxides

We have experiences in the design and syntheses of intramolecular phosphine borane Lewis pair [68]. Previous studies suggest that incorporating a Lewis acid and a Lewis base into a single molecule through a covalent linker imparts distinctive properties, especially exhibiting synergistic effect in small molecules activation and their transformation [72].

We therefore hypothesized that phosphines [68] or phosphonium halides could serve as valuable organic Lewis bases, capable of covalently linking with a borane Lewis acid, making them potential candidates for the ROP of epoxides. Additionally, due to the larger atomic radius of phosphorus, bulkier groups can be attached as compared to nitrogen. This variation could help adjust the confined space in which intermediates and monomers interact, potentially lowering the reaction energy barrier and enhancing the synergistic effect. Therefore, we synthesized a new phosphonium-bridged bisborane organo-initiator/catalyst (**5**) (Figure 9.12). ROP of PO initiated by this system offered several key advantages: (i) living polymerization characteristics were maintained within a temperature range of −30 to 0 °C; (ii) a very high monomer/initiator molar ratio (up to 30 000 : 1) and extremely high molecular weights ($M_n > 10^5$ g mol^{-1}) were achieved; (iii) water was introduced as an effective chain transfer reagent, resulting in the formation of α-hydroxy/ω-OH PPO; (iv) the refined structure of the bis-borane catalyst was ideal for ROP of PO, with both boron centers simultaneously stabilizing the transition state during the ROP process. Under optimized conditions, TOFs of up to 3300 h^{-1} were achieved, nearly double the rate of a similar ammonium diborane catalyst. Remarkably, at very low catalyst loadings (PO/**5** = 10,000 : 1), molecular weights of up to M_n = 490 kg mol^{-1} with 97% conversion in six hours were observed [73]. Kinetic experiments revealed a zero-order dependence on PO concentration. Interestingly, similar to nitrogen-based catalysts, a synergistic effect was observed between the phosphonium and borane functionalities. When a diborane with a neutral methylene linker and Ph_4PBr (in a binary setup) was used, monomer conversion was significantly slower compared to the use of catalyst **5**. The conversion of BO and AGE also proceeded successfully, though monomer consumption was notably slower in both cases. To gain a deeper understanding of the role of "cation element P" in catalyst **5**, we compared it with the previously studied "cation element N" [74]. Interestingly, calculations indicated that the activation energy for the initiation process with catalyst **5** was lower compared to its nitrogen analogue (17.3 kcal mol^{-1} vs. 19.7 kcal mol^{-1}). This is attributed to the larger radius of the phosphorus atom, which creates a more spacious environment for the coordination and ROP of PO. Additionally, the ROP of PO using catalyst 5 exhibited a lower Gibbs energy (9.9 kcal mol^{-1}) compared to its nitrogen counterpart (12.5 kcal mol^{-1}), due to a minimal change in entropy during PO coordination. Furthermore, the presence of the phosphonium structure reduced the energy of the unoccupied p orbital of the boron atom, thereby enhancing the Lewis acidity of catalyst **5**.

Kinetic experiments showed that the ROP of PO exhibited zero-order kinetics with respect to PO concentration and first-order dependence on catalyst **5** in bulk at −20 °C. ^{1}H NMR analysis of the resulting PPO indicated that the nucleophilic ring-opening of electron-rich epoxides predominantly occurs at the methene position rather than the methine, due to the steric effects of the substituents. The region-regularity of PPO was confirmed by $^{13}C\{^1H\}$ NMR spectroscopy, which revealed head-to-tail linkages and the absence of head-to-head and tail-to-tail enchainments, indicating a controlled and fully region-specific monomer insertion process. Based on the $^{11}B\{^1H\}$ NMR spectra, polymer microstructure, end group analysis, and control experiments, we propose that an intramolecular synergistic effect involving the two boron atoms, the electropositive P^+, and the Br^- anion is crucial to the ROP of epoxides. An intramolecular P^+-assisted S_N2 ROP mechanism is proposed (see Figure 9.12). Initially, the Br^- ion creates a dynamic association/dissociation equilibrium between the two boron atoms with the assistance of the central P^+ through electrostatic interactions. When PO is introduced to catalyst **5**, the Lewis basic PO molecules preferentially coordinate with the electron-deficient boron atoms. Subsequently, the nucleophilic Br^- ion attacks the less hindered carbon of the PO ring, forming an alkoxide-involved tetra-coordinated boron center. This results in the highly region-selective ring-opening reaction. The alkoxide anion then attacks the next activated PO molecule on the adjacent boron atom, assisted by electrostatic interactions with the central P^+, leading to chain propagation. This mechanism is consistent with the $^{11}B\{^1H\}$ NMR spectra of the ROP of PO with catalyst **5**, where alkoxide-coordinated boron (δ = 0.93 ppm) was observed. Through intramolecular P^+-assisted S_N2 ring-opening and interactions with adjacent boron atoms, the active species completes the catalytic cycle, resulting in the PPO product.

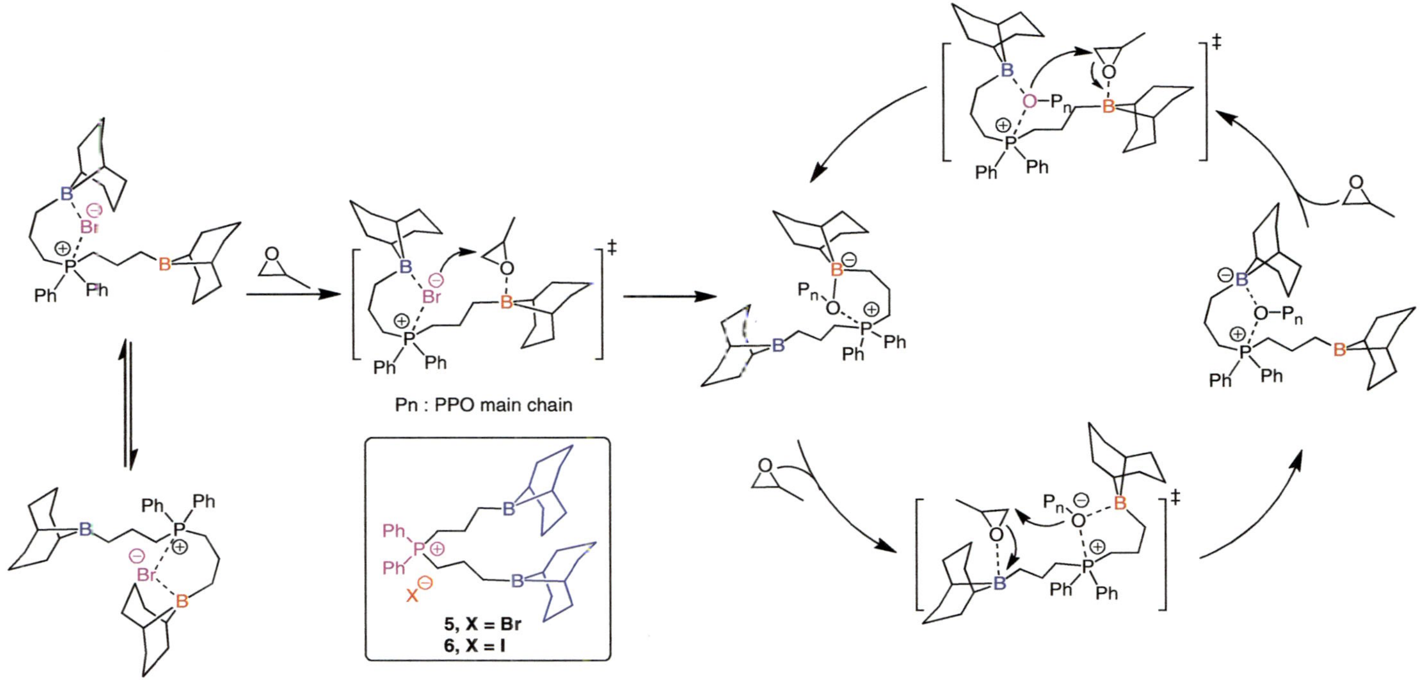

Figure 9.12 Chemical structure of phosphonium-bridged bisborane initiators/catalysts **5–6** and proposed mechanism for the ROP of PO with **5**.

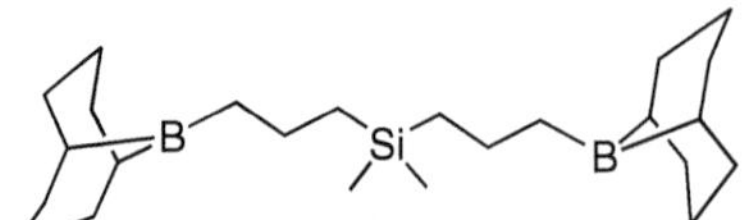

7

Figure 9.13 The chemical structure of silicon-centered organoborane.

Subsequently, Wu and coworkers used a different precursor to synthesize **5** and further demonstrated that phosphonium-bridged bisborane showed superior performance than its ammonium analogues. [75]

9.4.3 Silicon-Based Borane Lewis Pair System

In addition to the ammonium and phosphonium linked borane catalysts for the ROP of PO. In 2023, Wu and colleagues developed a binary organoborane catalyst system, utilizing a silicon-centered organoboron compound as the electrophile and pairing it with an ionic organic ammonium salt as the initiator [76]. The silicon-centered alkyl borane/ammonium salt binary catalyst (Figure 9.13) demonstrated significantly higher activity than its bifunctional counterpart in the ROP of propylene oxide. At a catalyst loading of 0.01 mol%, it exhibited activity 7.3 times greater than the bifunctional catalyst, and at an extremely low loading of 0.002 mol%, the activity was 15.3 times higher. Investigations including ^{19}F NMR, control experiments, and theoretical calculations revealed that the central silicon atom possesses a suitable electron density and a larger intramolecular cavity. These features enhance monomer coactivation and the delivery of propagating chains, resulting in a superior intramolecular synergistic effect compared to the bifunctional catalyst. This study suggested an intramolecular Si-assisted S_N2 ROP mechanism with a dual-pathway initiation mode to explain the exceptionally high activity of this binary catalytic system, challenging the traditional belief that binary Lewis acid/nucleophilic catalysts have low activity due to increased entropy. However, one limitation of such system is the polymerizations have to be performed at low temperatures (≤0 °C) to ensure its synergistic effect.

9.4.4 Macromolecular Organoborane Mediated ROP of Epoxides

In most instances, borane Lewis acids have been employed as small molecule catalysts, whether as binary or single-component organocatalysts for polymerization. However, there have been few reports focusing on incorporating a single borane unit into macromolecular boranes to create macromolecular Lewis acids for polymerization [77]. Nonetheless, some studies have shown significantly enhanced activity by developing macromolecular catalysts. Recently, Wang, Liu, and colleagues integrated aluminum porphyrin as a catalyst into a polymer backbone, guided by the principle "unity makes strength." They demonstrated that the resulting macromolecular catalyst exhibited multiple synergistic effects, achieving both high activity and selectivity in the ROCOP of CO_2 and epoxides [78–82].

Given the crucial role of borane elements as Lewis acids, we introduced the concept of "aggregation-induced enlargement effects" by chemically linking multiple

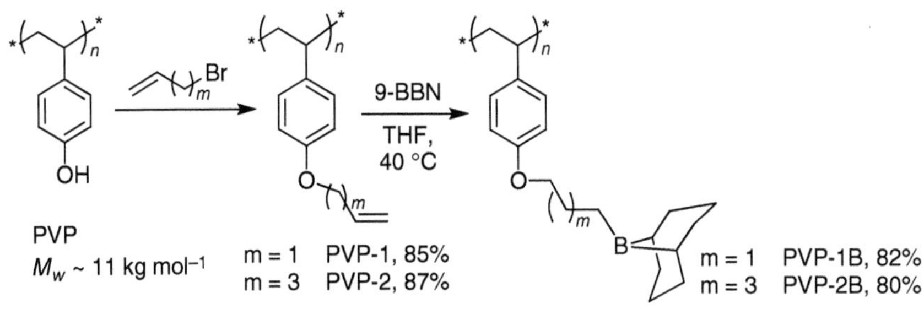

Figure 9.14 Synthetic route of **PVP-1B** and **PVP-2B**.

borane elements within a nanoscopic confinement to create macromolecular Lewis acids. We anticipated several advantages from this approach. The spatial confinement is expected to significantly increase the concentration of borane Lewis active sites beyond the average due to local confinement. Since boranes activate monomers through supramolecular interactions, the concentration of activated monomers around the chain propagation center will also increase, leading to enhanced activity. Additionally, the local aggregation of Lewis acids may improve tolerance to air and high concentrations of chain transfer agents (CTA) by facilitating rapid proton exchange between the growing anion and CTA.

We employed this strategy to synthesize macromolecular boranes and evaluate their tolerance to moisture and oxygen. Additionally, the catalytic performance of the designed macromolecular borane was tested for the ROP of PO under ambient conditions. The macromolecular boranes were synthesized through the hydroboration of a polymer containing repeating C=C double bonds. Poly(4-vinylphenol) (PVP) was selected as the candidate due to its commercial availability, range of molecular weights (M_ns), and reactive phenol groups, which allow for the introduction of pendant C=C groups. Following the synthetic route shown in Figure 9.14, two macromolecular boranes, **PVP-1B** and **PVP-2B**, were successfully synthesized in good yield. Based on the M_n of **PVP**, the number of repeating units in **PVP** was estimated to be 36 [83].

The most notable advantage of the synthesized macromolecular boranes is their exceptional tolerance to moisture and oxygen, as demonstrated by both the "standing test in air" and ROP of PO experiments. The strong moisture and oxygen tolerance of **PVP-1B** can be attributed to several factors: (i) Supramolecular interactions between boron and oxygen donors: These interactions contribute to the compact molecular structure of **PVP-1B**, which reduces the exposure of boron atoms to moisture and oxygen, thereby minimizing the risk of oxidation or hydrolysis. (ii) Increased molecular size: The polymeric nature of the borane compounds results in larger and bulkier molecules, which hinders the diffusion of moisture and oxygen into the polymer solution, limiting their interaction with the borane units. (iii) Hydrophobic properties: **PVP-1B** exhibits hydrophobic characteristics, making it insoluble in water, which further minimizes the absorption of moisture into the polymer solution, reducing the likelihood of hydrolysis. Additionally, the stability of **PVP-1B** may also be supported by the supramolecular interaction between the deficient p orbital of the borane centers and the phenol moieties.

To assess the effectiveness of **PVP-1B** and **PVP-2B**, tetrabutylammonium chloride (TBACl) and tetrabutylammonium succinate (TBASA) were selected as monofunctional and bifunctional initiators, respectively, for the bulk ROP of PO. Under conditions of $[PO]_0/[TBACl]_0/[B]_0$ = 10000/1/10 (where [B] represents the molar amount of boron centers in the polymer), control experiments showed that **PVP-1B** catalyzed the bulk ROP of PO to achieve an 88% conversion within 2.5 hours. Under the same conditions, **PVP-1B** was more active than **PVP-2B**, which achieved a 67% conversion. The dilution effect was not significant at a much higher feeding ratio of $[M]_0/[I]_0/[B]_0$ = 10000/1/10. The conversions of PO catalyzed by **PVP-1B** and **PVP-2B** reached 96% and 88%, respectively, within 2.5 hours. These polymerization rates were significantly higher than those of our previously reported **PBB-Br**, which achieved only a 30% PO conversion within six hours[73]. Remarkably, the turnover frequency (TOF) for **PVP-1B**-catalyzed PO polymerization at 25 °C reached up to 27,900 h^{-1} when the feeding ratio of $[M]_0/[I]_0/[B]_0$ was set at 10000/1/10. Additionally, **PVP-1B** proved to be an efficient catalyst for controlled ROP of PO, even in the presence of large amounts of CTA ($[CTA]_0/[\textbf{PVP-1B}]_0$ = 1800, with one borane tolerating 50 equivalents of CTA). As a result, a wide range of PPO molecular weights (1.0–23.4 kg mol^{-1}) could be produced with high productivity (TOF = 1980 h^{-1}).

To illustrate the polymerization process of PO catalyzed by **PVP-1B**/TBACl, a schematic diagram was proposed (Figure 9.15). In this study, the rapid polymerization of PO was clearly attributed to **PVP-1B**, which, while sharing the same active site structure as **B2**, operated in a distinct manner. The alkoxide end, with its strong nucleophilicity, preferentially attacked the neighboring activated PO, driven by the reduced enthalpy and stabilized by the formation of a stable anionic transition state, where two borane units cooperatively localized the epoxide and alkoxide ends. Despite the alkoxide' strong nucleophilicity and poor leaving ability, the comb-shaped structure of **PVP-1B** created multiple vacant coordination sites near the alkoxide, facilitating the accumulation of activated epoxides. Moreover, the intramolecular cavities formed by the flexible comb-shaped polymer reduced the distance between the PO and the alkoxide end. This proximity promoted the relay between the borane-stabilized alkoxide active chain end and borane-activated PO, significantly enhancing the ROP of PO.

This work demonstrates that integrating multiple borane units within a polymer at the macromolecular level offers superior catalytic performance compared to constructing binary or bi(multi)functional systems at the molecular level. This approach paves the way for the development of functional polyethers.

9.4.5 Poly(alkylene oxide) Structures: Innovative Polyether Architectures

The ammonium-based bifunctional 9-BBN-initiated ROP of epoxides exhibits high reactivity with living polymerization behavior, even at low catalyst loading (5 ppm), and allows for the synthesis of polyethers with molecular weights exceeding one

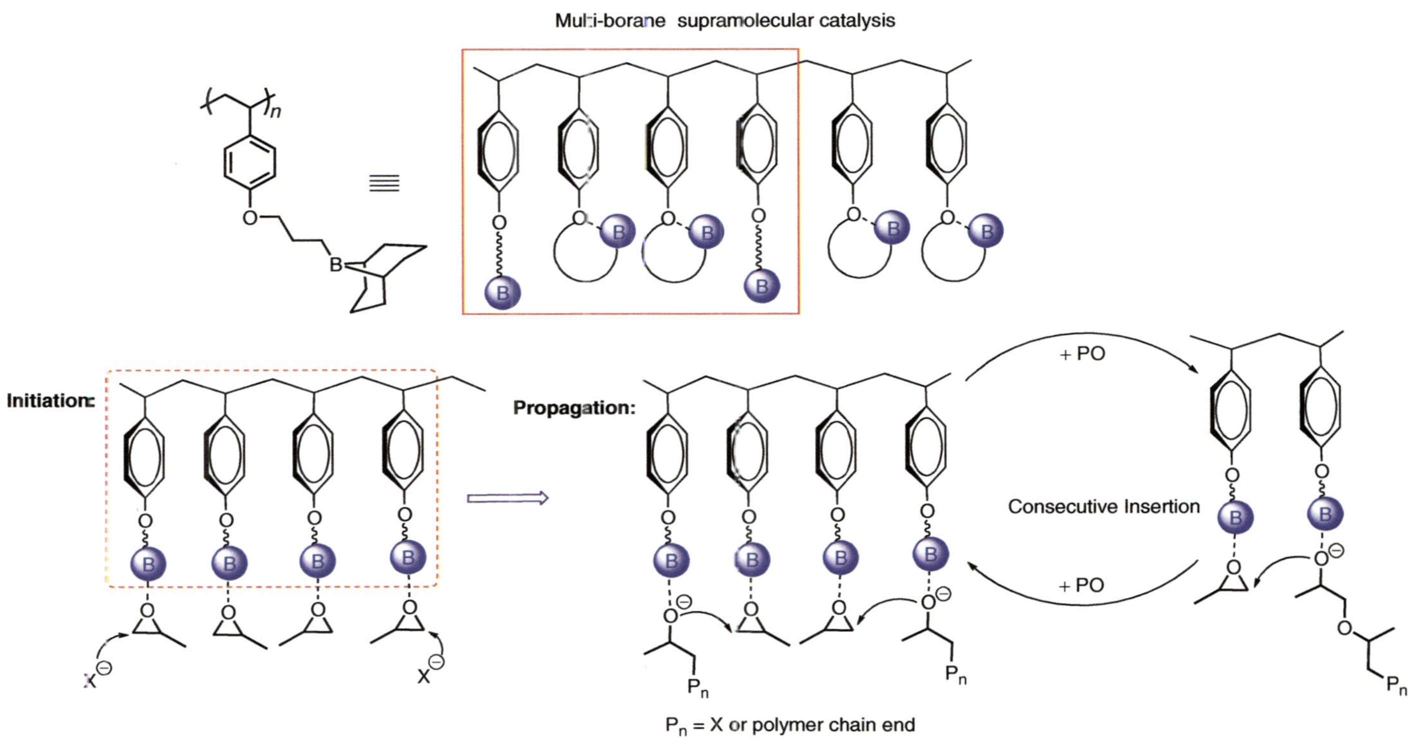

Figure 9.15 A schematic diagram proposed for macromolecular Lewis acid catalyzed ROP of PO.

million grams per mole (>10^6 g mol^{-1}) [74]. However, the generated polyethers featuring α-halide/ω-OH end groups, which significantly limit the kinds of end groups and architectures of the generated polyethers.

We developed the first example of an intramolecular phosphonium bisborane Lewis pair, which functioned as both a dual initiator and catalyst, achieving high activity and living characteristics in PO polymerization. Notably, well-defined α-OH/ω-OH PPO was produced using water as a chain transfer agent (CTA). Remarkably, Lewis pair **5** demonstrated good tolerance to water, even at a molar ratio of H_2O/**5** = 200/1. By using water as a cost-effective and renewable chain transfer reagent, various molecular weights (M_ns) of α-hydroxy/ω-OH PPO can be generated, depending on the feeding ratio of PO to H_2O [73]. The controlled nature of the polymerization in the presence of water was attributed to a reversible and rapid chain transfer reaction. We suggest that the further use of various protonic chain transfer reagents to produce other functionalized PPOs, as well as PPO-based block copolymers and elastomers, holds significant potential.

Subsequently, Wu and colleagues demonstrated that their organoboron initiator/catalyst **3** exhibited good tolerance to certain chain transfer agents, including 1,2-propanediol (PD), ethylene glycol (EG), and allyl alcohol (AA). To ensure rapid chain transfer in the presence of these CTAs, most reactions were conducted at 0 °C. By carefully adjusting the amount of CTAs, PPO and PEO polyols with tailored molecular weights (M_ns) ranging from 1.0 to 8.2 kg mol^{-1} and narrow dispersities ($Đ < 1.10$) were efficiently produced on a scale of several hundred grams. They also explored the synthesis of block copolymers composed of PPO and PEO, which are widely used in medicine, surfactants, and micelle formation due to the molecular-scale segregation and aggregation of the different blocks [15]. Triblock copolymers of PEO/PPO, such as PEO-*b*-PPO-*b*-PEO (Pluronic) and PPO-*b*-PEO-*b*-PPO (Pluronic R), were prepared in a straightforward manner by sequential monomer addition (Figure 9.16). The well-defined topology of these block copolymers was confirmed by gel permeation chromatography (GPC) and diffusion-ordered spectroscopy (DOSY). Their unimodal distribution and controlled molecular weight make them competitive with commercial analogs produced by BASF. Unsaturated polyethers are widely used in water-reducing agents and specialty rubbers. Therefore, polymerization of PO with 2-allyloxyethanol as a CTA and telomerization of PO with 6 wt% allyl glycidyl ether (a commercial PPO elastomer, Parel 58, by Hercules Inc.) were performed to generate unsaturated polyethers, which were further characterized by GPC, MALDI-TOF-MS, and NMR [71]. Such work broadens architecture of the generated polyethers. However, the ammonium based bifunctional 9-BBN system has limited endurance to CTA (≤50) and monomer loading (≤10,000). Further work needs to improve the tolerance to CTA, monomer loading as well as reducing the catalyst loading.

Building on their experience with borinane-based ammonium salts in the alternating copolymerization of CO_2 with cyclic ethers [84], Feng and colleagues synthesized bifunctional borinane compounds **B2**, **B3**, and **B4**, each bearing ammonium salts with 2, 3, and 4 functional groups, respectively (Figure 9.17) [70].

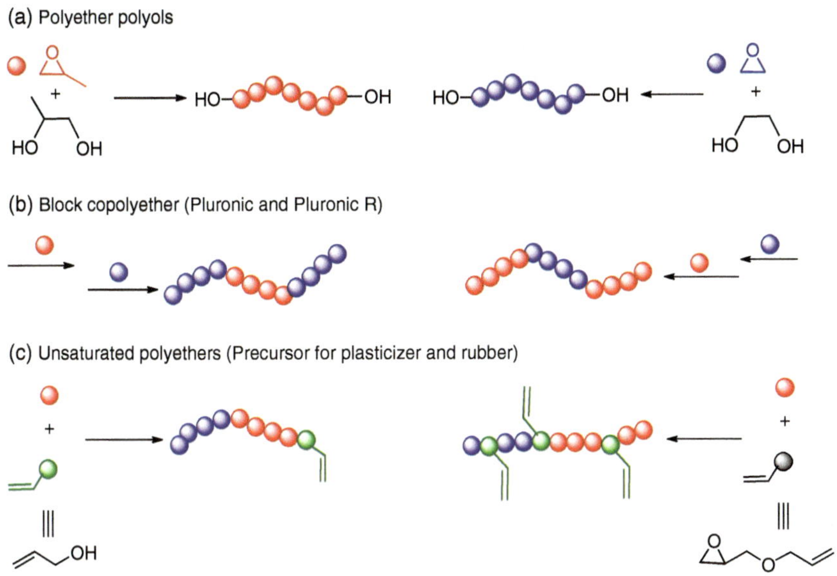

Figure 9.16 Different polyethers generated by 9-BBN ammonium borane system.

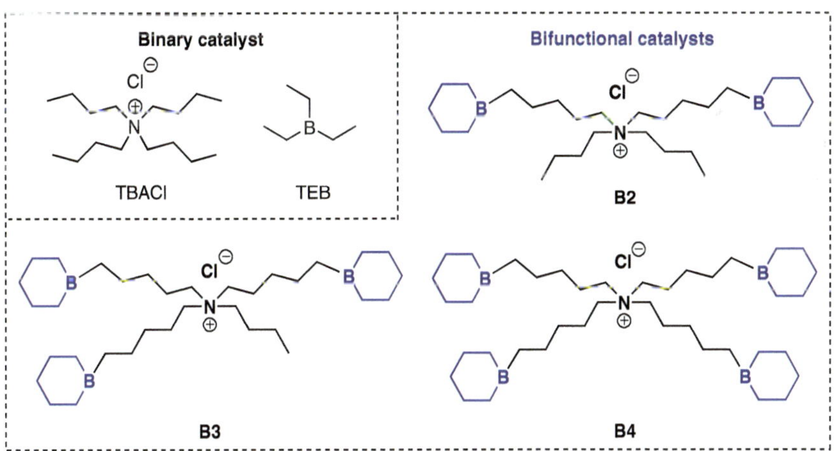

Figure 9.17 The chemical structures of bifunctional borinane compounds **B2**, **B3**, and **B4**.

Catalyst **B4** exhibited exceptional activity in the ROP of PO, surpassing any previously reported catalysts. The turnover frequency (TOF) for **B4** reached $1.8 \times 10^5\ h^{-1}$. Following this, **B4** was employed for the ROP of various epoxides in the presence of chain transfer agents (CTAs), e.g. H_2O, BnOH(benzyl alcohol), TFE(2,2,2-trifluoroethanol), PEG4OH([polyethylene glycol]$_4$-OH), PG(1,2-propanediol), HEA(2-hydroxyethyl acrylate), DBnOH(1,4-benzenedimethanol), DBnSH(1,4-benzenedimenethiol), and prepared α-functional low molar mass polyethers. They also explored the catalytic limit of **B4**. Even at an extremely low concentration of 2 ppm, **B4** demonstrated significant catalytic activity in the polymerization of PO, achieving a turnover frequency (TOF) of 18 125 h^{-1} when DBnOH was used as the chain transfer agent (CTA). Remarkably, with a catalyst loading as low as 2 ppm, **B4** retained high activity when using water as the chain transfer agent under the same conditions, although the polymerization rate decreased compared to DBnOH, with the TOF only reducing by half. Further investigations into the capabilities of **B4** were conducted by reducing the catalyst concentration to parts per billion (ppb) for the ring-opening polymerization (ROP) of ethylene oxide (EO) (2×10^7 equiv. EO loading, 250 000 equiv. CTA loading). In this setup, **B4** achieved turnover number (TON) and TOF values of 9×10^6 and 1.9×10^5 h^{-1}, respectively, establishing it as one of the most effective catalysts for the synthesis of polyethylene oxide (PEO) telechelics. This efficiency is highlighted by the ability of 1 g of **B4** to synthesize over 606.0 kg of PEO within 48 hours using a CTA, a performance unmatched by other catalysts. The successful integration of CTA motifs into PEO diols was confirmed through NMR spectroscopy. Further tests were performed using a variety of epoxide monomers with water as the CTA. It was noted that monomers with longer chains, such as butylene oxide (BO) and octene oxide (OO), polymerized at rates similar to PO. However, the polymerization rates of phenyl glycidyl ether (PGE) and allyl glycidyl ether (AGE) were slower, likely due to interactions between the extra oxygen in PGE and the double bond in AGE with the boron centers in **B4**. Despite these variations, the molecular weightsof the resulting diols were consistent with predictions, exhibiting a narrow distribution and free from high molar mass impurities.

In our previous research, we demonstrated that an intramolecular phosphonium bisborane Lewis pair, **5**, functions effectively as both a dual initiator and catalyst, achieving high activity and living characteristics in PO polymerization. We successfully produced well-defined α-OH/ω-OH PPO using water as the chain transfer agent (CTA) [73]. Moreover, our results showed that the precise control over M_n of the resulting polyether, with relatively low $Đ$, was due to the rapid, quantitative, and reversible chain transfer reaction between the active species and water molecules. Building on this, we utilized the synergistic effects of **5** to investigate the immortal ring-opening polymerization (iROP) of PO with functionalized CTAs of varying *p*Ka values, allowing us to introduce diverse functionalities to the produced PPOs for the first time. We generated both α-OH/ω-OH telechelic PPOs and a variety of α-functional/ω-OH heterotelechelic PPOs with predictable M_ns (1.3–10.1 kg mol^{-1}) and narrow polydispersities (1.03–1.09). The α-functional/ω-OH heterotelechelic PPOs included groups such as α-benzyl, allyl, amino, $SiMe_3$-protected alkynyl, furfuryl, (meth)acrylate, azide, (protected) carboxylate acid, (protected) amine, and nitro (Figure 9.18). The selection of alcohols with suitable substituents was essential for developing an effective immortal catalytic system. DFT calculations supported our experimental observations, revealing that the chain transfer to the CTA process is nearly barrierless (0.8 kcal mol^{-1}) and thermodynamically favorable compared to chain propagation. The Gibbs energy barrier for initiation is 4.9 kcal mol^{-1} lower than that for propagation. These findings enhance our understanding of

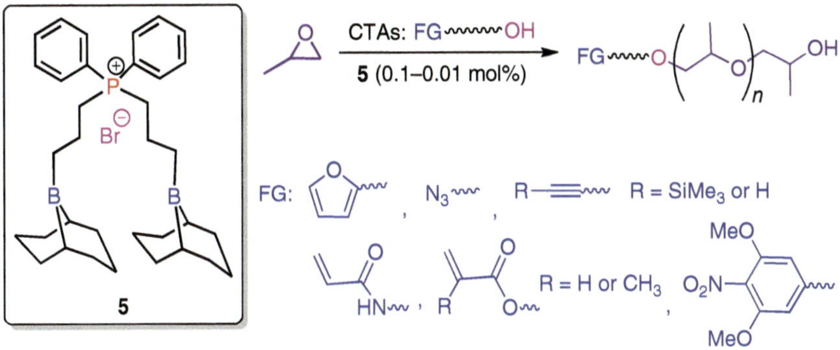

Figure 9.18 Syntheis of diverse α,ω-difunctionalized PPOs catalyzed by **5**.

the initiation, chain transfer, and propagation processes during the iROP of PO catalyzed by **5**. Additionally, well-defined AB or ABABAB block copolymers can be synthesized by adding mixtures of monomers such as PO, BO, and AGE in a single step. This organocatalyzed approach to α,ω-difunctionalized PPOs in an immortal manner not only greatly expands the range of PPO architectures but also offers a promising and more advantageous methodology compared to previous approaches [85].

9.5 Chiral Organoborane Mediated Asymmetric ROP of Epoxides

In 2022, Naumann and coworker detailed the use of organoboron catalysts for the stereoselective polymerization of racemic propylene oxide (*rac*-PO) [86]. Inspired by the epoxide homopolymerization mechanism, which incorporates two equivalents of Et_3B, and influenced by Du's research on chiral diboranes for hydrogenation, [87] they employed axially chiral linkers. Initiating with readily available diols such as 1,1′-bi-2-naphthol (BINOL), a straightforward process of etherification followed by hydroboration was used to produce the diboranes **8** and **9** (Figure 9.19). These catalysts achieved a moderate level of stereoselectivity, with isotactic diad placements (m) up to 88%, across a broad temperature range from −36 °C to 50 °C.

Interestingly, the choice of organobase (DBU, ^{t}Bu-P_2, or a specific NHO) did not affect the stereoselectivity, while the introduction of steric hindrance, particularly a phenyl group in compound **9** significantly improved stereoselectivity but reduced the rate of polymerization. Solvent dilution using THF or toluene were effective, likely by reducing non-selective bimolecular interactions [87]. The molecular weights reached up to 120 kg mol^{-1}, with well-defined properties confirmed by GPC and MALDI-TOF-MS, which identified end groups exclusively originating from the initiator, benzyl alcohol (BnOH). Kinetic studies with an enantiopure monomer revealed that (R)-**8** preferentially polymerized (S)-PO (and (S)-BO), while (S)-**8** favored the opposite enantiomers. Interestingly, similar selectivity was observed

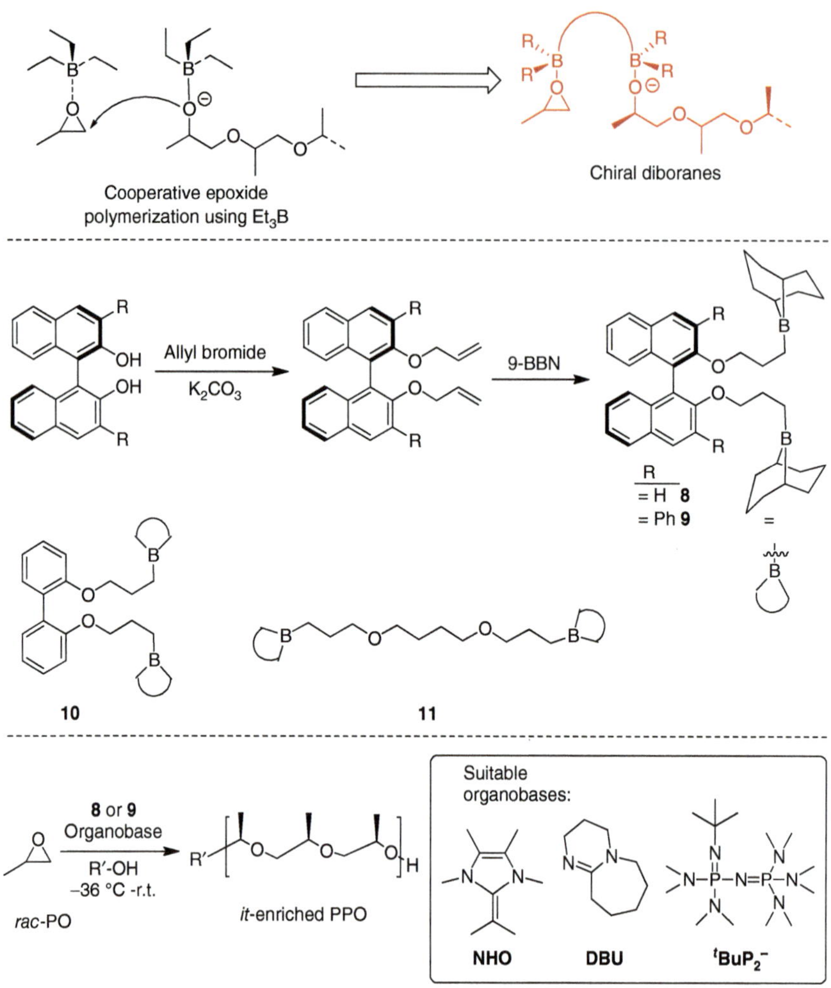

Figure 9.19 Rationale for the research (top), two-step synthetic pathway (middle) and polymerization procedure (bottom) for the production of *it*-PPO.

with racemic-**8**, indicating a strong association between the growing polymer chain end and the diborane catalyst. This association likely reduces the frequency of exchange of growing polyether chains between catalysts.

This study indicates a promising direction for simplifying catalyst synthesis if further research confirms these trends, marking a significant advancement in the field of stereoselective polymerization traditionally dominated by metal-based catalysts [38]. Metal-based catalysts achieve isotactic triad placement, *mm* rates over 99% for PO [22], the functional group tolerance exhibited by organoboron catalysts like **8** and **9** is a crucially differentiating factor of great practical importance [86]. These organoboron catalysts facilitate the introduction of blocks of isotactic polypropylene oxide (it-PPO) with high precision when used with polycaprolactone or polylactide as macroinitiators, a strategy not feasible with traditional metal-based systems. This "polyester-first" approach offers significant advantages for creating tacticity-tuned polymeric additives. Unlike the bifunctional catalysts numbered **1–7**, **8** and **9** do not rely on a functional linker setup, yet they still enable the exploration of new structural and property relationships within diborane-based catalysts. Further insights into structure–property relationships for diborane catalysts are still emerging. Control experiments with non-chiral (**10**) and linear (**11**) variants have begun to clarify potentially efficient catalyst configurations for epoxide homopolymerization. The biphenyl-based catalyst (**10**) produces PO up to fifty times faster than its linear counterpart (**11**), suggesting that the biphenyl backbone significantly enhances polymerization rate due to better preorganization, akin to observations made for bifunctional catalysts reported by Lin et al. [88].

To advance the development of organoboron catalysts that can rival top metal-based catalysts for the stereoselective ROP of epoxides, it is important to create a wider range of chiral organoboron catalysts. Additionally, a thorough examination of the relationship between catalyst structure and catalytic performance is necessary. In 2023, Wu and colleagues synthesized various organoboron catalyst systems, including both a binary diborane/ammonium salt system and bifunctional organoboron compounds (Figure 9.20), for the stereoselective ROP of epoxides [89]. They investigated how different catalytic structures and reaction conditions influence ROP stereoselectivity. Their study found that moderate steric hindrance in catalysts can enhance the tacticity of the resulting polyethers, while excessive steric hindrance significantly reduces catalytic activity. Temperature affects catalytic activity noticeably but has minimal impact on the tacticity of the polyethers. Appropriate solvent choices can improve the stereo-regularity of polyethers. The polyethers produced exhibited moderate isotacticity, ranging from 62.1% m to 84.8% m.

9.6 Conclusion and Outlook

The introduction and development of triethylborane (Et_3B) and subsequent borane catalysts have significantly propelled advancements in epoxide homo- and copolymerization. These borane catalysts offer a metal-free, sustainable approach to synthesizing polyethers and other copolymers, characterized by ultra-low catalyst loadings that eliminate the need for catalyst removal. This feature is especially beneficial in multifunctional or multiborane setups due to their high activity, allowing for usage in catalytic amounts. These advances have led to remarkable improvements in turnover numbers (TON), turnover frequencies (TOF), and achievable molecular weightsin epoxides polymerizations.

A distinctive advantage of borane catalysis is its functional group tolerance, due to the moderate Lewis acidity of the boranes. This enables the dual role of monomer activation and chain-end capping, facilitating the polymerization of sensitive, highly functionalized monomers. Such capabilities allow for the creation of advanced polymer architectures, like poly(ether-ester) multiblocks, functionalized polyethers, and

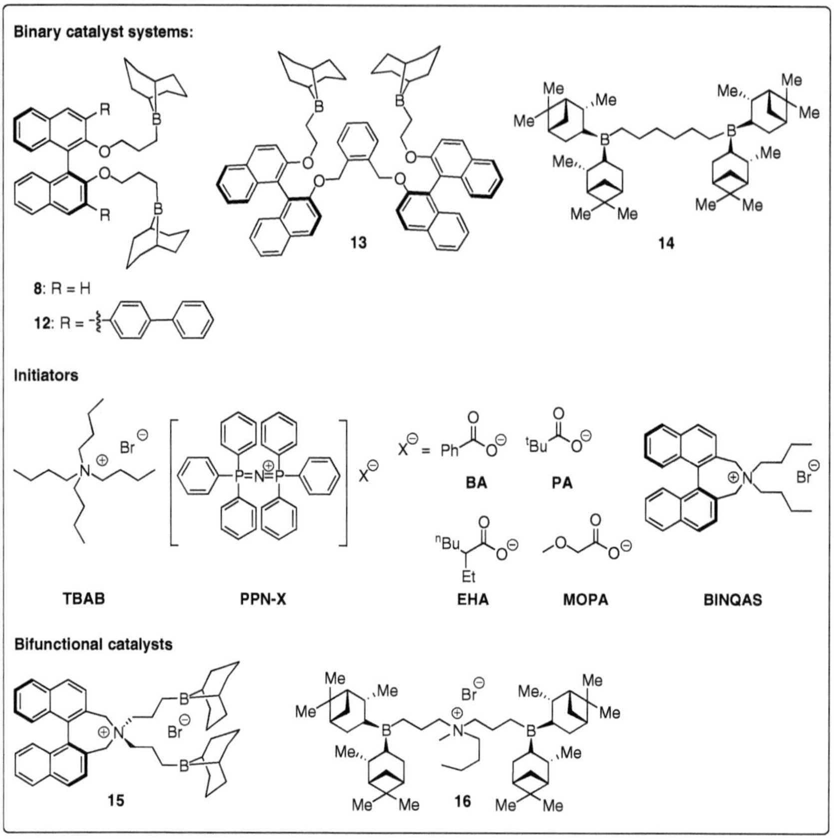

Figure 9.20 The binary and bifunctional organoboron compounds employed in this study.

even the first instances of organocatalytic, stereoselective polymerization of epoxides, although selectivity here still needs enhancement.

Additionally, the potential for structural or compositional changes of alkylboranes during the catalytic cycle opens new possibilities for novel chemoselectivities. This aspect, along with the conceptual principle to Lewis pair polymerization, suggests fertile grounds for further advances. Inspiration from classical organometallic chemistry, such as the Co(III)-Salen complexes, indicates the importance of optimally spaced Lewis acidic centers for effective catalysis, a principle that applies to both metal-based and organocatalytic systems. Our viewpoints for areas poised for significant breakthroughs in borane-catalyzed epoxide polymerization include: expanding of monomer scope to include other *O*-heterocyclic monomers (oxiranes with electron withdrawing group and other functional groups) and beyond *O*-heterocyclic monomers to diversify the kinds of polyethers; understanding of mechanism of borane deactivation and structural changes during polymerization; theoretically addressing the roles of Lewis acidity on the monomer activation and polymerization mechanisms; stereoseletive polymerization by boron based organocatalyst; developing high active immobilized and recyclable catalytic systems.

References

1 Krebs, R.E. (2006). *The History and Use of our earth's Chemical Elements: A Reference Guide*. New York: Bloomsbury Publishing USA.

2 Hall, D.G. (2005). Structure, properties, and preparation of boronic acid derivatives. Overview of their reactions and applications. In: *Boronic Acids: Preparation and Applications in Organic Synthesis and Medicine*, 1–99.

3 Sana, M., Leroy, G., and Wilante, C. (1991). Enthalpies of formation and bond energies in lithium, beryllium, and boron derivatives. A theoretical attempt for data rationalization. *Organometallics* 10 (1): 264–270.

4 Li, Z., Tan, B., Zhang, Q. et al. (2022). Process modeling of boron isotopes separation by cross-current and counter-current solvent extraction. *Ind. Eng. Chem. Res.* 61 (49): 18121–18126.

5 Coca, A. (2016). *Boron Reagents in Synthesis*, vol. 1236. Washington, DC: American Chemical Society.

6 Lewis, G.N. (1923). *Valence and the Structure of Atoms and Molecules*. Chemical Catalog Company, Incorporated.

7 Laurence, C. and Gal, J.-F. (2009). *Lewis Basicity and Affinity Scales: Data and Measurement*. Chichester, UK: John Wiley & Sons.

8 Olah, G.A., Prakash, G.K.S., Sommer, J., and Molnar, A. (2009). *Superacid Chemistry*. Hoboken, NJ: John Wiley & Sons.

9 Christe, K.O., Dixon, D.A., McLemore, D. et al. (2000). On a quantitative scale for Lewis acidity and recent progress in polynitrogen chemistry. *J. Fluorine Chem.* 101 (2): 151–153.

10 Gutmann, V. (1976). Solvent effects on the reactivities of organometallic compounds. *Coordin. Chem. Rev.* 18 (2): 225–255.

11 Beckett, M.A., Strickland, G.C., Holland, J.R., and Varma, K.S. (1996). A convenient nmr method for the measurement of Lewis acidity at boron centres: correlation of reaction rates of Lewis acid initiated epoxide polymerizations with Lewis acidity. *Polymer* 37 (20): 4629–4631.

12 Childs, R.F., Mulholland, D.L., and Nixon, A. (1982). The Lewis acid complexes of α, β-unsaturated carbonyl and nitrile compounds. A nuclear magnetic resonance study. *Can. J. Chem.* 60 (6): 801–808.

13 Luo, L. and Marks, T.J. (1999). Ziegler–Natta catalyst activation. Thermodynamic and kinetic aspects of metallocenium ion-pair formation, dissociation, and structural reorganization. *Top. Catal.* 7: 97–106.

14 Brocas, A.-L., Mantzaridis, C., Tunc, D., and Carlotti, S. (2013). Polyether synthesis: from activated or metal-free anionic ring-opening polymerization of epoxides to functionalization. *Prog. Polym. Sci.* 38 (6): 845–873.

15 Herzberger, J., Niederer, K., Pohlit, H. et al. (2016). Polymerization of ethylene oxide, propylene oxide, and other Alkylene oxides: synthesis, novel polymer architectures, and bioconjugation. *Chem. Rev.* 116 (4): 2170–2243.

16 Levene, P.A. and Walti, A. (1927). On condensation products of propylene oxide and of glycidol. *J. Biol. Chem.* 75 (1): 325–336.

17 Dudev, T. and Lim, C. (1998). Ring strain energies from ab initio calculations. *J. Am. Chem. Soc.* 120 (18): 4450–4458.

18 Darensbourg, D.J. and Chung, W.-C. (2013). Relative basicities of cyclic ethers and esters. Chemistry of importance to ring-opening co- and terpolymerization reactions. *Polyhedron* 58: 139–143.

19 Odian, G. (2004). *Principles of Polymerization*. Hoboken, NJ: John Wiley & Sons.

20 Fevre, M., Pinaud, J., Gnanou, Y. et al. (2013). N-heterocyclic carbenes (NHCs) as organocatalysts and structural components in metal-free polymer synthesis. *Chem. Soc. Rev.* 42 (5): 2142–2172.

21 Klein, R. and Wurm, F.R. (2015). Aliphatic Polyethers: classical polymers for the 21st century. *Macromol. Rapid Commun.* 36 (12): 1147–1165.

22 Longo, J.M., Sanford, M.J., and Coates, G.W. (2016). Ring-opening copolymerization of epoxides and cyclic anhydrides with discrete metal complexes: structure-property relationships. *Chem. Rev.* 116 (24): 15167–15197.

23 Naumann, S. and Dove, A.P. (2016). N-heterocyclic carbenes for metal-free polymerization catalysis: an update. *Polym. Int.* 65 (1): 16–27.

24 Dove, A., Sardon, H., and Naumann, S. (2018). *Organic Catalysis for Polymerisation*. Royal Society of Chemistry.

25 Ryzhakov, D., Printz, G., Jacques, B. et al. (2021). Organo-catalyzed/initiated ring opening co-polymerization of cyclic anhydrides and epoxides: an emerging story. *Polym. Chem.* 12 (20): 2932–2946.

26 Billouard, C., Carlotti, S., Desbois, P., and Deffieux, A. (2004). Controlled high-speed anionic polymerization of propylene oxide initiated by alkali metal Alkoxide/Trialkylaluminum systems. *Macromolecules* 37 (11): 4038–4043.

27 Imbrogno, J., Ferrier, R.C. Jr., Wheatle, B.K. et al. (2018). Decoupling catalysis and chain-growth functions of mono(μ-alkoxo)bis(alkylaluminums) in epoxide polymerization: emergence of the N–Al adduct catalyst. *ACS Catal.* 8 (9): 8796–8803.

28 Labbé, A., Carlotti, S., Billouard, C. et al. (2007). Controlled high-speed anionic polymerization of propylene oxide initiated by onium salts in the presence of Triisobutylaluminum. *Macromolecules* 40 (22): 7842–7847.

29 Carlotti, S., Labbé, A., Rejsek, V. et al. (2008). Living/controlled anionic polymerization and copolymerization of Epichlorohydrin with Tetraoctylammonium bromide–Triisobutylaluminum initiating systems. *Macromolecules* 41 (19): 7058–7062.

30 Gervais, M., Brocas, A.-L., Cendejas, G. et al. (2010). Synthesis of linear high molar mass Glycidol-based polymers by monomer-activated anionic polymerization. *Macromolecules* 43 (4): 1778–1784.

31 Labbé, A., Brocas, A.-L., Ibarboure, E. et al. (2011). Selective ring-opening polymerization of Glycidyl methacrylate: toward the synthesis of cross-linked (co)polyethers with Thermoresponsive properties. *Macromolecules* 44 (16): 6356–6364.

32 Widger, P.C.B., Ahmed, S.M., Hirahata, W. et al. (2010). Isospecific polymerization of racemic epoxides: a catalyst system for the synthesis of highly isotactic polyethers. *Chem. Commun.* 46 (17): 2935–2937.

33 Widger, P.C.B., Ahmed, S.M., and Coates, G.W. (2011). Exploration of Cocatalyst effects on a bimetallic cobalt catalyst system: enhanced activity and Enantioselectivity in epoxide polymerization. *Macromolecules* 44 (14): 5666–5670.

34 Horbatenko, Y., Pérez, J.P., Hernández, P. et al. (2014). Reaction mechanisms for the formation of mono- and Dipropylene glycol from the propylene oxide hydrolysis over ZSM-5 zeolite. *J. Phys. Chem. C.* 118 (38): 21952–21962.

35 Torchilin, V.P. (2001). Structure and design of polymeric surfactant-based drug delivery systems. *J. Controlled Release* 73 (2): 137–172.

36 Childers, M.I., Vitek, A.K., Morris, L.S. et al. (2017). Isospecific, chain shuttling polymerization of propylene oxide using a bimetallic chromium catalyst: a new route to Semicrystalline polyols. *J. Am. Chem. Soc.* 139 (32): 11048–11054.

37 Hirahata, W., Thomas, R.M., Lobkovsky, E.B., and Coates, G.W. (2008). Enantioselective polymerization of epoxides: a highly active and selective catalyst for the preparation of Stereoregular Polyethers and Enantiopure epoxides. *J. Am. Chem. Soc.* 130 (52): 17658–17659.

38 Childers, M.I., Longo, J.M., Van Zee, N.J. et al. (2014). Stereoselective epoxide polymerization and copolymerization. *Chem. Rev.* 114 (16): 8129–8152.

39 Zhang, J., Wang, L., Liu, S., and Li, Z. (2022). Synthesis of diverse polycarbonates by Organocatalytic copolymerization of CO_2 and epoxides: from high pressure and temperature to ambient conditions. *Angew. Chem. Int. Ed.* 61 (4): e202111197.

40 Eßwein, B., Steidl, N.M., and Möller, M. (1996). Anionic polymerization of oxirane in the presence of the polyiminophosphazene base t-Bu-P_4. *Macromol. Rapid Commun.* 17 (2): 143–148.

41 Rexin, O. and Mülhaupt, R. (2002). Anionic ring-opening polymerization of propylene oxide in the presence of phosphonium catalysts. *J. Polym. Sci. Part A: Polym. Chem* 40 (7): 864–873.

42 Raynaud, J., Ottou, W.N., Gnanou, Y., and Taton, D. (2010). Metal-free and solvent-free access to α,ω-heterodifunctionalized poly (propylene oxides) by N-heterocyclic carbene-induced ring opening polymerization. *Chem. Commun.* 46 (18): 3203–3205.

43 Naumann, S., Thomas, A.W., and Dove, A.P. (2015). N-heterocyclic olefins as Organocatalysts for polymerization: preparation of well-defined poly (propylene oxide). *Angew. Chem. Int. Ed.* 54 (33): 9550–9554.

44 Balint, A., Papendick, M., Clauss, M. et al. (2018). Controlled preparation of amphiphilic triblock-copolyether in a metal-and solvent-free approach for tailored structure-directing agents. *Chem. Commun.* 54 (18): 2220–2223.

45 Lindner, R., Lejkowski, M.L., Lavy, S. et al. (2014). Ring-opening polymerization and copolymerization of propylene oxide catalyzed by N-heterocyclic Carbenes. *ChemCatChem* 6 (2): 618–625.

46 Kanazawa, A., Kanaoka, S., and Aoshima, S. (2013). Concurrent cationic vinyl-addition and ring-opening copolymerization using $B(C_6F_5)_3$ as a catalyst: copolymerization of vinyl ethers and isobutylene oxide via crossover propagation reactions. *J. Am. Chem. Soc.* 135 (25): 9330–9333.

47 Asenjo-Sanz, I., Veloso, A., Miranda, J.I. et al. (2014). Zwitterionic polymerization of glycidyl monomers to cyclic polyethers with $B(C_6F_5)_3$. *Polym. Chem.* 5 (24): 6905–6908.

48 Zhang, D., Boopathi, S.K., Hadjichristidis, N. et al. (2016). Metal-free alternating copolymerization of CO_2 with epoxides: fulfilling “green” synthesis and activity. *J. Am. Chem. Soc.* 138 (35): 11117–11120.

49 Zhang, D.-D., Feng, X., Gnanou, Y., and Huang, K.-W. (2018). Theoretical mechanistic investigation into metal-free alternating copolymerization of CO_2 and epoxides: the key role of triethylborane. *Macromolecules* 51 (15): 5600–5607.

50 Boopathi, S.K., Hadjichristidis, N., Gnanou, Y., and Feng, X. (2019). Direct access to poly (glycidyl azide) and its copolymers through anionic (co-) polymerization of glycidyl azide. *Nat. Commun.* 10 (1): 293.

51 Jia, M., Hadjichristidis, N., Gnanou, Y., and Feng, X. (2019). Monomodal ultrahigh-molar-mass polycarbonate homopolymers and diblock copolymers by anionic copolymerization of epoxides with CO_2. *ACS Macro Lett.* 8 (12): 1594–1598.

52 Patil, N.G., Boopathi, S.K., Alagi, P. et al. (2019). Carboxylate salts as ideal initiators for the metal-free copolymerization of CO_2 with epoxides: synthesis of well-defined polycarbonates diols and polyols. *Macromolecules* 52 (6): 2431–2438.

53 Zhang, C.J., Duan, H.Y., Hu, L.F. et al. (2018). Metal-free route to precise synthesis of poly (propylene oxide) and its blocks with high activity. *ChemSusChem* 11 (24): 4209–4213.

54 Chen, Z., Yang, J.-L., Lu, X.-Y. et al. (2019). Triethyl borane-regulated selective production of polycarbonates and cyclic carbonates for the coupling reaction of CO_2 with epoxides. *Polym. Chem.* 10 (26): 3621–3628.

55 Hong, M., Chen, J., and Chen, E.Y.X. (2018). Polymerization of polar monomers mediated by main-group Lewis acid–base pairs. *Chem.Rev.* 118 (20): 10551–10616.

56 Walther, P., Krauß, A., and Naumann, S. (2019). Lewis pair polymerization of epoxides via Zwitterionic species as a route to high-molar-mass Polyethers. *Angew. Chem. Int. Ed.* 58 (31): 10737–10741.

57 Chen, Y., Shen, J., Liu, S. et al. (2018). High efficiency organic Lewis pair catalyst for ring-opening polymerization of epoxides with chemoselectivity. *Macromolecules* 51 (20): 8286–8297.

58 Vogler, C. and Naumann, S. (2020). A simplified approach for the metal-free polymerization of propylene oxide. *RSC Adv.* 10 (71): 43389–43393.

59 Song, Q., Zhao, J., Zhang, G. et al. (2020). N-heterocyclic carbene/Lewis acid-mediated ring-opening polymerization of propylene oxide. Part 2: toward dihydroxytelechelic polyethers using triethylborane. *Eur. Polym. J.* 134: 109839–109845.

60 Chen, Y., Liu, S., Zhao, J. et al. (2019). Chemoselective polymerization of epoxides from carboxylic acids: direct access to esterified polyethers and biodegradable polyurethanes. *ACS Macro. Lett.* 8 (12): 1582–1587.

61 Markus, F., Vogler, C., Bruckner, J.R., and Naumann, S. (2021). Ordered mesoporous carbons via self-assembly of tailored block Copolyethers for pore size-dependent applications. *ACS Appl. Nano Mater.* 4 (4): 3486–3492.

62 Vogler, C., Naumann, S., and Bruckner, J.R. (2022). Predictive design of ordered mesoporous silica with well-defined, ultra-large mesopores. *Mol. Syst. Des. Eng.* 7 (10): 1318–1326.

63 Meng, Y., Gu, D., Zhang, F. et al. (2006). A family of highly ordered mesoporous polymer resin and carbon structures from organic–organic self-assembly. *Chem.Mater.* 18 (18): 4447–4464.

64 Naumann, S. (2021). Strategies for pore-diameter control in mesoporous carbons derived from organic self-assembly processes. *Org. Mater.* 3 (2): 283–294.

65 Liu, S., Liu, L., Zhou, Y. et al. (2022). Selective ring-opening polymerization of glycidyl esters: a versatile synthetic platform for glycerol-based (co) polyethers. *Polym. Chem.* 13 (24): 3650–3659.

66 Liu, S., Bai, T., Ni, K. et al. (2019). Biased Lewis pairs: a general catalytic approach to ether-Ester block copolymers with unlimited ordering of sequences. *Angew. Chem. Int. Ed.* 58 (43): 15478–15487.

67 Lidston, C.A.L., Severson, S.M., Abel, B.A., and Coates, G.W. (2022). Multifunctional catalysts for ring-opening copolymerizations. *ACS Catal.* 12 (18): 11037–11070.

68 Wang, X., Kehr, G., Daniliuc, C.G., and Erker, G. (2014). Internal adduct formation of active intramolecular C4-bridged frustrated phosphane/borane Lewis pairs. *J. Am. Chem. Soc.* 136 (8): 3293–3303.

69 Yang, G.-W., Zhang, Y.-Y., and Wu, G.-P. (2021). Modular organoboron catalysts enable transformations with unprecedented reactivity. *Acc. Chem. Res.* 54 (23): 4434–4448.

70 Chen, C., Gnanou, Y., and Feng, X. (2022). Borinane boosted Bifunctional Organocatalysts for ultrafast ring-opening polymerization of cyclic ethers. *Macromolecules* 55 (23): 10662–10669.

71 Qi, H., Xie, R., Yang, G.-W. et al. (2022). Rational optimization of Bifunctional Organoboron catalysts for versatile Polyethers via ring-opening polymerization of epoxides. *Macromolecules* 55 (20): 9081–9090.

72 Stephan, D.W. (2021). Diverse uses of the reaction of frustrated Lewis pair (FLP) with hydrogen. *J. Am. Chem. Soc.* 143 (48): 20002–20014.

73 Wang, X.W., Hui, J.W., Shi, M.M. et al. (2022). Exploration of the synergistic effect in a one-component Lewis pair system: serving as a dual initiator and catalyst in the ring-opening polymerization of epoxides. *ACS Catal.* 12 (14): 8434–8443.

74 Yang, G.-W., Zhang, Y.-Y., Xie, R., and Wu, G.-P. (2020). High-activity Organocatalysts for polyether synthesis via intramolecular ammonium cation assisted SN 2 ring-opening polymerization. *Angew. Chem. Int. Ed.* 59: 16910–16917.

75 Zhang, Y.-Y., Lu, C., Yang, G.-W. et al. (2022). Mechanism-inspired upgradation of phosphonium-containing organoboron catalysts for epoxide-involved copolymerization and homopolymerization. *Macromolecules* 55 (15): 6443–6452.

76 Xu, C.K., Yang, G.W., Lu, C., and Wu, G.P. (2023). A binary silicon-centered Organoboron catalyst with superior performance to that of its Bifunctional analogue. *Angew. Chem. Int. Ed.* 62 (48): e202312376.

77 Yolsal, U., Horton, T.A.R., Wang, M., and Shaver, M.P. (2021). Cyclic ether triggers for polymeric frustrated Lewis pair gels. *J. Am. Chem. Soc.* 143 (33): 12980–12984.

78 Cao, H., Zhang, R., Zhou, Z. et al. (2021). On-demand transformation of carbon dioxide into polymers enabled by a comb-shaped metallic oligomer catalyst. *ACS Catal.* 12 (1): 481–490.

79 Kuang, Q., Zhang, R., Zhou, Z. et al. (2023). A supported catalyst that enables the synthesis of colorless CO2-polyols with ultra-low molecular weight. *Angew. Chem. Int. Ed.* 62 (35): e202305186.

80 Zhang, R., Kuang, Q., Cao, H. et al. (2023). Unity makes strength: constructing polymeric catalyst for selective synthesis of CO_2/epoxide copolymer. *CCS Chem.* 5 (3): 750–760.

81 Zhuo, C., Cao, H., Wang, X. et al. (2023). Polymeric aluminum porphyrin: controllable synthesis of ultra-low molecular weight CO_2-based polyols. *Chin. Chem. Lett.* 34 (8): 108011–108015.

82 Zhuo, C., You, H., Gao, F. et al. (2023). Bottlebrush polymeric catalyst: boosting activity for CO_2/epoxide copolymerization. *Fuel* 333: 126434–126443.

83 Gu, Y., Kou, X., Wang, X., and Li, Z. (2024). Creating remarkably moisture-and air-stable macromolecular Lewis acid by integrating borane within the polymer chain: a highly active catalyst for homo (co) polymerization of epoxides. *Angew. Chem. Int. Ed.* 63 (7): e202318645.

84 Chen, C., Gnanou, Y., and Feng, X. (2022). Borinane-based organoboron catalysts for alternating copolymerization of CO_2 with cyclic ethers: improved productivity and facile recovery. *Polym. Chem.* 13 (45): 6312–6321.

85 Wang, X., Shi, M., Hui, J. et al. (2023). Controllable synthesis of Hetero-functionalized Polyethers via intramolecular Phosphonium Bisborane Lewis pair-catalyzed immortal ring-opening polymerization. *Macromolecules* 56 (11): 4030–4040.

86 Sirin-Sariaslan, A. and Naumann, S. (2022). Chiral diboranes as catalysts for the stereoselective organopolymerization of epoxides. *Chem. Sci.* 13 (36): 10939–10943.

87 Liu, Y. and Du, H. (2013). Chiral dienes as "ligands" for borane-catalyzed metal-free asymmetric hydrogenation of imines. *J. Am. Chem. Soc.* 135 (18): 6810–6813.

88 Schaefer, J., Zhou, H., Lee, E. et al. (2022). Tertiary and quaternary Phosphonium borane Bifunctional catalysts for CO2/epoxide copolymerization: a mechanistic investigation using in situ Raman spectroscopy. *ACS Catal.* 12 (19): 11870–11885.

89 Fang, Y.-B., Qi, H., Li, B. et al. (2024). Stereoselective polymerization of epoxides using organoboron catalysts. *Polym. Chem.* 13: 1297–1302.

10

Organocatalyst for Ring-Opening Polymerization of Cyclosiloxanes Toward Polysiloxanes

Na Zhao[1], *Yuetao Liu*[2], *and Zhibo Li*[1,2]

[1]*Qingdao University of Science and Technology, Key Laboratory of Biobased Polymer Materials, Shandong Provincial Education Department; College of Polymer Science and Engineering, 53 Zhengzhou Road, Qingdao 260042, China*
[2]*Qingdao University of Science and Technology, College of Chemical Engineering, 53 Zhengzhou Road, Qingdao 266042, China*

10.1 Introduction

Silicone materials are important hybrid organic/inorganic materials with unique properties compared to common polymer materials. They have been widely used as pharmaceuticals and medical devices, high temperature lubricants, adhesives, insulator materials, sealants, and so on [1, 2]. Polysiloxanes represent a pivotal class of silicon-based polymers characterized by a backbone composed of alternating silicon and oxygen atoms. The advantageous properties of silicone materials are derived from the unique characteristics of the silicon-oxygen bond. The relatively large bond angles (ranging from 140° to 180°) and longer bond lengths (approximately 1.64 Å) allow easy rotation of the substituents, endowing polysiloxanes with greater flexibility than carbon-based polymers [3]. This, in turn, contributes to their unique properties, particularly their exceptionally low glass transition temperature (T_g). Conversely, the high bond dissociation energy (BDE) of the silicon-oxygen bond, approximately 460 kJ mol^{-1}, underpins the remarkable thermal and oxidative stability of polysiloxanes. The characteristics of the bonds may vary with changes in the substituents on the silicon atoms. With the diversity of organic substituents, polysiloxanes have broad applications for their different physicochemical characteristics and operating temperature ranges [2, 4, 5].

Generally, two principal strategies exist for the synthesis of high molecular weight (MW) linear polysiloxanes. The first one is hydrolytic polycondensation of reactive dichlorodialkylsilanes to synthesize oligomeric linear or cyclic silicone prepolymers. The second one is the condensation polymerization of linear oligomer siloxanes or the ring-opening polymerization (ROP) of cyclosiloxanes [1, 4]. The ROP process is recognized as the predominant method for synthesizing most of the silicon-containing polymers. The preference is due to the superior controllability of ROP compared to hydrolytic polycondensation, enabling synthesizing high MW polysiloxanes with narrow dispersity [6–11].

Organocatalysts in Polymer Chemistry: Synthesis and Applications, First Edition. Edited by Zhibo Li.

Both in academia and industry, the long-standing challenge of the ROP of cyclosiloxanes is achieving improved control over the polymerization process to suppress the backbiting reactions occurring at the propagating chain ends [7, 11]. This undesirable process always produces systematic contamination of 10–15% oligocyclosiloxanes of different ring sizes (4–7 membered rings) that must be removed from the material after polymerization. The reason for forming large-size cyclic by-products is the thermodynamic competition of backbiting reaction with chain propagation during high-temperature ROP of cyclic monomers.

Different catalysts can be used for the ROP of cyclosiloxanes via anionic or cationic mechanism. In most instances, both anionic and cationic ROP lead to the same equilibrium characterized by a polysiloxane/oligocyclosiloxane mixture in a ratio of approximately 85 : 15 [11]. Nevertheless, the mechanisms underlying the attainment of equilibrium differ, giving rise to two distinct pathways for ROP that are either kinetically or thermodynamically controlled. The thermodynamically controlled ROP, known as equilibrium polymerization, is entropy-driven and is typically employed when the equilibrium polymer yield is substantially high. The kinetically controlled ROP, known as non-equilibrium polymerization, is enthalpy-driven and has to be quenched before the attainment of equilibrium [1, 7]. The former process simultaneously generates both linear polymers and oligocyclosiloxanes, whereas the latter initially forms linear polymers, and then attains equilibrium after polymerization (Figure 10.1) [1].

Cationic ROP of cyclosiloxanes, the mechanisms of which are quite complex [12], is particularly employed for monomers containing reactive functional groups like Si-H or Si-Cl, which exhibit instability in the ROP system catalyzed by strong bases. Strong protic acids, including sulfonic acid, sulfuric acid, and perchloric acid, are commonly employed as catalysts. The cationic polymerization of cyclosiloxanes

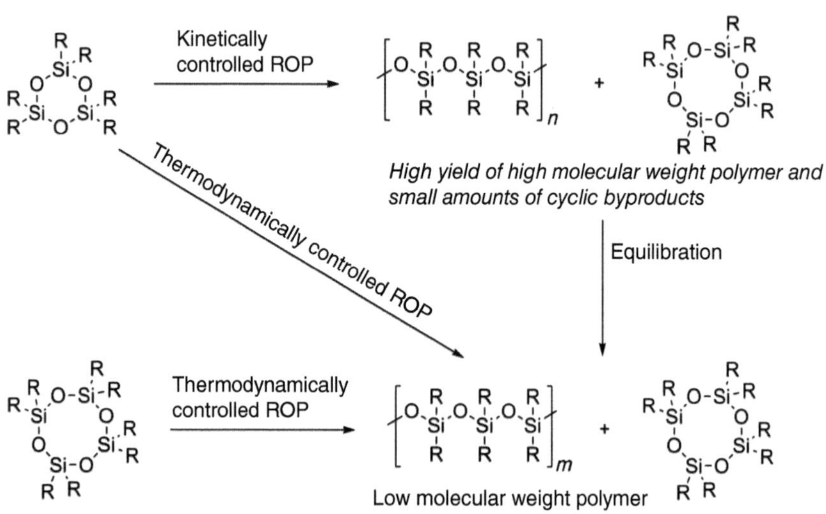

Figure 10.1 Different routes of thermodynamically controlled polymerization of cyclosiloxanes and kinetically controlled polymerization of cyclosiloxanes. Source: Köhler et al. [1]/Royal Society of Chemistry.

using different acid catalysts has been comprehensively summarized by several reviews and the relevant contents will not be discussed here [7, 13].

Anionic ROP of cyclosiloxanes is commonly catalyzed by strong bases to synthesize high MW polysiloxanes. Hence, a strong base is required during the initiation step to generate silanolate anions, which serve as the active centers in the propagation reaction [12]. Over the past two decades, organocatalysis has gained increased attention in the anionic ROP of cyclic monomers, including cyclosiloxanes with different ring sizes and substituents [14–19]. Organocatalysts, such as tetraalkylammonium hydroxide, phosphazene bases [10, 20–26], NHCs [27, 28], phosphorus ylides [11, 29], and guanidines [8, 9, 30] have proven efficient for the anionic ROP of cyclosiloxanes. Compared with the traditional metal-based catalysts, organocatalysts exhibited better activity and controllability toward the ROP of cyclosiloxanes. This chapter focuses on the anionic ROP of cyclosiloxanes initiated by various organocatalysts. Particularly, phosphazene-catalyzed anionic ROP of cyclosiloxanes with different substituents was emphasized in this section.

10.2 Anionic ROP of Cyclosiloxanes

Anionic ROP of cyclic siloxanes includes the breaking of the silicon-oxygen bond in the cyclics, followed by the propagation to form a polymer chain. The initiation step requires strong bases, including organometallic bases, organic bases, and inorganic bases, to initiate the ring-opening reaction of cyclic siloxanes. Silanolate anions are generated and serve as active centers of propagation (Figure 10.2, equation 1). The most commonly used catalysts for anionic ROP of cyclic siloxanes are alkali metals hydroxides (such as potassium hydroxide), tetraalkylammonium hydroxide, and tetraalkylphosphonium salts. The catalytic efficiency significantly increases with the increasing size of the counterion: $Li^+ < Na^+ < K^+ < Rb^+ < Cs^+ \sim Et_4N^+ \sim Et_4P^+$ [20, 31]. The increased polymerization activities derived from the less electrophilic bulky cations and weaker ion pairing led to the increased concentration of free anions in the polymerization system. The chain growth is reversible, with backbiting reactions occurring and leading to cyclosiloxanes of varying sizes (Figure 10.2, equation 2). The polymer could also be attacked by the terminal silanolate (as a nucleophile) in the growing chain of another polymer, leading to inter-chain rearrangements with the formation of a new silanolate (Figure 10.2, equation 3). In order to realize the non-equilibrium anionic ROP of cyclosiloxanes, the reaction has to be quenched immediately upon achieving the maximum monomer conversion [32].

10.3 Phosphazene-Catalyzed Anionic ROP of Cyclosiloxanes

Phosphazene bases are non-nucleophilic Brønsted bases with extremely strong basicity. Schwesinger and coworkers developed and synthesized a series of phosphazene bases with varied structures and different numbers of P=N bonds [33].

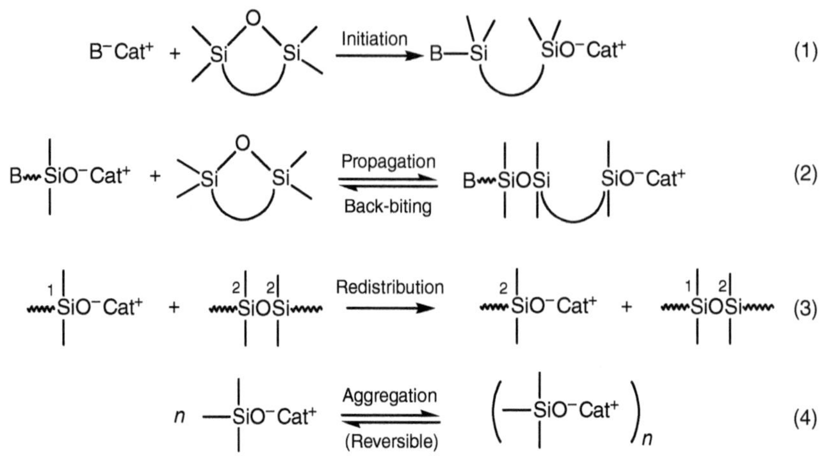

Figure 10.2 The mechanism for the anionic ROP of cyclosiloxanes. Source: Dubois et al. [7]/John Wiley & Sons.

Owing to the high basicity and non-nucleophilic properties of phosphazene bases, they could convert protic compounds into nucleophiles through deprotonating or activating weak nucleophilic species, thereby enabling the efficient ROP of cyclosiloxanes [15, 17, 18]. Recently, several phosphazene bases were applied and developed as effective catalysts for the ROP of different cyclosiloxanes (Figure 10.3) [10, 23, 24, 34–36].

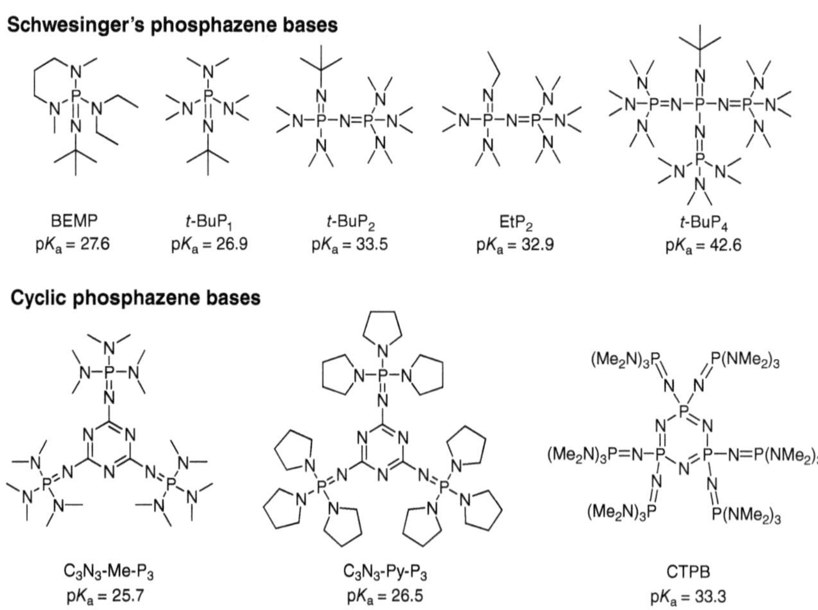

Figure 10.3 Representative Schwesinger's and cyclic phosphazene bases and their pK_a values determined in acetonitrile.

10.3.1 Schwesinger's Phosphazene Bases

In 1995, Möller et al. reported the homopolymerization of octamethylcyclotetrasiloxane (D_4) catalyzed by the extremely strong, non-ionic phosphazene base *t*-BuP_4 [20, 21]. Methoxide of phosphazenium formed by deprotonation of methanol (MeOH) by *t*-BuP_4. The protonated form of *t*-BuP_4 served as a counterion in the ROP of D_4. Extremely fast polymerization rate was observed both in bulk and in toluene solutions. The reaction reached equilibrium almost instantaneously for bulk ROP and within approximately 1 min in toluene solution (1.37 M), affording polydimethylsiloxanes (PDMSs) with MW up to 440 kg mol^{-1}. However, because of the backbiting reaction, the product consisted of linear high MW PDMS (up to 80%), macrocyclics, and small rings.

Dow Corning then published a series of patents about applying phosphazene as catalysts for ROP of cyclosiloxanes in 1997 [37–39]. It was found that phosphazenes with at least three P atoms (including *t*-BuP_4) were preferred ones. Even with reinforcing fillers in the polymerization system, the ROP could be accomplished within a very short time. The results indicated that phosphazene could react with H_2O and/or carbon dioxide to generate hydroxide or carbonate salt of phosphazene that was insoluble in the polymerization system. Consequently, the ROP was deactivated. The deactivation is reversible, the activity could be recovered through depressurizing, heating, or passing an inert gas into the polymerization system. This in return offered the possibility to modulate or control the polymerization by external stimulus.

Later in 1998, Clarson and coworkers used *t*-BuP_4/MeOH to achieve the efficient ROP of tetraphenyltetramethylcyclotetrasiloxane (P_4^{Me}) under ambient conditions [40]. The *t*-BuP_4/MeOH system showed a high polymerization rate of up to 1.6 min^{-1}, which was much faster than the conventional potassium hydroxide-catalyzed system. Equilibrium MWs of the ROP in bulk (1.76 M) and in toluene (0.957 M) were 249.0 and 122.0 kg mol^{-1}, respectively, suggesting a dramatic effect of dilution on ring-chain equilibration.

Afterward, Hupfield and Taylor used a mixture of *t*-BuP_4 and water/silanol for the ROP of D_4 [26]. The active hydroxide [*t*-BuNHP$^+$(NMe_2)$_3$P=N–)$_3$][OH^-] or silanolate [*t*-BuNHP$^+$(NMe_2)$_3$P=N–)$_3$] [Me_2SiO^-] was generated *in situ* (Figure 10.4). They demonstrated ROP of D_4 could be accomplished within very short reaction times at extremely low catalyst loading. They also found that CO_2/water could react with *t*-BuP_4 to form a bicarbonate material, which turned out to be a poor catalyst for ROP of D_4. However, as shown in Figure 10.5, the bicarbonate would decompose back to the active catalyst and then catalyze the ROP.

With the lower pK_a value than *t*-BuP_4, EtP_2 was not applied alone in the ROP of cyclosiloxanes. In combination with lithium cation, EtP_2 provided alternative ways to activate the initiating sites and catalyzed ROP of cyclosiloxanes, such as hexamethylcyclotrisiloxane (D_3) and hexaethylcyclotrisiloxane (D_3^{Et}) [21].

10.3.2 Phosphazenium Hydroxide P_5OH

In 1999, Hupfield and Taylor disclosed that tetrakis[tris(dimethylamino) phosphoranylidenamino] phosphazenium hydroxide (P_5OH), successfully prepared from the

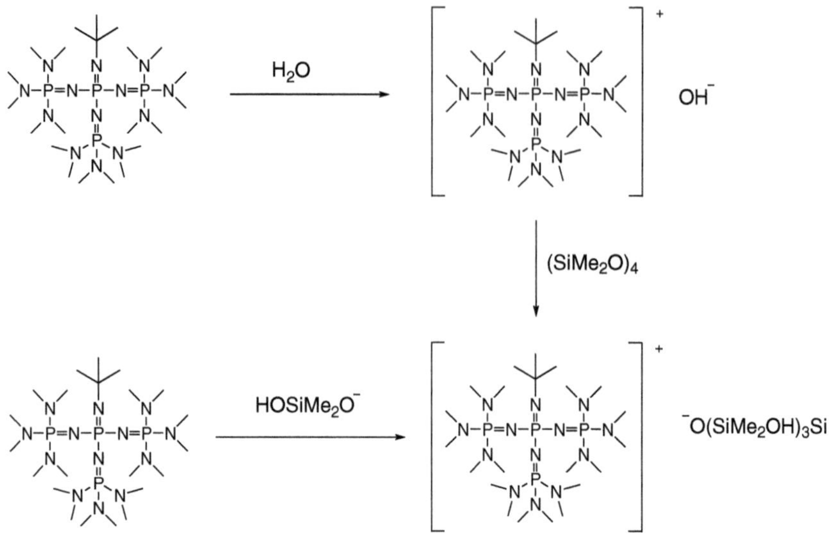

Figure 10.4 Phosphazene base t-BuP_4 conversion to active hydroxide or silanolate catalyst. Source: Hupfield and Taylor [26]/Springer Nature.

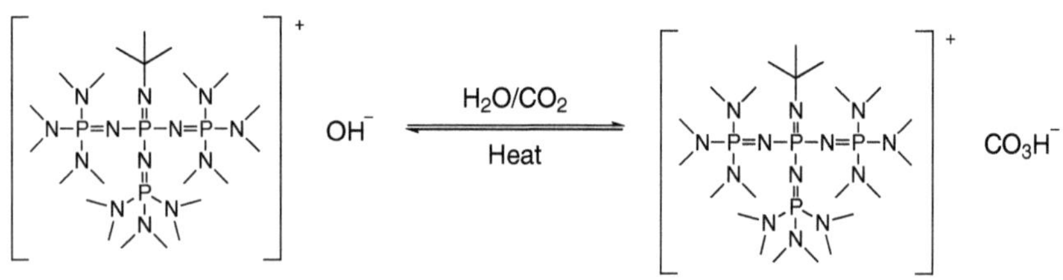

Figure 10.5 Reaction of t-BuP_4 hydroxide with CO_2/H_2O [26].

precursor tetrakis[tris(dimethylamino) phosphoranylidenamino] phosphazenium chloride (P_5Cl) through a straightforward anion exchange process (Figure 10.6), could catalyze the ROP of D_4 at elevated temperature (≥100 °C) [26]. Remarkably, within a few minutes, PDMSs with high M_n of 51.2 kg mol^{-1} could be obtained [26]. P_5OH exhibited sluggish catalytic behavior at reduced catalyst concentrations and lower reaction temperatures, which might be a reflection of basicity differences with t-BuP_4.

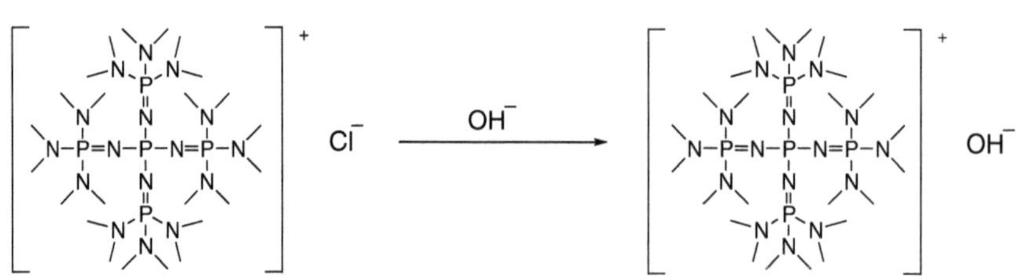

Figure 10.6 Conversion of P_5Cl to P_5OH by anion exchange [26].

10.3.3 Cyclic Trimeric Phosphazene Base (CTPB)

In 2017, we developed and synthesized a novel phosphazene base with a cyclic structure, acronymically referred to as cyclic trimeric phosphazene base (CTPB) (Figure 10.3), possessing a moderate basicity (pK_a = 33.3 in MeCN) [34]. CTPB has a spherical structure with a diameter of about 1.4 nm, which has a similar molecular size as t-BuP$_4$ but is larger than that of t-BuP$_2$. Surprisingly, the pK_a value of CTPB is lower than t-BuP$_4$, probably due to the limitation of homologation for this system.

We found that CTPB exhibited remarkable room-temperature activity and efficiency for D_4 ROP within minutes [23]. The polymerizations proceeded fast, yielding polymers with exceptionally high M_n up to 1353 kg mol^{-1}. Additionally, CTPB proved to be a competent catalyst for the ROP of D_4 and octaphenylcyclotetrasiloxane (P_4) under a mild reaction condition. Diphenylsiloxane content of the resulting copolymers ranged from 6.1 to 64.0 mol%. Surprisingly, there was neither a T-branch nor a Q-branch in the copolysiloxanes (Figure 10.7), demonstrating the good controllability of CTPB in diphenyl polysiloxane synthesis. The T and Q units refer to silicon atoms bonded to three and four oxygen atoms, respectively [2].

Subsequently, we demonstrated the successful CTPB-mediated ROP of D_4 and 2,4,6,8-tetravinyl-2,4,6,8-tetramethylcyclotetrasiloxane (V_4) at ambient temperature [41]. With a remarkably low CTPB loading (0.02 mol%), a series of linear PDMSs with different MWs and vinylsiloxane contents were successfully prepared (Figure 10.8). The vinylsiloxane contents in the copolysiloxanes could be adjusted from 8.2 to 16.1 mol% by simply tuning the feeding ratio of methyl and vinyl monomers, and the MWs could be regulated by the proportion of D_4, V_4, and end capping agents.

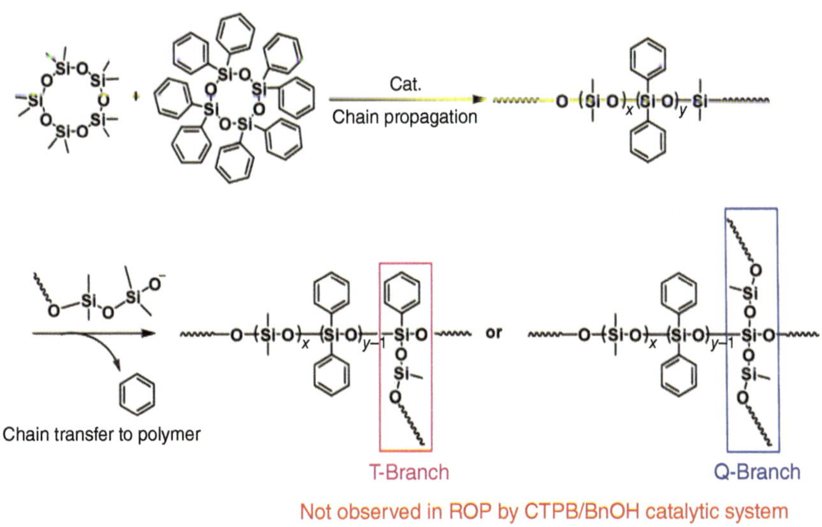

Figure 10.7 Possible T and Q branches in the ROP of D_4 and P_4. Source: Shi et al. [23]/Royal Society of Chemistry.

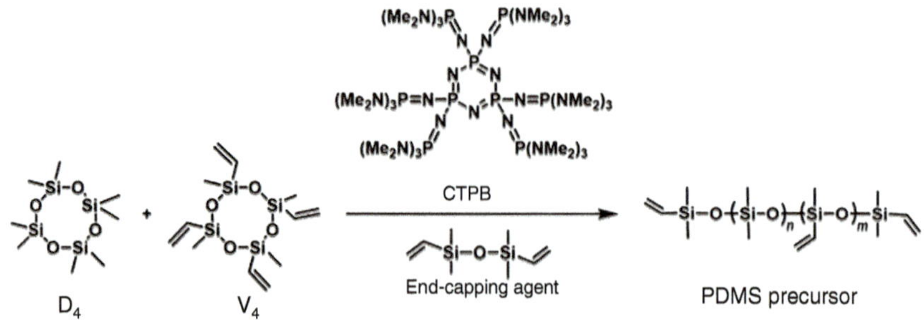

Figure 10.8 CTPB-catalyzed copolymerization of methyl and vinyl cyclosiloxane monomers. Source: Shi et al. [41]/Royal Society of Chemistry.

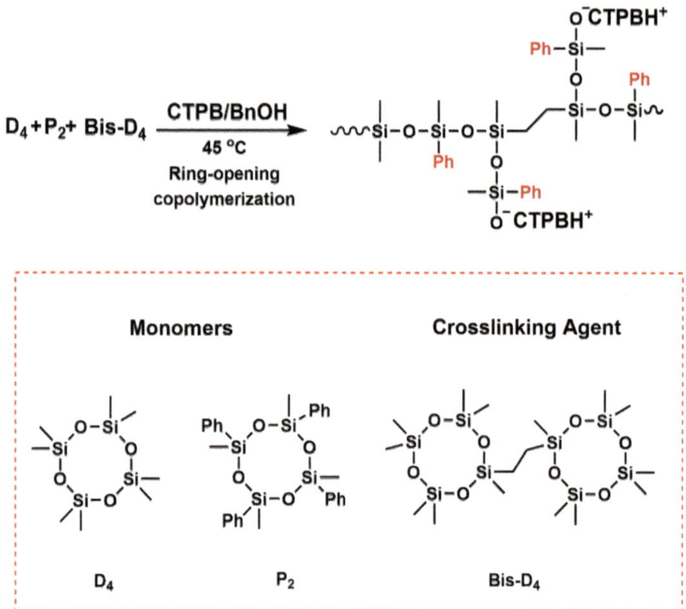

Figure 10.9 Synthetic route of *p*-PDMS prepolymer via CTPB-catalyzed ROP. Source: Xia et al. [36]/John Wiley & Sons.

Later, we synthesized phenyl-containing prepolymers (*p*-PDMS) for elastomers through one-pot anionic ROP of D_4 and 2,4,6,8-tetramethyl-2,4,6,8-tetraphenylcyclotetrasiloxane (P_2) [36]. As shown in Figure 10.9, the anionic copolymerization of D_4, P_2, and the cross-linker bis(heptamethylcyclotetrasiloxanyl) ethane (Bis-D_4) mediated by CTPB/BnOH proceeded at 45 °C, resulting in a pellucid low viscosity prepolymer for the following fabrication of micropillar adhesives.

Most recently, we prepared polydiethylsiloxane (PDES) (Figure 10.10) [24] with high M_n of 404.0 kg mol^{-1} in impressive yields via CTPB-catalyzed ROP of D_3^{Et} under ambient condition. Furthermore, random poly(dimethylsiloxane-*co*-diethylsiloxane)s (PMESs) with diethylsiloxane content ranging from 20 to 87 mol% was

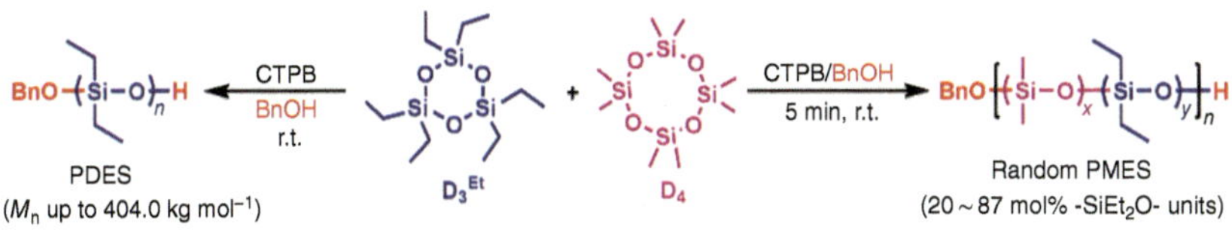

Figure 10.10 Synthesis of PDES and PMES copolymers.

conveniently synthesized by CTPB-catalyzed copolymerization of D_3^{Et} and D_4 within five minutes. The similar reactivity ratio ($r_E = 1.04$, $r_D = 0.89$) indicated the similar reactivity of dimethyl and diethyl cyclosiloxane monomers. The exceptional efficiency of CTPB and characteristics of random polymerization of D_3^{Et} and D_4 were thus proved.

10.3.4 Trisphosphazene Base – C_3N_3-Me-P_3

Relatively weaker phosphazene bases usually lead to slower polymerization rate during ROP of cyclosiloxane process. However, they offer improved control over the polymerization, making them attractive for precision syntheses. It is well-established that the ring strain of cyclosiloxanes significantly influences polymerization rate constants, with the observed order being: $D_6 < D_5 < D_4 \ll D_3$. Among these, cyclic trisiloxanes exhibit remarkably high reactivity because of their planar structure and higher ring strain. Thus, we applied the less basic, bulky trisphosphazene bases C_3N_3-Me-P_3 and C_3N_3-Py-P_3, as exhibited in Figure 10.11, to initiate the ROP of D_3 under ambient condition [10]. This ROP proceeded in a fast and kinetically controlled manner. Moreover, linear PDMSs with various end groups were prepared via C_3N_3-Me-P_3-mediated polymerization of D_3, followed by the end-capping with a range of organochlorosilanes.

10.3.5 Guanidinophosphazene – HTGCP

Recently, Zhao and coworkers developed a guanidinophosphazene superbase with medium basicity ($pK_a = 27.2$ in acetonitrile), named as 2,2,4,4,6,6-hexa(tetramethyl-guanidyl)-1,3,5,2λ^5,4λ^5,6λ^5-cyclictriphosphazene (HTGCP) [25]. As depicted in Figure 10.12, HTGCP showed remarkable catalytic performance in ROP of D_3 in toluene under mild conditions, giving PDMS with MWs of 10 ~ 1136 kg mol^{-1} within 120 min. In addition, HTGCP can also prepare high MW copolymer by the copolymerization of D_4 and V_4, but with wide MW distribution.

10.4 Guanidines-Catalyzed Anionic ROP of Cyclosiloxanes

The high basicity of guanidine superbases derived from their conjugated acids exhibited good stability [42, 43]. Among which, bicyclic guanidines

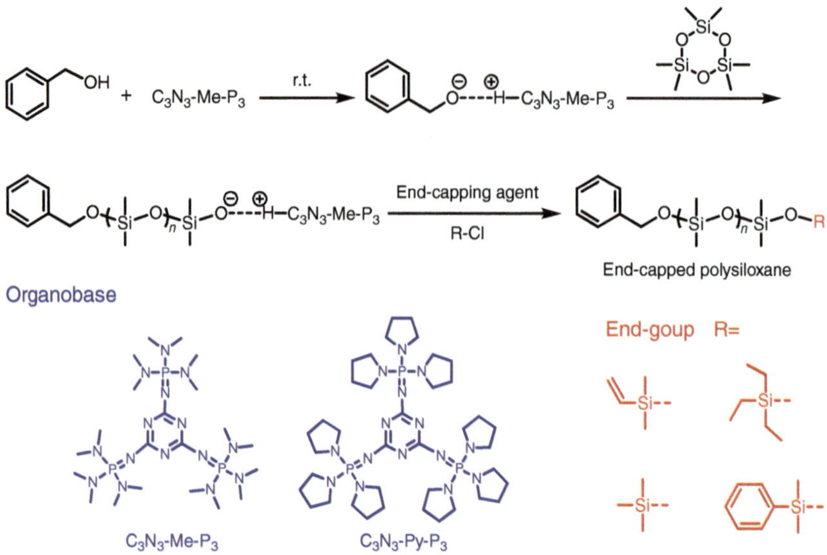

Figure 10.11 C_3N_3-Me-P_3-mediated polymerization of D_3. Source: Shi et al. [10]/American Chemical Society.

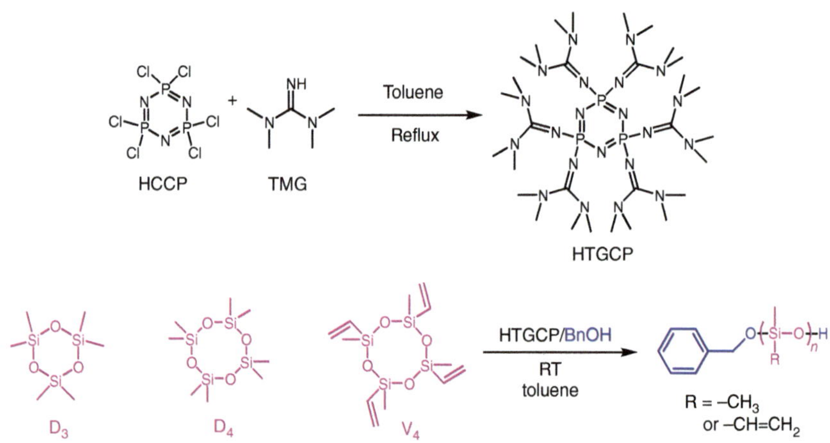

Figure 10.12 Structure of HTGCP and the ROP of cyclic siloxanes. Source: Duan et al. [25]/Royal Society of Chemistry.

such as 1,5,7-triazabicycle[4.4.0]dec-5-ene (TBD) and its *N*-methyl derivative 7-methyl-1,5,7-triazabicyclo[4.4.0]dec-5-ene (MTBD) have proven to be efficient to catalyze various organic reactions and polymerizations [15–17].

In 2006, Hedrick, Waymouth, and coworkers reported the preparation of poly(carbosiloxanes) with well-defined MW and terminal structure [30]. Adopting alcohol as an initiator, TBD (pK_a = 26 in MeCN) effectively catalyzed the ROP of 2,2,5,5-tetramethyl1-oxa-2,5-disilacyclopentane (TMOSC, Figure 10.13). However, under identical conditions, the slightly less basic MTBD (pKa = 25.5 in MeCN)

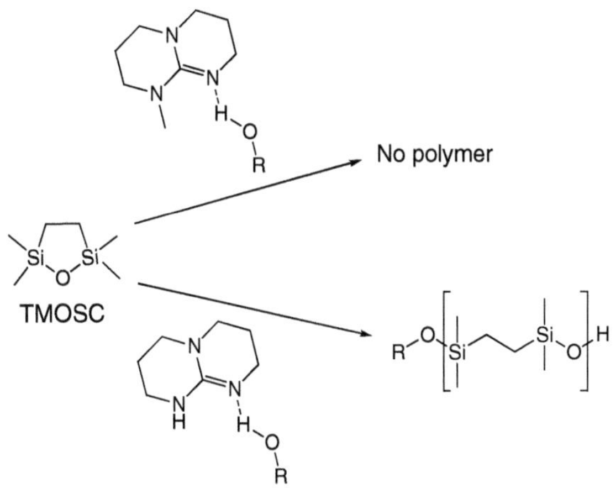

Figure 10.13 TBD-catalyzed ROP of carbosiloxane. Source: Lohmeijer et al. [30]/American Chemical Society.

exhibited no catalytic activity at all. This observation suggests that basicity is just one of the important factors for apparent catalytic activity. It is assumed that the hydrogen bonding between organobase and alcohols or silanols played an important role in the ring-opening reaction.

In 2018, Fuchise, Sato and Shimada et al. reported the guanidine-catalyzed controlled/living ROP of D_3. The monocyclic guanidine 1,3-trimethylene-2-*n*-propylguanidine (TMnPG) could catalyze the ROP of D_3 to produce various polysiloxanes with controlled terminal structures [8]. The MWs of the resulting polysiloxanes can be controllable from 2.64 to 102.3 kg mol^{-1} with rather narrow dispersity. Given the controllable characteristic, the statistical and triblock copolymers can also be conveniently prepared via ROP of D_3 with $D_3^{Me,Vi}$ or D_3^{2Ph} (Figure 10.14). End-capping of the polymer was achieved by a substitution reaction of functional chlorosilanes. The results of kinetic studies indicated that polymerization proceeded via an initiator/end-chain activation mechanism. Water was activated by TMnPG and then initiated the polymerization. Silanols in the chain ends were active centers for the propagation reaction.

In the past few years, this group has made significant contributions to the controlled ROP of cyclic siloxanes using guanidine catalysts. A series of studies have been reported to detail the syntheses and characterizations of polysiloxanes catalyzed by guanidines [9, 44–47]. One of the key innovations in their research is the utilization of non-equilibrium ROP to prepare end-functionalized polysiloxanes. This approach could precisely control the MWs and their dispersity as well as the terminal structure of the resulting polysiloxanes. The controlled/living ROP of D_3 and $D_3^{Me,Vi}$ catalyzed by TMnPG and other guanidines afforded various asymmetric linear PDMS and polymethylvinylsiloxane (PMVS) products with controlled M_n, narrow dispersity, and precisely controlled end groups [44]. Sequential ROP of D_3 and $D_3^{Me,Vi}$ could give PDMS/PMVS diblock copolymer in a controlled manner.

Subsequently, they reported the controlled/living ROP of 1,3,5-triphenyl-1,3,5-tri(ptolyl)cyclotrisiloxane (PT3) catalyzed by TMnPG, utilizing water or silanols as initiators (Figure 10.15) [45]. By employing linear symmetric silanol end-capped PDMS

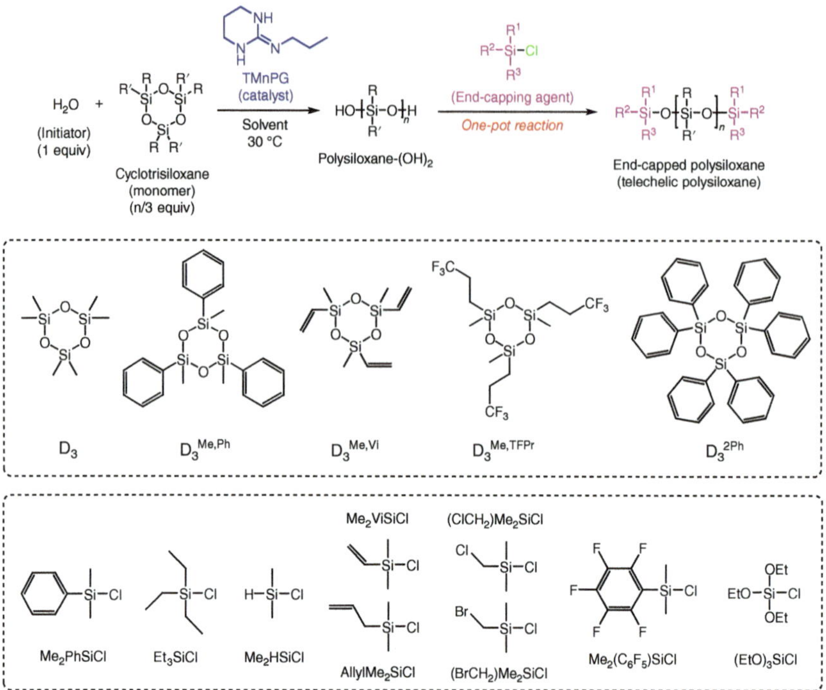

Figure 10.14 TMnPG-mediated polymerization of cyclic trisiloxanes.

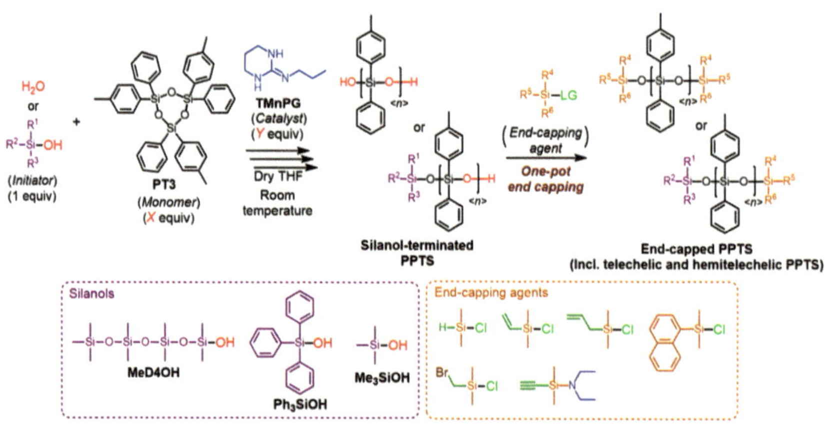

Figure 10.15 TMnPG-catalyzed ROP of PT3. Source: Fuchise et al. [45]/Royal Society of Chemistry.

as a macromolecular initiator, PPTS/PDMS triblock copolymers could be precisely prepared by the ROP of PT3 and the two-step polymerization of D_3 and PT3.

Monofunctional cyclotrisiloxanes could be polymerized via guanidines-catalyzed ROP adopting water or a silanol as an initiator [46]. This method can be conveniently employed to synthesize polysiloxanes bearing allyl group, 3-chloropropyl group, and

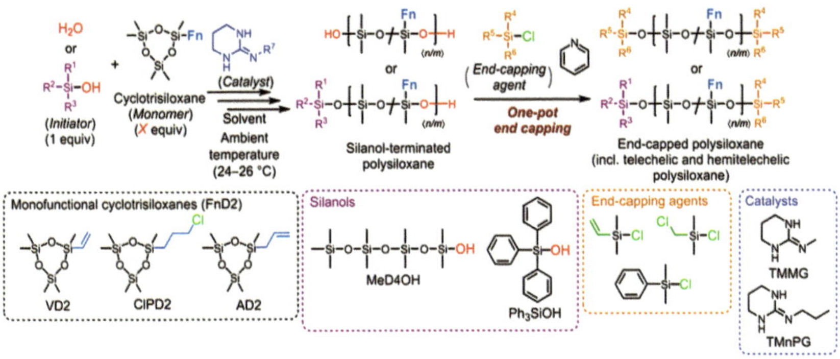

Figure 10.16 Guanidines-catalyzed polymerization of monofunctional cyclic trisiloxanes. Source: [46]/American Chemical Society.

vinyl group in the side chains (Figure 10.16). (3-Cyanopropyl)pentamethylcyclotrisiloxane (CNPD2) could also be polymerized using TMMG as the catalyst, giving linear polysiloxanes with 3-cyanopropyl groups randomly distributed in side chains (Figure 10.17) [47].

Linear polysiloxanes bearing mono- or di-substituted alkynyl groups can be precisely synthesized via guanidine-catalyzed controlled/living ROP of cyclic trisiloxanes with different substituents, employing alkynyl(amino)silanes as end-capping agents [9]. Polysiloxanes with both symmetrical and asymmetrical end-group structures, such as PDMS, poly[dimethylsiloxane-*co*-allyl(methyl)siloxane], and poly[dimethylsiloxane-*co*-(3-cyanopropyl)methylsiloxane], were synthesized accordingly (Figure 10.18). The precise synthesis of di-terminated PDMSs with bromoaryl groups, BrPh-PDMS-PhBr, was obtained through the ROP of D_3 using BrPhSi-Me_2Cl as end-capping agents (Figure 10.19) [48].

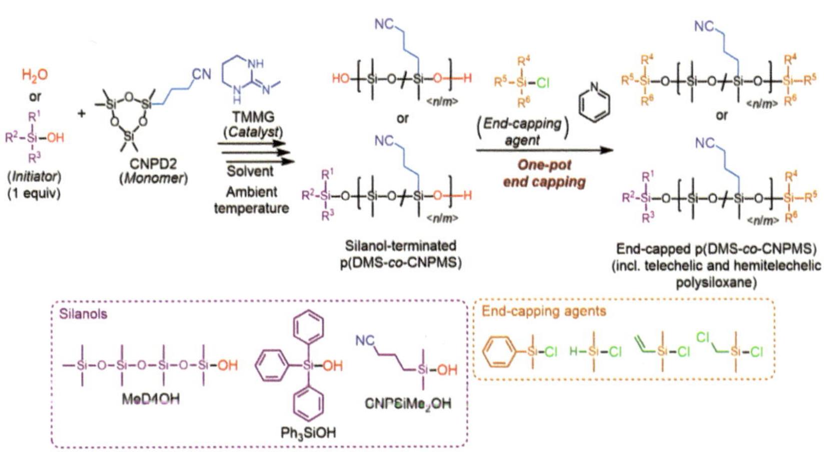

Figure 10.17 TMMG-catalyzed polymerization of CNPD2. Source: [47]/American Chemical Society.

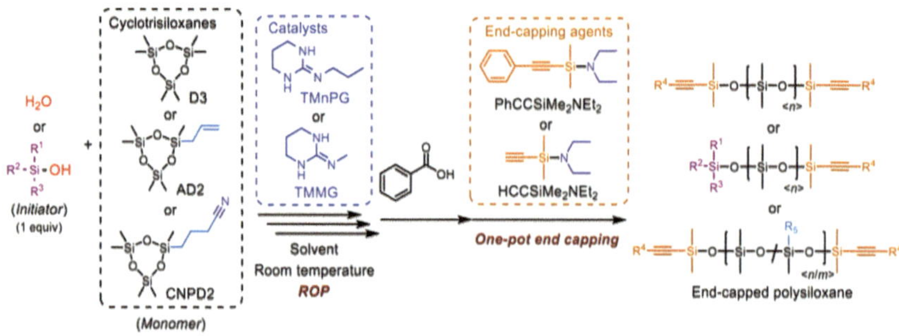

Figure 10.18 Preparation of alkynylsilyl end-capped polysiloxanes with well-defined structure. Source: [9]/American Chemical Society.

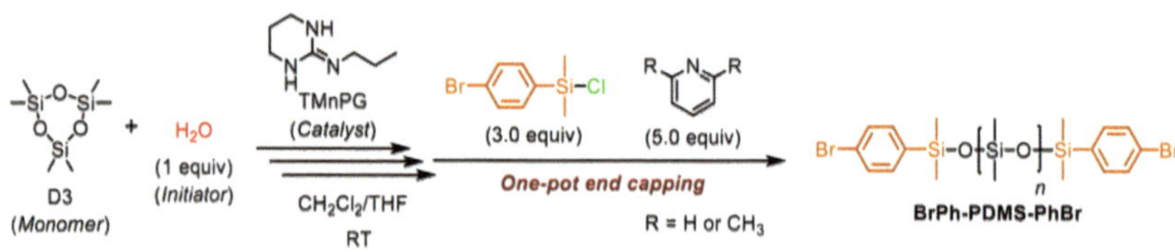

Figure 10.19 Preparation of PDMS bi-terminated with bromoaryl groups. Source: Sato et al. [48]/with permission of ELSEVIER.

10.5 Phosphorus Ylides-Catalyzed Anionic ROP of Cyclosiloxanes

The utilization of alcohols, specifically those deprotonated by phosphorus ylides, as initiators for ROP of D_4 has been proposed due to their unique thermal labile nature. This characteristic simplifies their removal from the final polymer, ensuring a cleaner and more controlled polymerization process. In 2003, Baceiredo and Mignani et al. demonstrated that phosphorus ylides could be adopted as catalysts for the ROP of cyclosiloxanes due to their basic properties [29]. Phosphonium alcoholates (Figure 10.20), formed by treating ylides using an excess amount of MeOH, could catalyze the ROP of D_4 under ambient condition. The polymerization rate is significantly influenced by the character and ratio of the initiator. Moreover, the polymer separation/isolation was simplified by using ylides as the catalyst.

Recently, Kato and coworkers made a milestone contribution to the ROP of cyclosiloxanes. They found that the backbiting side reactions could be completely inhibited by the coordination of alcohol to the active center of the propagating chain. They also pointed out that phosphonium cations worked as a self-quenching catalyst. Upon the loss of coordinating alcohols, the polymerization could be stopped before the onset of the backbiting [11]. The synergy of these effects enables

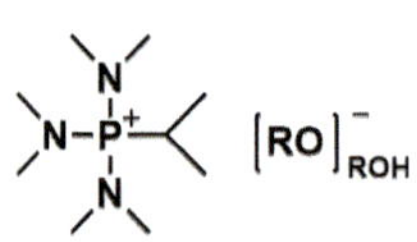

Figure 10.20 Phosphonium alcoholates for the polymerization of cyclic siloxanes.

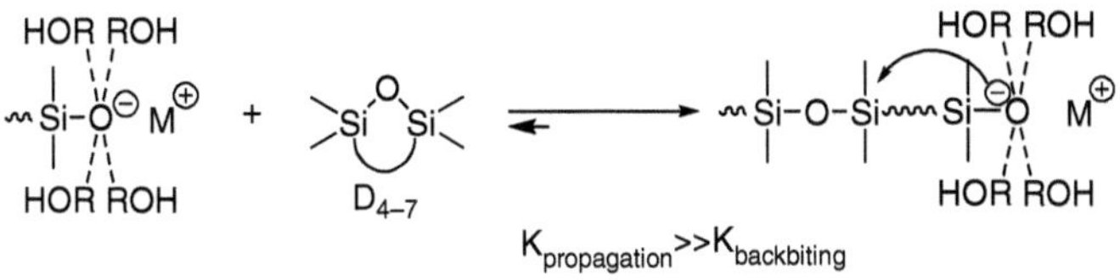

Figure 10.21 Propagation/backbiting equilibrium modified by the effect of coordinating ROH. Source: Shi et al. [11]/American Association for the Advancement of Science.

a thermodynamically controlled polymerization of D_4 without producing undesirable cyclic oligosiloxane by-products (Figure 10.21). This contribution addressed a long-standing challenge in silicone synthesis that the presence of 10% to 15% cyclosiloxanes in the ROP polymer is caused by the backbiting process.

10.6 NHC-Catalyzed Anionic ROP of Cyclosiloxanes

Over the past 20 years, *N*-heterocyclic carbenes (NHCs) have garnered growing attention in the field of organic catalysis. These neutral carbon atoms exhibit a six-electron valency, with four electrons participating in σ-bond formation, while the remaining two electrons are localized on the central carbon atom. The robust stability of NHCs stems from the interaction of the two lone pair electrons on the nitrogen with the empty p orbital at the carbene center. These distinctive properties endow NHCs with dual functionality as powerful ligands for transition metals and organic catalysts in polymer synthesis [17].

In 2006, Hedrick and coworkers reported that NHC could also catalyze the ROP of cyclic carbosiloxane TMOSC (Figure 10.13) [30]. NHC-2 (Figure 10.22a) with higher basicity ($pK_a = 35.7$ in acetonitrile) showed the highest activity, and a high monomer conversion of 99% was achieved within 1 min to give a polymer with low dispersity of 1.19, whereas NHC-1 (Figure 10.22a, $pK_a = 28.2$ in acetonitrile) only afforded 80% conversion after 30 min. However, the activation of terminal silanols via hydrogen bonding with NHC-1 and NHC-2 was sufficient to induce transesterification of the polymer backbone. Polymer scrambling (Figure 10.22b) took place after complete monomer conversion and the polydispersities increased gradually.

Figure 10.22 (a) Organocatalysts for ROP of TMOSC and (b) scrambling products. Source: Lohmeijer et al. [30]/American Chemical Society.

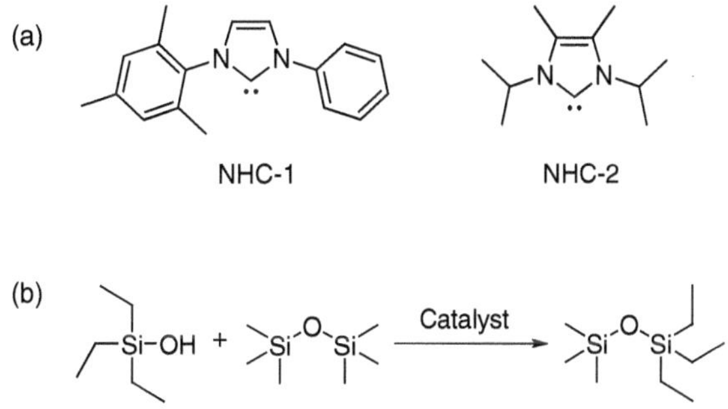

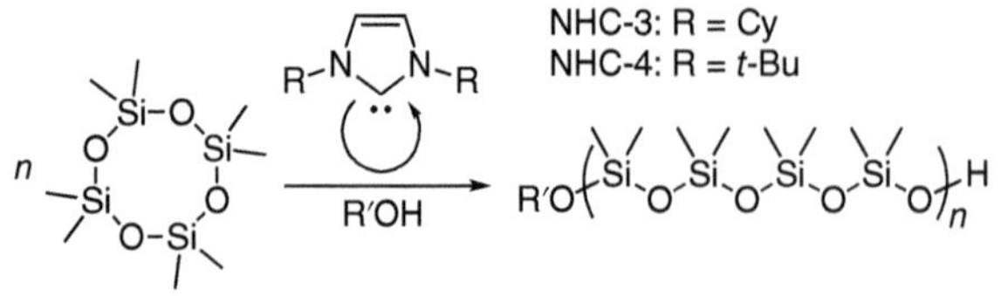

Figure 10.23 NHC-catalyzed polymerization of cyclosiloxanes. Source: Jain and Malik [28]/with permission of ELSEVIER.

Afterward, Baceiredo et al. found that NHC-3/NHC-4 can catalyze the ROP of D_4 at 80 °C with alcohol as the initiator (Figure 10.23) [27]. The polymerization condition was neutral. The by-products are moisture-sensitive catalysts and volatile initiators, which can be easily removed. By changing the feeding ratio of the alcohol initiator, polysiloxanes with different M_n can be conveniently synthesized.

10.7 Conclusion

In this chapter, we summarized the recent progress of the organocatalyzed ROP of cyclosiloxanes toward polysiloxanes. Through this strategy, a range of functional cyclic siloxanes, including those containing dimethylsiloxane, diethylsiloxane, diphenylsiloxane, methylvinylsiloxane, methylphenylsiloxane, as well as monosubstituted alkynyl and 3-cyanopropyl groups can be successfully polymerized. These reactions occur under mild conditions, resulting in the formation of polysiloxanes with high M_n.

Organic superbases, especially strong phosphazene bases t-BuP_4 and CTPB, exhibited extraordinary catalytic activity in the anionic ROP of cyclosiloxanes. Notably, equilibrium could be achieved within a matter of minutes, even with extremely low catalyst loadings. However, side reactions including backbiting and redistribution were unavoidable in these efficient catalytic systems, leading to the relatively wide distribution of the resulting polysiloxanes.

The kinetically controlled anionic ROP of cyclic siloxanes leverages the high reactivity of ring-strained cyclic trisiloxanes as monomers, allowing the polymerization to be stopped before backbiting occurs. Phosphazene bases with lower basicity and guanidines could catalyze the ROP of D_3 and other cyclotrisiloxanes in a kinetically controlled manner. These catalytic systems enable synthesizing polysiloxanes with controlled M_n, narrow dispersity, and precise terminal groups under mild conditions.

Specifically, phosphonium cations could work as a self-quenching catalyst triggered by the loss of coordinating alcohols. Thus, the thermodynamically controlled ROP of D_4 without cyclic by-products was achieved for the first time. This contribution was groundbreaking that solved the long-standing problem of cyclic oligosiloxanes associated with silicone synthesis.

Research on organocatalyzed ROP of cyclosiloxanes began in the 1990s and boomed again in the last decade. The remarkable progress in this field has demonstrated the potential of organocatalysts for ROP of cycloxiloxanes as a sustainable approach to silicone production. It is unequivocal that organocatalyzed ROP of cyclosiloxanes will keep providing sustainable solutions to societal needs in the silicone industry.

References

1 Köhler, T., Gutacker, A., and Mejía, E. (2020). Industrial synthesis of reactive silicones: reaction mechanisms and processes. *Org. Chem. Front.* 7 (24): 4108–4120.

2 Ryan, K.M., Drumm, A.D., Martin, C.E. et al. (2020). Synthetic methods and applications of functional and reactive silicone polymers. In: *Reactive and Functional Polymers Volume One*, 301–328. Springer Nature.

3 Yilgör, E. and Yilgör, I. (2014). Silicone containing copolymers: synthesis, properties and applications. *Prog. Polym. Sci.* 39 (6): 1165–1195.

4 Ackermann, J., Wrobel, D., Hoffmann, U. et al. (2011). Rubber, 7. Synthesis by polyaddition, polycondensation, and other mechanisms. In: *Ullmann's Encyclopedia of Industrial Chemistry*. Wiley-VCH Verlag GmbH & Co. KGaA.

5 Ren, Z.J. and Yan, S.K. (2016). Polysiloxanes for optoelectronic applications. *Prog. Mater. Sci.* 83: 383–416.

6 Jones, R.G., Ando, W., and Chojnowski, J. (2000). *Silicon-Containing Polymers*. Springer Science+Business Media Dordrecht.

7 Dubois, P., Coulembier, O., and Raquez, J.-M. (2009). *Handbook of Ring-Opening Polymerization*. Wiley-VCH Verlag GmbH & Co. KGaA.

8 Fuchise, K., Igarashi, M., Sato, K., and Shimada, S. (2018). Organocatalytic controlled/living ring-opening polymerization of cyclotrisiloxanes initiated by water with strong organic base catalysts. *Chem. Sci.* 9 (11): 2879–2891.

9 Fuchise, K., Sato, K., and Igarashi, M. (2021). Precise synthesis of linear polysiloxanes end-functionalized with alkynylsilyl groups by organocatalytic ring-opening polymerization of cyclotrisiloxanes. *Macromolecules* 54 (12): 5765–5773.

10 Shi, J., Liu, Z., Zhao, N. et al. (2022). Controlled ring-opening polymerization of hexamethylcyclotrisiloxane catalyzed by trisphosphazene organobase to well-defined poly(dimethylsiloxane)s. *Macromolecules* 55 (7): 2844–2853.

11 Shi, L., Boulègue-Mondière, A., Blanc, D. et al. (2023). Ring-opening polymerization of cyclic oligosiloxanes without producing cyclic oligomers. *Science* 381 (6661): 1011–1014.

12 Penczek, S., Cypryk, M., Duda, A. et al. (2007). Living ring-opening polymerizations of heterocyclic monomers. *Prog. Polym. Sci.* 32 (2): 247–282.

13 Hendrick, T.C., Parbhoo, B.M., and White, J.M. (1989). Comprehensive polymer science vol. 4. In: *Comprehensive Polymer Science* (ed. G. Allen). Pergamon Press.

14 Kamber, N.E., Jeong, W., Waymouth, R.M. et al. (2007). Organocatalytic ring-opening polymerization. *Chem. Rev.* 107 (12): 5813–5840.

15 Kiesewetter, M.K., Shin, E.J., Hedrick, J.L., and Waymouth, R.M. (2010). Organocatalysis: opportunities and challenges for polymer synthesis. *Macromolecules* 43 (5): 2093–2107.

16 Dove, A.P. (2012). Organic catalysis for ring-opening polymerization. *ACS Macro Lett.* 1 (12): 1409–1412.

17 Ottou, W.N., Sardon, H., Mecerreyes, D. et al. (2016). Update and challenges in organo-mediated polymerization reactions. *Prog. Polym. Sci.* 56: 64–115.

18 Liu, S., Ren, C., Zhao, N. et al. (2018). Phosphazene bases as organocatalysts for ring-opening polymerization of cyclic esters. *Macromol. Rapid Commun.* 39 (24): e1800485.

19 Walsh, D.J., Hyatt, M.G., Miller, S.A., and Guironnet, D. (2019). Recent trends in catalytic polymerizations. *ACS Catal.* 9 (12): 11153–11188.

20 Molenberg, A. and Moller, M. (1995). A fast catalyst system for the ring-opening polymerization of cyclosiloxanes. *Macromol. Rapid Commun.* 16 (6): 449–453.

21 Esswein, B., Molenberg, A., and Möller, M. (1996). Use of polyiminophosphazene bases for ring-opening polymerizations. *Macromol. Symp.* 107 (International Symposium on Ionic Polymerization, 1995): 331–340.

22 Molenberg, A. and Möller, M. (1997). Polymerization of cyclotrisiloxanes by organolithium compounds and P2-Et basea. *Macromol. Chem. Phys.* 198: 717–726.

23 Shi, J., Zhao, N., Xia, S. et al. (2019). Phosphazene superbase catalyzed ring-opening polymerization of cyclotetrasiloxane toward copolysiloxanes with high diphenyl siloxane content. *Polym. Chem.* 10 (17): 2126–2133.

24 Liu, Z., Shi, J., Zhao, N., and Li, Z. (2022). Fast synthesis of high molecular weights polydiethylsiloxanes and random poly(dimethylsiloxane-*co*-diethylsiloxane) copolysiloxanes via cyclic trimeric phosphazene base catalyzed ring-opening (co)polymerization. *Eur. Polym. J.* 173: 111280.

25 Duan, W., Li, Z., Yang, Z. et al. (2023). Synthesis of guanidinophosphazene superbase for the facile preparation of high molecular weight polysiloxane under mild conditions. *Polym. Chem.* 14 (2): 172–182.

26 Hupfield, P.C. and Taylor, R.G. (1999). Ring-opening polymerization of siloxanes using phosphazene base catalysts. *J. Inorg. Organomet. Polym.* 9 (1): 17–34.

27 Rodriguez, M., Marrot, S., Kato, T. et al. (2007). Catalytic activity of *N*-heterocyclic carbenes in ring opening polymerization of cyclic siloxanes. *J. Organomet. Chem.* 692 (4): 705–708.

28 Jain, I. and Malik, P. (2021). *N*-heterocyclic carbene complexes in ring opening polymerization. *Eur. Polym. J.* 150: 110412.

29 Bessmertnykh, A., Ben, F., Baceiredo, A., and Mignani, G. (2003). Anionic ring-opening polymerization of cyclic organosiloxanes using phosphorus ylides as strong non-ionic bases. *J. Organomet. Chem.* 686 (1-2): 281–285.

30 Lohmeijer, B.G.G., Dubois, G., Leibfarth, F. et al. (2006). Organocatalytic living ring-opening polymerization of cyclic carbosiloxanes. *Org. Lett.* 8 (21): 4683–4686.

31 Hurd, D.T., Osthoff, R.C., and Corrin, M.L. (1954). The mechanism of the base-catalyzed rearrangement of organopolysiloxanes. *J. Am. Chem. Soc.* 76 (1): 249–252.

32 Bellas, V., Iatrou, H., and Hadjichristidis, N. (2000). controlled anionic polymerization of hexamethylcyclotrisiloxane: model linear and miktoarm star co- and terpolymers of dimethylsiloxane with styrene and isoprene. *Macromolecules* 33 (19): 6993–6997.

33 Schwesinger, R., Schlemper, H., Hasenfratz, C. et al. (1996). Extremely strong, uncharged auxiliary bases: monomeric and polymer-supported polyaminophosphazenes (P2-P5). *Liebigs Ann.* 7: 1055–1081.

34 Zhao, N., Ren, C., Li, H. et al. (2017). Selective ring-opening polymerization of non-strained gamma-butyrolactone catalyzed by a cyclic trimeric phosphazene base. *Angew. Chem. Int. Ed.* 56 (42): 12987–12990.

35 Wang, L., Zhang, J., Zhao, N. et al. (2020). Synthesis of tris-phosphazene bases with triazine as core and their applications for efficient ring-opening alternating copolymerization of epoxide and anhydride: notable effect of basicity and molecular size. *ACS Macro Lett.* 9 (9): 1398–1402.

36 Xia, S., Chen, Y., Tian, J. et al. (2021). Superior low-temperature reversible adhesion based on bio-inspired microfibrillar adhesives fabricated by phenyl containing polydimethylsiloxane elastomers. *Adv. Funct. Mater.* 31 (26): 2101143.

37 Remy, B., John, C., William, H., and Richard, T. (1997). *Polymerisation of Cyclosiloxanes*. European Patent. Patent number: EP0860461A2.

38 Brian, H. and Richard, T. (1997). *Polymerisation of Cyclosiloxanes in the Presence of Fillers*. European Patent. Patent number: EP0860459A2.

39 Remy, B. and Richard, T. (1997). *Process for Making Siloxane Compositions*. European Patent. Patent number: EP0860460A2

40 Van Dyke, M.E. and Clarson, S.J. (1998). Reaction kinetics for the anionic ring-opening polymerization of tetraphenyltetramethylcyclotetrasiloxane using a fast initiator system. *J. Inorg. Organomet. Polym.* 8 (2): 111–117.

41 Shi, J., Zhao, N., Yan, D. et al. (2020). Design of a mechanically strong and highly stretchable thermoplastic silicone elastomer based on coulombic interactions. *J. Mater. Chem. A* 8 (12): 5943–5951.

42 Kovačević, B. and Maksić, Z.B. (2001). Basicity of some organic superbases in acetonitrile. *Org. Lett.* 3 (10): 1523–1526.

43 Ishikawa, T. (2009). *Superbases for Organic Synthesis: Guanidines, Amidines, Phosphazenes and Related Organocatalysts*. Wiley.

44 Fuchise, K., Kobayashi, T., Sato, K., and Igarashi, M. (2020). Organocatalytic ring-opening polymerization of cyclotrisiloxanes using silanols as initiators for the precise synthesis of asymmetric linear polysiloxanes. *Polym. Chem.* 11 (48): 7625–7636.

45 Fuchise, K., Sato, K., and Igarashi, M. (2021). Organocatalytic controlled/living ring-opening polymerization of 1,3,5-triphenyl-1,3,5-tri-*p*-tolylcyclotrisiloxane for the precise synthesis of fusible, soluble, functionalized, and solid poly[phenyl (*p*-tolyl)siloxane]s. *Polym. Chem.* 12: 5178–5190.

46 Fuchise, K., Sato, K., and Igarashi, M. (2021). Precise synthesis of side-chain-functionalized linear polysiloxanes by organocatalytic ring-opening polymerization of monofunctional cyclotrisiloxanes. *Macromolecules* 54 (11): 5204–5217.

47 Fuchise, K., Sato, K., and Igarashi, M. (2021). Precise synthesis of linear polysiloxanes with a polar side-chain structure by organocatalytic controlled/living ring-opening polymerization of (3-cyanopropyl)pentamethylcyclotrisiloxane. *Polym. Chem.* 12 (22): 3321–3331.

48 Sato, K.-I., Hemmi, Y., Kato, A. et al. (2022). Precise synthesis of α,ω-chain-end-functionalized poly(dimethylsiloxane) with bromoaryl groups for incorporation in naphthalene-diimide-based *N*-type semiconducting polymers. *Polymer* 252: 124934.

11

Organic Lewis Pair in Polymer Synthesis

Mingqian Wang, Zhiqiang Ding, Bin Wang, and Yuesheng Li

Tianjin University, Tianjin Key Laboratory of Composite & Functional Materials, School of Materials Science and Engineering, Peiyang Park Campus, No.135 Yaguan Road, Haihe Education Park, Tianjin 300350, China

11.1 Lewis Pair Catalytic Ring-Opening Polymerization of Cyclic Esters

Aliphatic polyesters, such as poly(lactic acid) (PLA) and poly(ε-caprolactone) (PCL), have attracted increasing attention in packaging and biomedical applications due to their intriguing physical properties, biocompatibility, and environmentally friendly properties [1]. Ring-opening polymerization (ROP) of cyclic esters, which provides a high degree of molecular weight (MW) and polydispersity, is a promising approach for the promising approach for the efficient synthesis of aliphatic polyesters [2]. Catalysts play an increasingly important role in ROP, because the polymerization behavior and structure of polyesters are highly dependent on catalysts and initiators [3]. The development of catalysts for ROP has allowed exquisite control over chain growth, stereochemistry, and even the topological structure of polyesters [4].

The concept of Lewis pairs (LPs) describes the combination and interaction of a Lewis acid and a Lewis base (LB). The reaction between a Lewis acid and base often results in the formation of neutral classical Lewis adducts (CLAs), which significantly quench their individual reactivity [5]. In 2006, Stephan et al. constructed sterically hindered LPs that prevent the interaction of Lewis acid and base, preserving the individual reactivity capable of activating small molecule substrates [5d]. These pairs are called frustrated Lewis pairs (FLP). In 2010, Chen and colleagues were the first to use FLPs as dual catalysts to mediate vinyl-addition polymerization, resulting in high MW polymers [6]. This discovery opened up a new area in polymer synthesis known as Lewis pair polymerization (LPP). Now, the LPP has been extended to the ROP for polyester synthesis. In this chapter, we will introduce the ROP of lactide and related cyclic esters, with a special focus on the pivotal role of organocatalysts in the ROP process, highlighting their unique advantages. The mechanism of ROP based on LP systems will be introduced first, and then it will be

Organocatalysts in Polymer Chemistry: Synthesis and Applications, First Edition. Edited by Zhibo Li.

presented according to different organic LP system types. The use of LP has greatly expanded the application fields of polyesters.

11.2 Mechanisms for Lewis Pair-Catalyzed ROP of Cyclic Esters

The ROP of cyclic esters using LPs has been extensively studied [7]. The combination of a Lewis acid and base not only enhances the polymerization activity but also improves selectivity. The main driving force behind ROP of small- and medium-ring lactones is the release of ring strain in the monomer, which is minimal compared to the driving force of olefin polymerization (C-C bond formation). As a result, these polymerizations are often less active for less strained monomers, and it is not uncommon for these reactions to occur over the course of hours or even days. Therefore, the active species is exposed for extended periods, allowing it to participate in unwanted side reactions. One such reaction is chain-to-chain transesterification, which is known to cause dispersity broadening and scrambling of block copolymers (BCPs). This reaction is thermoneutral and similar in character to propagation. To prevent transesterification and produce well-defined polymers or BCPs, the active species must be designed to selectively favor propagation over chain-to-chain transesterification. This requires inhibiting transesterification completely.

The specific mechanisms can become complicated depending on the type of LPs used. Generally, there are two main mechanisms: zwitterionic and proton-mediated dual catalysis. In zwitterionic mode (see **Figure** 11.1(a)), chain initiation occurs through nucleophilic attack of LBs on cyclic esters coordinated with Lewis acids, generating zwitterionic active species for chain propagation. Different topological polyesters can be produced due to the existence of intramolecular or separated ion-pairing, which relies on the interaction modes of LPs [8]. In dual catalysis mode (see **Figure** 11.1(b)), monomer activation is achieved through the cooperative coordination of carbonyl by Lewis acid and active chain end by electron donation of LB to the hydroxyl proton. This process requires the use of three components: alcohols, Lewis acids, and bases.

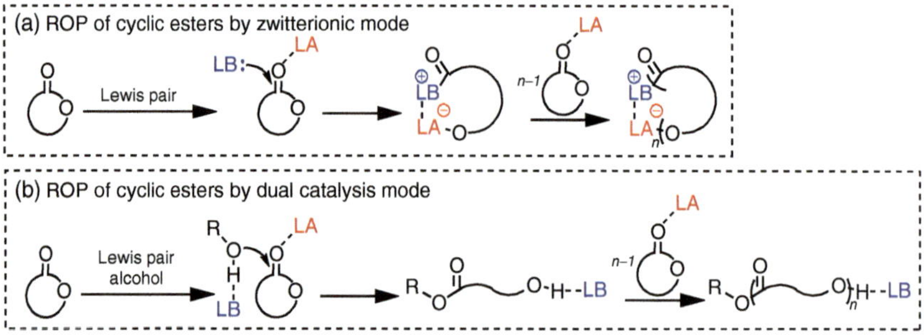

Figure 11.1 Mechanism of LPP of cyclic esters operating in different modes.

11.3 Lewis Pairs Used for ROP of Cyclic Ester

11.3.1 Organoboron-Based Lewis Pair

The initial attempt of the ROP of δ-VL and *rac*-lactide by $B(C_6F_5)_3$/phosphine (P*t*Bu_3, PCy_3) or amine (2,6-lutidine, *N,N*-dimethylaniline) FLPs was made by Stephan and colleagues in 2011 [7a]. However, ring opening of δ-VL and ring contraction of *rac*-LA were observed without any formation of polyesters. The mechanistic rationale for the observed ring opening of the lactone is as follows. The C – O bond of the lactone was activated by $B(C_6F_5)_3$, prompting a nucleophilic attack by the amine or phosphine LB to generate the zwitterionic species. The ring-contracted product was proposed to be initiated by deprotonation of $B(C_6F_5)_3$-activated *rac*-lactide by the organic base, presumably due to the steric demand of the methyl groups of *rac*-lactide that precluded the nucleophilic attack of the sterically encumbered organic base. The resulting carbanion then undergoes a nucleophilic attack by the carboanion to the carbonyl C, thereby affording the ring-contracted product. Zhu and Chen discovered a method for quantitatively converting *meso*-lactide, a byproduct of L-lactide production, into *rac*-lactide via rapid epimerization catalyzed by the $B(C_6F_5)_3$/1,4-diazabicyclo[2.2.2]octane (DABCO) LP, which forms a stable CLA (Figure 11.2) [9]. The use of the relatively nonpolar solvent toluene was essential for this rapid and quantitative epimerization process, as it enabled the precipitation of the resulting *rac*-lactide from the solution once formed due to solubility differences. These differences constantly disrupt the thermodynamic equilibrium, ensuring the completion of epimerization. Even with a low catalyst loading of 0.01 mol%, 95.4% *meso*-to-*rac* lactide conversion was accomplished in just 5 min (TOF = 114,480 h^{-1}). They also developed a highly enantioselective bifunctional chiral organic catalyst (Figure 11.2), which combines three essential elements (β-isocupreidine core, thiourea functionality,

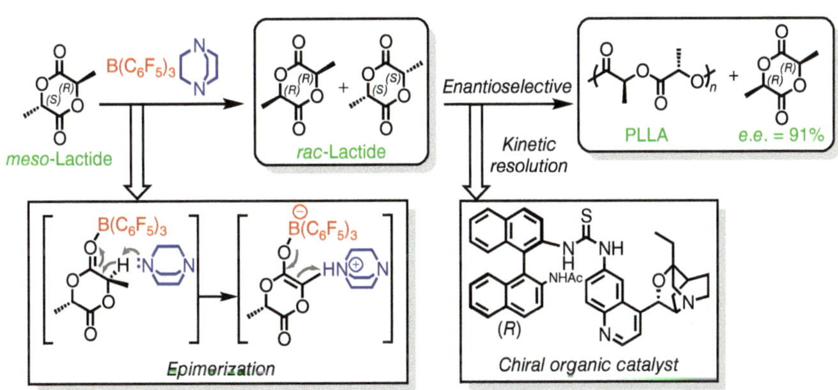

Figure 11.2 Epimerization of *meso*-lactide into *rac*-lactide catalyzed by the B/N LP and subsequent kinetic resolution polymerization of the resulting *rac*-lactide into PLLA and optically resolved D-lactide.

and binaphthylamine framework) into a single molecule that can asymmetrically activate both the monomer (by the chiral thiourea) and the propagating hydroxyl species (by the chiral amine). In the context of optimized conditions for kinetic resolution polymerization of *rac*-lactide, this bifunctional chiral organic catalyst exhibited a high degree of stereoselectivity, preferentially polymerizing L-lactide and kinetically resolving D-lactide with a stereoselectivity factor of 53 and an ee of 91% at 50.6% monomer conversion. Furthermore, the epimerization and kinetic resolution polymerization can be coupled into a one-pot process, which effectively transforms *meso*-lactide into isotactic poly(L-lactide) (PLLA) and D-lactide.

The scope of the LP-mediated ROP has been expanded to other types of cyclic monomers. Yang et al. reported the utilization of borane/amine LPs for the controlled ROP of *N*-carboxyanhydrides (NCAs) to synthesize well-defined polypeptides [10], opening up a new opportunity for the application of LP-mediated polymerizations. Control runs demonstrated that the ROP of γ-benzyl-L-glutamate (Glu-NCA) by the LB aniline alone was completed in 1 h (Glu-NCA/LB = 50/1, RT), resulting in the corresponding polypeptide with a relatively broad dispersity ($Đ$ = 1.63). This outcome aligns with expectations for a normal amine mechanism. In contrast, the bulky borane LA, bis[2,4,6-tris (trifluoromethyl)phenyl]boron fluoride ($Fmes_2BF$), was unable to initiate the polymerization. The dispersity ($Đ$ = 1.63) is consistent with a normal amine mechanism. In contrast, the bulky borane LA, bis[2,4,6-tris(trifluoromethyl)phenyl]boron fluoride ($Fmes_2BF$), was unable to initiate the polymerization. However, when $Fmes_2BF$ and aniline were combined, the ROP became controlled, producing the polypeptide with a measured M_n (7.5–12.0 kg mol^{-1}) close to the calculated value and a narrow dispersity. However, the polymerization time was extended to 8 h in order to achieve a 99% monomer conversion (TOF = 6.3 h^{-1}, Glu-NCA/LA/LB = 50/1/1). In contrast to the conventional mechanism of LPP, where LA activates the monomer through coordination, it was postulated that, in this instance, the bulky borane LA captures the terminal amino group of the propagating chain to form a novel LP intermediate (Figure 11.3), which is crucial for the inhibition of side reactions. The nucleophilicity of the aniline initiator and the terminal amine of the propagation chain end is diminished by the interaction with LA, resulting in a reduced rate of polymerization but a

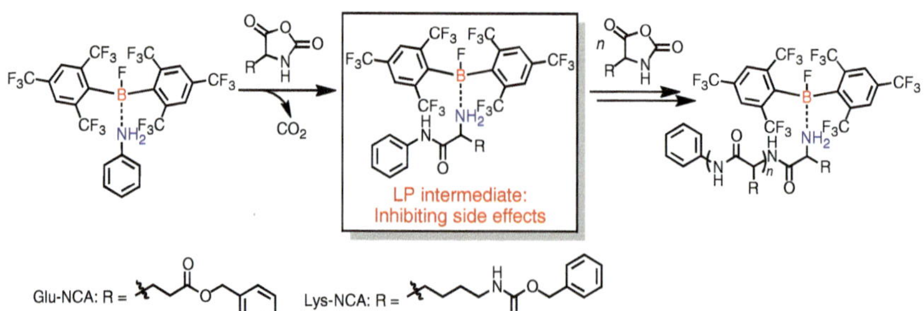

Figure 11.3 Proposed mechanism for the controlled ROP of NCAs by B/N LP.

more controlled process than that observed in aniline-mediated ROP. Furthermore, the controlled ROP of ε-benzyl-L-lysine *N*-carboxyanhydride (Lys-NCA) was also achieved by the $Fmes_2BF$/aniline LP (TOF = 9.9 h^{-1}, M_n = 11.9 kg mol^{-1}, $Đ$ = 1.22). In a similar manner, the bis[2,4,6-tris(trifluoromethyl)phenyl]boron hydride ($Fmes_2BH$)/aniline pair demonstrated a low activity for the ROP of Glu-NCA (TOF = 2.5 h^{-1}). The polymer was produced with a broad dispersity ($Đ$ = 1.53), while the $B(C_6F_5)_3$/aniline pair was totally inactive, highlighting the importance of choosing a sterically and electronically balanced LA relative to the pairing LB for achieving an efficient and controlled LPP. The use of alternative primary amines, including *p*-methoxyaniline, *p*-bromoaniline, and benzylamine, in conjunction with $Fmes_2BF$ also resulted in controlled polymerizations (TOF = 3.1–5.6 h^{-1}, M_n = 6.4–11.6 kg mol^{-1}, $Đ$ = 1.24–1.29).

11.3.2 Organoaluminum-Based Lewis Pair

Chen et al. initially employed Al-based LPs consisting of highly acidic $Al(C_6F_5)_3$ LA and $PtBu_3$ LB for the ROP of ε-CL, albeit with low activity (TOF = 23.2 h^{-1}, 0.125 mol% catalyst), producing PCL with a MW of 74.0 kg mol^{-1} and a dispersity index of 2.76 [11]. Recently, Nakayama and colleagues described the controlled ROP of L-lactide by employing $Al(C_6F_5)_3$·THF/$PMes_3$ or PPh_3 in the presence of BnOH to produce PLA with relatively low dispersity (M_n = 7.2–25 kg mol^{-1}, $Đ$ = 1.11–1.46, TOF <4.0 h^{-1}) [12]. The ratio of ideal/LA/LB/BnOH was 100/1/1/1, and the reaction temperature was 100 °C. The reaction time was 24 hours. The ROP of ε-CL by this LP system was faster (TOF < 170 h^{-1}) but less controlled (M_n = 4.1–10 kg mol^{-1}, $Đ$ = 1.41–1.67). Dagorne et al. employed adducts of NHC (IMes, IDipp, I*t*Bu) with group 13 metal alkyl (MR_3; M = Al, Ga, In; R = Me, Et, C_6F_5) as catalysts for the ROP of *rac*-lactide [13]. Controlled ROP was achieved when adducts of IMes with AlR_3 were used as catalysts (R = Me, TOF = 100 h^{-1}; R = Et, TOF = 24.7 h^{-1}) at room temperature (RT) (*rac*-lactide/LP adduct = 100, M_n = 13–15 kg mol^{-1}, $Đ$ = 1.31–1.40, I^* = 90.3–107.1%), whereas the more sterically bulky adducts IDipp-AlR_3 (R = Me, Et) adducts required heating (90 °C) for polymerization to occur, and exhibited significantly reduced activity and control during the ROP (TOF = 3.9–4.5 h^{-1}, M_n = 5.1–8.2 kg mol^{-1}, $Đ$ = 1.32, I^* = 52.8–76.5%). It is noteworthy that the IMes–$AlMe_3$ adduct led to the selective production of imidazolium-ended linear PLA, whereas the IDipp–$AlMe_3$ adduct resulted in the formation of cyclic PLA. Heating a linear PLA obtained by IMes–$AlMe_3$ after complete conversion can quantitatively convert it into the cyclic topology, which is attributed to the thermally induced intramolecular cyclization. The IMes–$Al(C_6F_5)_3$ adduct with the more acidic LA exhibited a much lower activity (TOF = 0.63 h^{-1}), presumably due to the stronger Al – NHC interaction, which thereby disfavored the dissociation of the adduct. No polymerization activity was observed at RT in the case of I*t*Bu–$AlMe_3$, and poor control of the ROP process was observed when IMes–$GaMe_3$ and IMes–$InMe_3$ were utilized. The turnover frequency (TOF) ranged from 1.8 to 10.0 h^{-1}, and the dispersity ($Đ$) ranged from 2.22 to 2.54.

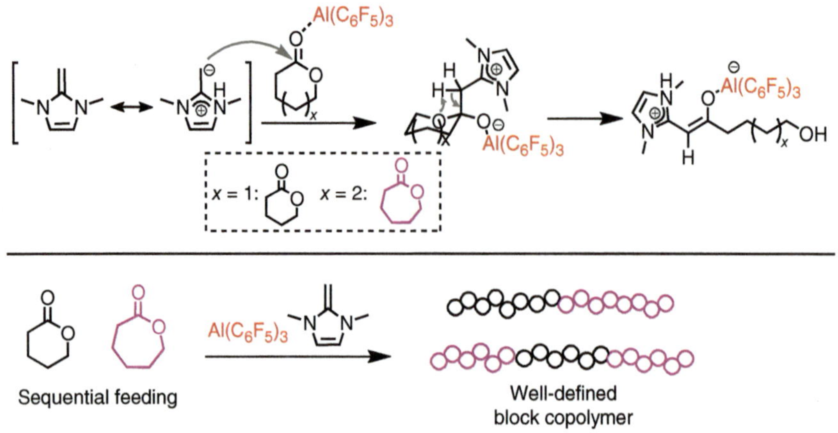

Figure 11.4 Proposed initiation mechanism for the living ROP of δ-VL or ε-CL by NHO/$Al(C_6F_5)_3$ LPs.

Zhang and colleagues explored ROP of lactones by bulky organoaluminum and *N*-heterocyclic olefins. *N*-heterocyclic olefins (NHOs) with exocyclic methylene underwent an unexpected initiation process in which, during the initial ring opening, a proton is transferred from the NHO methylene carbon to the terminal alkoxy to form a neutral amine (Figure 11.4) [8a]. This molecule is still nucleophilic enough to attack the next monomer via the carbonyl conjugated to the enamine, thus initiating a living polymerization. Interestingly, this initiation sequence achieves a near-ideal efficiency (106%) and causes a living polymerization where transesterification is completely avoided. On the other hand, the dimethyl-substituted NHO, which cannot undergo proton transfer, results in broad MW distribution. The ability to shut down the transesterification side reaction in this LPP enabled the synthesis of well-defined BCPs with M_n as high as $855\ \mathrm{kg\,mol^{-1}}$ and $Đ$ as low as 1.02. The mechanism by which transesterification occurs is interesting and twofold. First, the LA-stabilized alkoxide is not a strong enough nucleophile to attack in-chain carbonyls unless the in-chain carbonyls are activated by an LA. It is reasonable to assume that in this LPP system, the LA $Al(C_6F_5)_3$ has a higher affinity for free monomer carbonyls than for in-chain carbonyls, but if this were the only relevant argument, one would suspect that after complete conversion, the free LA would then be allowed to begin activating the in-chain carbonyls. This does not seem to be the case due to the lack of transesterification following the chain extension and copolymerization experiments. Thus, it seems likely that there is a steric factor that precludes the development of inter- and intra-chain-to-chain transition states. The living characteristics of this LPP system, cleanly confirmed by the linear growth of M_n values with an increase of conversions and monomer-to-initiator ratio as well as the chain extension experiments, enabled the synthesis of well-defined di- and triblock copolymers regardless of the monomer addition order (PCL-*b*-PVL, PCL-*b*-PVL-*b*-PCL) with narrow dispersity ($Đ < 1.15$).

11.3.3 Organozinc-Based Lewis Pair

An LPP system comprising a cationic zinc complex supported by the diaminophenolate (NNO) ligand and a tertiary amine was employed by Guillaume, Bourissou, Carpentier, and colleagues to facilitate the efficient and controlled ROP of *rac*-lactide (Figure 11.5) [14]. In the presence of neopentyl alcohol (*neo*-PentOH) as the initiator, near-quantitative monomer conversion was realized to atactic PLA with a MW (M_n) up to 14.5 kg mol^{-1} and dispersity ($Đ$) values ranging between 1.2 and 1.4. This was achieved in 3 h using *rac*-lactide, *neo*-PentOH, $[\{NNO\}Zn]^+$ $[B(C_6F_5)_4]^-$, and 1,2,2,6,6-pentamethylpiperidine (PMP) in a ratio of 100/1/0.5/0.5, with a TOF of approximately 33 h^{-1}. No polymerization was observed when the polymerization was conducted in the presence of *neo*-PentOH with only $[\{NNO\}Zn]^+$ $[B(C_6F_5)_4]^-$ or PMP. The authors postulated that $[\{NNO\}Zn]^+$ $[B(C_6F_5)_4]^-$/amine LA/LB LP cooperatively activates the monomer and initiates/propagates alcohol. In situ formation of zinc alkoxides that initiated the polymerization was ruled out, as no reaction occurred in an equivalent mixture of $[\{NNO\}Zn]^+$ $[B(C_6F_5)_4]^-$/PMP/*neo*-PentOH. The polymerization activity exhibited a positive correlation with the basicity of the amine. An increase in the amount of alcohol relative to the catalytic amounts of $[\{NNO\}Zn]^+$ $[B(C_6F_5)_4]^-$/amine LP permitted the simultaneous growth of multiple polymer chains per metal center without affecting either the control or the rate of polymerization. This phenomenon was comparable to the "immortal" ROP processes by organometallic catalysts. The $[\{NNO\}Zn]^+$ $[B(C_6F_5)_4]^-$/amine LP system was also found to be an effective catalyst for the ROP of trimethylene carbonate (TMC), with no significant impact of the nature of the anion (e.g., $\{H_2N[B(C_6F_5)_3]_2\}^-$ and $[EtB(C_6F_5)_3]^-$) on the catalyst performance [15]. However, the change of the amine did follow the same trend observed in the ROP of *rac*-lactide.

A series of NHC-$Zn(C_6F_5)_2$ neutral LP adducts (Figure 11.6) were also investigated for the ROP of β-BL, *rac*-lactide, and TMC [16]. In the presence of such monomers, NHC-$Zn(C_6F_5)_2$ adducts can be dissociated to generate free NHC

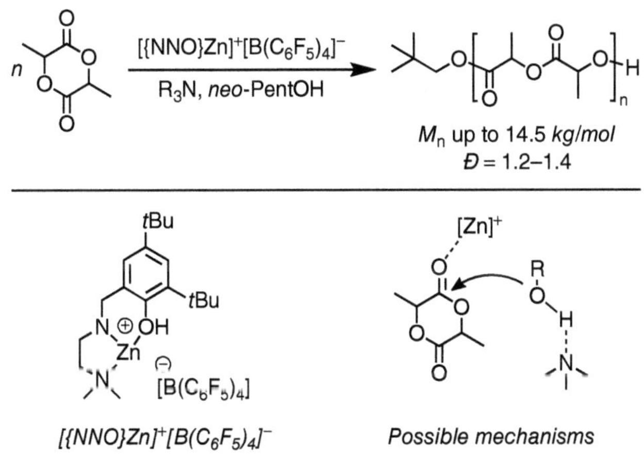

Figure 11.5 Proposed initiation mechanism for the controlled ROP of *rac*-lactide by cationic zinc complex LPs.

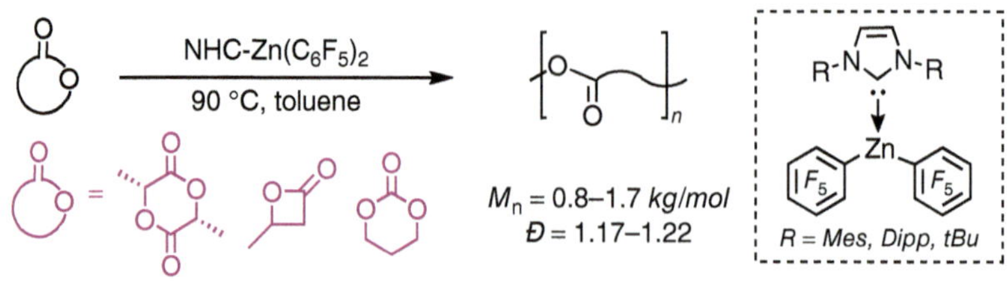

Figure 11.6 ROP of β-BL, *rac*-lactide, and TMC was mediated by NHC-Zn(C_6F_5)$_2$ LPs.

and Zn(C_6F_5)$_2$, which promote the ROP of β-BL was attempted, but the resulting activity was low (TOF = 2.7–4.0 h^{-1}, β-BL/Zn = 100, 90 °C, toluene). Furthermore, the M_n values of the resultant polymers (M_n = 0.8–1.7 kg mol^{-1}, $Đ$ = 1.17–1.22) were much lower than the theoretical values. The presence of crotonate/H chain ends indicated that the NHC acted as a base, abstracting the α-H of Zn(C_6F_5)$_2$$^-$ activated β-BL, which led to the alkyl cleavage and subsequent polymerization. Additionally, the NHC – Zn(C_6F_5)$_2$ adducts were observed to undergo ROP of *rac*-lactide (TOF = 5 h^{-1}, M_n = 3.8–9.2 kg mol^{-1}, $Đ$ = 1.96–2.13). TMC (TOF = 50 h^{-1}, M_n = 14.1–23.9 kg mol^{-1}, $Đ$ = 1.39–1.93) at RT (monomer/Zn = 100) afforded polymers with an ill-controlled MW. In contrast to β-BL polymerization, the obtained polytrimethylene carbonate (PTMC) contained MeO- and $HOCH_2CH_2CH_2O$-carbonate chain ends, which were derived from the methanolysis of an imidazolium chain end. This suggested that the polymerization was initiated by the nucleophilic attack of the NHC.

In 2013, Bourissou and Amgoune reported the synthesis of cyclic polyesters from lactide or ε-CL by combining Zn(C_6F_5)$_2$ with an amine or phosphine (Figure 11.7) [17]. It was discovered that weakly interacting Zn(C_6F_5)$_2$/PMP LP was more active toward L-lactide polymerization in 2-methyltetrahydrofuran (Me-THF) [18] at 65 °C (TOF = 5.7 h^{-1}, L-lactide/LA/LB = 30/1/1), resulting in PLA with a relatively narrow distribution. This LPP system exhibited a pronounced solvent dependence, with polymerization proceeding at a markedly reduced rate in toluene or THF relative to that in Me-THF. The formed PLA was found to be contaminated by poly(THF) in THF, while no polymerization occurred in CH_2Cl_2. The absence of polymer formation when PMP or Zn(C_6F_5)$_2$ was used alone at 65 °C after 12 h indicates that

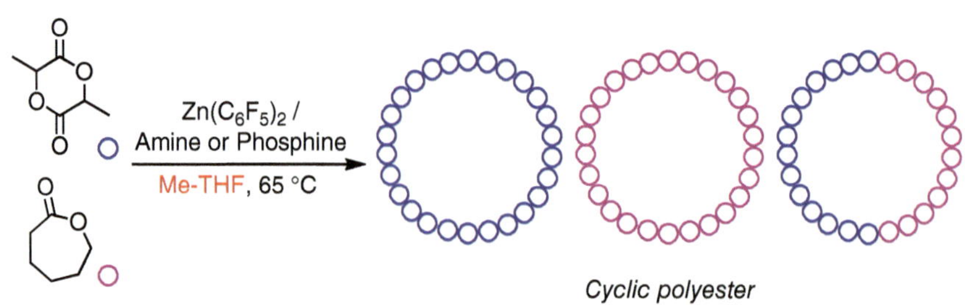

Figure 11.7 Synthesis of cyclic polyesters from lactide or ε-CL by Zn(C_6F_5)$_2$ with N/P LPs.

the cooperative activation of the LP is crucial to the polymerization process. The use of more nucleophilic LBs led to more efficient polymerization, with the TOF value following the order of PPh_3 (7.0 h^{-1}), DMAP (13.7 h^{-1}), and $PnBu_3$ (14.9 h^{-1}) ($M_n \leq 50.6$ kg mol^{-1}, $Đ$ = 1.1–1.6). Of greater interest is the observation that the resulting PLA obtained by $Zn(C_6F_5)_2$/PMP has a cyclic architecture, as corroborated by the absence of end groups at any MW of the PLA. Additionally, the MW of PLA exhibited a linear increase with monomer conversion, while the D value exhibited a slight increase from 1.1 to 1.5 over time.

The ROP mechanism, based on a $Zn(C_6F_5)_2$-based LP, was further developed by Wang et al. for the synthesis of cyclic polymers [7b]. The combination of $Zn(C_6F_5)_2$ with organic bases, including 2,2-dimethyl-1,3-propanediol (DMAP), imidazole (IMes), dibutyl urea (DBU), and 7-methyl-1,5,7-triazabicyclo[4.4.0]decane (MTBD), The 5-ene (MTBD) afforded LPs with varying degrees of LA-LB interactions, thus modulating the ROP of L-lactide. A direct relationship between polymerization activity and the degree of interaction of the LP can be observed. For instance, the pair comprising the most interacting $Zn(C_6F_5)_2$/DMAP exhibited the lowest activity (TOF = 46.3 h^{-1}, L-lactide/LA/LB = 50/1/1, toluene, 80 °C). In contrast, the pair comprising the least interacting $Zn(C_6F_5)_2$/IMes exhibited moderate activity (TOF = 92.6 h^{-1}). Finally, the pair comprising the least interacting $Zn(C_6F_5)_2$/DBU and $Zn(C_6F_5)_2$/MTBD exhibited the highest activity (TOF = 277.8 h^{-1}). Moreover, the interaction between LA and LB was found to be weakened at high temperatures or in solvents with a high dielectric constant, resulting in further enhanced polymerization activity. The analysis of the unquenched polymerization mixture produced by $Zn(C_6F_5)_2$/MTBD pair revealed that the polymerization proceeded via zwitterionic species through the bifunctional activation of the LP. Ring opening of lactone was initiated by an attack to the activated lactone monomer by an LB, followed by ring opening of the lactone to an LA-stabilized alkoxide. The propagation facilitated by $Zn(C_6F_5)_2$ follows a coordination-insertion mechanism, distinguishing it from typical LPP bimetallic addition. However, LB/$Zn(C_6F_5)_2$ pairs can still be used to polymerize cyclic esters in a controlled manner.

It is noteworthy that the intramolecular ion-pairing interaction of zwitterionic species completely inhibits transesterification, as indicated by low dispersity even at prolonged reaction time, which allows for sequential addition of a second comonomer for the formation of cyclic BCPs. Building upon the cyclic zwitterionic species mechanism, Wang and Li subsequently reported the synthesis of long-chain aliphatic polyesters with high MW through the ROP of ω-pentadecalactone (PDL) using the $Zn(C_6F_5)_2$/DBU pair (Figure 11.8) [8b]. The researchers were able to obtain cyclic and linear block copolyesters through a sequential addition strategy and/or subsequent nucleophilic substitution by bulky and electron-deficient alcohols, thanks to the absence of transesterification. Additionally, an effective method for the production of linear diblock copolymers, namely *l*-poly(PDL-*b*-CL) and *l*-poly(PDL-*b*-LA), was developed. This involved the sequential polymerization of PDL and ε-CL or L-lactide, followed by nucleophilic substitution by alcohol after full

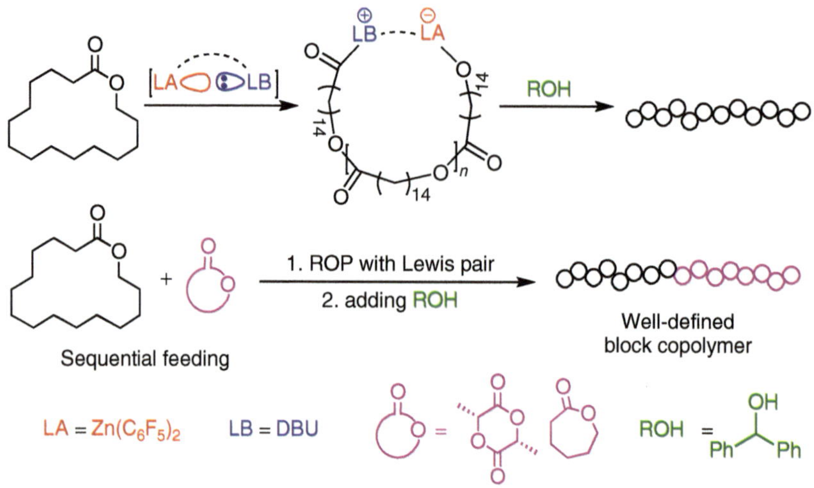

Figure 11.8 Synthesis of "polyethylene-like" copolyester by a cascade of ring-opening polymerization and subsequent nucleophilic substitution.

monomer conversion. It has been demonstrated that bulky and electron-deficient Ph_2CHOH was the most suitable alcohol for exclusively attacking the acylazolium species in the polymer chain, rather than the ester group, thereby leading to the formation of the linear polymer without random chain scission.

11.3.4 Metal Halide-Based LPs

Buchmeiser, Naumann, and colleagues also employed isolated NHC/metal halide LP adducts (**1–4**, Figure 11.9) as the latent precatalysts for the ROP of ε-CL [19]. One of these (**1**) has already been shown to act as a latent catalyst in the successful synthesis of polyurethanes. The high catalytic activity of this system is attributed to the cooperative polymerization mechanism, where the free carbene and $SnCl_2$ formed by thermal dissociation serve as a base for deprotonation of polyol and an LA for activation of the isocyanate, respectively [20]. The polymerization activity is found to be strongly dependent on the nature of the LA and dissociation ability of the NHC-LA adducts. No polymer was generated at RT, with the exception of **2**, due to the weak coordination of the NHC to the Mg center, which resulted in the formation of a small amount of the free NHC to initiate the polymerization. At relatively high temperatures (70 °C, ε-CL/BnOH/LP = 280/2/1, neat, 5 h), all of these adducts were active. The weakly bound **2** polymerized more rapidly than the 28, which in turn was more efficient than the more strongly associated **1**. Further increasing the temperature to 90 and 130 °C, which facilitated the dissociation of the adduct, accelerated the polymerization for all the NHC-LA adducts (TOF ≈ 560 h^{-1}, 15 min). In the case of **4**-mediated polymerization, a high temperature of 130 °C was necessary to achieve quantitative conversion, whereas the polymerization at 70 °C was only marginally active (TOF = 2.5 h^{-1}).

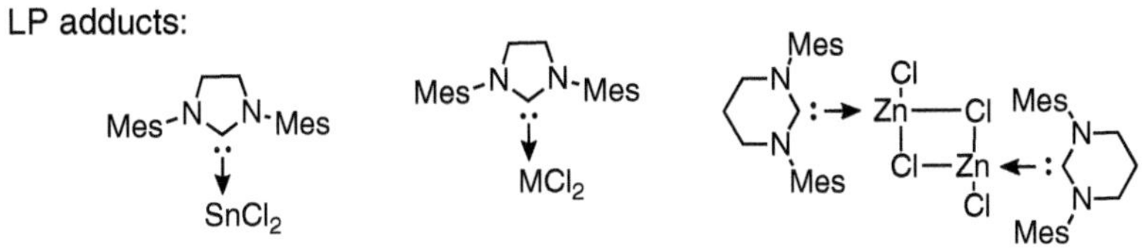

Figure 11.9 NHC – metal chloride adducts employed for the ROP of cyclic esters.

Dove, Naumann, and colleagues described the combination of simple metal halides (MgX_2 [X = Cl, Br, I], YCl_3, $AlCl_3$, etc.) with NHCs (*s*-IMes (1,3-dimesitylimidazolin-2-ylidene) as the ligands for the ROP of cyclic esters, with particular reference to the use of PDL as a renewable monomer. The authors describe the use of simple metal halides [MgX_2 (X = Cl, Br, I), YCl_3, $AlCl_3$, etc.] with NHCs [*s*-IMes (1,3-dimesitylimidazolin-2-ylidene], IMes, and IiPr] and organic bases (DBU, DMAP) as the ligands for the ROP of cyclic esters [21], particularly macrolactone PDL, a renewable monomer that has attracted increasing interest [22]. The properties of the corresponding polymer (PPDL) were found to be similar to those of linear low-density polyethylene [22f]. This study demonstrated that the isolated NHC-$MgCl_2$ adduct **2** and the mixture of *s*-IMes and $MgCl_2$ displayed an identical catalytic effect for the polymerization of PDL. In contrast, the use of YCl_3 as the LA led to a conversion approximately half as high as observed with $MgCl_2$, while $AlCl_3$ resulted in an extremely low conversion. The addition of $FeCl_3$, $ZnCl_2$, BPh_3, and $Bi(OTf)_3$ did not induce any polymerization. For different magnesium halide salts, MgI_2 exhibited the highest polymerization activity, which was explained by its cationic character resulting from the dissociation of one iodide ligand. Other LBs, such as IMes, IiPr, DBU, and DMAP, were also found to be active in the presence of MgX_2. The order of activity was observed to be $MgX_2 > YCl_3 \gg AlCl_3$ and $MgI_2 > MgBr_2 > MgCl_2$ in every case, regardless of the nature of the nucleophile. This indicates that the LA-monomer activation/interaction is the decisive factor in this catalytic system. The combination of MgI_2 and DBU was found to be the most active catalytic system for PDL polymerization, with a TOF of 308 h^{-1}. As a widely proposed mechanism in organopolymerization, LP-mediated polymerization proceeded through nucleophilic ring opening of LA-activated PDL by either activated alcohol (initiator or OH-terminated growing polymer chain, Figure 11.1(b)) or by the NHC itself (Figure 11.1(a)). It is noteworthy that the MgI_2/DMAP LP system exhibited monomer selectivity. This phenomenon was observed when the polymerization gradually lost its activity when the nature of the monomer shifted more distantly from the original PDL. After two hours, the conversion of ε-CL was 95%, while that of δ-VL was 38%, and the conversion of *rac*-lactide was 5% and that of β-BL was 0%. In the case of copolymerization of ε-CL and δ-VL (1/1), only MgI_2 and YCl_3 coupled with DMAP resulted in the formation of an isolated polymer. MgI_2, when coupled with DMAP, resulted in the formation of an isolated polymer after two hours with a 52% conversion, while YCl_3, when coupled with DMAP, resulted in the formation of an isolated polymer after 10 minutes with a 58%

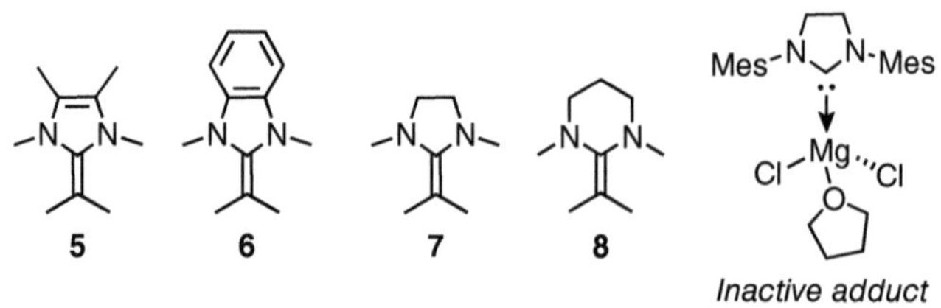

Figure 11.10 Structures of NHOs employed in this LPP study and inactive NHO-metal halide adduct.

conversion. In the same year, Waymouth and Chang discovered that the addition of an appropriate amount of LiCl (1 M) to the $I^iPr(Me)$-mediated ROP of δ-VL resulted in a more controlled polymerization (TOF = 532.5 h^{-1}) with a higher I^* (108.0–130.0). The polymerization yielded a 30.6% conversion, a predictable MW ($M_n \leq 5.46$ kg mol^{-1}), and narrow dispersities ($Đ$ = 1.08–1.26), in stark contrast to the polymerization by $I^iPr(Me)$ alone (TOF = 540.0 h^{-1}, M_n = 19.6 kg mol^{-1}) [23]. The polymerization yielded a MW of 0.6 kg mol^{-1}, a dispersity of 1.35, and an intrinsic viscosity of 36.7%. The presence of LiCl also influenced the topology of the resultant poly(δ-valerolactone) (PVL), with the formation of linear chains observed at high LiCl concentrations, while cyclic PVLs were isolated in the absence of LiCl. Naumann et al. employed NHOs (**5–8**, Figure 11.10) as the LB, in combination with simple metal halides (MgI_2, $MgCl_2$, YCl_3, ZnI_2, $AlCl_3$) as the LA, to achieve controlled homo- and copolymerization of ε-CL and δ-VL at RT. In the polymerization of ε-CL, the polymerization activity was determined by the LA: MgI_2 (TOF $\approx$ 36 h^{-1}) > YCl_3 > ZnI_2 $\approx$ $AlCl_3$ > $MgCl_2$, while in the δ-VL polymerization, ZnI_2 (TOF $\approx$ 36 h^{-1}) has gained reactivity in relation to YCl_3 and MgI_2.

This LPP system was also successful for copolymerization of equimolar ε-CL and δ-VL (NHO/BnOH/metal halide/δ-VL/ε-CL = 1/2/5/100/100), where the most rapid monomer consumption was observed for MgI_2 and the slowest for $AlCl_3$. Additionally, the metal halide is also the decisive factor for the copolymerization selectivity. MgI_2, $MgCl_2$, ZnI_2, and $AlCl_3$ favored the incorporation of δ-VL, whereas YCl_3 showed a preference for ε-CL. In all cases, relatively narrow dispersity values were observed ($Đ$ = 1.05–1.37) for the obtained polymers, with a linear correlation of M_n and conversion, despite the moderate I* value of ~50%. In contrast, NHO **5** alone promoted uncontrolled polymerization, and NHOs **6–8** were inactive with respect to the ROP of ε-CL or δ-VL. These results demonstrated that suitable metal halides not only activate the monomer to facilitate the ring-opening process but also suppress side reactions (transesterification and enolization [24]) arising from a reduction of the free, active NHO in solution as a result of competitive formation of inactive NHO-metal halide adduct (Figure 11.10). It is also conceivable that the activation of the main chain ester group by the metal halide is weaker than that of the lactone, which would lead to a higher selectivity of ring opening over transesterification. More recently, Naumann et al. successfully extended the above NHO (**5, 7–8**)/metal halide (MgI_2, $MgCl_2$, YCl_3, ZnI_2, LiCl) LP systems to homopolymerization of PDL, copolymerization of PDL with five-, six-, and seven-membered lactones (γ-BL, δ-VL, ε-CL), and copolymerization of γ-BL with δ-VL and ε-CL [25]. With an NHO/BnOH/LA/PDL ratio of 1/2/5/100 at 110 °C, all three NHOs paired with metal

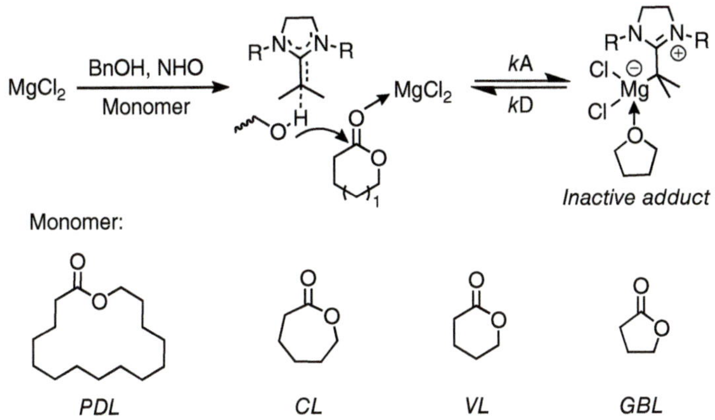

Figure 11.11 Proposed equilibrium between dissociated NHO/metal halide pairs (active for polymerization) and NHO – metal halide adducts (inactive for polymerization).

halides were able to mediate PDL. Homopolymerization (conversion = 85–97%, 4 h, $MgI_2 > MgCl_2 > YCl_3 > ZnI_2$) yielded PPDL with M_n = 10–40 kg mol^{-1} and relatively broad dispersity of 1.46–1.79, due to the existence of transesterification at high temperature.

A variety of copolymers were produced by employing this LP system, with a fixed monomer feed ratio of 1/1 (Figure 11.11). In comparison to the homopolymerization of PDL, copolymerizations of PDL with ε-CL proceeded at a slower rate (conversion = 54–97%, 4 h) and produced copolymers with a broader dispersity (M_n = 10.9–28.1 kg mol^{-1}, $Đ$ = 1.72 2.10, 100 °C, THF, pressure tube). A fully randomized copolymer with a 50% PDL content was achieved through the transesterification of NHO 6/MgI_2. Subsequently, δ-VL/PDL copolymerization was conducted at 110 °C in toluene. It was observed that the conversion of δ-VL was relatively sluggish, with less than 50% conversion achieved within four hours. In contrast, YCl_3 exhibited a conversion of 78% within the same time frame. However, PDL incorporation was largely disfavored. A fully randomized δ-VL/PDL copolymer with a 50% PDL content was also accessible by NHO **5**/YCl_3 (M_n = 23.0 kg mol^{-1}, $Đ$ = 1.53). A temperature step was introduced, during which the reaction was held at 50 °C for 8 h to promote conversion of δ-VL. Subsequently, the temperature was increased to 110 °C to support PDL incorporation by transesterification. Conversely, it is also possible to copolymerize PDL with thermodynamically opposite "non-strained" γ-BL [3, 26], with the yield (15–50%) and γ-BL content (5–22%) varying depending on the LA used. Similarly, γ-BL/ε-CL and γ-BL/δ-VL copolymers (M_n = 2.0–6.9 kg mol^{-1}, $Đ$ = 1.09–2.40) can be produced by this LP system. Most recently, Naumann and colleagues employed NHO/LiCl LPs for γ-BL polymerization, affording polymers with different topologies (linear, cyclic, or a mixture of cyclic and linear polymers) which can be tuned by the structures of NHOs. However, the activities (TOF = 0.2–1.6 h^{-1}) and MWs (M_n = 4.7–7.8 kg mol^{-1}, $Đ$ = 1.40–1.80) of the resulting polymers were relatively low [27]. In a related study, Naumann et al. also investigated this LPP system theoretically [28]. Their findings indicated that the formation of NHO/metal halide adducts occurs through the coordination of the exocyclic carbon atom of NHO to the metal center, rather than the side-on coordination. The tendency to form adducts followed the order of $ZnCl_2 > MgCl_2 > LiCl$. The saturated NHO backbones and the introduction of substituents on the exocyclic olefinic carbon can disfavor the adduct formation. When the lactone coordinates to a metal halide ($MgCl_2$), the most likely binding mode is via the carbonyl oxygen. It was also demonstrated that in a metal-free setting, an initiating alcohol cannot be deprotonated by NHO. In contrast, in the presence of $MgCl_2$, the same process is exothermic with a low barrier.

References

1 (a) Chen, X., McCarthy, S.P., and Gross, R.A. (1997). Synthesis, characterization, and epoxidation of an aliphatic polycarbonate from 2, 2-(2-pentene-1, 5-diyl) trimethylene carbonate (cHTC) ring-opening polymerization. *Macromolecules* 30: 3470–3476. (b) Drumright, R.E., Gruber, P.R., and Henton, D.E. (2000). Polylactic acid technology. *Adv. Mater.* 12: 1841–1846. (c) Gross, R.A. and Kalra, B. (2002). Biodegradable polymers for the environment. *Science* 297: 803–807. (d) Hillmyer, M.A. and Tolman, W.B. (2014). Aliphatic polyester block polymers: renewable, degradable, and sustainable. *Acc. Chem. Res.* 47: 2390–2396. (e) Zhu, Y., Romain, C., and Williams, C.K. (2016). Sustainable polymers from renewable resources. *Nature* 540: 354–362.

2 Tschan, M.J.L., Brulé, E., Haquette, P., and Thomas, C.M. (2012). Synthesis of biodegradable polymers from renewable resources. *Polym. Chem.* 3: 836–851.

3 Hong, M. and Chen, E.Y.X. (2016). Completely recyclable biopolymers with linear and cyclic topologies via ring-opening polymerization of γ-butyrolactone. *Nat. Chem.* 8: 42–49.

4 (a) Dechy-Cabaret, O., Martin-Vaca, B., and Bourissou, D. (2004). Controlled ring-opening polymerization of lactide and glycolide. *Chem. Rev.* 104: 6147–6176. (b) Kamber, N.E., Jeong, W., Waymouth, R.M. et al. (2007). Organocatalytic ring-opening polymerization. *Chem. Rev.* 107: 5813–5840. (c) O'Keefe, B.J., Hillmyer, M.A., and Tolman, W.B. (2001). Polymerization of lactide and related cyclic esters by discrete metal complexes. *J. Chem. Soc. Dalton Trans.* 2215–2224.

5 (a) Körte, L.A., Warner, R., Vishnevskiy, Y.V. et al. (2015). Intramolecular pyridine-based frustrated Lewis-pairs. *Dalton Trans.* 44: 9992–10002. (b) Stephan, D.W. (2008). "Frustrated Lewis pairs": a concept for new reactivity and catalysis. *Org. Biomol. Chem.* 6: 1535–1539. (c) Stephan, D.W. (2015). Frustrated lewis pairs. *J. Am. Chem. Soc.* 137: 10018–10032. (d) Welch, G.C., Juan, R.R.S., Masuda, J.D., and Stephan, D.W. (2006). Reversible, metal-free hydrogen activation. *Science* 314: 1124–1126.

6 Zhang, Y., Miyake, G.M., and Chen, E.Y.-X. (2010). Alane-Based Classical and Frustrated Lewis Pairs in Polymer Synthesis: Rapid Polymerization of MMA and Naturally Renewable Methylene Butyrolactones into High-Molecular-Weight Polymers. *Angew. Chem. Int. Ed.* 49: 10158–10162.

7 (a) Kreitner, C., Geier, S.J., Stanlake, L.J.E. et al. (2011). Ring openings of lactone and ring contractions of lactide by frustrated Lewis pairs. *Dalton Trans.* 40: 6771–6777. (b) Li, X.-Q., Wang, B., Ji, H.-Y., and Li, Y.-S. (2016). Insights into the

mechanism for ring-opening polymerization of lactide catalyzed by Zn (C 6 F 5) 2/organic superbase Lewis pairs. *Cat. Sci. Technol.* 6: 7763–7772.

8 (a) Wang, Q., Zhao, W., He, J. et al. (2017). Living Ring-Opening Polymerization of Lactones by N-Heterocyclic Olefin/Al(C6F5)3 Lewis Pairs: Structures of Intermediates Kinetics, and Mechanism. *Macromolecules* 50: 123–136. (b) Wang, B., Pan, L., Ma, Z., and Li, Y. (2018). Ring-opening polymerization with lewis pairs and subsequent nucleophilic substitution: a promising strategy to well-defined polyethylene-like polyesters without transesterification. *Macromolecules* 51: 836–845.

9 Zhu, J.-B. and Chen, E.Y.X. (2015). From meso-Lactide to Isotactic Polylactide: Epimerization by B/N Lewis Pairs and Kinetic Resolution by Organic Catalysts. *J. Am. Chem. Soc.* 137: 12506–12509.

10 Zhang, H., Nie, Y., Zhi, X. et al. (2017). Controlled ring-opening polymerization of α-amino acid N-carboxy-anhydride by frustrated amine/borane Lewis pairs. *Chem. Commun.* 53: 5155–5158.

11 Zhang, Y., Miyake, G.M., John, M.G. et al. (2012). Lewis pair polymerization by classical and frustrated Lewis pairs: acid, base and monomer scope and polymerization mechanism. *Dalton Trans.* 41: 9119–9134.

12 Nakayama, Y., Kosaka, S., Yamaguchi, K. et al. (2017). Controlled ring-opening polymerization of l-lactide and ε-caprolactone catalyzed by aluminum-based L ewis pairs or L ewis acid alone. *J. Polym. Sci. A Polym. Chem.* 55: 297–303.

13 Schnee, G., Bolley, A., Hild, F. et al. (2017). Group 13 metal (Al, Ga, In) alkyls supported by N-heterocyclic carbenes for use in lactide ring-opening polymerization catalysis. *Catal. Today* 289: 204–210.

14 Piedra-Arroni, E., Brignou, P., Amgoune, A. et al. (2011). A dual organic/organometallic approach for catalytic ring-opening polymerization. *Chem. Commun.* 47: 9828–9830.

15 Brignou, P., Guillaume, S.M., Roisnel, T. et al. (2012). Discrete Cationic Zinc and Magnesium Complexes for Dual Organic/Organometallic Catalyzed Ring-Opening Polymerization of Trimethylene Carbonate. *Chem. Eur. J.* 18: 9360–9370.

16 Schnee, G., Fliedel, C., Avilés, T., and Dagorne, S. (2013). Neutral and Cationic N-Heterocyclic Carbene Zinc Adducts and the BnOH/Zn (C6F5) 2 Binary Mixture–Characterization and Use in the Ring-Opening Polymerization of β-Butyrolactone, Lactide, and Trimethylene Carbonate. *Eur. J. Inorg. Chem.* 2013: 3699–3709.

17 Piedra-Arroni, E., Ladavière, C., Amgoune, A., and Bourissou, D. (2013). Ring-opening polymerization with Zn(C6F5)2-based lewis pairs: original and fficient approach to cyclic polyesters. *J. Am. Chem. Soc.* 135: 13306–13309.

18 Aycock, D.F. (2007). Solvent applications of 2-methyltetrahydrofuran in organometallic and biphasic reactions. *Org. Process Res. Dev.* 11: 156–159.

19 Naumann, S., Schmidt, F.G., Frey, W., and Buchmeiser, M.R. (2013). Protected N-heterocyclic carbenes as latent pre-catalysts for the polymerization of ε-caprolactone. *Polym. Chem.* 4: 4172–4181.

20 Bantu, B., Pawar, G.M., Decker, U. et al. (2009). CO2 and SnII Adducts of N-Heterocyclic Carbenes as Delayed-Action Catalysts for Polyurethane Synthesis. *Chem. Eur. J.* 15: 3103–3109.

21 Naumann, S., Scholten, P.B.V., Wilson, J.A., and Dove, A.P. (2015). Dual catalysis for selective ring-opening polymerization of lactones: Evolution toward simplicity. *J. Am. Chem. Soc.* 137: 14439–14445.

22 (a) Bouyahyi, M. and Duchateau, R. (2014). Metal-based catalysts for controlled ring-opening polymerization of macrolactones: high molecular weight and well-defined copolymer architectures. *Macromolecules* 47: 517–524. (b) Jasinska-Walc, L., Bouyahyi, M., Rozanski, A. et al. (2015). Synthetic principles determining local organization of copolyesters prepared from lactones and macrolactones. *Macromolecules* 48: 502–510. (c) Ladelta, V., Bilalis, P., Gnanou, Y., and Hadjichristidis, N. (2017). Ring-opening polymerization of ω-pentadecalactone catalyzed by phosphazene superbases. *Polym. Chem.* 8: 511–515. (d) Myers, D., Witt, T., Cyriac, A. et al. (2017). Ring opening polymerization of macrolactones: high conversions and activities using an yttrium catalyst. *Polym. Chem.* 8: 5780–5785. (e) Todd, R., Tempelaar, S., Lo Re, G. et al. (2015). Poly(ω-pentadecalactone)-b-poly(l-lactide) Block Copolymers via Organic-Catalyzed Ring Opening Polymerization and Potential Applications. *ACS Macro Lett.* 4: 408–411. (f) van der Meulen, I., Gubbels, E., Huijser, S. et al. (2011). Catalytic ring-opening polymerization of renewable macrolactones to high molecular weight polyethylene-like polymers. *Macromolecules* 44: 4301–4305. (g) Wilson, J.A., Hopkins, S.A., Wright, P.M., and Dove, A.P. (2014). 'Immortal'ring-opening polymerization of ω-pentadecalactone by Mg (BHT) 2 (THF) 2. *Polym. Chem.* 5: 2691–2694. (h) Wilson, J.A., Hopkins, S.A., Wright, P.M., and Dove, A.P. (2015). Synthesis of ω-pentadecalactone copolymers with independently tunable thermal and degradation behavior. *Macromolecules* 48: 950–958. (i) Wilson, J.A., Hopkins, S.A., Wright, P.M., and Dove, A.P. (2015). Synthesis and postpolymerization modification of one-pot ω-pentadecalactone block-like copolymers. *Biomacromolecules* 16: 3191–3200.

23 Chang, Y.A. and Waymouth, R.M. (2015). Ion pairing effects in the zwitterionic ring opening polymerization of δ-valerolactone. *Polym. Chem.* 6: 5212–5218.

24 Naumann, S., Thomas, A.W., and Dove, A.P. (2016). Highly polarized alkenes as organocatalysts for the polymerization of lactones and trimethylene carbonate. *ACS Macro Lett.* 5: 134–138.

25 Walther, P. and Naumann, S. (2017). N-Heterocyclic olefin-based (co) polymerization of a challenging monomer: Homopolymerization of ω-pentadecalactone and its copolymers with γ-butyrolactone, δ-valerolactone, and ε-caprolactone. *Macromolecules* 50: 8406–8416.

26 Hong, M. and Chen, E.Y.-X. (2016). Towards truly sustainable polymers: a metal-free recyclable polyester from biorenewable non-strained γ-butyrolactone. *Angew. Chem. Int. Ed.* 55: 4188–4193.

27 Walther, P., Frey, W., and Naumann, S. (2018). Polarized olefins as enabling (co) catalysts for the polymerization of γ-butyrolactone. *Polym. Chem.* 9: 3674–3683.

28 Meisner, J., Karwounopoulos, J., Walther, P. et al. (2018). The Lewis Pair Polymerization of Lactones Using Metal Halides and N-Heterocyclic Olefins: Theoretical Insights. *Molecules* 23: 432.

12

Organocatalysts for Radical Polymerization

Chenyu Wu[1], Qiang Ma[2], Baoshan Hou[1], Zhilei Wang[1], Zikuan Wang[3], and Saihu Liao[2]

[1] *Shandong University, Institute of Frontier Chemistry, School of Chemistry and Chemical Engineering, 72 Binhai Road, Qingdao 266237, China*
[2] *Xiamen University, State Key Laboratory of Physical Chemistry of Solid Surfaces, College of Chemistry and Chemical Engineering, 422 Siming South Road, Xiamen 361005, China*
[3] *Max-Planck-Institut für Kohlenforschung, Department of Theoretical Chemistry, Kaiser-Wilhelm-Platz 1, 45470 Mülheim an der Ruhr, Germany*

12.1 Photocontrolled Organocatalyzed Radical Polymerization

Radical polymerization represents a key type of polymerization technique, being useful in a wealth of application scopes, which range from aeronautics to daily uses [1–16]. This method includes a generated carbon radical that triggers a sequence of monomer additions, giving rise to consecutive growth of polymer chains. As per mechanisms of radical generation, radical polymerizations can be categorized differently. Almost exclusively, organocatalysts are utilized in conjunction with photocatalysis, contributing to photocatalyzed reversible deactivation radical polymerization (photo-RDRP) [17], a recently emerging approach featuring a myriad of advantages, viz. pseudo-linear kinetics with temporal and spatial control, compatibility to various monomers and solvents, as well as tolerance to ambient conditions such as air exposure and room temperature [18–25]. In photo-RDRP, the performance is typically measured by temporal precision, apparent propagation rate constant (k_p^{app}), initiator efficiency (I^*), and dispersity of molecular weights ($Đ$), at given conditions (viz. monomer, initiator, formulation, solvent, temperature, light, and environment) [26].

A catalytic cycle of photo-RDRP begins with carbon-heteroatom bond cleavage of initiator R–X (R is a carbon-based initiating moiety and X is a functional moiety), referred to as the activation step, followed by radical polymerization; finally, the catalytic cycle is completed by a deactivation step, forming a new carbon-heteroatom bond between the propagating chain radical and the functional moiety [26–28]. According to the nature of activation/deactivation steps, catalytic mechanisms include oxidative quenching pathway (OQP, Scheme 12.1A), triplet energy transfer

Organocatalysts in Polymer Chemistry: Synthesis and Applications, First Edition. Edited by Zhibo Li.

pathway (TET, Scheme 12.1B), reductive quenching pathway (RQP, Scheme 12.1C), and oxygen-mediated reductive quenching pathway (O-RQP, Scheme 12.1D).

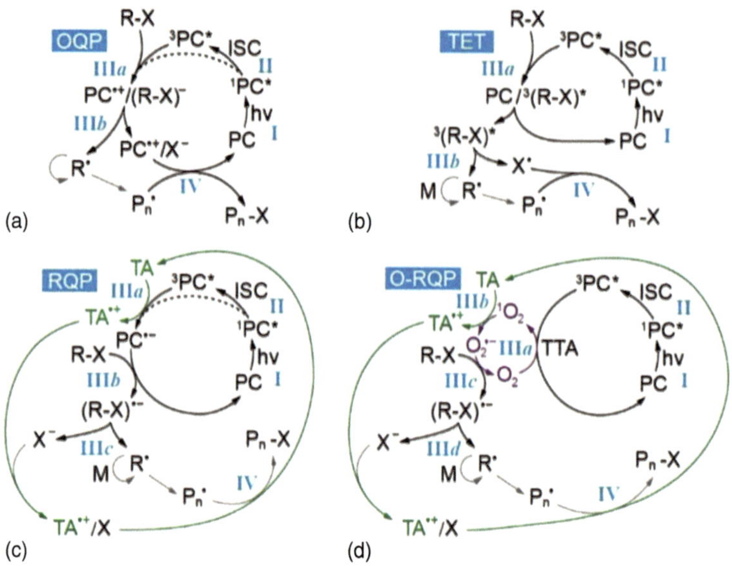

Scheme 12.1 Typical Catalytic Cycles of Photo-RDRP.

OQP: oxidative quenching pathway. TET: triplet energy transfer. RQP: reductive quenching pathway. O-RQP: oxygen-mediated reductive quenching pathway. PC: photocatalyst, herein referred to as the organocatalyst in photo-RDRP. TA: tertiary amine, or other sacrificial reductants. R-X: initiator, with R being the functional group and X being the end group. M: monomer. P: polymer chain. ISC: intersystem crossing. TTA: triplet-triplet annihilation. I: photon absorption. II: excited-state evolutions. III: activation. IV: deactivation.

In the following context, OQP is mainly considered since it is free of co-catalysts and therefore of many unwanted side reactions. On the other hand, chemical structures together with the resulting geometric and electronic structures of the PC, determine the properties, and ultimately affect the performance of photo-RDRP. These delicate reasonings are structure-property-performance (SPP) relationships, inferred as the underlying principles for the rational design of the PC.

SPP relationships, as detailed in Sec. 12.3, have been extensively investigated in recent years. They are theoretically rooted in the mechanistic aspects and are useful in designing PCs with specific properties by selecting designated geometrical and chemical structures. This includes a variety of chromophore cores and moieties, as illustrated in Scheme 12.2 and Scheme 12.3.

Compared to metal-containing PCs, organocatalysts are purely constructed using a vast number of available organic blocks, which allows for a great variety of possible structures (Scheme 12.2 and 12.3), such as polynuclear aromatic hydrocarbons [29, 30], phenothiazines (Scheme 12.2a) [31], dihydrophenazines (Scheme 12.2b)

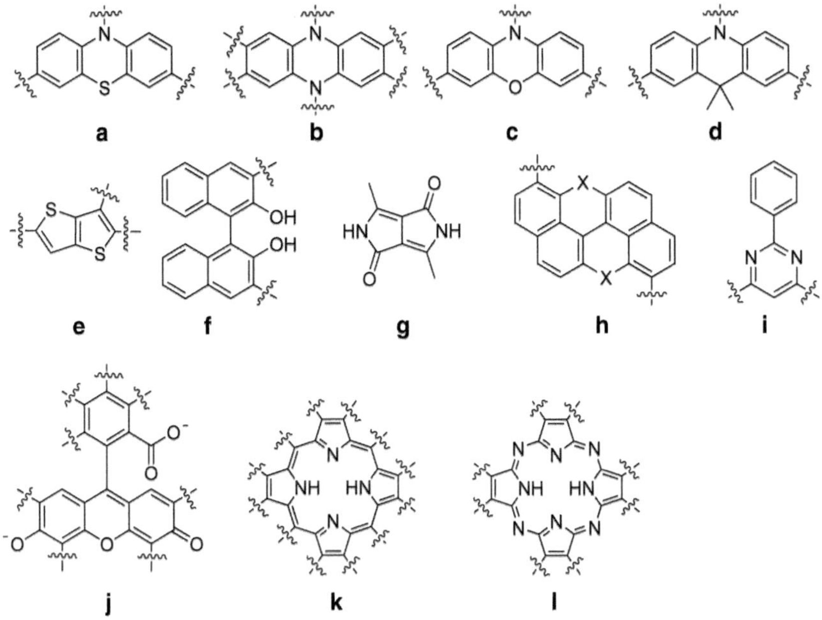

Scheme 12.2 Typical Chromophore Cores in Organocatalysts Used for Photo-RDRP.

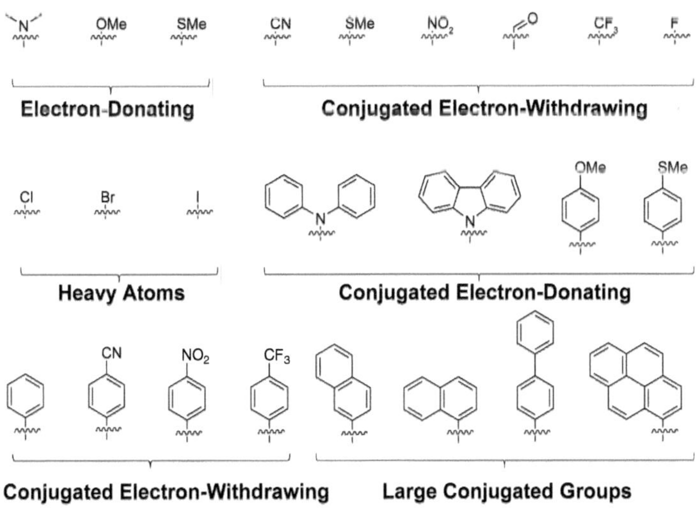

Scheme 12.3 Typical Functional Groups in Organocatalysts Used for Photo-RDRP.

[32], phenoxazines (Scheme 12.2c) [33], dimethylacridines (Scheme 12.2d) [34], thienothiophenes (Scheme 12.2e) [35], 1,1′-bisnaphthols (Scheme 12.2f) [36], *N*-unsubstituted diketopyrrolopyrroles (Scheme 12.2g) [37], heteroatom-doped polycyclic arenes [38, 39] (Scheme 12.2h), 2-phenylpyrimidines (Scheme 12.2i) [40], xanthenes (Scheme 12.2j) [41], porphyrins (Scheme 12.2k) [42], phthalocyanines (Scheme 12.2l) [43], and donor-acceptor (D–A) scaffolds [44, 45].

12.2 Organocatalyzed Atom Transfer Radical Polymerization (O-ATRP)

12.2.1 Mechanism of O-ATRP

The catalytic cycle of O-ATRP can proceed via either an OQP (Scheme 12.1A) or RQP (Scheme 12.1C), depending on the catalytic species used [26, 46]. Most commonly, it follows an OQP mechanism. However, RQP mechanisms are also possible in the presence of additional reductants [47].

Upon photon absorption, the ground state photocatalyst (PC) is excited to the S_n state, which quickly relaxes to the S_1 state ($^1PC^*$). From there, it may undergo intersystem crossing (ISC) to reach the long-lived T_1 state ($^3PC^*$). Both $^1PC^*$ and $^3PC^*$ can reduce the initiator R–X (activating O-ATRP), generating a carbon-centered radical capable of initiating radical polymerization, along with a halide ion (X^-, such as Br^- or Cl^-), and the radical cation $PC^{\bullet+}$. After several monomer additions to the growing chain, the $PC^{\bullet+}/X^-$ pair reacts with the propagating radical ($R^\bullet$), producing a dormant halogen-capped polymer (P_n-X) and regenerating the ground state PC (deactivating O-ATRP). The polymer chain, P_n-X, can be reactivated by $^3PC^*$ for further monomer propagation, followed by recapping with X^-.

12.2.2 Polynuclear Aromatic Hydrocarbons

Although some polynuclear aromatic hydrocarbons (PAHs) (Scheme 12.4) were explored and used as PCs in O-ATRP of methyl methacrylate (MMA), only moderate control ($Đ > 1.3$) was demonstrated in earlier work [29, 30]. Additionally, the utilization of PAH, such as anthracene 1b (Scheme 12.4), under UV light

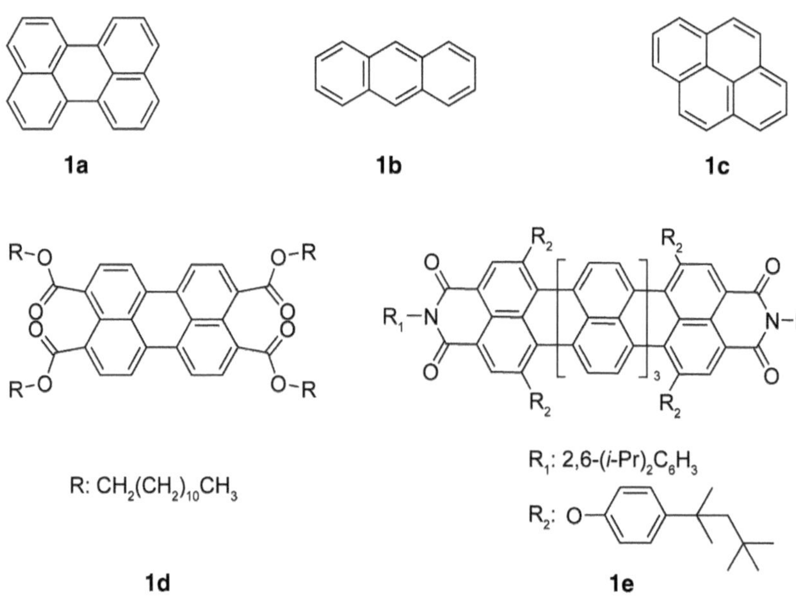

Scheme 12.4 Unfunctionalized PAHs and Functionalized PAHs.

exposure may cause side reactions during the polymerization process. Therefore, understanding the role of PAHs remains significant.

The first example of a PAH used as a PC was perylene **1a** (Scheme 12.4), reported by Miyake and Theriot in 2014 [30]. This compound was utilized for the polymerization of MMA and other monomers under white light. Although monomer conversion exhibits first-order kinetics during irradiation, the polymer's molecular weight (MW) does not continue to increase after the initial irradiation period but rather decreases.

Subsequently, similar PAHs such as anthracene **1b** (Scheme 12.4) and pyrene **1c** (Scheme 12.4) were explored by Yagci in 2016 [29]. Although anthracene proved to be more efficient in the radical generation step, detailed experiments showed that the anthracene/alkyl halide initiating system produces polymers through a free radical mechanism. This is attributed to the transfer of propagating sites to unstable positions on the anthracene ring. The resulting polymers exhibit a wide range of MWs (M_n = 4.1–19.1 kDa) and moderate dispersities ($Đ \sim 1.4$–1.5). As compared to the excited state of anthracene, that of pyrene undergoes electron transfer with the alkyl halide initiator at a slower rate. However, in the case of the pyrene/alkyl halide system, polymerizations yielded polymer products with $Đ$ down to 1.37 and well-reserved chain ends. This was validated through spectroscopic analysis, along with chain extension tests.

In addition to changing the number of aromatic rings to modify the conjugation length, some functionalized PAHs have also been reported. In 2018, Zhou and He reported **1d** (Scheme 12.4) [48] with moderate MW distributions ($M_w/M_n < 1.5$) at catalyst loading from 12.5 to 125 ppm. Increased conjugation yields **1e** (Scheme 12.4), with near-infrared (NIR) absorption, at $\lambda_{max} = 870$ nm [49], using ppm levels of **1e** (1–20 ppm) to produce polymers with M_w/M_n around 1.50.

12.2.3 Phenothiazine Derivatives

In 2014, Hawker and his team introduced phenothiazine **2a** (Scheme 12.5) as the first PC for O-ATRP, demonstrating precise control over MW (1.30–15.4 kDa) and MW distribution ($Đ \sim 1.18$–1.32). But strictly speaking, the light source they employed is not visible light (380 nm LED) [31]. Since then, many phenothiazine derivatives and variants have been gradually discovered. Matyjaszewski and his research group revisited the core mechanism of O-ATRP, expanding the use of organic photocatalysts to cover a more extensive selection of phenothiazine-based compounds, including π-extended derivatives like **2a-2f** (Scheme 12.5) [50–52].

Chen's team recently developed a new series of *N*-aryl-phenothiazine photocatalysts (**2a**, **2k-2n**, Scheme 12.5) for O-ATRP, employing aryl sulfonyl halide initiators and activating the process with white or purple lights [53]. This method enables the synthesis of polyacrylates/polymethacrylates and polyacrylamides/polymethacrylamides with a wide variety of heteroaryl or aryl chain ends, eliminating concerns about metal contamination, yet poor polymerization control was evident from $Đ > 1.5$. To enhance the performance, a continuous flow technique was combined with O-ATRP, and investigated using multiple derivatives **2a-2e**, **2i-2j** (Scheme 12.5) [54]. Accelerated polymerization and better

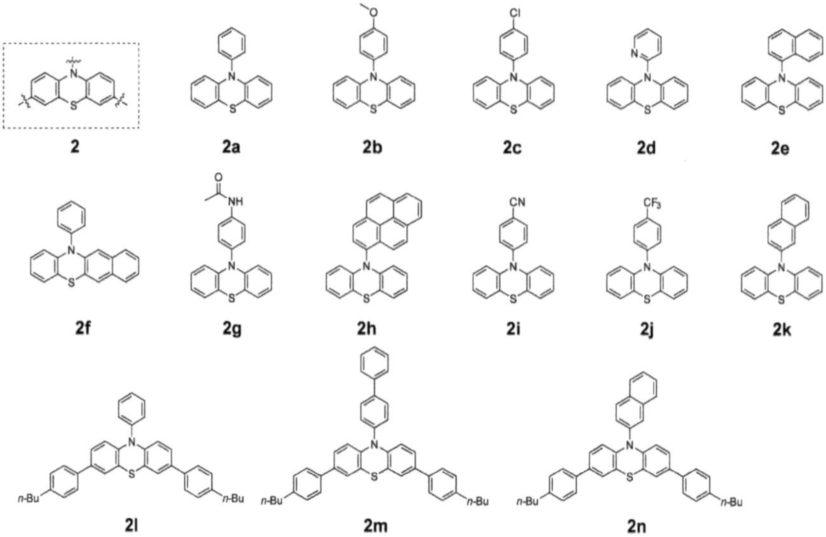

Scheme 12.5 Phenothiazine Derivatives.

$Đ \sim 1.13$–1.58 were achieved. Johnson introduced a straightforward thermally responsive organo-hydrogel, featuring covalently attached **2g**. This Gel-PTH system demonstrated the ability to control polymerization by toggling temperature, light, and catalyst presence [51].

12.2.4 Dihydrophenazine Derivatives

The first generation of dihydrophenazine **3a**-**3f** (Scheme 12.6) was reported by Miyake in 2016 [55]. To begin with, four *N*-aryl dihydrophenazines with various substituents (**3a**-**3d**), incorporating electron-withdrawing/-donating properties, were synthesized and studied. 5,10-diphenyl-5,10-dihydrophenazine **3a** showed control of M_n = 11.9 kDa, $Đ$ = 1.55, and I^* = 74.5%, which is moderate. However, PCs substituted by ligands with electron-withdrawing properties demonstrated better control, producing polymers with $Đ$ as low as 1.17 and 1.33 in the case of **3c** and **3d**, respectively, compared to PCs substituted by electron-donating ligands ($Đ$ of 1.47 for 3b), although the initiator efficiency (I^*) was lower than 60%. Moreover, PCs **3e** and **3f** (2-naphthyl and 1-naphthyl substitution) were found to exhibit better control in terms of $Đ < 1.10$ and $I^* > 50\%$.

As research progressed, it was found that the first generation of the dihydrophenazine family was prone to side reactions associated with alkyl core substitution (AkCS), affecting polymerization performance. Moreover, a typical catalyst loading of 1000 ppm was still necessary for this dihydrophenazine family in good polymerization control. To address this issue, Miyake et al. synthesized new core-substituted diaryl dihydrophenazine PCs (**3g**-**3p**, Scheme 12.6) that effectively work at 5–100 ppm loading in O-ATRP for synthesizing polymers with $Đ \sim 1.09$–1.42 and I^* = 93–106%.

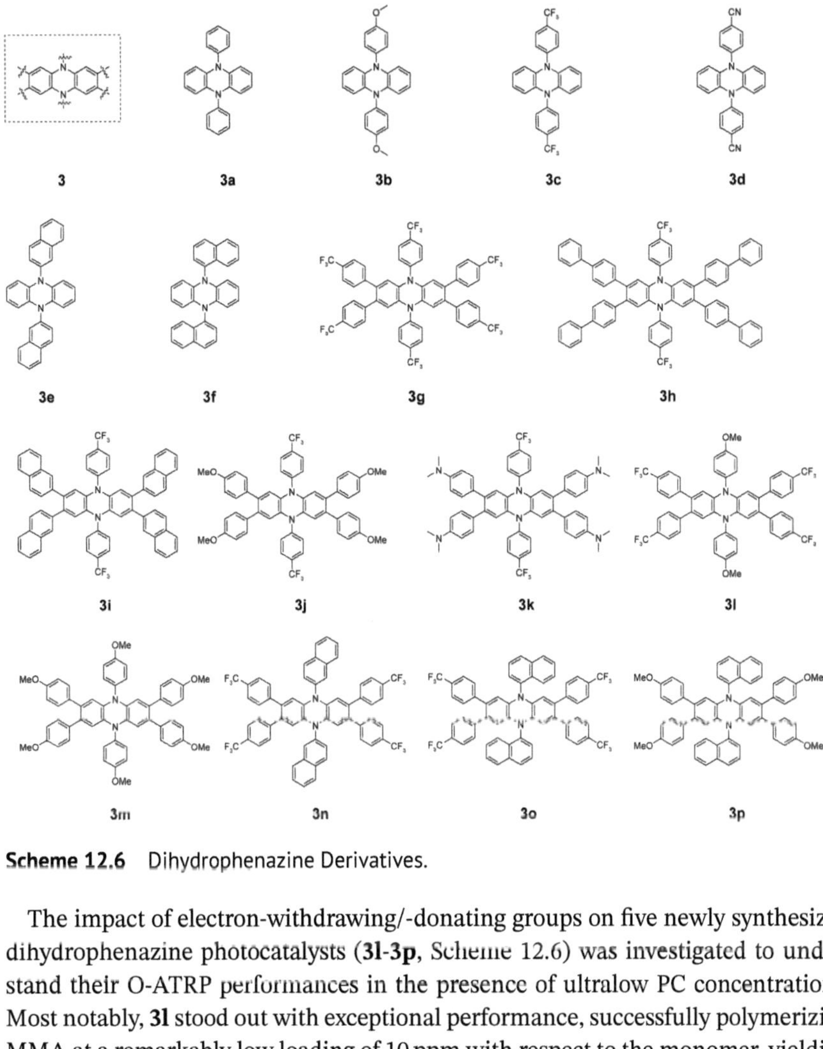

Scheme 12.6 Dihydrophenazine Derivatives.

The impact of electron-withdrawing/-donating groups on five newly synthesized dihydrophenazine photocatalysts (**3l-3p**, Scheme 12.6) was investigated to understand their O-ATRP performances in the presence of ultralow PC concentrations. Most notably, **3l** stood out with exceptional performance, successfully polymerizing MMA at a remarkably low loading of 10 ppm with respect to the monomer, yielding polymers with $Đ = 1.33$ and an outstanding I^* up to 102% [56].

12.2.5 Phenoxazine Derivatives

In 2016, researchers synthesized a series of *N*-aryl phenoxazines (**4a-4c**, **4f-4g**, Scheme 12.7) [57]. While **4a** and **4b** displayed moderate control over polymerization in O-ATRP of MMA – resulting in $M_n = 7.2$ kDa, $Đ = 1.48$, $I^* = 137\%$ for 4a, and $M_n = 6.5$ kDa, $Đ = 1.45$, $I^* = 85.5\%$ for **4b**. Compounds **4c** and **4g**, with *N*-naphthyl substitutions, exhibited much stronger control. **4c** achieved $M_n = 8.8$ kDa, $Đ = 1.22$, and $I^* = 92.6\%$, while **4g** delivered $M_n = 10.8$ kDa, $Đ = 1.11$, and $I^* = 73.3\%$. Further modification of **4c** by attaching 4-biphenyl groups at positions 3 and 7 led to **4f**, which demonstrated a red shift in absorption, thus enabling activity in the visible spectrum. Following this, the team designed a system leveraging orthogonal

π-systems and an intermediate charge transfer (CT) excited state to enhance ISC yields (Φ_{ISC}) in phenoxazine-based compounds **4e** and **4f**. Computational analysis was also performed on **4d** and **4h**-**4p** to identify additional candidates for further testing [58].

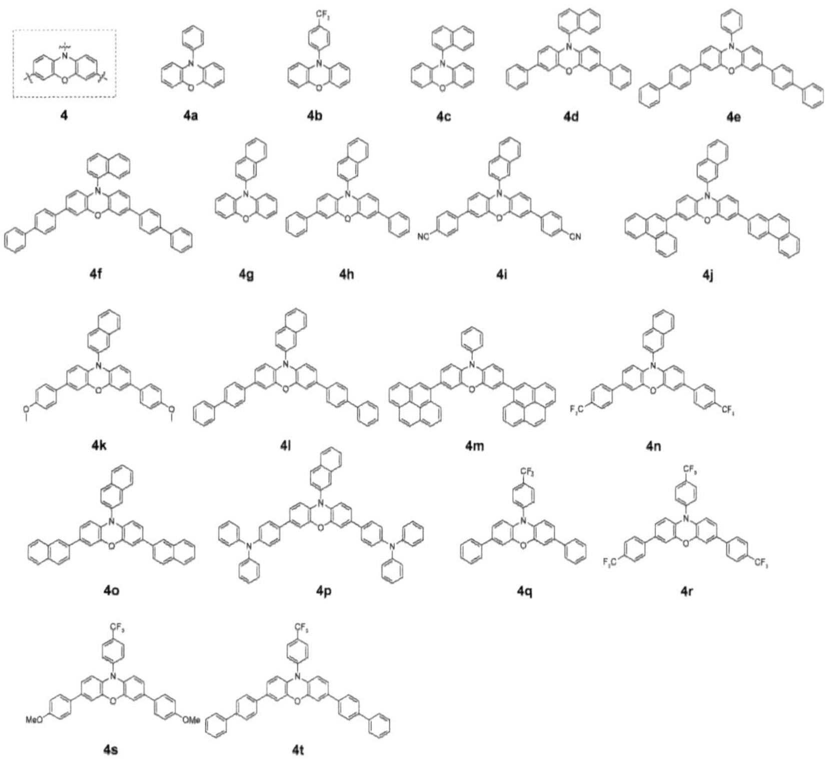

Scheme 12.7 Phenoxazine Derivatives.

From the work of Son's group, also by further core modifications of *N*-trifluoromethylphenyl phenoxazine, the new phenoxazine derivatives **4q**-**4t** were evaluated for O-ATRP in both experimental and theoretical studies [59]. Especially, combination of **4t** and EBPA as the PC and initiator, respectively, enables successful O-ATRP with a CFL bulb ($Đ = 1.23$, $I^* = 74\%$), white LEDs ($Đ = 1.47$, $I^* = 71\%$), or sunlight ($Đ = 1.28$, $I^* = 49\%$).

12.2.6 Dimethylacridine Derivatives

Miyake's group presented a series of dimethylacridines **5a**-**5g** (Scheme 12.8) as new PCs, enabling controlled O-ATRP of acrylates for the first time [60]. The PCs feature $PC^{•+}$ capable of strong oxidation for effective deactivation reactions. Among PCs tested in both batch and flow systems, **5f** achieved the lowest $Đ = 1.53$, while **5g** exhibited the highest $Đ = 4.93$ in batch. A similar trend was observed in flow, with **5f** reaching $Đ = 1.35$ and **5g** reaching $Đ = 4.57$. Further testing with additional acrylate monomers led to the successful preparation of the *p*(n-BA)-*b*-*p*(EA)-*b*-*p*(t-BA) triblock copolymer using **5f**. Ma et al. reported dimethyl dihydroacridines **5h-5m** (Scheme 12.8) based on the same backbone for O-ATRP [61]. Despite the similar structure with phenothiazine, phenoxazine, and phenazine, the use of 9,9-dimethylacridine has been uncommon due to the instability of its oxidative state.

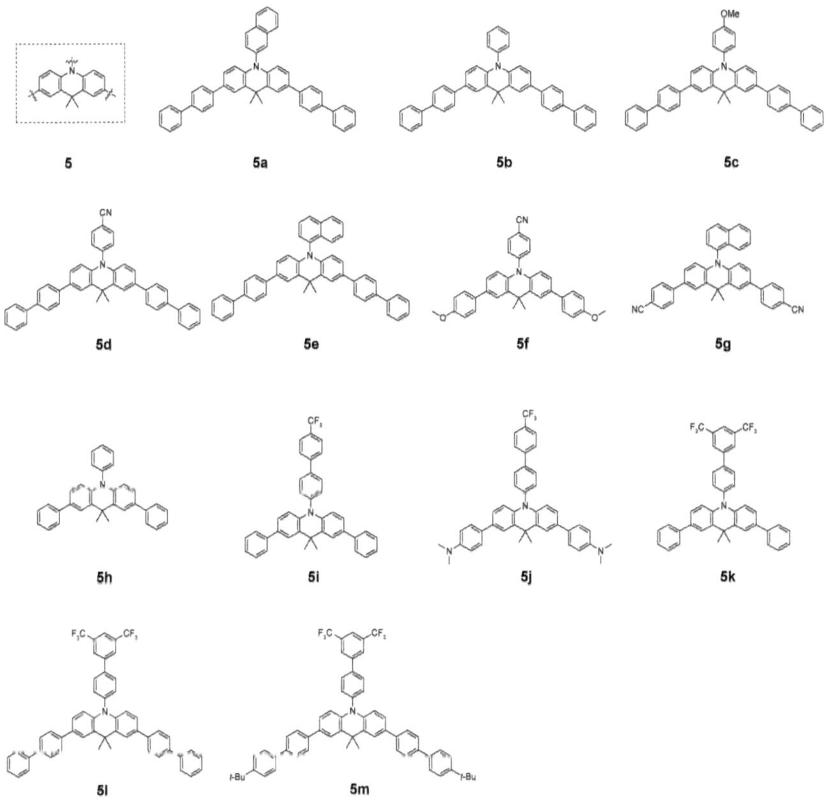

Scheme 12.8 Dimethylacridine Derivatives.

12.2.7 Organic Donor–Acceptor Scaffolds

Kim et al. developed a computational strategy for molecular design and screening, aimed at directly constructing organic PCs with donor-acceptor (D–A) architectures, drawing from an extensive library of commonly used donor and acceptor building blocks [44, 62]. Dating back to 2016, the first novel carbazole-based PC, **6a** (Scheme 12.9), was first introduced into O-ATRP by Cheng and Zhu et al. featuring high catalytic efficiency with 5 ppm under blue lights [63]. Only the control of the product (MW, M_n = 23.4 kDa and MW distribution, $Đ$ = 1.50) is unsatisfactory and only moderately controlled is realized. This progress represents a significant advance toward highly efficient O-ATRP. Kwon et al. discovered an organic photoredox catalyst, 4DP-IPN **6b** (Scheme 12.9), which can control O-ATRP of MMA at sub-ppm level of loading [44].

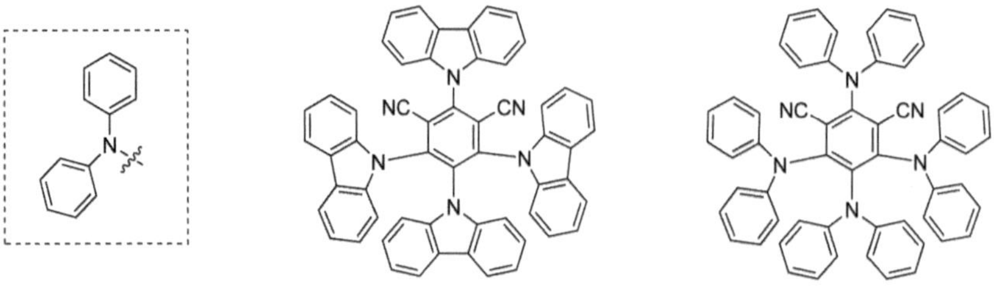

6 **6a** **6b**

Scheme 12.9 Donor-acceptor Scaffolds.

12.2.8 Heteroatom-Doped Anthanthrene Derivatives

In 2021, Liao et al. proposed a novel approach to catalyst design, centered around the use of heteroatom-doped polycyclic arenes (HDPA) to accelerate the discovery of new PCs [64]. This strategy leads to great success, which can be illustrated by the fact that different heteroatom-doping photocatalysts have been applied in controlled radical polymerization (O-ATRP and PET-RAFT) and cationic polymerization, respectively [65, 66]. First, oxygen-doped anthanthrene "**7a-7d**" (Scheme 12.10) was proven to be a desirable series of PCs strongly absorbing in the visible range ($\varepsilon_{450\,\mathrm{nm}} > 20\,000\,\mathrm{M}^{-1}\ \mathrm{cm}^{-1}$), with catalyst loadings as low as ppm/sub-ppm (0.05–10 ppm), allowing for metal-free, controlled polymerization

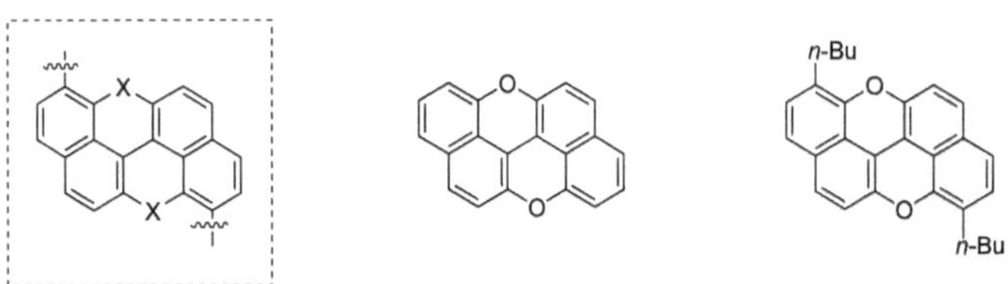

7 **7a** **7b**

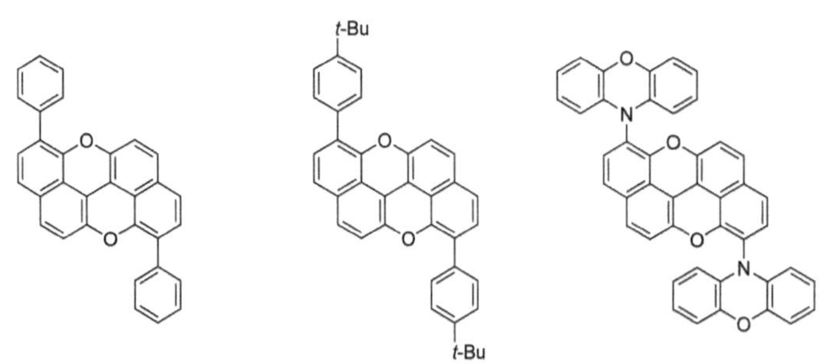

7c **7d** **7e**

Scheme 12.10 Heteroatom-doped Anthanthrene Derivatives.

reactions in the visible range and even in sunlight. An impressive feature is that the visible-light absorption strength is nearly ten times greater than the commonly used organometallic photocatalyst *fac*-Ir(ppy)$_3$ (ε_{458nm} = 2450 M^{-1} cm^{-1}). Additionally, this framework is a rare instance of a non-*N*-heterocyclic organic catalyst, without CT characteristics, capable of mediating ATRP with high control and low $Đ < 1.20$ at ppm-level catalyst concentrations.

Lately, based on ODA parent structure, a new oxygen-doped anthanthrene (ODA)-derivated catalysts **7e** (Scheme 12.10) was reported by Jaewoo Lee et al. [67]. Using this PC under UV light, they synthesized amphiphilic polyvinyl chloride (PVC) via metal-free ATRP and developed a highly permeable PVC-based microfiltration (MF) membrane. This membrane efficiently separates microplastics (MP), with the amphiphilic PVC functioning as both a pore-forming agent during phase separation and a hydrophilic modifier after solidification. The research provides promising practical approaches to tackle the issue of MP pollution.

12.2.9 Other Classes

In 2017, O-ATRP with thienothiophene derivatives **8a-8d** (Scheme 12.11) as PCs has been investigated by Yagci's group, namely **8a, 8b, 8c,** and **8d** [35].

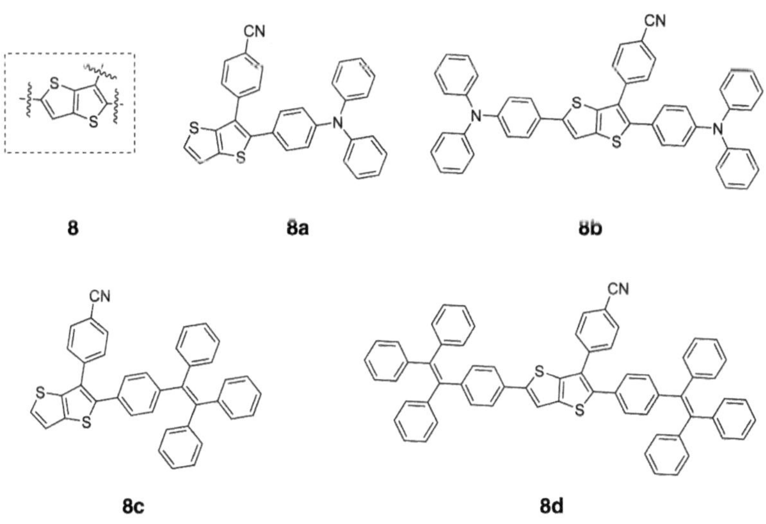

Scheme 12.11 Thienothiphene Derivatives.

In 2019, Liao et al. identified **9** (Scheme 12.12) as a novel class of organic PCs for O-ATRP. Under visible light, this system achieved good control over M_n, maintained a low $Đ$, and ensured high chain-end fidelity [36]. The series **9a-9f** (Scheme 12.12) exhibits high modifiability, allowing for the adjustment of its photophysical properties by introducing various substituents at the 3,3′-positions, enabling O-ATRP [36]. The "living" nature of the polymerization system was clearly demonstrated through kinetic analysis, chain extension, and precise "on–off" light-control experiments.

Furthermore, a dual-function photoredox phosphoric acid catalyst was designed using the same backbone, enabling O-ATRP to be proceeded orthogonally with ring-opening polymerization (ROP) [68].

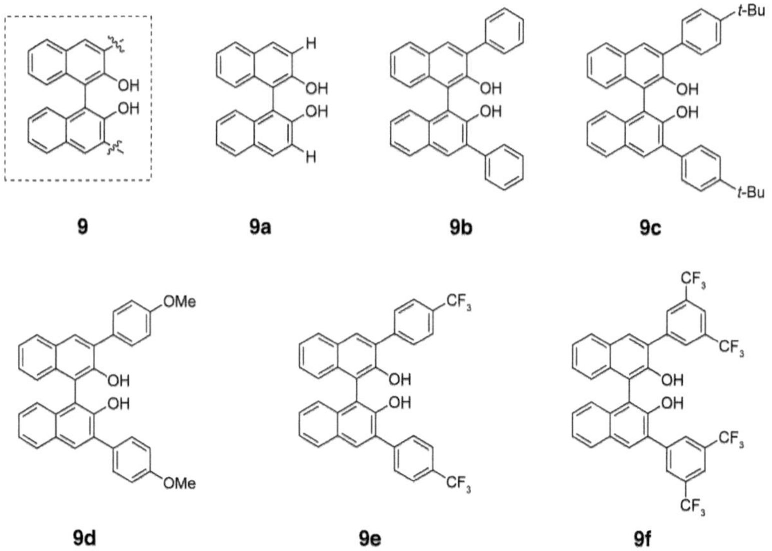

Scheme 12.12 1,1'-Bisnaphthol Derivatives.

In 2020, by integrating theoretical design and experimental validation, a novel class **10** (Scheme 12.13) was shown to be highly effective for photocontrolled O-ATRP of both MMA and styrene [69]. Using this catalyst under white lights yielded polymers with cross-linked characteristics, via an OQP mechanism. Following this, Kang's group developed analogues **10a-10d** (Scheme 12.13) [37].

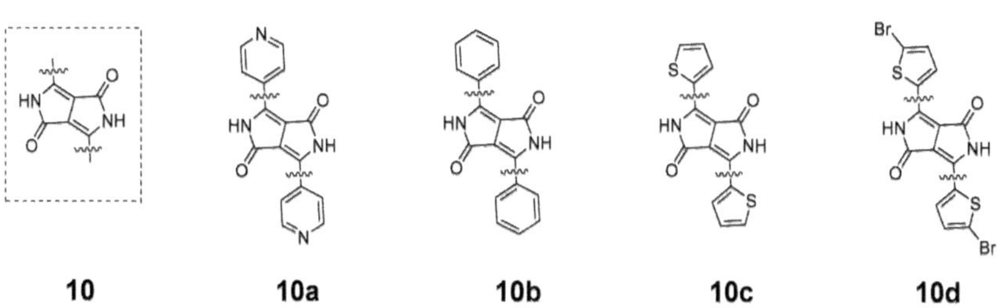

Scheme 12.13 *N*-unsubstituted Diketopyrrolopyrrole Derivatives.

PCs based on the 2-phenylpyrimidine framework are shown in Scheme 12.14. These PCs **11a-11c** were successfully applied to the O-ATRP of methacrylic monomers, producing methacrylic polymers with *Đ* below 1.3 at 50 ppm loading [40], as well as effective synthesis of block copolymers.

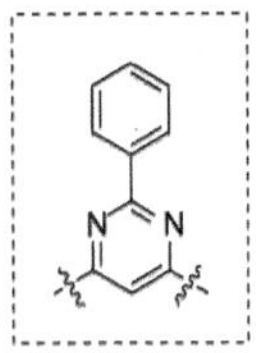

11

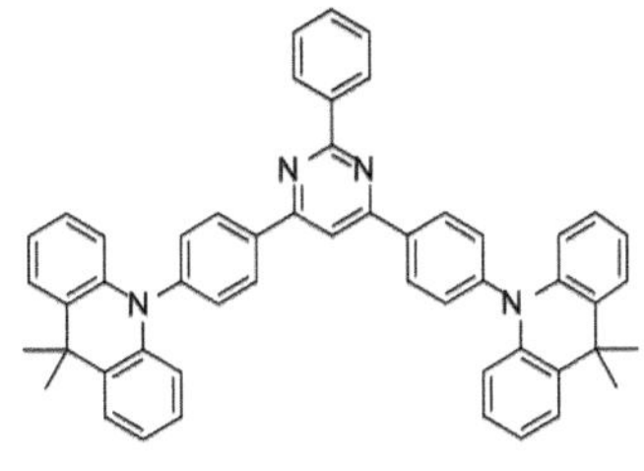

11a

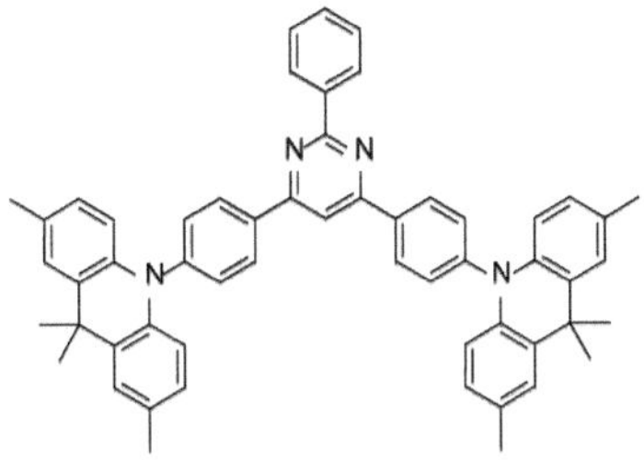

11b

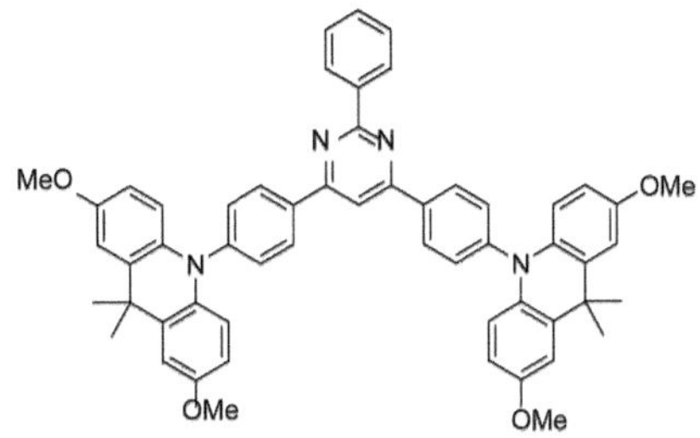

11c

Scheme 12.14 2-Phenylpyrimidine Derivatives.

12.3 Photoinduced Electron/Energy Transfer Reversible Addition-Fragmentation Chain Transfer (PET-RAFT)

12.3.1 Mechanism of PET-RAFT

As with O-ATRP, the PET-RAFT polymerization can also operate via multiple different mechanisms. In the simplest case, where only the PC is present as recoverable catalytic species, OQP and TET are both possible [26, 70, 71]. Such an uncertainty of which mechanism dominates PET-RAFT is unique, which stems from the fact that the initiator, specifically an RAFT agent, is a chemical with conjugated core structure. Indeed, the conjugated nature of the RAFT agent can accommodate low-energy excited states, making TET a possible channel in activation. Generally speaking, whether PET-RAFT predominately operates via a certain mechanism remains an open question to date, but several certainties exist:

(1) For OQP, the excited PC loses an electron to generate $PC^{\bullet+}$ which can form an ion-pair $PC^{\bullet+}/RAFT^{\bullet-}$ as with O-ATRP. This feature has not been validated by experimental results in the exact context of PET-RAFT. But if a PC-RAFT agent complex exists, whether OQP or TET happens is only a matter of charge distribution within the PC-RAFT agent complex.
(2) The possibility of TET expands the scope of available PCs, which provides additional factors to tune a PC in PET-RAFT to its optimum.

(3) In all PET-RAFT systems, rapid cessation of polymerization is observed upon removal of light irradiation [72], meaning that the uncertainty regarding whether an OQP or TET mechanism is operative only has an importance in theoretical understanding rather than actual implementation.

While having a homolytic bond cleavage step just like that in photoiniferter systems [73], the only systems that have been experimentally validated to adopt a TET mechanism, are those with Ir-/Ru-based PCs, corroborated by transient absorption spectroscopy as well as theoretical calculations [74, 75]. On the contrary, Falvey and coworkers challenged the capability of many RAFT agents in possessing sufficient triplet excitation energies for homolytic bond cleavage, based on laser flash photolysis [76], from which they concluded that only OQP is possible for the discussed cases, encompassing both dithiobenzoates and trithiocarbonates. However, it is deemed that whether TET or OQP dominates in different cases still need cross-validation from both Stern-Volmer/spectroscopy experiments and reaction path calculations [77].

One typical feature of PET-RAFT is that the MW homogeneity is optimized by chain transfer reactions, which makes the apparent propagation rate constant (k_p^{app}) to be the dominant performance of concern. In practice, Konkolewicz et al. demonstrated that variations of light intensities, chemical compositions, as well as geometries of reaction vessels exert very limited impact on the polymerization control of PET-RAFT systems [78].

12.3.2 Porphyrin Derivatives

12c (Scheme 12.15), although technically a transition-metal complex, does not display the typical metal-to-ligand charge transfer (MLCT) states due to zinc's inertness. Instead, its excitation remains localized within the structure upon light absorption [71]. Boyer and his team investigated the potential of **12c** as a photocatalyst for PET-RAFT polymerization for various monomers such as acrylates, acrylamides, and styrene. **12c** was found to enhance oxygen tolerance in these reactions by enabling triplet-triplet annihilation (TTA) that generates singlet oxygen ($^1O_2$), which is then quenched by either monomers or solvents [79, 80]. Furthermore, Boyer's group introduced modified versions of **12c** that are water-soluble, allowing for its use in aqueous PET-RAFT polymerization [9, 81].

In their latest work, Boyer and Wu introduced a porphyrin-based PC, **12f** (Scheme 12.15). **12f** outperformed **12c** as a PC, achieving a k_p^{app} of 0.138 min^{-1}, compared to 0.118 min^{-1} for **12c** [82]. This increased efficiency is likely due to **12f**'s higher extinction coefficient around 600 nm.

The introduction of benzo groups on each pyrrole ring of **12c** leads to the formation of **12g** (Scheme 12.15). This structural modification not only extends the conjugation length of the system but also breaks the accidental degeneracy between the a_{1u} and a_{2u} orbitals, the two highest occupied orbitals of the molecule [83]. Consequently, **12g** exhibits significant hyperchromism and bathochromic shift in the

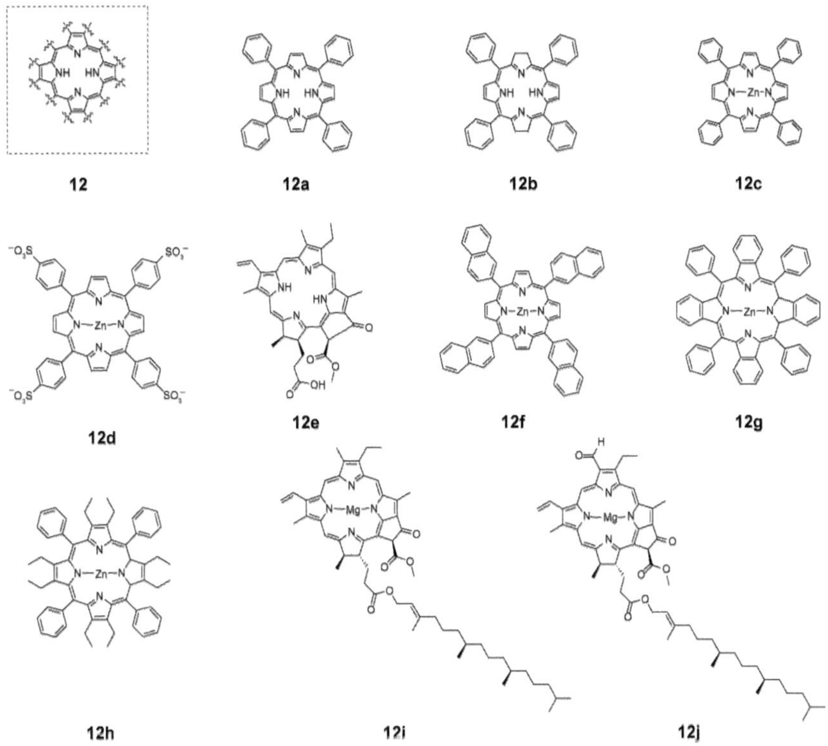

Scheme 12.15 Porphyrin Derivatives.

Q-band, without intensifying the less practically useful short-wavelength absorption peaks. Compared to **12c**, **12g** demonstrated a 49% improvement in k_p^{app}.

Holten and colleagues explored porphyrin derivatives with conformational distortions, revealing that the core-twisting of the macrocycle, caused by steric interactions from bulky peripheral substituents, has a significant impact on the S_1 state [84]. Compound **12h** (Scheme 12.15) is a porphyrin dye featuring core-twisting, with its planar equivalent being **12c**. As reported, the experimentally determined structure of **12h** demonstrates a saddle-shaped, nonplanar macrocycle, driven by steric hindrance between ethyl and phenyl groups [85, 86]. Such structural variation results in a substantial red shift in light absorption of ~85 nm, from **12c** [42] to **12h** [87].

12i, [88] **12j**, [89] and **12e** (Scheme 12.15), [90] are among natural PCs studied for PET-RAFT systems. Both **12i** and **12j** are effective PCs, yet **12e** showed limited compatibility, functioning effectively with dithiobenzoates for methacrylate polymerization but failing with acrylates when trithiocarbonates were used. Also, the free-base **12a** efficiently absorbed red light and showed similar activity, working well with dithiobenzoates but not with trithiocarbonates [42]. A significant advancement was made by Zhang and colleagues, who developed a reduced porphyrin **12b**. This new PC proved effective in PET-RAFT polymerization of both acrylates and methacrylates, with strong absorption extending into the far-red to near-infrared (NIR) spectrum [91].

12.3.3 Fluorescein Derivatives

The Boyer's group identified fluorescein derivatives **13a-13g** (Scheme 12.16) as efficient PET-RAFT PCs [92, 93]. Among them, **13f** emerged as a sensitive, dual-responsive (pH/light) organic PC for the polymerization of acrylates, acrylamides, methacrylates, and methacrylamides [93]. Additionally, **13g**, featuring four bromine and four iodine atoms, was developed to maximize oxygen tolerance, achieving the highest level of oxygen tolerance among synthetic fluorescein derivatives [92].

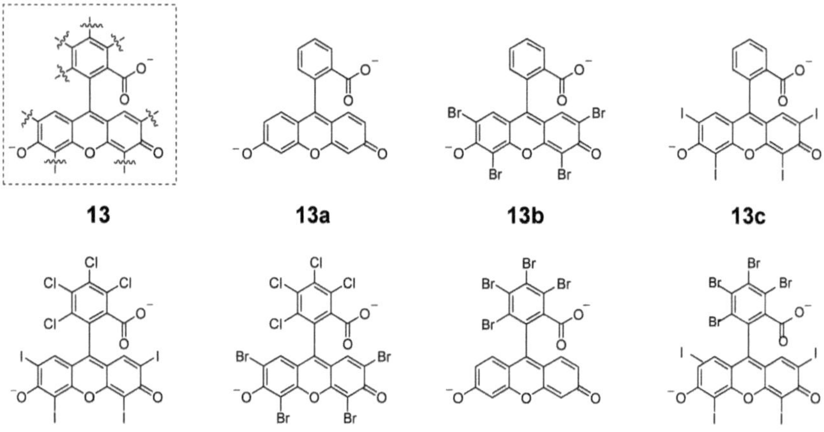

13 13a 13b 13c 13d 13e 13f 13g

Scheme 12.16 Fluorescein Derivatives.

12.3.4 Phthalocyanine Derivatives

Phthalocyanine derivatives **14a-14e** (Scheme 12.17) are employed in PET-RAFT polymerization via O-RQP, whose absorption peaks cover wavelengths from orange to near-infrared light [47]. For example, the implementation of **14e** expanded PET-RAFT polymerization to deep-red (~700 nm) and near-infrared (780 nm) light.

12.3.5 Other Classes

Several other classes of PCs have also been utilized in PET-RAFT polymerization, many of which originate from those used in O-ATRP. This overlap is due to the chain transfer mechanism in PET-RAFT, which introduces an additional level of control over MW. As a result, PCs that perform well in O-ATRP are also well-suited for PET-RAFT polymerization, leveraging similar characteristics for effective polymerization processes.

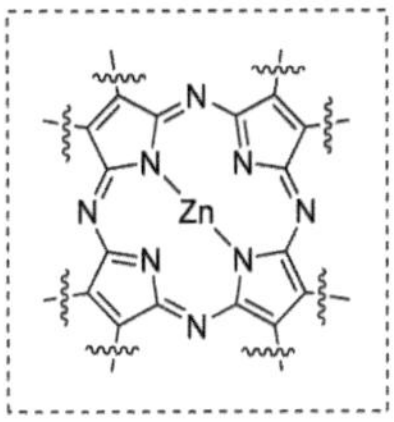

14

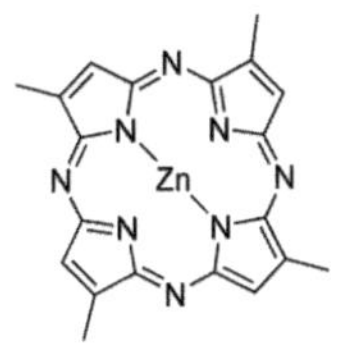

14a

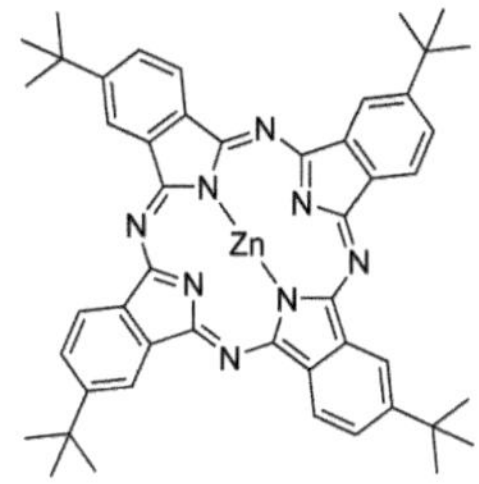

14b

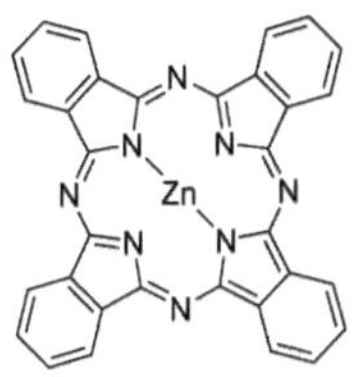

14c

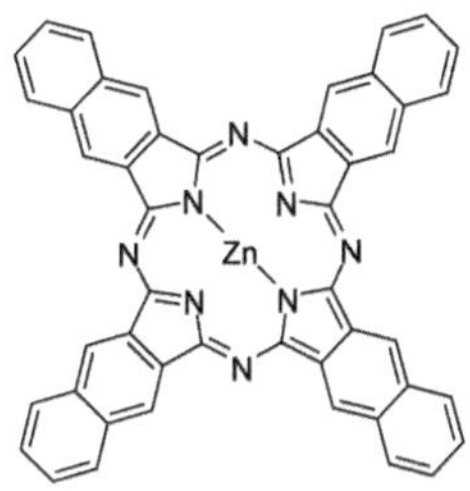

14d

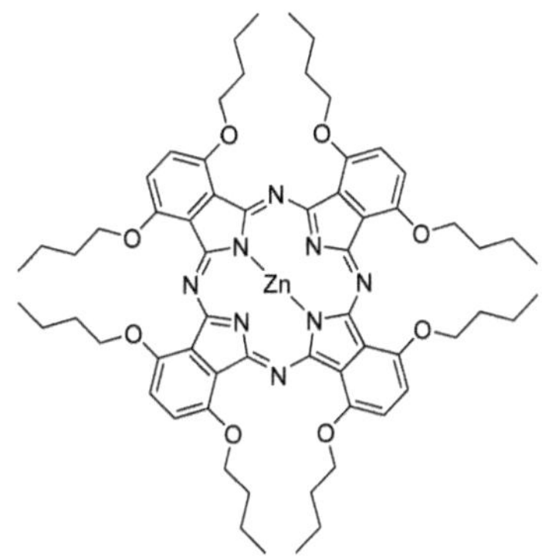

14e

Scheme 12.17 Phthalocyanine Derivatives.

12.4 Structure-Property-Performance Relationships

By virtue of these diverse structures of organocatalysts, a great variety of properties are tunable [26, 94], including light absorption (e.g. maximum absorption wavelength, λ_{max}, molar extinction coefficient at λ_{max}, ε_{max}, and full width at half maximum, FWHM; detailed in Sec. 12.4.2), excited-state evolution (e.g. ISC rate constant, k_{ISC}, triplet state lifetime, τ_T, and triplet quantum yield, Φ_T; detailed in Sec. 12.4.3), along with activation and deactivation efficiencies represented by energy barriers, reaction rates, or alternatively reduction/oxidation potentials (detailed in Secs. 12.4.4 and 12.4.5).

12.4.1 Structure–Property-Relationships

From the above analysis of photo-RDRP performances, it is deduced that only a few properties control all these performances. The key properties include ε_{max} at λ_{max} in photon absorption (Sec. 12.4.2), Φ_T and τ_T in excited-state evolution (Sec. 12.2.3), as well as activation and deactivation energy barriers (Sec. 4.4 and 4.5) in the following steps. The relationships between properties and performances are either empirically inferred or theoretically concluded, which are more intuitive compared to the structure–property relationships that determine the properties themselves.

Indeed, chemical and geometric structures of a molecule determine its electronic structures of all possible states, from which all kinds of molecular properties can be derived. The question is then, how ligands and the chromophore core should be chosen, so as to bring forth the designated properties. Decisive principles to this subject, namely the rational design of PCs, can be deduced as structure–property relationships as will be manifested in theories, mechanistic formulas, and equations presented in Secs. 12.4.2–12.4.5.

12.4.2 Photon Absorption

When exposed to light, the PC engages in a resonant interaction with the electromagnetic field E_0 of the light wave (angular frequency ω). In the ultraviolet and visible wavelength ranges, this interaction is dominated by the contribution of the electric dipole moment operator $\hat{\mu}$, which triggers the absorbance of a photon into the PC through an electronic transition. The perturbation operator for this transition can be represented as $\hat{H}' = -\hat{\mu} \cdot E_0 \cos(\omega t)$.

12.4.2.1 Theory for Absorption Spectrum

To analyze this transition, the point of departure is the Fermi golden rule [95, 96] for the rate of photon absorption:

$$\sigma_{\mathrm{abs}}(\omega) = \frac{4\pi^2\omega}{3\hbar c}\sum_{v,v'} P_{iv}(T)\ |\langle \Phi_f \Theta_{fv'}\ |\ \hat{\mu}|\Phi_i \Theta_{iv}\rangle|^2 \delta(\overline{\omega}_{fi} - \omega) \tag{12.1}$$

$$= \frac{4\pi^2\omega}{3\hbar c}|\langle \Phi_f|\hat{\mu}\ |\Phi_i\rangle|^2 \rho_{\mathrm{abs}}(\omega), \tag{12.2}$$

$$\rho_{\mathrm{abs}}(\omega) = \sum_{v,v'} P_{iv}(T)\ |\langle \Theta_{fv'}|\Theta_{iv}\rangle|^2 \delta(\overline{\omega}_{fi} - \omega), \tag{12.3}$$

$$\overline{\omega}_{fi} = \omega_{fi} + \omega_{fv'} - \omega_{iv}, \omega_{xy} = \frac{E_{xy}}{\hbar}, E_{xy} = E_x - E_y. \tag{12.4}$$

Here, Φ_i and Φ_f represent the initial and final electronic states of energies E_i and E_f, respectively, whereas $\{\Theta_{iv}\}$ and $\{\Theta_{fv'}\}$ are the corresponding vibrational states, respectively; ω is the angular frequency of the absorbed photon. The parametrical dependence of the transition dipole moment $langle\Phi_f|\hat{\mu}|\Phi_i\rangle$ on the nuclear coordinates has been neglected when going from Eq. (12.1) to Eq. (12.2), in accordance with the Condon approximation. The expression $\rho_{\mathrm{abs}}(\omega)$ in Eq. (12.3) is then the Franck-Condon weighted density of states (FCWD) in which $P_{iv}(T)$ is the Boltzmann distribution function for the vibrational states of the initial electronic state; $\langle \Theta_{fv'}|\Theta_{iv}\rangle$ is the overlap between the vibrational wave functions of the final and initial electronic states, whereas the δ function plays the role of resonance condition (i.e. the transitions could occur only when $\omega \approx \overline{\omega}_{fi}$).

The FCWD $\rho_{\mathrm{abs}}(\omega)$ is generally a very complicated function but whose main feature can be understood from the displaced multi-mode harmonic oscillator model (DHO), which reads [97, 98]:

$$\rho_{\mathrm{abs}}(\omega) = \frac{1}{2\pi}\int_{-\infty}^{\infty} dt e^{it(\omega_{fi}-\omega)-g(t)}, \tag{12.5}$$

$$g(t) = \sum_k S_k \left[(2\overline{n}_k + 1) - (\overline{n}_k + 1)e^{it\omega_k} - \overline{n}_k e^{-it\omega_k}\right], \tag{12.6}$$

$$S_k = \frac{\omega_k d_k^2}{2\hbar}, \tag{12.7}$$

$$\overline{n}_k = \frac{1}{e^{\beta\hbar\omega_k} - 1}, \beta = \frac{1}{k_B T}, \tag{12.8}$$

where $\{-d_k = Q_{fk} - Q_{ik}\}$ are the displacements of the normal coordinates $\{Q_{fk}\}$ of the final state relative to those $\{Q_{ik}\}$ of the initial state; ω_k is the angular frequency of one of the vibration modes k. $\{S_k\}$ Eq. (12.7) are the dimensionless Huang-Rhys factors characterizing the coupling strength between the electronic and nuclear degrees of freedom. $g(t)$ Eq. (12.6) is the line shape function. It is instructive to look at the simplest case of $\rho_{abs}(\omega)$ with $T = 0$ (i.e. $\overline{n}_k = 0$) and only one vibration mode of frequency ω_0:

$$\rho_{abs}(\omega) = \frac{1}{2\pi}\int_{-\infty}^{\infty} dt e^{it(\omega_{fi}-\omega)-S[1-e^{it\omega_0}]}, S = \frac{\omega_0 d^2}{2\hbar} \tag{12.9}$$

$$= \frac{1}{2\pi}e^{-S}\sum_{n=0}^{\infty}\frac{S^n}{n!}\int_{-\infty}^{\infty} dt e^{it(\omega_{fi}-\omega+n\omega_0)} \tag{12.10}$$

$$= e^{-S}\sum_{n=0}^{\infty}\frac{S^n}{n!}\delta(\omega_{fi} - \omega + n\omega_0), \tag{12.11}$$

which shows that the spectrum is a vibrational progression accompanying the electronic transition at ω_{fi}, separated by ω_0 with a Poisson distribution of intensities. It is clear that the absorption peak appears at the electronic energy gap ω_{fi} in the weak coupling regime ($S < 1$), whereas the absorption peak occurs at $n \approx S$ in the strong coupling regime ($S > 1$). Therefore, S can in the latter case be interpreted as the mean number of vibrational quanta excited in the final electronic state. This is just the Franck-Condon principle, stating that transition intensities are dictated by the vertical overlap between nuclear wave functions of the initial and final electronic states.

12.4.2.2 Full Width at Half Maximum

It is seen from Eq. (12.7) that a larger displacement d brings about a larger S, thus broadening the absorption peak. The displacement d, more generally speaking, is how much the equilibrium geometries in the ground and excited states differ from each other. It follows that rigid molecules (whose equilibrium geometries in the ground and excited states are close) usually exhibit narrower absorption peaks, and thereby smaller FWHM of the absorption peak, than flexible or soft molecules [47, 83].

On the other hand, for molecules with significantly different excited-state and ground-state characters, such as those exhibiting a low-lying CT state, the corresponding absorption and emission peaks have notably large FWHMs, which is due to the different equilibrium geometries from distinct wavefunctions of the two states. Since CT states can be routinely constructed from organic D–A scaffolds, this is a rather effective strategy for broadening absorption peaks in organic PCs [55, 99]. In practice, a broad peak with large FWHM is beneficial for light absorption from wide spectrum light sources (such as solar light and compact fluorescent lamp).

12.4.2.3 Maximum Absorption Wavelength

The energy $\hbar\omega$ of the absorbed photon has to closely approach the energy gap $\hbar(\omega_{fi} + n\omega_0)$ (i.e. $\omega \approx \omega_{fi} + n\omega_0$) for some $n \geq 0$ to yield appreciable $\alpha_{abs}(\omega)$, as revealed in Eqs. (12.2), (12.4) and (12.5). As ω_0 is usually much smaller than ω_{fi}, and $\rho_{abs}(\omega)$ decays factorially when n is large (Eq. (12.11)), this means that ω has to be close to ω_{fi} in order to have efficient light absorption. The maximum of the peak appears at $\lambda_{max} = \frac{2\pi c}{\omega_{fi}+n\omega_0}$, where $n = 0$ at $S < 1$ (weak coupling regime) or $n \approx S$ at $S > 1$ (strong coupling regime). As such, λ_{max} is directly related to $\omega_{fi} = E_f - E_i$. Since the excitation energy $\hbar\omega_{fi}$ is determined by the orbitals from and to which the electron is excited, λ_{max} can be tuned by raising or lowering the energies of the relevant orbitals, mostly being π and π* in organocatalysts [83, 91, 100–103]. Choosing ligands that allow easy manipulation of π and π* orbital energies is pivotal to λ_{max} manipulation [47, 104].

In general, a short absorption wavelength brings the PC to a high-lying state, resulting in higher reactivity, while long absorption wavelengths are desired for the efficient utilization of long-wavelength light in the light source (especially when the light source is solar light), and the resulting low-lying excited states may lead to better reaction selectivity. However, it should be noted that the energy available to the PC for subsequent reactions is smaller than the absorbed photon energy, due to excited-state geometry relaxation. For example, for absorption and emission related to CT excited state, the emission peak is usually much more bathochromic compared to the absorption peak, marking a considerable loss of the excitation energy as heat before it can be utilized in subsequent reactions. Such energy waste is unwanted as to photocatalysis.

12.4.2.4 Molar Extinction Coefficient

As to the molar extinction coefficient ε_{max} at the maximum absorption wavelength λ_{max}, it has a formula of:

$$\varepsilon_{max} \propto \frac{f_{i\to f}}{\Delta\overline{\nu}_{FWHM}} \tag{12.12}$$

$$f_{i\to f} = \frac{4m_e\pi c}{3e^2\hbar\lambda_{max}} |\langle\Phi_f|\hat{\mu}|\Phi_i\rangle|^2 \tag{12.13}$$

$$\langle\Phi_f|\hat{\mu}|\Phi_i\rangle = \int \Phi_f^*(r)\,\hat{\mu}\,\Phi_i(r)dr \tag{12.14}$$

where $f_{i\to f}$ is the oscillator strength, which is correlated to the probability of the electronic transition under light; $\Delta\overline{\nu}_{FWHM}$ is the full width at the half maximum of the absorption peak; m_e is the mass of an electron and e is the charge of an electron; r is the position vector representing any point in the space. $\langle\Phi_f|\hat{\mu}|\Phi_i\rangle$ is the electronic transition dipole moment, which is an integral in the whole space shown as Eq. (12.14). It is obvious that the magnitude of ε_{max} is proportional to the square of $|\langle\Phi_f|\hat{\mu}|\Phi_i\rangle|$. The increase of the latter value, particularly in transitions such as

$S_0 \rightarrow S_1$, is associated with a higher spatial overlap of the orbitals which the excited electron transits in between. Meanwhile, since $|\langle\Phi_f|\hat{\mu}|\Phi_i\rangle|$ is not only proportional to the wavefunction overlap but also the dipole operator $\hat{\mu}$, delocalized orbitals give larger transition dipole moments than localized orbitals for a given level of orbital overlap.

Usually, for a typical $\pi \rightarrow \pi^*$ excitation, larger π conjugation of the chromophore scales up the distribution of the transition density $\rho_{i\rightarrow f}(r) = \Phi_f^*(r)\,\Phi_i(r)$, rendering the amplitude of its dipole moment $|\langle\Phi_f|\hat{\mu}|\Phi_i\rangle|$ larger and hence enhancing ε_{max}. Moreover, to achieve a large value of $|\langle\Phi_f|\hat{\mu}|\Phi_i\rangle|$ and consequently a high ε_{max}, it is essential to avoid spatial-symmetry-forbidden transitions.

It is noted that, under the Born-Oppenheimer approximation, electronic structures are analyzed at fixed molecular geometries. However, interconversion between electronic states, such as absorption of light by photoexcitation, is ultimately accompanied by nuclear motions. It is noted that in Eqs. (12.2) and (12.13), the $\langle\Phi_f|\hat{\mu}|\Phi_i\rangle$ is still vibrationally independent. This is not accurate, especially in situations involving notable vibrational contributions. The most viable solution is to account for the Herzberg-Teller effect by including the first-order derivative of $\langle\Phi_f|\hat{\mu}|\Phi_i\rangle$ with respect to the normal mode coordinate [105]. The Herzberg-Teller effect is particularly significant when photoexcitation is symmetry-forbidden. In such cases, nuclear vibrations cause variations in the symmetric structure, enabling photoinduced electronic transitions that are otherwise forbidden at the equilibrium structure. Consequently, introducing more degrees of freedom in the ligands of the chromophore facilitates the absorption of light by dark states. Actually, the Herzberg-Teller effect also applies to the overall wave function $\langle\Phi_f\Theta_{fv'}|\hat{\mu}|\Phi_i\Theta_{iv}\rangle$, not just the electronic part $\langle\Phi_f|\hat{\mu}|\Phi_i\rangle$. Thus, the first-order derivative of $\langle\Phi_f\Theta_{fv'}|\hat{\mu}|\Phi_i\Theta_{iv}\rangle$ with respect to the normal mode coordinate can be separated into $\langle\Phi_f|\hat{\mu}|\Phi_i\rangle$ and the thermal vibrational correlation function, which has a different formula for the Herzberg-Teller effect. This effect is ubiquitous in a wide variety of systems, especially in macrocyclic molecules known for their strong Herzberg-Teller effect.

In macrocyclic molecules, such as porphyrins and phthalocyanines, excited-state degeneracy arising from molecular orbital degeneracy leads to additional structure–property relationships. Specifically, metalloporphyrins suffer from a low-absorption Q-band due to the cancellation of the transition dipole moments associated to the equally contributed $a_{1u} \rightarrow e_g(1)$ and $a_{2u} \rightarrow e_g(2)$ configurations in one of the degenerate states in the Q-band, and that for $a_{1u} \rightarrow e_g(2)$ and $a_{2u} \rightarrow e_g(1)$ in another. This cancellation results from two factors: (i) the accidental degeneracy of a_{1u} and a_{2u}, and (ii) the different orientations of the degenerate $e_g(1)$ and $e_g(2)$. To break this cancellation and enable strong Q-band absorption, the most viable approach is to introduce ligands to the α- and β-positions to tune the energy levels of a_{1u} and a_{2u}, eliminating the accidental degeneracy and allowing one configuration to dominate in the two states of the Q-band [83]. Another viable approach is to reduce one or a pair of opposite peripheral double bonds in the pyrrole, which moves only one of $e_g(1)$ and $e_g(2)$ up, again causing only one configuration to dominate in the two states of the Q-band [106].

12.4.3 Excited-State Evolution

12.4.3.1 Quantum Yield for Long-Lived State

When a photon is absorbed, as depicted in Scheme 12.18, it enables the transition from S_0 to a singlet excited state, S_n. When $n > 1$, this is usually followed by a rapid internal conversion (IC), which typically relaxes the system to the most stable singlet excited-state S_1. Subsequently, S_1 may undergo ISC to transition into a triplet state T_n, thus reaching the triplet manifold.

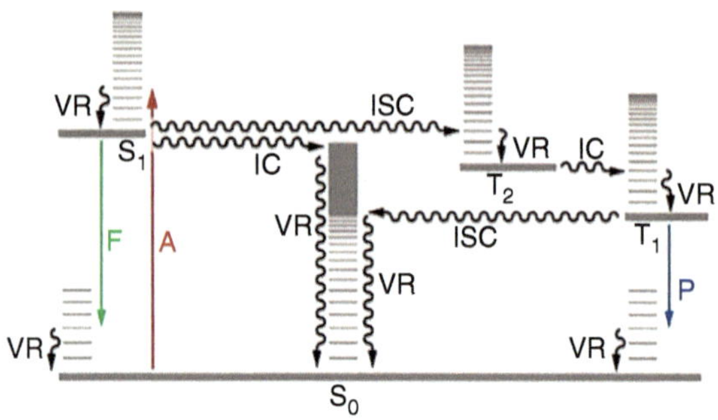

Scheme 12.18 Jablonski Diagram of Excited-State Evolutions.
A: absorption. F: fluorescence. P: phosphorescence. IC: internal conversion. ISC: intersystem crossing. VR: vibrational relaxation.

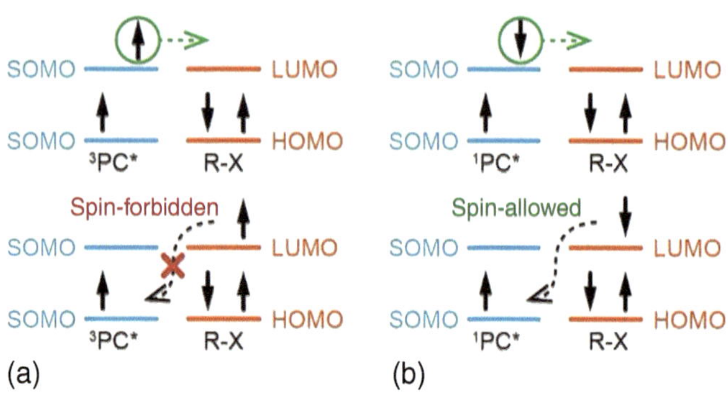

Scheme 12.19 Representations for Electron Transitions in Molecular Orbitals.
Note: (a) Electron transfers from $^3PC^*$ to R–X (above) where the back electron transfer is spin-forbidden (below). (b) Electron transfer from $^1PC^*$ to R–X (above) where the back electron transfer is spin-allowed (below). HOMO: highest occupied molecular orbital. LUMO: lowest unoccupied molecular orbital. SOMO: singly occupied molecular orbital.

In dilute solutions or under low PC concentrations, the excited state of PC must have a sufficient lifetime to diffuse through the solution and interact with the initiator. This interaction is crucial for activating polymerization via bimolecular mechanisms [26, 77, 107, 108]. T_1, which is accessed via ISC, is generally more stable than other excited states. This stability arises from the spin-forbidden nature of the transition between T_1 and the ground state S_0, preventing side reactions that would otherwise return the species to a dormant state, as illustrated in Scheme 12.19A. Considering the vital role of the long-lived T_1 in photo-RDRP, particularly in catalytic cycles that require the T_1 state such as TET in Scheme 12.1B and O-RQP in Scheme 12.1D, achieving a significant population of T_1 turns out to be a premise. As indicated in Scheme 12.18, this enhanced quantum yield of T_1 (Φ_T) is achieved by enhancing ISC of $S_1 \rightarrow T_n$ (followed by IC of $T_n \rightarrow T_1$), and in the meantime suppressing IC and fluorescence of $S_1 \rightarrow S_0$. In this vein, fluorescence, IC, and ISC are analyzed individually in the following sections. The underlying goal of these sections is to identify strategies for maximizing Φ_T [26, 44].

Conversely, the transition from S_1 to S_0 is spin-allowed, as illustrated in Scheme 20B, enabling deexcitation to the ground state either before or after electron transfer, through IC, fluorescence, and back electron transfer, respectively. These unproductive pathways reduce the importance of electron transfer from the S_1 state compared to that from the T_1 state. Nevertheless, a high concentration of PC allows massive encounters between PC and initiator before the S_1 state of the PC decays into the T_1 state, such that S_1 may exert a significant impact when the activation reaction is fast enough to fit within the bounds of the S_1 lifetime; however, this is only possible for scenarios where triplet states are not mandatory, as with OQP and RQP.

12.4.3.2 Theory for Fluorescence

The FGR [93, 94] for the emission cross-section $\sigma_{em}(\omega)$ reads

$$\sigma_{em}(\omega) = \frac{4\omega^3}{3\hbar c^3} |\langle \Phi_f | \hat{\mu} | \Phi_i \rangle|^2 \rho_{em}(\omega), \tag{12.15}$$

$$\rho_{em}(\omega) = \sum_{v,v'} P_{iv}(T) |\langle \Theta_{fv'} | \Theta_{iv} \rangle|^2 \delta(\overline{\omega}_{fi} + \omega) \tag{12.16}$$

under the Condon approximation. Again, $\hat{\mu}$ is here the electric dipole moment operator. The fluorescence rate constant k_F is defined as

$$k_F = \int_0^\infty \sigma_{em}(\omega) d\omega. \tag{12.17}$$

It is clear that $\sigma_{em}(\omega)$ differs from $\sigma_{abs}(\omega)$ in Eq. (12.2) only in the dependency on ω. In particular, replacing $-\omega$ with $+\omega$ in Eqs. (12.5), (12.9)–(12.11), (12.19), and (12.21) will give rise to the corresponding expressions for $\rho_{em}(\omega)$ under the DHO approximation. The magnitude of transition dipole moment $\langle \Phi_f | \hat{\mu} | \Phi_i \rangle$ is expected to be appreciable when the two electronic wavefunctions have a large absolute overlap. The magnitude of the $S_1 \rightarrow S_0$ electric transition dipole moment $|\langle \Phi_{S_0} | \hat{\mu} | \Phi_{S_1} \rangle|$ as the perturbation matrix element becomes larger with stronger mixing of electronic wave functions of S_1 and S_0, which happens when the spatial overlap of the pair of orbitals that are dominant in the electronic transition becomes better. Overall, an increased fluorescence decay rate constant $k_{f,\, i\rightarrow f}$ resulting from a larger value of $\alpha_{em,\, i\rightarrow f}(\omega)$ can be attained by ensuring substantial spatial overlap between the orbitals during the electronic transition. Similar to absorption, the fluorescence process also exhibits the Herzberg-Teller effect, where symmetry-forbidden transitions can occur with fluorescence due to nuclear motions. Additionally, breaking the accidental degeneracy between a_{1u} and a_{2u} or the degeneracy between $e_g(1)$ and $e_g(2)$ can enable Q-band fluorescence just as it does for absorption [109, 110].

12.4.3.3 Theory for Internal Conversion

Nonradiative IC occurs due to the interaction between electronic and nuclear motions, known as vibronic coupling. Typically, this vibronic coupling is represented by the Born-Oppenheimer coupling operator $\hat{H}'_{BO}$, which symbolizes the perturbation resulting from the deviation from the Born-Oppenheimer approximation [111]:

$$\hat{H}'_{BO}(\Phi_i \Theta_{iv}) \approx \sum_n -\hbar^2 \frac{\partial \Theta_{iv}}{\partial Q_n} \frac{\partial \Phi_i}{\partial Q_n}. \tag{12.18}$$

Here, Q_n represents the normal coordinates associated with the nth vibrational mode.

Consequently, the rate constant for IC under the Condon approximation can be articulated as follows:

$$k_{IC,iv \to f} = 2\pi\hbar^3 \left| \left\langle \Phi_f \middle| \sum_n \frac{\partial}{\partial Q_n} \middle| \Phi_i \right\rangle \right|^2 \rho_{IC},$$

$$\rho_{IC} = \sum_{v'} \left| \left\langle \Theta_{fv'} \middle| \sum_n \frac{\partial}{\partial Q_n} \middle| \Theta_{iv} \right\rangle \right|^2 \delta(\overline{\omega}_{fi}). \tag{12.19}$$

The vibronic coupling term $\left\langle \Phi_f \middle| \sum_n \frac{\partial}{\partial Q_n} \middle| \Phi_i \right\rangle$ represents the overlap of Φ_f with the first derivative of Φ_i with respect to all of the normal modes Q_n. Based on and Hellmann-Feynman theorem [112–115], this term is expressed with a complete basis set as follows:

$$\left\langle \Phi_f \middle| \sum_n \frac{\partial}{\partial Q_n} \middle| \Phi_i \right\rangle = \frac{\left\langle \Phi_f^0 \middle| \left(\sum_n \frac{\partial \hat{V}}{\partial Q_n} \right)_0 \middle| \Phi_i^0 \right\rangle}{E_f^0 - E_i^0} \tag{12.20}$$

In this context, Φ_i^0 refers to the electronic wave function Φ_i at the equilibrium structure of the ground state, while E_i^0 denotes the electronic energy of Φ_i^0. The term $\hat{V}$ represents the operator for the Coulomb interaction potential between electrons and the nucleus.

With regard to the $S_1 \to S_0$ IC, the rate constant can be expressed as

$$k_{IC} = \frac{2\pi\hbar^3}{\left|\Delta E^0_{S_0-S_1}\right|^2} \left| \left\langle \Phi^0_{S_0} \middle| \left(\sum_n \frac{\partial \hat{V}}{\partial Q_n} \right)_0 \middle| \Phi^0_{S_1} \right\rangle \right|^2 \rho_{IC} \tag{12.21}$$

where $\Delta E^0_{S_0-S_1}$ is the vertical energy gap between S_1 and S_0 at the equilibrium structure of the ground state S_0. A larger energy gap $\Delta E^0_{S_0-S_1}$ results in a lower rate constant $k_{IC,S_1 \to S_0}$ according to Eq. (12.21), which makes $k_{IC,S_1 \to S_0}$ more responsive to changes in $\Delta E^0_{S_0-S_1}$, compared to the fluorescence rate constant $k_{f,S_1 \to S_0}$.

The squared magnitude of the vibronic coupling matrix element $\left|\left\langle \Phi^0_{S_0} \left| \left(\sum_n \frac{\partial \hat{V}}{\partial Q_n}\right)_0 \right| \Phi^0_{S_1} \right\rangle\right|^2$, increases with enhanced mixing of the electronic wave functions of S_1 and S_0. This increase is due to the improved spatial overlap of the dominant orbital pair involved in the electronic transition. Moreover, $\frac{\partial \hat{V}}{\partial Q_n}$ signifies the force driving the conversion of potential energy, $\hat{V}$, along the mode Q_n into vibrational energies. Consequently, flexible molecules such as those bearing twisting phenyl rings, have stronger nuclear motions and tend to exhibit a larger $\left|\left\langle \Phi^0_{S_0} \left| \left(\frac{\partial \hat{V}}{\partial Q_p}\right)_0 \right| \Phi^0_{S_1} \right\rangle\right|^2$ and a higher IC rate constant $k_{IC,S_1 \to S_0}$.

12.4.3.4 Theory for Intersystem Crossing

For ISC scenarios involving electronic transitions between singlet and triplet manifolds, the perturbation operator, denoted as $\hat{H}'$, is specifically defined as the spin-orbit coupling (SOC) operator $\hat{H}_{SO}$ [116]:

$$\hat{H}_{SO} = \frac{1}{2m_e^2c^2} \sum_n \sum_j \frac{Z_n^{eff}}{|\hat{r}_{jn}|^3} \hat{L}_{jn} \cdot \hat{s}_j$$
$$\hat{L}_{jn} = \hat{r}_{jn} \times \hat{p}_j \tag{12.22}$$

In this framework, Z_n^{eff} represents the effective nuclear charge of the nth nucleus, considering the two-electron screening effect. The term $\hat{L}_{jn}$ is the orbital angular momentum operator of the jth electron relative to the nth nucleus, while $\hat{r}_{jn}$ denotes the position vector of the jth electron relative to the nth nucleus. Additionally, $\hat{p}_j$ is the linear momentum operator, and $\hat{s}_j$ represents the spin angular momentum operator of the jth electron.

The rate model for ISC, as per the Fermi golden rule, is formulated as follows [117]:

$$k_{ISC} = \frac{2\pi}{\hbar} |\langle \Phi_f | \hat{H}_{SO} | \Phi_i \rangle|^2 \rho_{ISC}$$
$$\rho_{ISC} = \sum_{v'} \left| \langle \Theta_{fv'} | \Theta_{iv} \rangle \right|^2 \delta(\overline{\omega}_{fi}) \tag{12.23}$$

Analysis of Eq. (12.22) reveals that the well-known heavy atom effect on the SOC operator $\hat{H}_{SO}$ is attributed not only to the larger atomic number Z_n when heavier atoms are present in the molecule [92], but also to the fact that the average distance $|\hat{r}_{jn}|$ (the distance of the jth electron relative to the nth nucleus) tends to decrease, bringing electrons closer to the nucleus. As a result, the strength of SOC is approximately proportional to Z_n^4, considering the $\frac{Z_n^{eff}}{|\hat{r}_{jn}|^3}$ term.

The scalar product $\hat{L}_{jn} \cdot \hat{s}_j$ quantifies the interaction between the magnetic spin moment of the jth electron and the magnetic moment generated by its orbital motion around the nth nucleus. While $\hat{s}_j$ influences the spin component and $\hat{L}_{jn}$ affects the spatial component of the electronic wave function, their scalar product, $\hat{L}_{jn} \cdot \hat{s}_j$ can mix wavefunctions with different spin quantum numbers, in a way that conserves total angular momentum. Therefore, the SOC operator $\hat{H}_{SO}$, comprising the sum of these scalar products, is capable of changing the multiplicity of the molecular system and enabling ISC. This mechanism underlies the El-Sayed rule [118], which typically states that effective electronic transitions between S_n and T_n are only observed when there is a change in molecular orbital type, such as from $^1(\pi, \pi^*)$ to $^3(n, \pi^*)$.

To summarize, there are three strategies for rational design of SOC matrix elements [119]. First, the inclusion of heavier atoms increases the squared magnitude of the SOC matrix element $|\langle\Phi_{T_n}|\hat{H}_{SO}|\Phi_{S_1}\rangle|^2$, thereby enhancing $k_{ISC,S_1\rightarrow T_n}$. Second, the El-Sayed rule, which is derived from the characteristics of $\hat{H}_{SO}$, stipulates that a change in molecular orbital types during ISC enhances $k_{ISC,S_1\rightarrow T_n}$, when the number of heavier atoms is fixed. Last, spatial proximity of the orbitals involved in the transitions yields better orbital overlaps, and thus also contributes to an increase in the SOC matrix element and, consequently, a higher $k_{ISC,S_1\rightarrow T_n}$, in direct analogy of how better orbital overlaps increase the transition dipole moments and vibronic coupling vectors.

For the ISC process between S_1 and T_n, the rate constant $k_{ISC,S_1\rightarrow T_n}$ is given by:

$$k_{ISC} = \frac{2\pi}{\hbar}|\langle\Phi_{T_n}|\hat{H}_{SO}|\Phi_{S_1}\rangle|^2\rho_{ISC}. \tag{12.24}$$

ρ_{ISC} shares the same mathematical form with ρ_{IC} and is thus similarly influenced by the energy gap law, where a smaller energy gap $\Delta E_{S_1\rightarrow T_n}$ leads to a faster ISC process and a higher $k_{ISC,S_1\rightarrow T_n}$. This provides another way of tuning $k_{ISC,S_1\rightarrow T_n}$ in addition to tuning the SOC matrix elements [44].

12.4.4 Activation

In principle, both $^1PC^*$ and $^3PC^*$ generated from Steps I and II can serve as the reactant in the activation step III (Scheme 12.1). However, $^3PC^*$ should possess a long lifetime in order to encounter the initiator [77, 107, 108]. This section deals with $^3PC^*$ since it usually bears a much longer lifetime.

The collision of $^3PC^*$ with an initiator R-X, electron donor (e.g. tertiary amine [TA]) or O_2 forms an exciplex, which can occur in different forms symbolized by $PC^{\bullet+}/(R\text{-}X)^{\bullet-}$, $PC/^3(R\text{-}X)^*$, $PC^{\bullet-}/(TA)^{\bullet+}$, or $^3PC^*/^3(O_2)^*$. The forms of ($PC^{\bullet+}/(R\text{-}X)^{\bullet-}$, $PC^{\bullet-}/(TA)^{\bullet+}$) are obtained by electron transfer, while the rest two are obtained by TET. The four forms correspond to four different catalytic cycles as shown in Scheme 12.1. As shown in Sec. 12.2.1, it is obvious that the activation for OQP represents the most straightforward mechanism in photo-RDRP, involving just a PC and an initiator R–X in the catalytic cycle. As a result, the activation in this pathway primarily involves interactions between the PC* and R-X. In this vein, the activation for OQP is analyzed individually in the following sections. The underlying goal of these sections is to identify SPP relationships for maximizing activation efficiency.

Taking the activation step of OQP for example, the PC* and R–X (which are usually present in rather low concentrations) must diffuse through the solution to encounter and then engage in a single electron transfer reaction. This process requires diffusion prior to the reaction, making the corresponding rate constant dependent on the concentrations of both PC* and R-X, thereby classifying it as a second-order reaction. Empirically, the Rehm-Weller equation provides an effective means to express this second-order electron transfer rate constant [77, 120]:

$$k_{\text{act}} = \frac{k_f K_C k_{\text{ET}}}{K_C k_{\text{ET}} + k_f} \tag{12.25}$$

The diffusion rate constant k_f is predominantly influenced by the viscosity of the solvent, as the diffusion of the reactants involves displacing solvent molecules to move through the liquid. When focusing solely on the impact of solvent viscosity, k_f can be expressed as:

$$k_f \approx \frac{8RT}{3\eta} \tag{12.26}$$

Here, η represents the viscosity of the solvent, R is the ideal gas constant, and T denotes the temperature.

The equilibrium constant K_C is particularly noteworthy, as it reflects the stability of the encounter complex PC*/R–X within the equilibrium PC* + R–X ↔PC*/R–X. It is defined by the formula:

$$K_C = \frac{k_f}{k_d} \tag{12.27}$$

where k_d is the dissociation rate constant of the encounter complex PC*/R–X. A higher value of K_C indicates a stronger binding between PC* and R–X within PC*/R–X, signifying greater resistance to dissociation.

The rate constant k_{ET} is commonly expressed using the first-order Marcus equation for dissociative electron transfer:

$$k_{\text{ET}} = \frac{k_B T}{h} \exp\left[-\frac{\Delta G^{\ddagger}_{\text{ET}}}{k_B T}\right] \tag{12.28}$$

$$\Delta G^{\ddagger}_{\text{ET}} = \frac{\left(\Delta G^0_{\text{ET}} + \lambda\right)^2}{4\lambda} \tag{12.29}$$

$$\lambda = \lambda_o + \text{BDE}, \tag{12.30}$$

$$\Delta G^0_{\text{ET}} = eE^0(\text{PC}^{\bullet} + /\text{PC}^*) - eE^0(\text{R}-\text{X}/\text{R}-\text{X}^{\cdot-}) - w_C \tag{12.31}$$

where k_b is the Boltzmann constant, h denotes the Planck constant, and T represents the temperature. $\Delta G^{\ddagger}_{\text{ET}}$ is the Gibbs energy barrier for the electron transfer reaction and ΔG^0_{ET} signifies the overall Gibbs energy change of the reaction. λ_o stands for the outer-sphere reorganization energy and BDE refers to the bond dissociation energy of R–X (which in the present reaction is a good approximation to the inner-sphere reorganization energy, since the largest structural change during the electron transfer is the rupture of the R–X bond). $E^0(\text{PC}^{\bullet+}/\text{PC}^*)$ is the oxidation potential of PC*, while $E^0(\text{R}-\text{X}/\text{R}-\text{X}^{\cdot-})$ is the reduction potential of R–X. The term e is the fundamental charge and w_C represents a Coulombic work term associated to the electrostatic interaction between $\text{PC}^{\bullet+}$ and $\text{R}-\text{X}^{\cdot-}$ after electron transfer.

According to Eq. (12.31), it becomes evident that the activation efficiency, especially for PCs that share the same chromophore core combined with the same initiator R-X, can be characterized by $E^0(PC^{\bullet+}/PC^*)$ [26, 77]. This is based on the fact that PCs with very similar chemical and/or geometric structures usually maintain similar binding strengths between PC* and R-X.

Based on the descriptions above, we propose strategies for achieving efficient activation. These strategies can be summarized as follows: (i) enhancing the binding strength between PC* and R-X, which encourages the formation of their exciplex and speed up the subsequent dissociative electron transfer process. (ii) Reducing the energy barrier for the reaction $PC^*/R\text{-}X \rightarrow PC^{\bullet+}/X^- + R^\bullet$, facilitating faster electron transfer, or alternatively, lowering the excited-state potential $E^0(PC^{\bullet+}/PC^*)$, which gives an approximate measure of the energy barrier given the same R–X and PCs sharing the same chromophore core [77].

12.4.5 Deactivation

The deactivation step in OQP is also a reaction within PC and R-X. In the activation step, the dissociated X group acquires a formal negative charge due to the electron transfer to the R–X bond, while the R group remains neutral and becomes a radical $R^\bullet$. This results in the formation of an anion, X^-. This anion can pair with the electron-deficient $PC^{\bullet+}$ to create an ion pair, $PC^{\bullet+}/X^-$. Such ion pairing plays a crucial role in the deactivation step [52, 121, 122]. In this phase, $PC^{\bullet+}/X^-$ can interact as a single entity with the growing polymer chain $P_n^\bullet$ initiated by $R^\bullet$. This interaction enables a bimolecular (two-body) reaction, represented as $PC^{\bullet+}/X^- + P_n^\bullet \rightarrow PC + P_n\text{-}X$, instead of an unfavorable three-body reaction, $PC^{\bullet+} + X^- + P_n^\bullet \rightarrow PC + P_n\text{-}X$. As such, maintaining the stability of the ion-pair $PC^{\bullet+}/X^-$ is important, as it is essential for effective deactivation and, consequently, plays a significant role in enhancing the performance of photo-RDRP [46, 121, 123]. It is important to emphasize that the entity $PC^{\bullet+}/X^-$ should not be regarded solely as $PC^{\bullet+}$ and X^-; rather, its electron distribution is more delocalized. In simpler terms, this deactivating species can exist in a state between $PC^{\bullet+}/X^-$ and $PC/X^\bullet$. This delocalization characteristic aids in facilitating the deactivation reaction, as the deactivation step involves a complex interplay of back electron transfer and bond reformation. Similar to activation, where electron transfer precedes bond dissociation, deactivation is initiated by back electron transfer followed by bond reformation. The two processes are intricately intertwined to a certain extent, contingent on the nature of both the X moiety of R–X and the PC.

Based on the descriptions above, we propose methods for achieving efficient deactivation. These methods can be summarized as follows: (i) Enhancing the stability of ion-pair $PC^{\bullet+}/X^-$, which plays a crucial role in ensuring efficient deactivation. (ii) Reducing the energy barrier for the back electron transfer reaction of $PC^{\bullet+}/X^-$ to $R^\bullet$, in analogy with the activation step. (iii) Lowering the potential $E^0(PC^{\bullet+}/PC)$ to provide a bigger thermodynamic driving force for deactivation [77].

12.4.6 Property-Performance Relationships

Scheme 12.1 illustrates the catalytic mechanisms of OQP, TET, RQP, and O-RQP, all enclosing typical steps, viz. photon absorption (Step I, as detailed in Sec. 12.4.2), evolution of excited states (Step II, as detailed in Sec. 12.4.3), activation of polymerization (Step III, as detailed in Sec. 12.4.4), and eventually deactivation of polymerization (Step IV, as detailed in Sec. 12.4.5). After deactivation, the catalytic cycle is completed, restoring all compounds to a dormant state that pauses the chain growth started after activation.

OQP is the most common mechanism for photo-RDRP. It is simple yet robust, composed of the PC and R–X only. Such a mechanism has an activation step involving donating an electron from the excited PC (PC*) to R-X, resulting in R–X bond cleavage in a process known as dissociative electron transfer. TET is a counterpart to dissociative electron transfer, but differs in that the activation step uses electron-exchange triplet/Dexter energy transfer. Hitherto, only R–X with salient conjugation, such that they possess sufficiently stable low-lying triplet excited states, can undergo such a mechanism. Another requirement for TET is the population of a triplet excited-state $^3PC^*$. RQP, on the other hand, has an activation step that includes an electron transfer from the TA to PC*, followed by dissociative electron transfer from the charged $PC^{\bullet-}$ to R–X (coupled by dissociation of the latter). Finally, O-RQP is a new pathway for photo-RDRP which uses oxygen (O_2) as a co-catalyst [47]. It involves transferring an electron from a TA to the photosensitized oxygen ($^1O_2$) by the aid of triplet-TTA between $^3PC^*$ and O_2, thus facilitating the generated superoxide $O_2^{\bullet-}$ to react with R–X and break its bond.

k_p^{app}, a constant that measures the kinetics of photo-RDRP, is a product of the steady-state concentration of the propagating radicals and the secondary rate constant of the addition of the propagating chain radical onto the monomer; while the latter is basically fixed for a given combination of monomer, solvent, and viscosity, the former depends heavily on the PC and initiator. While the activation step leads to a net increase in the propagating radical concentration, the deactivation step caps the propagating radical with an end group and reduces its concentration. Therefore, activation and deactivation reach an equilibrium under irradiation, yielding an equilibrated population of propagating radicals. As such, k_p^{app} is determined by the relative rate constant of activation versus deactivation. For photo-RDRP incorporating reversible chain transfer, the chain transfer intermediate temporarily captures a propagating radical, thus the lifetime of the chain transfer intermediate also has an influence on k_p^{app}. In a word, the key characteristics of a PC that govern k_p^{app} includes ε_{max}, Φ_T, τ_T, the activation energy barrier associated with activation efficiency, the deactivation energy barrier associated with deactivation efficiency, along with the dissociation energy barrier of the chain transfer intermediate characterizing the latter's stability.

Temporal precision, characterized by how fast the polymerization stops after cessation of irradiation, is the symbolic character of photocontrolled polymerization. The population of propagating radicals decreases after ceasing the irradiation; hence the deactivation efficiency is the key factor. $Đ$ is an indicator of the chain length homogeneity, which is the weight-averaged MW divided by the number-averaged MW (i.e. M_w/M_n). For controlled polymerization, deactivation must be more efficient than activation, that is, propagating radicals need effective restoration to guarantee all R–X grow alternatingly and only a few monomers are added to a given propagating radical before the latter is deactivated. Moreover, low radical concentration at a time will prevent accumulation of radicals that trigger bi-radical termination.

I^* is an indicator of how much of R–X is effectively converted to polymer chains at the beginning of polymerization. I^* is defined as the theoretical number-average MW divided by the experimentally determined number-average MW (i.e. $M_{n,theo}/M_{n,exp} \times 100\%$). I^* is affected by multiple factors, such as ε_{max}, Φ_T, τ_T, activation energy barrier, as well as the stability of PC and R-X.

12.5 Outlook

Leveraging the vast diversity of organocatalysts and the established and growing knowledge of SPP relationships, several emerging directions are becoming hotspots in the field of photo-RDRP based on organocatalysts.

(1) NIR-light-regulated polymerization. Developing PCs that absorb beyond 800 nm can improve depth penetration for applications within tissues and materials. Strategies include customizing PCs, modifying initiators, and exploring two-photon energetics, photochemistry-electrochemistry coupling, and direct T_1 excitation.
(2) Solubility of large π-conjugation PCs. Extending π conjugation improves photophysical properties but increases hydrophobicity, posing challenges for 3D printing and biomedical applications. Future designs must balance these properties and solubility.
(3) Computer-guided PC and initiator design. Computational methods like machine learning can accelerate the discovery of new PC-initiator systems, enabling a broader range of chemistries and complex multifunctional materials.
(4) Orthogonal polymerization. Designing PCs with non-overlapping absorptions allows multiple polymerization systems to be independently activated, facilitating complex polymer synthesis and surface patterning with different wavelengths.
(5) Additive manufacturing. Photoinduced additive manufacturing benefits from light-based methods for higher resolution and speed. PCs must deactivate oxygen, catalyze ultrafast polymerization, and enable chain extension and surface processing for improved 3D networks.
(6) Biocompatible PCs for bioapplications. For bio-related materials, PCs must be nontoxic, biocompatible, and water-soluble, with high Φ_T to counteract oxygen exposure. Biodegradability or photobleaching is also desired to minimize residual PC impact.

References

1 Judzewitsch, P.R., Zhao, L., Wong, E.H.H., and Boyer, C. (2019). High-Throughput Synthesis of Antimicrobial Copolymers and Rapid Evaluation of Their Bioactivity. *Macromolecules* 52 (11): 3975–3986. https://doi.org/10.1021/acs.macromol.9b00290.

2 Judzewitsch, P.R., Nguyen, T.K., Shanmugam, S. et al. (2018). Towards sequence-controlled antimicrobial polymers: effect of polymer block order on antimicrobial activity. *Angew. Chem., Int. Ed.* 57 (17): 4559–4564. https://doi.org/10.1002/anie.201713036.

3 Niu, J., Lunn, D.J., Pusuluri, A. et al. (2017). Engineering live cell surfaces with functional polymers via cytocompatible controlled radical polymerization. *Nat. Chem.* 9 (6): 537–545. https://doi.org/10.1038/nchem.2713.

4 Tucker, B.S., Coughlin, M.L., Figg, C.A., and Sumerlin, B.S. (2017). Grafting-from proteins using metal-free PET-RAFT polymerizations under mild visible-light irradiation. *ACS Macro Lett.* 6 (4): 452–457. https://doi.org/10.1021/acsmacrolett.7b00140.

5 Yeow, J., Shanmugam, S., Corrigan, N. et al. (2016). A polymerization-induced self-assembly approach to nanoparticles loaded with singlet oxygen generators. *Macromolecules* 49 (19): 7277–7285. https://doi.org/10.1021/acs.macromol.6b01581.

6 Szymanski, J.K. and Pérez-Mercader, J. (2016). Direct optical observations of vesicular self-assembly in large-scale polymeric structures during photocontrolled biphasic polymerization. *Polym. Chem.* 7 (47): 7211–7215. https://doi.org/10.1039/c6py01497f.

7 Albertsen, A.N., Szymanski, J.K., and Pérez-Mercader, J. (2017). Emergent properties of giant vesicles formed by a polymerization-induced self-assembly (PISA) reaction. *Sci. Rep.* 7. https://doi.org/10.1038/srep41534.

8 Xu, S.H., Zhang, T., Kuchel, R.P. et al. (2020). Gradient polymerization induced self-assembly: a one-step approach. *Macromol. Rapid Commun.* 41 (1): https://doi.org/10.1002/marc.201900493.

9 Xu, S.H., Yeow, J., and Boyer, C. (2018). Exploiting wavelength orthogonality for successive photoinduced polymerization-induced self-assembly and photo-crosslinking. *ACS Macro Lett.* 7 (11): 1376–1382. https://doi.org/10.1021/acsmacrolett.8b00741.

10 Poelma, J.E., Fors, B.P., Meyers, G.F. et al. (2013). Fabrication of complex three-dimensional polymer brush nanostructures through light-mediated living radical polymerization. *Angew. Chem., Int. Ed.* 52 (27): 6844–6848. https://doi.org/10.1002/anie.201301845.

11 Fors, B.P., Poelma, J.E., Menyo, M.S. et al. (2013). Fabrication of unique chemical patterns and concentration gradients with visible light. *J. Am. Chem. Soc.* 135 (38): 14106–14109. https://doi.org/10.1021/ja408467b.

12 Pester, C.W., Narupai, B., Mattson, K.M. et al. (2016). Engineering surfaces through sequential stop-flow photopatterning. *Adv. Mater.* 28 (42): 9292–9300. https://doi.org/10.1002/adma.201602900.

13 Discekici, E.H., Pester, C.W., Treat, N.J. et al. (2016). Simple benchtop approach to polymer brush nanostructures using visible-light-mediated metal-free atom transfer radical polymerization. *ACS Macro Lett.* 5 (2): 258–262. https://doi.org/10.1021/acsmacrolett.6b00004.

14 Chen, M., Cu, Y.W., Singh, A. et al. (2017). Living additive manufacturing: transformation of parent gels into diversely functionalized daughter gels made possible by visible light photoredox catalysis. *ACS Cent. Sci.* 3 (2): 124–134. https://doi.org/10.1021/acscentsci.6b00335.

15 Zhang, Z.H., Corrigan, N., Bagheri, A. et al. (2019). A versatile 3D and 4D printing system through photocontrolled raft polymerization. *Angew. Chem., Int. Ed.* 58 (50): 17954–17963. https://doi.org/10.1002/anie.201912608.

16 Jung, K., Corrigan, N., Ciftci, M. et al. (2020). Designing with light: advanced 2D, 3D, and 4D materials. *Adv. Mater.* 32 (18): https://doi.org/10.1002/adma.201903850.

17 Chen, M., Zhong, M.J., and Johnson, J.A. (2016). Light-controlled radical polymerization: mechanisms, methods, and applications. *Chem. Rev.* 116 (17): 10167–10211. https://doi.org/10.1021/acs.chemrev.5b00671.

18 Dadashi-Silab, S., Doran, S., and Yagci, Y. (2016). Photoinduced electron transfer reactions for macromolecular syntheses. *Chem. Rev.* 116 (17): 10212–10275. https://doi.org/10.1021/acs.chemrev.5b00586.

19 Corrigan, N., Ciftci, M., Jung, K.W., and Boyer, C. (2021). Mediating reaction orthogonality in polymer and materials science. *Angew. Chem., Int. Ed.* 60 (4): 1748–1781. https://doi.org/10.1002/anie.201912001.

20 Corrigan, N. and Boyer, C. (2019). 100th anniversary of macromolecular science viewpoint: photochemical reaction orthogonality in modern macromolecular science. *ACS Macro Lett.* 8 (7): 812–818. https://doi.org/10.1021/acsmacrolett.9b00292.

21 Fu, C.K., Xu, J.T., and Boyer, C. (2016). Photoacid-mediated ring opening polymerization driven by visible light. *Chem. Commun.* 52 (44): 7126–7129. https://doi.org/10.1039/c6cc03084j.

22 Kottisch, V., Michaudel, Q., and Fors, B.P. (2017). Photocontrolled interconversion of cationic and radical polymerizations. *J. Am. Chem. Soc.* 139 (31): 10665–10668. https://doi.org/10.1021/jacs.7b06661.

23 Xiong, X.H., Xue, L.L., and Cui, J.X. (2018). Phototriggered growth and detachment of polymer brushes with wavelength selectivity. *ACS Macro Lett.* 7 (2): 239–243. https://doi.org/10.1021/acsmacrolett.7b00989.

24 Hiltebrandt, K., Pauloehrl, T., Blinco, J.P. et al. (2015). λ-orthogonal pericyclic macromolecular photoligation. *Angew. Chem., Int. Ed.* 54 (9): 2838–2843. https://doi.org/10.1002/anie.201410789.

25 Hiltebrandt, K., Kaupp, M., Molle, E. et al. (2016). Star polymer synthesis *via* λ-orthogonal photochemistry. *Chem. Commun.* 52 (60): 9426–9429. https://doi.org/10.1039/c6cc03848d.

26 Wu, C., Corrigan, N., Lim, C.H. et al. (2022). Rational design of photocatalysts for controlled polymerization: effect of structures on photocatalytic activities. *Chem. Rev.* 122 (6): 5476–5518. https://doi.org/10.1021/acs.chemrev.1c00409.

27 Lorandi, F., Fantin, M., and Matyjaszewski, K. (2022). Atom transfer radical polymerization: a mechanistic perspective. *J. Am. Chem. Soc.* 144 (34): 15413–15430. https://doi.org/10.1021/jacs.2c05364.

28 Lee, Y., Boyer, C., and Kwon, M.S. (2023). Photocontrolled RAFT polymerization: past, present, and future. *Chem. Soc. Rev.* 52 (9): 3035–3097. https://doi.org/10.1039/D1CS00069A. 10.1039/D1CS00069A.

29 Allushi, A., Jockusch, S., Yilmaz, G., and Yagci, Y. (2016). Photoinitiated metal-free controlled/living radical polymerization using polynuclear aromatic hydrocarbons. *Macromolecules* 49 (20): 7785–7792. https://doi.org/10.1021/acs.macromol.6b01752.

30 Miyake, G.M. and Theriot, J.C. (2014). Perylene as an organic photocatalyst for the radical polymerization of functionalized vinyl monomers through oxidative quenching with alkyl bromides and visible light. *Macromolecules* 47 (23): 8255–8261. https://doi.org/10.1021/ma502044f.

31 Treat, N.J., Sprafke, H., Kramer, J.W. et al. (2014). Metal-free atom transfer radical polymerization. *J. Am. Chem. Soc.* 136 (45): 16096–16101. https://doi.org/10.1021/ja510389m.

32 Theriot, J.C., Lim, C.H., Yang, H. et al. (2016). Organocatalyzed atom transfer radical polymerization driven by visible light. *Science* 352 (6289): 1082–1086. https://doi.org/10.1126/science.aaf3935.

33 Pearson, R.M., Lim, C.H., McCarthy, B.G. et al. (2016). Organocatalyzed atom transfer radical polymerization using *N*-aryl phenoxazines as photoredox catalysts. *J. Am. Chem. Soc.* 138 (35): 11399–11407. https://doi.org/10.1021/jacs.6b08068.

34 Buss, B.L., Lim, C.H., and Miyake, G.M. (2020). Dimethyl dihydroacridines as photocatalysts in organocatalyzed atom transfer radical polymerization of acrylate monomers. *Angew. Chem., Int. Ed.* 59 (8): 3209–3217. https://doi.org/10.1002/anie.201910828.

35 Kutahya, C., Allushi, A., Isci, R. et al. (2017). Photoinduced metal-free atom transfer radical polymerization using highly conjugated thienothiophene derivatives. *Macromolecules* 50 (17): 6903–6910. https://doi.org/10.1021/acs.macromol.7b01335.

36 Ma, Q., Zhang, X., Ji, L., and Liao, S. (2019). BINOLs as visible light photocatalysts for metal-free atom transfer radical polymerization. *Polym. Chem.* 10 (48): 6662–6668. https://doi.org/10.1039/C9PY01370A. 10.1039/C9PY01370A.

37 Yang, L., Ngo, H.M., Luo, Y. et al. (2022). Specific active sites of organo-photocatalysts for photo-atom transfer radical polymerization: a combined experimental and theoretical study of N-unsubstituted diketopyrrolopyrrole. *J. Phys. Chem. C* 126 (51): 21576–21584. https://doi.org/10.1021/acs.jpcc.2c04920.

38 Yang, Z., Xiao, W., Zhang, X., and Liao, S. (2022). Organocatalytic cationic degenerate chain transfer polymerization of vinyl ethers with excellent temporal control. *Polym. Chem.* 13 (19): 2776–2781. https://doi.org/10.1039/D2PY00134A. 10.1039/D2PY00134A.

39 Yang, Z., Chen, J., and Liao, S. (2022). Monophosphoniums as effective photoredox organocatalysts for visible light-regulated cationic RAFT polymerization. *ACS Macro Lett.* 11 (9): 1073–1078. https://doi.org/10.1021/acsmacrolett.2c00418.

40 Polgar, A.M., Huang, S.H., and Hudson, Z.M. (2022). Donor modification of thermally activated delayed fluorescence photosensitizers for organocatalyzed atom transfer radical polymerization. *Polym. Chem.* 13 (26): 3892–3903. https://doi.org/10.1039/D2PY00470D. 10.1039/D2PY00470D.

41 Wu, C.Y., Corrigan, N., Lim, C.H. et al. (2019). Guiding the design of organic photocatalyst for PET-RAFT polymerization: halogenated xanthene dyes. *Macromolecules* 52 (1): 236–248. https://doi.org/10.1021/acs.macromol.8b02517.

42 Shanmugam, S., Xu, J., and Boyer, C. (2015). Exploiting metalloporphyrins for selective living radical polymerization tunable over visible wavelengths. *J. Am. Chem. Soc.* 137 (28): 9174–9185. https://doi.org/10.1021/jacs.5b05274.

43 Wu, C.Y., Jung, K., Ma, Y.T. et al. (2021). Unravelling an oxygen-mediated reductive quenching pathway for photopolymerization under long wavelengths. *Nat. Commun.* 12 (1): https://doi.org/10.1038/s41467-020-20640-z.

44 Singh, V.K., Yu, C., Badgujar, S. et al. (2018). Highly efficient organic photocatalysts discovered via a computer-aided-design strategy for visible-light-driven atom transfer radical polymerization. *Nat. Catal.* 1 (10): 794–804. https://doi.org/10.1038/s41929-018-0156-8.

45 Huang, Z., Gu, Y., Liu, X. et al. (2017). Metal-free atom transfer radical polymerization of methyl methacrylate with ppm level of organic photocatalyst. *Macromol. Rapid Commun.* 38 (10): 1600461. https://doi.org/10.1002/marc.201600461 (accessed 2024/06/13).

46 Corbin, D.A. and Miyake, G.M. (2022). Photoinduced organocatalyzed atom transfer radical polymerization (O-ATRP): precision polymer synthesis using organic photoredox catalysis. *Chem. Rev.* 122 (2): 1830–1874. https://doi.org/10.1021/acs.chemrev.1c00603.

47 Wu, C., Jung, K., Ma, Y. et al. (2021). Unravelling an oxygen-mediated reductive quenching pathway for photopolymerization under long wavelengths. *Nat. Commun.* 12 (1): 478. https://doi.org/10.1038/s41467-020-20640-z.

48 Wang, G.-X., Lu, M., Zhou, M.-J. et al. (2018). Photo-induced ATRP of MMA under blue light irradiation in the presence of 3,4,9,10-tetra-(12-alkoxycarbonyl)-perylene as a photocatalyst. *Iran. Polym. J.* 27 (1: 43–48. https://doi.org/10.1007/s13726-017-0583-4.

49 Zeng, L.-L., Xie, W.-Y., Yang, C.-X. et al. (2018). Photomediated atom transfer radical polymerization of MMA under long-wavelength light irradiation. *Iran. Polym. J.* 27 (11): 881–887. https://doi.org/10.1007/s13726-018-0661-2.

50 Pan, X.C., Lamson, M., Yan, J.J., and Matyjaszewski, K. (2015). Photoinduced metal-free atom transfer radical polymerization of acrylonitrile. *ACS Macro Lett.* 4 (2): 192–196. https://doi.org/10.1021/mz500834g.

51 Dadashi-Silab, S., Pan, X., and Matyjaszewski, K. (2017). Phenyl Benzo[b]phenothiazine as a visible light photoredox catalyst for metal-free atom transfer radical polymerization. *Chem. Eur. J.* 23 (25): 5972–5977. https://doi.org/10.1002/chem.201605574 (acccessed 2024/06/14).

52 Pan, X.C., Fang, C., Fantin, M. et al. (2016). Mechanism of photoinduced metal-free atom transfer radical polymerization: experimental and computational studies. *J. Am. Chem. Soc.* 138 (7): 2411–2425. https://doi.org/10.1021/jacs.5b13455.

53 Zhao, Y., Gong, H., Jiang, K. et al. (2018). Organocatalyzed photoredox polymerization from aromatic sulfonyl halides: facilitating graft from aromatic C–H bonds. *Macromolecules* 51 (3): 938–946. https://doi.org/10.1021/acs.macromol.8b00134.

54 Huang, W., Zhai, J., Hu, X. et al. (2020). Continuous flow photoinduced phenothiazine derivatives catalyzed atom transfer radical polymerization. *Eur. Polym. J.* 126: 109565. https://doi.org/10.1016/j.eurpolymj.2020.109565.

55 Theriot, J.C., Lim, C.-H., Yang, H. et al. (2016). Organocatalyzed atom transfer radical polymerization driven by visible light. *Science* 352 (6289): 1082–1086. https://doi.org/10.1126/science.aaf3935.

56 Price, M.J., Puffer, K.O., Kudisch, M. et al. (2021). Structure–property relationships of core-substituted diaryl dihydrophenazine organic photoredox catalysts and their application in O-ATRP. *Polym. Chem.* 12 (42): 6110–6122. https://doi.org/10.1039/D1PY01060C. 10.1039/d1py01060c.

57 Pearson, R.M., Lim, C.-H., McCarthy, B.G. et al. (2016). Organocatalyzed atom transfer radical polymerization using N-aryl phenoxazines as photoredox catalysts. *J. Am. Chem. Soc.* 138 (35): 11399–11407. https://doi.org/10.1021/jacs.6b08068.

58 McCarthy, B.G., Pearson, R.M., Lim, C.-H. et al. (2018). Structure–property relationships for tailoring phenoxazines as reducing photoredox catalysts. *J. Am. Chem. Soc.* 140 (15): 5088–5101. https://doi.org/10.1021/jacs.7b12074.

59 Park, G.S., Back, J., Choi, E.M. et al. (2019). Visible light-mediated metal-free atom transfer radical polymerization with N-trifluoromethylphenyl phenoxazines. *Eur. Polym. J.* 117: 347–352. https://doi.org/10.1016/j.eurpolymj.2019.05.023.

60 Buss, B.L., Lim, C.-H., and Miyake, G.M. (2020). Dimethyl dihydroacridines as photocatalysts in organocatalyzed atom transfer radical polymerization of acrylate monomers. *Angew. Chem. Int. Ed.* 59 (8): 3209–3217. https://doi.org/10.1002/anie.201910828 (acccessed 2023/08/26).

61 Liu, Y., Chen, Q., Tong, Y., and Ma, Y. (2020). 9,9-dimethyl dihydroacridine-based organic photocatalyst for atom transfer radical polymerization from modifying "unstable" electron donor. *Macromolecules* 53 (16): 7053–7062. https://doi.org/10.1021/acs.macromol.0c00377.

62 Song, Y., Kim, Y., Noh, Y. et al. (2019). Organic photocatalyst for ppm-level visible-light-driven reversible addition-fragmentation chain-transfer (RAFT) polymerization with excellent oxygen tolerance. *Macromolecules* 52 (15): 5538–5545. https://doi.org/10.1021/acs.macromol.9b00940.

63 Huang, Z., Gu, Y., Liu, X. et al. (2017). Metal-free atom transfer radical polymerization of methyl methacrylate with ppm level of organic photocatalyst. *Macromol. Rapid Commun.* 38 (10): https://doi.org/10.1002/marc.201600461.

64 Ma, Q., Song, J., Zhang, X. et al. (2021). Metal-free atom transfer radical polymerization with ppm catalyst loading under sunlight. *Nat. Commun.* 12 (1): 429. https://doi.org/10.1038/s41467-020-20645-8.

65 Yang, Z., Chen, J. et al. (2022). Monophosphoniums as effective photoredox organocatalysts for visible light-regulated cationic RAFT polymerization. *ACS Macro Lett.* 11 (9): 1073–1078. https://doi.org/10.1021/acsmacrolett.2c00418.

66 Zhang, X., Jiang, Y. et al. (2021). Metal-free cationic polymerization of vinyl ethers with strict temporal control by employing an organophotocatalyst. *J. Am. Chem. Soc.* 143 (17): 6357–6362. https://doi.org/10.1021/jacs.1c02500.

67 Minh Tran, H., Kwak, D., Lee, U. et al. (2024). A hydrophilic near dissolved organic matter microfiltration (NDOM MF) membrane prepared using multifunctional porogen synthesized via metal-free atom transfer radical polymerization for highly efficient microplastic removal. *Chem. Eng. J.* 480: 147564. https://doi.org/10.1016/j.cej.2023.147564.

68 Ma, Q., Jiang, Y., Lin, J. et al. (2022). Organocatalytic orthogonal ATRP and ring-opening polymerization using a single dual-function photocatalyst. *Polym. Chem.* 13 (29): 4284–4289. https://doi.org/10.1039/D2PY00633B. 10.1039/D2PY00633B.

69 Yang, L., Huang, Y., Peng, Y. et al. (2020). Pyridine-diketopyrrolopyrrole-based novel metal-free visible-light organophotoredox catalyst for atom-transfer radical polymerization. *Chem. A Eur. J.* 124 (6): 1068–1075. https://doi.org/10.1021/acs.jpca.9b10404.

70 Xu, J.T., Jung, K., Atme, A. et al. (2014). A robust and versatile photoinduced living polymerization of conjugated and unconjugated monomers and its oxygen tolerance. *J. Am. Chem. Soc.* 136 (14): 5508–5519. https://doi.org/10.1021/ja501745g.

71 Seal, P., Xu, J.T., De Luca, S. et al. (2019). Unraveling photocatalytic mechanism and selectivity in PET-RAFT polymerization. *Adv. Theory Simul.* 2 (6): https://doi.org/10.1002/adts.201900038.

72 Dolinski, N.D., Page, Z.A., Discekici, E.H. et al. (2019). What happens in the dark? Assessing the temporal control of photo-mediated controlled radical polymerizations. *J. Polym. Sci. Part A: Polym. Chem.* 57 (3): 268–273. https://doi.org/10.1002/pola.29247.

73 Otsu, T. (2000). Iniferter concept and living radical polymerization. *J. Polym. Sci. Part A: Polym. Chem.* 38 (12): 2121–2136. https://doi.org/10.1002/(sici)1099-0518(20000615)38:12<2121::Aid-pola10>3.0.Co;2-x.

74 Corrigan, N., Xu, J.T., Boyer, C., and Allonas, X. (2019). Exploration of the PET-RAFT initiation mechanism for two commonly used photocatalysts. *ChemPhotoChem* 3 (11): 1193–1199. https://doi.org/10.1002/cptc.201800182.

75 Christmann, J., Ibrahim, A., Charlot, V. et al. (2016). Elucidation of the key role of ru(bpy)$_{3\cdot 2+}$in photocatalyzed raft polymerization. *ChemPhysChem* 17 (15): 2309–2314. https://doi.org/10.1002/cphc.201600034.

76 Thum, M.D., Wolf, S., and Falvey, D.E. (2020). State-dependent photochemical and photophysical behavior of dithiolate ester and trithiocarbonate reversible addition-fragmentation chain transfer polymerization agents. *J. Phys. Chem. A* 124 (21): 4211–4222. https://doi.org/10.1021/acs.jpca.0c02678.

77 Wang, Z., Wu, C., and Liu, W. (2024). Toward the rational design of organic catalysts for organocatalyzed atom transfer radical polymerization. *Polymers* 16 (3): 323.

78 Kurek, P.N., Kloster, A.J., Weaver, K.A. et al. (2018). How do reaction and reactor conditions affect photoinduced electron/energy transfer reversible addition-fragmentation transfer polymerization? *Ind. Eng. Chem. Res.* 57 (12): 4203–4213. https://doi.org/10.1021/acs.iecr.7b05397.

79 Corrigan, N., Rosli, D., Jones, J.W.J. et al. (2016). Oxygen tolerance in living radical polymerization: investigation of mechanism and implementation in continuous flow polymerization. *Macromolecules* 49 (18): 6779–6789. https://doi.org/10.1021/acs.macromol.6b01306.

80 Xu, S.H., Ng, G., Xu, J.T. et al. (2017). 2-(methylthio)ethyl methacrylate: a versatile monomer for stimuli responsiveness and polymerization-induced self-assembly in the presence of air. *ACS Macro Lett.* 6 (11): 1237–1244. https://doi.org/10.1021/acsmacrolett.7b00731.

81 Shanmugam, S., Xu, J.T., and Boyer, C. (2016). Aqueous RAFT ohotopolymerization with oxygen tolerance. *Macromolecules* 49 (24): 9345–9357. https://doi.org/10.1021/acs.macromol.6b02060.

82 Zhang, Z., Yu, Y., Boyer, C. et al. (2024). Design and synthesis of a new indazole-decorated RAFT agent for highly efficient PET-RAFT polymerization. *Macromolecules* 57 (9): 4421–4429. https://doi.org/10.1021/acs.macromol.3c02315.

83 Wang, Z., Zhang, Z., Wu, C. et al. (2024). Pushing the Limit of Photo-Controlled Polymerization: Hyperchromic and Bathochromic Effects. *Molecules* 29 (10): 2377.

84 Gentemann, S., Medforth, C.J., Forsyth, T.P. et al. (1994). Photophysical properties of conformationally distorted metal-free porphyrins - investigation into the deactivation mechanisms of the lowest excited singlet-state. *J. Am. Chem. Soc.* 116 (16): 7363–7368. https://doi.org/10.1021/ja00095a046.

85 Senge, M.O. (2006). Exercises in molecular gymnastics—bending, stretching and twisting porphyrins. *Chem. Commun.* 3: 243–256. https://doi.org/10.1039/b511389j.

86 Sparks, L.D., Medforth, C.J., Park, M.S. et al. (1993). Metal dependence of the nonplanar distortion of octaalkyltetraphenylporphyrins. *J. Am. Chem. Soc.* 115 (2): 581–592. https://doi.org/10.1021/ja00055a030.

87 Zhang, L., Wu, C., Jung, K. et al. (2019). An oxygen paradox: catalytic use of oxygen in radical photopolymerization. *Angew. Chem. Int. Ed.* 58 (47): 16811–16814. https://doi.org/10.1002/anie.201909014 (accessed 2024/03/02).

88 Shanmugam, S., Xu, J.T., and Boyer, C. (2015). Utilizing the electron transfer mechanism of chlorophyll a under light for controlled radical polymerization. *Chem. Sci.* 6 (2): 1341–1349. https://doi.org/10.1039/c4sc03342f.

89 Shanmugam, S., Xu, J.T., and Boyer, C. (2016). Light-regulated polymerization under near-infrared/far-red irradiation catalyzed by bacteriochlorophyll a. *Angew. Chem., Int. Ed.* 55 (3): 1036–1040. https://doi.org/10.1002/anie.201510037.

90 Xu, J.T., Shanmugam, S., Fu, C.K. et al. (2016). Selective photoactivation: from a single unit monomer insertion reaction to controlled polymer architectures. *J. Am. Chem. Soc.* 138 (9): 3094–3106. https://doi.org/10.1021/jacs.5b12408.

91 Cao, H.L., Wang, G.C., Xue, Y.D. et al. (2019). Far-red light-induced reversible addition-fragmentation chain transfer polymerization using a man-made bacteriochlorin. *ACS Macro Lett.* 8 (5): 616–622. https://doi.org/10.1021/acsmacrolett.9b00320.

92 Wu, C., Corrigan, N., Lim, C.-H. et al. (2019). Guiding the design of organic photocatalyst for pet-raft polymerization: halogenated xanthene dyes. *Macromolecules* 52 (1): 236–248. https://doi.org/10.1021/acs.macromol.8b02517.

93 Wu, C., Chen, H., Corrigan, N. et al. (2019). Computer-guided discovery of a ph-responsive organic photocatalyst and application for ph and light dual-gated polymerization. *J. Am. Chem. Soc.* 141 (20): 8207–8220. https://doi.org/10.1021/jacs.9b01096.

94 Corrigan, N., Shanmugam, S., Xu, J.T., and Boyer, C. (2016). Photocatalysis in organic and polymer synthesis. *Chem. Soc. Rev.* 45 (22): 6165–6212. https://doi.org/10.1039/c6cs00185h.

95 Dirac, P.A.M. (1927). The quantum theory of the emission and absorption of radiation. *P R Soc Lond a-Conta* 114 (767): 243–265. https://doi.org/10.1098/rspa.1927.0039.

96 Orear, J., Fermi, E., Rosenfeld, A.H., and Schluter, R.A. (1950). *Nuclear Physics: A Course Given by Enrico Fermi at the University of Chicago*. University of Chicago Press.

97 Englman, R. and Jortner, J. (1970). The energy gap law for radiationless transitions in large molecules. *Mol. Phys.* 18 (2): 145–164. https://doi.org/10.1080/00268977000100171.

98 Lin, S.H., Chang, C.H., Liang, K.K. et al. (2002). Ultrafast dynamics and spectroscopy of bacterial photosynthetic reaction centers. *Adv. Chem. Phys.* 121: 1–88. https://doi.org/10.1002/0471264318.ch1.

99 Theriot, J.C., McCarthy, B.G., Lim, C.H., and Miyake, G.M. (2017). Organocatalyzed atom transfer radical polymerization: perspectives on catalyst design and performance. *Macromol. Rapid Commun.* 38 (13): https://doi.org/10.1002/marc.201700040.

100 Senge, M.O. (2006). Exercises in molecular gymnastics—bending, stretching and twisting porphyrins. *Chem. Commun.* 3: 243–256. https://doi.org/10.1039/B511389J. 10.1039/B511389J.

101 Adachi, M. and Nagao, Y. (2001). Design of near-infrared dyes based on π-conjugation system extension 2. Theoretical elucidation of framework extended derivatives of perylene chromophore. *Chem. Mater.* 13 (2): 662–669. https://doi.org/10.1021/cm000857v.

102 Ruiz-Morales, Y. (2002). HOMO–LUMO Gap as an index of molecular size and structure for polycyclic aromatic hydrocarbons (PAHs) and asphaltenes: a theoretical study. I. *Chem. A Eur. J.* 106 (46): 11283–11308. https://doi.org/10.1021/jp021152e.

103 Roznyatovskiy, V.V., Lee, C.-H., and Sessler, J.L. (2013). π-Extended isomeric and expanded porphyrins. *Chem. Soc. Rev.* 42 (5): 1921–1933. https://doi.org/10.1039/C2CS35418G. 10.1039/c2cs35418g.

104 Martynov, A.G., Mack, J., May, A.K. et al. (2019). Methodological survey of simplified TD-DFT methods for fast and accurate interpretation of UV–Vis–NIR spectra of phthalocyanines. *ACS Omega* 4 (4): 7265–7284. https://doi.org/10.1021/acsomega.8b03500.

105 Roy, P.P., Kundu, S., Makri, N., and Fleming, G.R. (2022). Interference between franck–condon and herzberg–teller terms in the condensed-phase molecular spectra of metal-based tetrapyrrole derivatives. *The Journal of Physical Chemistry Letters* 13 (32): 7413–7419. https://doi.org/10.1021/acs.jpclett.2c01963.

106 Shanmugam, S., Xu, J.T., and Boyer, C. (2015). Exploiting metalloporphyrins for selective living radical polymerization tunable over visible wavelengths. *J. Am. Chem. Soc.* 137 (28): 9174–9185. https://doi.org/10.1021/jacs.5b05274.

107 Rosokha, S.V. and Kochi, J.K. (2008). Fresh look at electron-transfer mechanisms via the donor/acceptor bindings in the critical encounter complex. *Acc. Chem. Res.* 41 (5): 641–653. https://doi.org/10.1021/ar700256a.

108 Di Savino, A., Foerster, J.M., La Haye, T. et al. (2020). Efficient encounter complex formation and electron transfer to cytochrome c peroxidase with an additional, distant electrostatic binding site. *Angew. Chem. Int. Ed.* 59 (51): 23239–23243. https://doi.org/10.1002/anie.202010006 (accessed 2023/08/26).

109 Peng, Q., Yi, Y.P., Shuai, Z.G., and Shao, J.S. (2007). Toward quantitative prediction of molecular fluorescence quantum efficiency: Role of duschinsky rotation. *J. Am. Chem. Soc.* 129 (30): 9333–9339. https://doi.org/10.1021/ja067946e.

110 Taniguchi, M. and Lindsey, J.S. (2021). Absorption and fluorescence spectral database of chlorophylls and analogues. *Photochem. Photobiol.* 97 (1): 136–165. https://doi.org/10.1111/php.13319 (acccessed 2024/03/07).

111 Worth, G.A. and Cederbaum, L.S. (2004). Beyond born-oppenheimer: molecular dynamics through a conical intersection. *Annu. Rev. Phys. Chem.* 55: 127–158. https://doi.org/10.1146/annurev.physchem.55.091602.094335.

112 Güttinger, P. (1932). Das verhalten von atomen im magnetischen drehfeld. *Z. Phys.* 73 (3): 169–184. https://doi.org/10.1007/BF01351211.

113 Hellmann, H. (1937). *Einführung in die quantenchemie.* F.Deuticke.

114 Pauli, W. *Die allgemeinen prinzipien der Wellenmechanik.* Julius Springer.

115 Feynman, R.P. (1939). Forces in molecules. *Phys. Rev.* 56 (4): 340–343. https://doi.org/10.1103/PhysRev.56.340.

116 Penfold, T.J., Gindensperger, E., Daniel, C., and Marian, C.M. (2018). Spin-vibronic mechanism for intersystem crossing. *Chem. Rev.* 118 (15): 6975–7025. https://doi.org/10.1021/acs.chemrev.7b00617.

117 Ma, H., Peng, Q., An, Z. et al. (2019). Efficient and long-lived room-temperature organic phosphorescence: theoretical descriptors for molecular designs. *J. Am. Chem. Soc.* 141 (2): 1010–1015. https://doi.org/10.1021/jacs.8b11224.

118 Elsayed, M.A. (1963). Spin-orbit coupling and radiationless processes in nitrogen heterocyclics. *J. Chem. Phys.* 38 (12): 2834–2838. https://doi.org/10.1063/1.1733610.

119 Penfold, T.J., Gindensperger, E., Daniel, C., and Marian, C.M. (2018). Spin-Vibronic Mechanism for Intersystem Crossing. *Chem. Rev.* 118 (15): 6975–7025. https://doi.org/10.1021/acs.chemrev.7b00617.

120 Lattke, Y.M., Corbin, D.A., Sartor, S.M. et al. (2021). Interrogation of O-ATRP activation conducted by singlet and triplet excited states of phenoxazine photocatalysts. *Chem. A Eur. J.* 125 (15): 3109–3121. https://doi.org/10.1021/acs.jpca.1c00855.

121 Lim, C.H., Ryan, M.D., McCarthy, B.G. et al. (2017). Intramolecular charge transfer and ion pairing in *N,N*-diaryl dihydrophenazine photoredox catalysts for efficient organocatalyzed atom transfer radical polymerization. *J. Am. Chem. Soc.* 139 (1): 348–355. https://doi.org/10.1021/jacs.6b11022.

122 Corbin, D.A., McCarthy, B.G., van de Lindt, Z., and Miyake, G.M. (2021). Radical cations of phenoxazine and dihydrophenazine photoredox catalysts and their role as deactivators in organocatalyzed atom transfer radical polymerization. *Macromolecules* 54 (10): 4726–4738. https://doi.org/10.1021/acs.macromol.1c00640.

123 Ryan, M.D., Theriot, J.C., Lim, C.H. et al. (2017). Solvent effects on the intramolecular charge transfer character of *N,N*-diaryl dihydrophenazine catalysts for organocatalyzed atom transfer radical polymerization. *J. Polym. Sci. Part A: Polym. Chem.* 55 (18): 3017–3027. https://doi.org/10.1002/pola.28574.

13

Organocatalyst-Catalyzed Non-Radical Photopolymerizations

Yun Liao and Saihu Liao

State Key Laboratory of Physical Chemistry of Solid Surfaces, Xiamen University, 422 Siming Road, Xiamen 361005, China

13.1 Introduction

Along with the development of organic and macromolecular synthesis in the past century, nature has offered synthetic chemists many inspirations for reaction development and catalyst design. In this regard, photosynthesis, as nature's photochemical process [1], has attracted numerous efforts of chemists to emulate this complex and sophisticated process to harness visible light to promote the transformation of molecules and materials. In fact, the utilization of light has also greatly pushed forward the advancement of polymer science in the past decade, including various aspects, such as photoinduced polymer synthesis, 3D printing, light-regulated transformation of polymers, photogated release of small molecules from polymer materials, polymeric photosensors, etc.

Light has been employed as stimuli for polymerization for a long time, and the past five decades have witnessed a significant advancement in the development and application of photoinitiated processes in both academia and industry [2–5]. Most of these processes involve an initiation of polymerization induced by light. Typically, photoinitiation relies on the activation of a specific photoinitiator/photoinitiating system or monomer under light irradiation. The excited species would then generate a reactive radical or ionic species to initiate the corresponding radical or ionic polymerization. In a traditional photoinitiated polymerization, the light regulation operates mainly at the initiation stage of polymerization, and normally exerts no regulation/control over the subsequent chain growth. This character is in contrast to the photo "controlled" polymerization, in which control on the chain growth by light (e.g. via the light on/off operation) is expected. In the past decade, the development of photocontrolled polymerizations has been greatly facilitated by the application of photocatalysts (PCs), in which a reversible photocatalytic activation of the dormant chain ends, coupling with deactivation of propagating chains in equilibrium, can be established to reach an on/off control on the polymerization process by light (Figure 13.1). As a result, the chain growth can be gated by switching on/off the

Organocatalysts in Polymer Chemistry: Synthesis and Applications, First Edition. Edited by Zhibo Li.

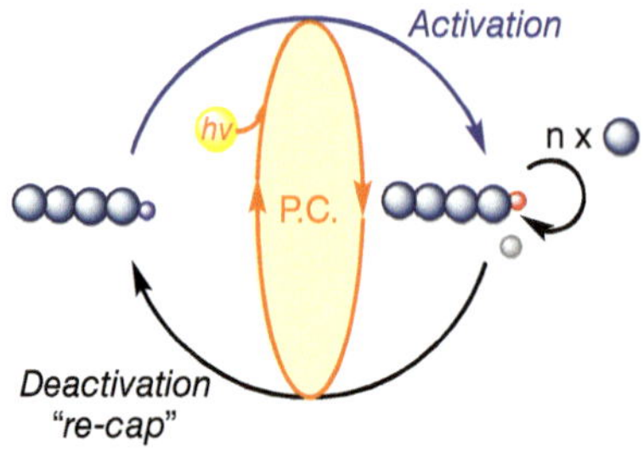

Figure 13.1 The application of photocatalyst (PC) for the development of photocontrolled polymerization.

light to alternate the corresponding periods of growth (after activation) and dormancy (after deactivation), and the rate of polymerization can also be regulated by the intensity of the light.

The integration of a PC has greatly expanded the types of photocontrolled polymerization, and a number of photocatalytic polymerizations have been established in the past decade, albeit most efforts concentrating on the development of photocontrolled radical polymerization such as photocontrolled atom transfer radical polymerization (ATRP) and photoinduced electron/energy transfer reversible addition-fragmentation chain transfer polymerization (PET-RAFT) polymerization as summarized in the previous chapter as well as other excellent reviews [6–9]. Nevertheless, the advance of PCs and photocatalytic strategies has also provided the opportunity to promote the development of photocontrolled polymerization beyond radical variants. In fact, a significant expansion of non-radical polymerizations under photocontrol has been accomplished in the past several years, including photocontrolled cationic polymerization, ring-opening polymerization (ROP) of lactones, ring-opening metathesis polymerization (ROMP), as well as several step-growth photopolymerizations based on photocatalytic addition/cycloaddition reactions, etc [10]. It is worth mentioning that a substantial portion of these advancements has benefited from the development of polymerization-oriented organic PCs and organocatalytic systems, such as highly oxidative organic PCs for cationic polymerizations and ROMP, reversible photoacid catalysts for the ROP of lactones, as well as redox and energy-transfer PCs for the step-growth polymerization via repeating radical addition or cycloaddition, which also forges the content of this chapter.

13.2 Organocatalysts for Photocontrolled Cationic Polymerization

Photoinitiated cationic polymerization relying on the strong acids generated from the photosensitive compounds, e.g. photoacid generators (PAGs), has been well-known, while the decomposition of PAGs is typically irreversible, which make the PAGs inherently unable to serve as a photo-reversible catalyst. In 2015, Perkowski, You, and Nicewicz disclosed an example of metal-free photoinitiated living cationic polymerization (LCP) of 4-methoxystyrene (*p*-MOS), in which pyrylium salts were employed as an organic PC to generate the key cationic species via a photoredox mechanism (Figure 13.2) [11]. The polymerization was carried

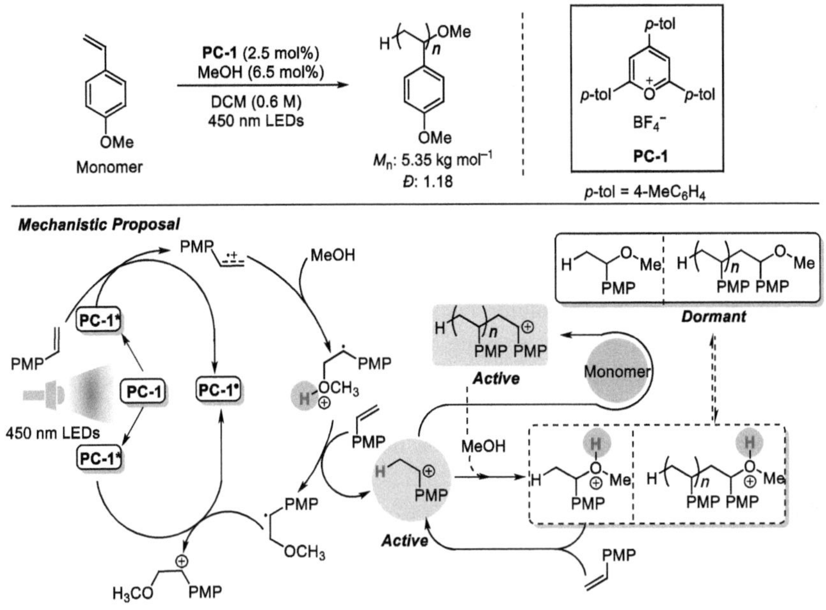

Figure 13.2 Visible light-initiated metal-free living cationic polymerization of 4-methoxystyrene in the presence of pyrylium-type photoredox catalyst.

out under the irradiation of bluc LEDs (γ = 450 nm), and methanol was added as a controlling agent, which could afford poly (*p*-methoxyl styrene) with controlled molecular weights and low dispersities (e.g. $M_{n,GPC}$ = 5.4 kg mol^{-1} and Đ 1.2).

According to the proposed polymerization mechanism, the photo-excited photocatalyst (**PC-1***) first oxidizes the monomer (*p* MOS) via single-electron transfer to generate a radical cationic species from monomer, which can subsequently be trapped by methanol (MeOH) to form a radical cationic adduct. The proton transfer from the cationic adduct to another *p*-MOS molecule could produce the key active cationic species (via protonation of monomer), which can then initiate the cationic polymerization and reach a chain growth. At the same time, after the loss of a proton, the adduct of *p*-MOS$^{\cdot+}$ and MeOH is converted to a radical intermediate, which can be further oxidized by the excited photocatalyst (**PC-1***) to generate cationic species and initiate the cationic polymerization. Authors suggested that this LCP proceeds via a mechanism similar to RAFT polymerization, in which the methoxy group serves as a chain transfer agent (CTA). The oxonium ion species generated from the reaction between a cationic chain and the chain end ether could reversibly fragment into another reactive cationic chain and a methoxy group-caped chain. The living character of this photoinitiated cationic polymerization was demonstrated in the reaction kinetics, plot of M_n vs monomer conversion, as well as chain extension experiments.

It is worth mentioning that Spokoyny and coworkers later on reported a photoinitiated cationic polymerization of *p*-MOS by using perfunctionalized icosahedral dodecaborate clusters, $B_{12}(OCH_2Ar)_{12}$ (Ar = Ph or C_6F_5), as a metal-free photooxidant. The authors showed that upon photoexcitation with visible light (blue LEDs), the clusters possess an excited-state reduction potential as high as 3 volts and can readily oxidize the styrene monomers and initiate their polymerization. Initiation was observed at cluster loadings as low as 0.005 mol% in both cases of electron-rich and electron-deficient styrene monomers, while the molecular weight distributions were relatively broad ($Đ$ = 1.6–5.8). Notably, upon photoexcitation, the $B_{12}(OCH_2C_6F_5)_{12}$ cluster could also promote the polymerization of isobutylene, a less activated olefin, giving a highly branched poly(isobutylene) product [12].

In the aforementioned examples, the organic PCs were mainly involved at the photoinitiation stage of the polymerization, while a photocontrol over the chain growth, e.g. temporal control by light, was elusive. Nevertheless, the introduction of photoredox catalysis into cationic polymerization provides a perspective for the regulation of chain growth by light. In 2016, a seminal contribution by Fors and coworkers unveiled that the combination of photoredox catalysts with RAFT reagents could realize a photocontrol on the cationic polymerization of vinyl ether monomers (Figure 13.3) [13]. In this photocatalytic system, 2,4,6-tris (*p*-methoxyphenyl)pyrylium tetrafluoroborate (**PC-2**) was used as the photoredox catalyst, in combination with a dithiocarbamate-type chain transfer agent (**CTA-1**) that showed a good mediation in the Brønsted acid (e.g. trifluoromethanesulfonic acid [TfOH])-catalyzed cationic RAFT polymerization [14]. This method could provide isobutyl vinyl ether (IBVE) polymers with low dispersity and predictable molecular mass in accord with the monomer/CTA molar ratios. The monomer scope was further examined with trithiocarbonate-type RAFT agent (**CTA-2**), and good control on molecular mass was observed in the polymerizations of a range of alkyl vinyl ethers, such as ethyl vinyl ether (EVE), *n*-propyl vinyl ether (PVE), *n*-butyl vinyl ether (BVE), and 2-chloroethyl vinyl ether (Cl-EVE). The synthesis of diblock copolymers via chain extension was also demonstrated with this method,

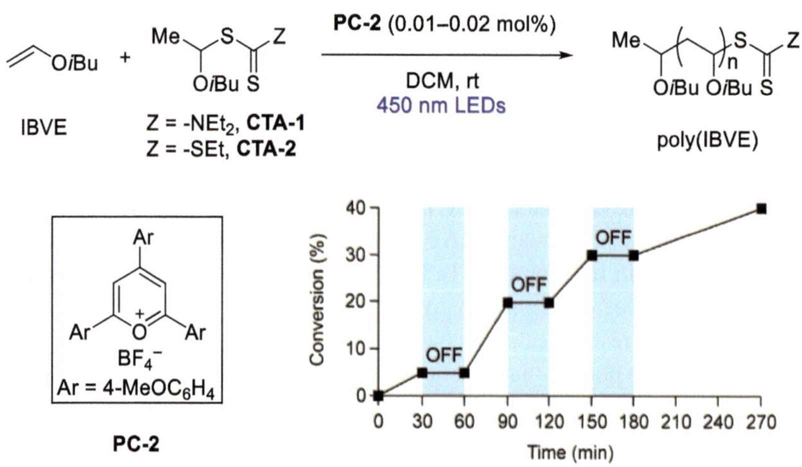

Figure 13.3 Photocontrolled cationic polymerization of vinyl ethers using pyrilium photocatalyst and RAFT agents. Kottisch et al., [13]/American Chemical Society.

suggesting a high chain end fidelity. The authors observed the rate of polymerization could be proportional directly to the light intensity, which suggests that the concentration of propagating cationic species in the reaction can be regulated by the external irradiation of light. As a result, a temporal control on the polymer chain growth by light was successfully realized. As shown in Figure 13.3, the on/off experiment with intermittent light irradiation could achieve good temporal control on the chain propagation, while at high conversion stages, monomer consumption during dark periods could not be completely avoided.

To obtain some insight about the polymerization mechanism, the authors performed photoluminescence quenching experiments with the PCs, and observed a stronger quenching of **PC-2*** by **CTA-2** than that by monomer IBVE. Based on these experimental results, they postulated that polymerization could be initiated by either oxidation of the CTA with excited photocatalysts (PC^{+*}) or oxidation of monomers followed by electron transfer from the CTA (Figure 13.4). Mesolytic fragmentation of the radical cation of CTA produces the key carbocation that would lead to the chain growth, while the presence of RAFT agent could enable the polymerization to proceed via a degenerative chain transfer (DCT) mechanism, thereby allowing for the synthesis of polymers with narrow dispersities and controlled molecular weights. Reduction of the radicals of CTA (•SCSZ) with the reduced photocatalyst (•PC) to the dithiocarbamate (for **CTA-1**) or trithiocarbonate (for **CTA-2**) anion could enable the regeneration of the PC and a concurrent deactivation of the active chain end via chain end capping. This final step closes the catalytic cycle and allows for the photoreversible formation of a cation via an activation-deactivation equilibrium, which is crucial to the realization of a temporal control over the chain growth, as demonstrated in the light on/off experiments.

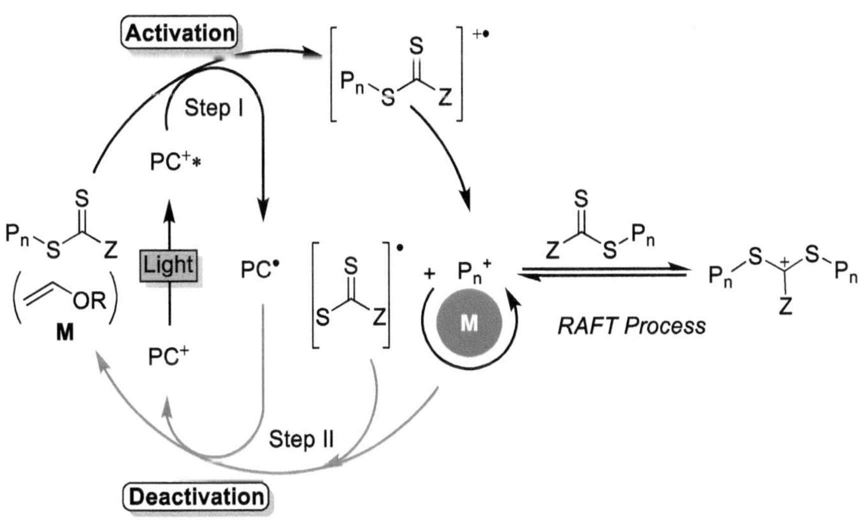

Figure 13.4 A mechanistic proposal for the photocontrolled cationic RAFT polymerization of vinyl ethers with photocatalyst (PC) and chain transfer agent.

To gain a deeper understanding of the mechanism of this light-regulated polymerization of vinyl ethers in the presence of PCs and CTA, the authors later further investigated the mechanism through a combination of different techniques, such as spectroscopies, ultraviolet–visible (UV–Vis) absorption, time-resolved photoluminescence, electron spin resonance (ERS), and electrochemistry [15]. In the photoluminescence quenching experiments, they discovered that quenching of **PC-2*** by **CTA-1** was much faster ($k_q = 7.52 \times 10^9\ M^{-1}\ s^{-1}$) than quenching by IBVE ($k_q = 1.23 \times 10^8\ M^{-1}\ s^{-1}$), revealing that direct oxidation of CTA by PC* is favored kinetically. Additionally, the electrochemical analysis indicated the onset of oxidation potential (+1.19 V vs saturated calomel electrode [SCE]) of **CTA-1** is also lower than the oxidation potential of IBVE monomer (+1.46 V vs SCE), providing further evidence that the oxidation of CTA is easier than monomer in this case. This result also demonstrated the excited PC (e.g. **PC-2***: +1.84 V vs SCE) could selectively oxidize the CTA rather than the monomer. To investigate the influence of the structure and redox potentials of PCs on the polymerization, they further explored a series of aryl pyrylium-type PCs with varying excited-state oxidation power, and discovered both the catalyst structure and excited-state oxidation capability play a key role in the polymerization (Figure 13.5). Even though some loss of temporal control at high monomer conversions was observed with these organic PCs, this study opened the door for the development of photocontrolled cationic RAFT polymerization. Later on, to pursue better temporal control over the polymerization, the authors turned to iridium complex-type PCs, which could afford a significantly enhanced/strict temporal control on the polymerization, while the iridium PCs are costly and also less active than pyrylium PCs [16].

To achieve effective and strict temporal control over the cationic polymerization, the development of a PC possessing enough photostability and appropriate redox potential is critical for establishing an effective activation-deactivation equilibrium. In 2021, Liao and coworkers introduced a class of bisphosphonium salts as a novel organic PC for photopolymerization [17], which could provide an excellent temporal control in the cationic polymerization of vinyl ethers (Figure 13.6). The bisphosphonium photocatalysts (**PC-3**) were developed following a design logic of heteroatom-doping of polycyclic arenes (HDPA) for organic PC development, which also inspired the development of oxygen-doped anthanthrene as a new class of organic PCs for ATRP [18]. Compared with the pyrylium salt catalyst (**PC-2**, $E^* = +1.84$ V vs SCE), the bisphosphonium PCs possess a highly oxidizing

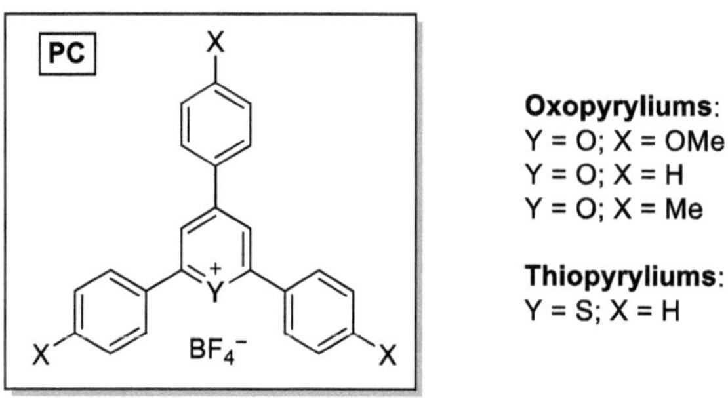

Figure 13.5 Pyrylium photocatalyst derivatives examined in the photocontrolled cationic polymerization of vinyl ethers.

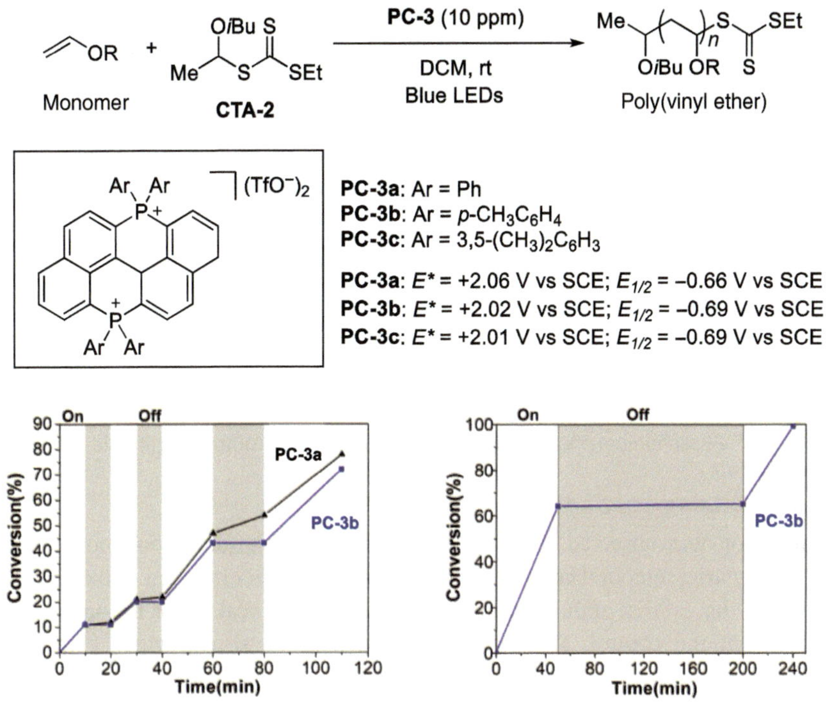

Figure 13.6 Bisphosphonium-mediated photocontrolled cationic polymerization and the temporal control in light on/off experiments.

excited state (E^*_{S1} = +2.02–2.06 V vs SCE) and also a more reducing state after accepting one electron, which, in principle, could afford a more effective deactivation. The bisphosphonium salts could readily promote the photocontrolled cationic polymerization of vinyl ethers at a low-ppm level of catalyst loading (as low as 2.5 ppm) under the irradiation of blue LEDs (λ = 460 nm). In the presence of **CTA-2** as an RAFT agent, the polymerization could afford poly(vinyl ether)s with controlled and predictable molecular mass and low dispersity. This type of bisphosphonium salts can be easily obtained via oxidative cyclization from the corresponding commercially available phosphine ligands [19], and the stability and redox potentials are tunable by varying the Ar groups.

To evaluate these PCs' performance in the temporal control, the authors performed the light on/off experiments with the three bisphosphonium PCs. **PC-3a** showed good temporal control in the early stage, but some loss of control at high conversions (>60%) was also observed, which may be ascribed to its instability under photoirradiation. Upon further optimization, they found introducing methyl substituents (**PC-3b** & **3c**) on the phenyl group could forge a PC with slightly less oxidizing excited states (E^* = +2.02 V vs SCE) and also a better photostability. Both PCs **PC-3b** and **PC-3c** exhibited a high catalytic efficiency (10–20 ppm catalyst loading) and could afford strict temporal control over the polymerization. Even prolonging the dark period (light off) to several hours, no detectable monomer

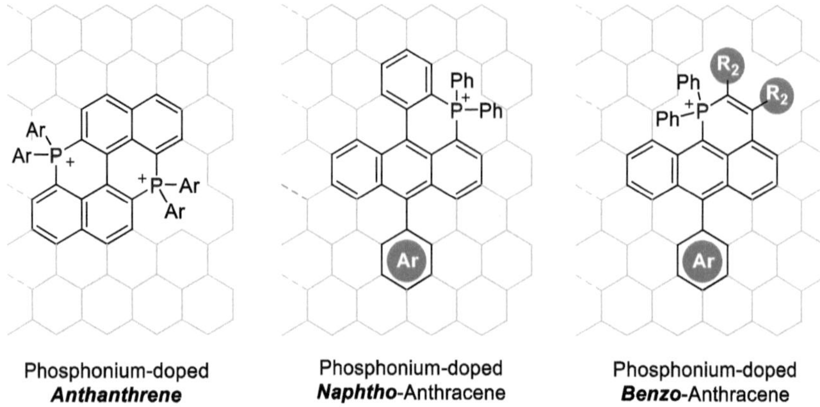

Figure 13.7 Phosphonium photocatalysts developed for photocontrolled cationic polymerization.

consumption was observed. Of note, **PC-3a** could also afford a polymerization with better/strict temporal control by light at lower temperature (e.g. –20 or 0 °C), suggesting the catalyst stability under light may be a critical factor for achieving a strict temporal control. Moreover, these organic PCs also provided excellent temporal control in the DCT cationic polymerization of vinyl ethers by employing thioacetal as a CTA [20].

The effectiveness and features observed with these phosphonium-type PCs encouraged the authors to search for new phosphonium frameworks with better accessibility and tunable redox potentials. In 2022, the same group reported a series of monophosphonium PCs, designed by monophosphonium-doping of naphtho-anthracene, for controlled cationic polymerization (Figure 13.7). This class of organic PCs showed a tunable, broad excited redox window. Even though the catalytic activity of the monophosphonium catalysts was in general lower than that of the bisphosphonium salts, they could also impose a strict temporal control over the polymerization of vinyl ethers under visible light [21]. Very recently, to develop more effective monophosphoium PCs, they prepared a range of monophosphoium salts by the assembly of alkynes and phosphines, which could deliver a higher catalytic efficiency under visible light in the cationic polymerization of vinyl ethers. Compared with the previous bis- and monophosphonium salts, these phosphonium salts possess more sites for modification and readily tunable redox potentials with a broader window. Notably, with these newly developed monophosphonium catalysts, the catalyst loading can be reduced to a low ppm level (e.g. 5 ppm) [22].

Acridinium salts are a class of organic photoredox catalysts widely employed in organic transformations, which showed high oxidizing capability upon excitation by visible light (E^{*}_{red} = ~ 2.06 V vs. SCE) [23, 24]. In 2022, Kamigaito and coworkers disclosed that acridinium salts could also serve as an effective PC for the cationic RAFT polymerization of vinyl ethers under the irradiation of visible light (Figure 13.8) [25]. In this study, they examined a range of acridinium salts with different substituents and counter-anions, which showed different catalytic activity in the model

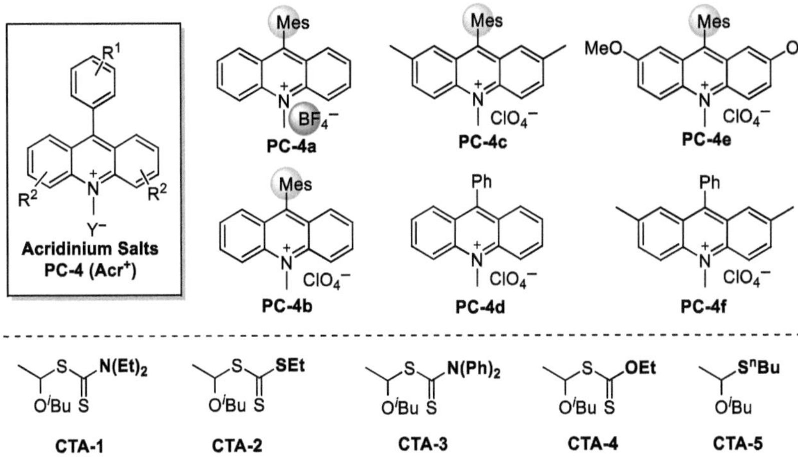

Figure 13.8 Acridinium photocatalysts for photocontrolled cationic polymerization of vinyl ethers.

polymerization of IBVE (in DCM, −40 °C). Among these catalysts, **PC-4d**, which possesses the highest reduction potential at its excited state, provided the fastest reaction, with the monomer completely consumed within four minutes. In the presence of acridinium catalysts with the same counter-anion (ClO_4^-), the polymerization rates were found in the following order: **PC-4e** < **4b** < **4c** < **4f** < **4d**. Further, they also investigated a series of RAFT agents and different polymerization rates were observed. The molecular weights of polymer products can be readily regulated by varying the ratios of monomer/RAFT agent. For example, when the ratio was set to 500/1, the polymerization could afford poly(IBVE) with high molecular weight ($M_n = 40.7\ \text{kg mol}^{-1}$) and low dispersity ($Đ$ 1.28).

The temporal control capability of the catalysts on the polymerization was examined in experiments by switching on/off the light irradiation, which showed good regulation by light through the whole process of polymerization, while a small amount of monomers were still consumed unavoidable in the dark periods (Figure 13.9). The **PC-4a**-catalyzed polymerization exhibited a good character of living polymerization, and the molecular mass of polymers was found to increase gradually and in a linear trend when plotted to the monomer conversion. It is worth mentioning that, Liao and coworkers recently found that an enhanced temporal control can be realized when the *N*-methyl group in acridinium salts was replaced by an aryl group. The 9-mesyl group also demonstrated to be crucial to the observed strict temporal control. The two substituents are suggested as being important to increase the catalysts' photostability to suppress the decomposition of the catalysts under light irradiation, which would lead to background polymerization in dark periods, thus resulting in a loss of photocontrol [26].

In 2022, Zhou, Zhang, and coworkers introduced tris(2,4-dimethoxyphenyl)methylium tetrafluoroborate (**PC-5**) as a PC and successfully developed a photocontrolled cationic polymerization of *p*-MOS under visible light irradiation by using phosphate

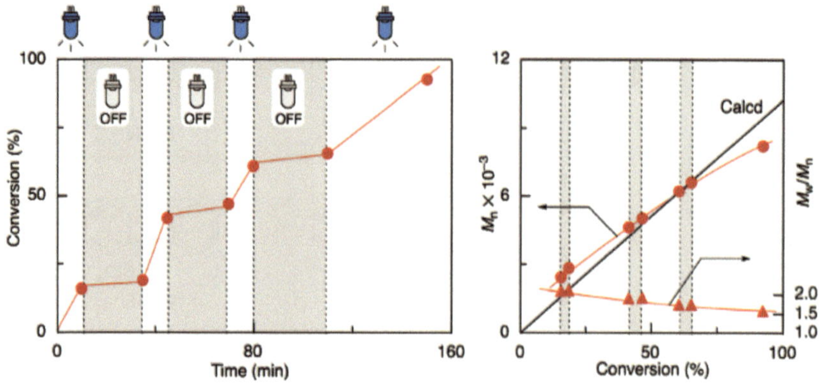

Figure 13.9 Temporal control evaluation of the photomediated cationic polymerization of IBVE with **CTA-2**/**PC-4a**, $[IBVE]_0$/$[\mathbf{PC\text{-}4a}]_0$/$[\mathbf{CTA\text{-}2}]_0$ 500/5.0/0.20 mM, in CH_2Cl_2 at −40 °C, by switching on/off the blue LED irradiation.

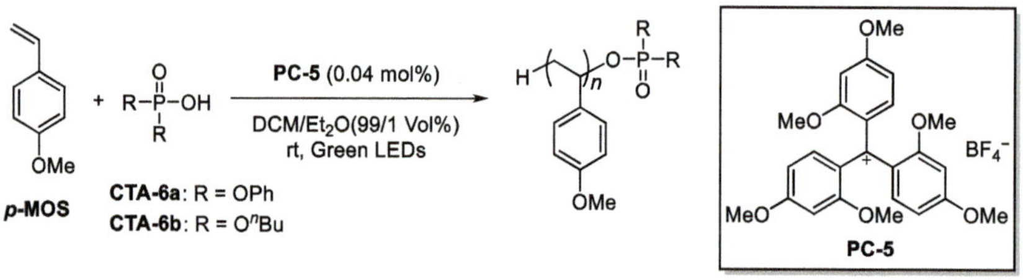

Figure 13.10 Tritylium photocatalysts for the photocontrolled cationic RAFT polymerization of *p*-MOS with phosphate as chain transfer agent. [27]/Springer Nature/CC BY 4.0.

as a CTA (Figure 13.10) [27]. The polymerization could prepare poly(*p*-MOS) with controlled molecular weights and low dispersity. Notably, an excellent temporal control was also achieved. The polymerization only proceeded under the irradiation of green light and ceased immediately after the light source was removed (Figure 13.11). Further, **PC-5** was also effective for the photopolymerization of vinyl ethers. A mechanism involving the single-electron oxidation of the phosphate CTA was proposed, and the phosphate serves as an RAFT agent to mediate the cationic polymerization, which could account for the observed low dispersity.

The aforementioned photocontrolled cationic polymerizations have achieved significant advances in terms of catalytic efficiency and photoregulation by developing organic PCs, while a photocontrolled cationic polymerization with a concurrent effective tacticity-regulation remains elusive. Based on their previous studies on the development of phosphonium-type photoredox catalysts for photocontrolled cationic polymerization [17, 21, 22] and stereoselective cationic polymerization with confined Brønsted acid catalysis [28, 29], Liao and coworkers recently introduced an asymmetric ion-pairing photoredox catalysis strategy, and successfully developed a photocontrolled cationic polymerization with simultaneous light regulation and stereo-chemical control (Figure 13.12). They replaced the small

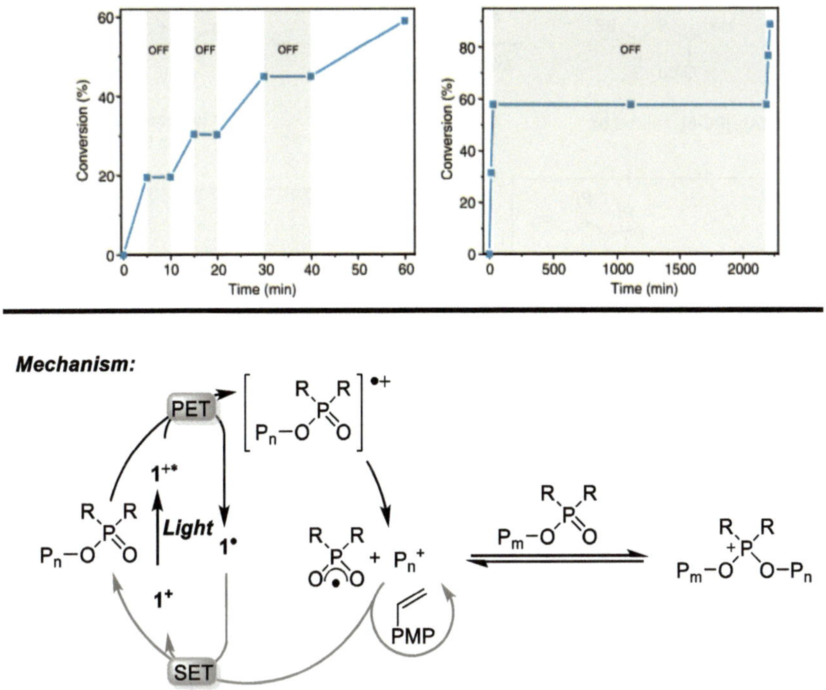

Figure 13.11 Temporal control and proposed mechanism for the photocontrolled living cationic polymerization of *p*-MOS with **PC-5** and phosphate CTA.

anions (e.g. $^{-}$OTf, BF_4^{-}) in the previously developed phosphonium catalysts with a sterically confined chiral anion (*A^{-}) via anion exchange, and thus obtained a series of bifunctional photoredox ion pair catalysts (e.g. **PC-6**: PC^{+}/*A^{-}). The photoredox active cation (PC^{+}) could activate the photocontrolled cationic polymerization by photooxidization of the RAFT chain end, while the bulky anion (*A^{-}) could deliver a stereo-chemical control in the monomer insertion step, thus realizing a tacticity-regulation. Remarkably, with these bifunctional ion pair photoredox catalysts, the cationic polymerization of vinyl ethers could proceed with excellent photocontrol and high stereoselectivity (up to 91% *m*) at a notably low catalyst loading of 50 ppm. **PC-6*** showed an apparent preference for CTA rather than monomer in the single-electron photooxidation process, as indicated by the photoluminescence quenching experiments, in which a strong quenching of **PC-6*** by CTA was observed. Further, the polymerization and chain growth can be effectively controlled by the light "on/off" operations. No apparent monomer conversion and chain growth were observed during the dark periods. These light on/off experiments also suggest the successful establishment of an effective activation-deactivation equilibrium with the ion pair PC, which is a prerequisite for temporal control.

Previously, the authors introduced a confined counter-anion strategy and realized a high stereoselectivity in the cationic polymerization by employing sterically confined Brønsted acid catalysts [28, 29]. Inspired by this success, they conceived

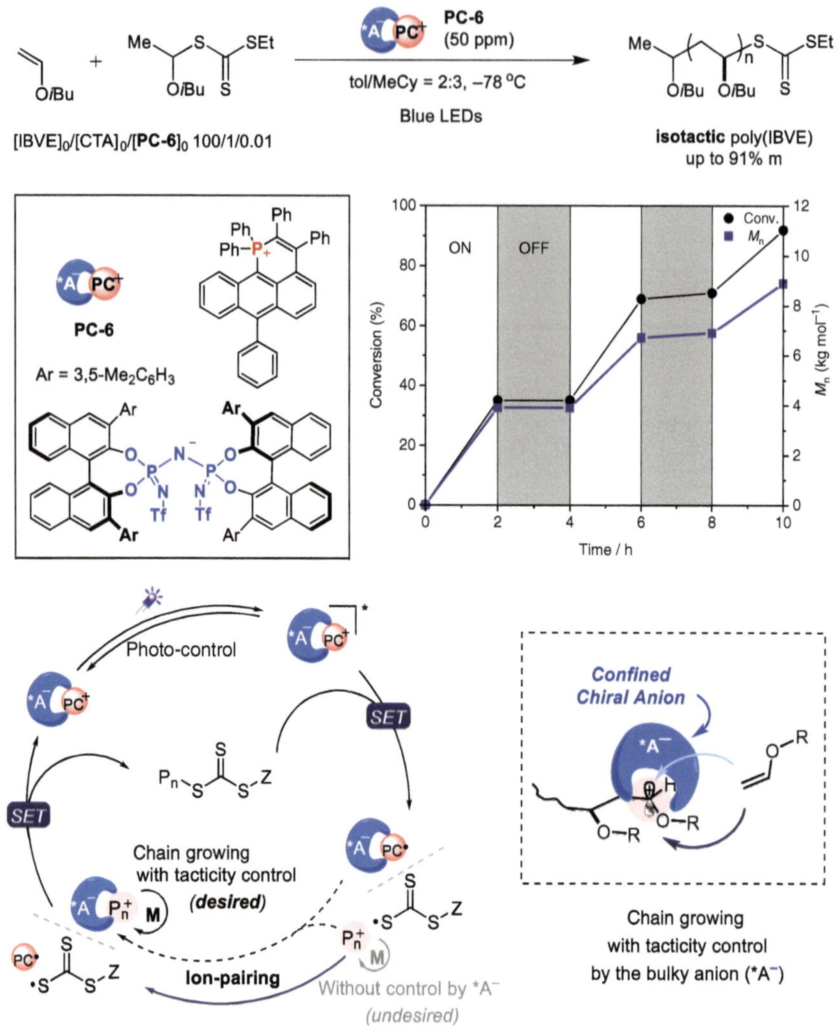

Figure 13.12 Stereoselective photocontrolled living cationic polymerization with ion pair photoredox catalyst **PC-6** and the temporal control. A mechanistic consideration for the stereoselective photocontrolled cationic polymerization with ion pair photoredox catalysts.

these ion pair photoredox catalysts ($PC^+/{*A^-}$) comprising of a bulky anion (${*A^-}$), the anion ${*A^-}$ can combine with P_n^+ during the polymerization via ion pairing by virtue of cation-anion attraction. If the re-combination of the anion ${*A^-}$ (first leave from the reduced PC [PC· / ${*A^-}$]) with the propagating P_n^+ via a cation/anion ion-paring process (the dotted line) is fast enough (Figure 13.12), a chain growth with stereo-chemical mediation on the monomer addition by the bulky anion ${*A^-}$ could become possible, by affecting the trajectory of the incoming vinyl ether to establish a face-selectivity (Figure 13.12, right). As a result, a cationic polymerization with a concurrent photocontrol and stereoselectivity becomes possible.

13.3 Organocatalysts for Photocontrolled Ring-Opening Polymerization

ROP of lactones is an important approach for the synthesis of polyesters, which can be promoted by acid or base catalysts. Accordingly, a number of organocatalysts, including organic acids, organic bases, and organic bifunctional catalysts, have been developed in the past two decades to achieve higher efficiency and better control over the ROP [30–33]. By virtue of the appealing regulation by light, photoinduced ROP technique has also attracted much research interest. In this regard, different photosensitive substances such as photoacid generators (PAGs) and photobase generators (PBGs) have been applied to the ROP, which would decompose to produce catalytic acid or base to initiate/promote the ROP upon exposure to light [34, 35]. For example, sulfonium salts used as PAGs have long been utilized in polymerization, which could undergo photolysis and release protons rapidly under UV irradiation (Figure 13.13). The strong Brønsted acid ($HSbF_6$) generated from the PAGs like triphenylsulfonium hexafluoroantimonate ($Ph_3S^+SbF_6^-$, under UV light, $\lambda = 365$ nm) can effectively initiate/catalyze the ROP of lactones [36, 37]. It is worth mentioning that, besides these classic PAGs, terarylene scaffolds that could undergo photochemical 6π-percyclization can also be employed for the development of photoacid generators [38, 39].

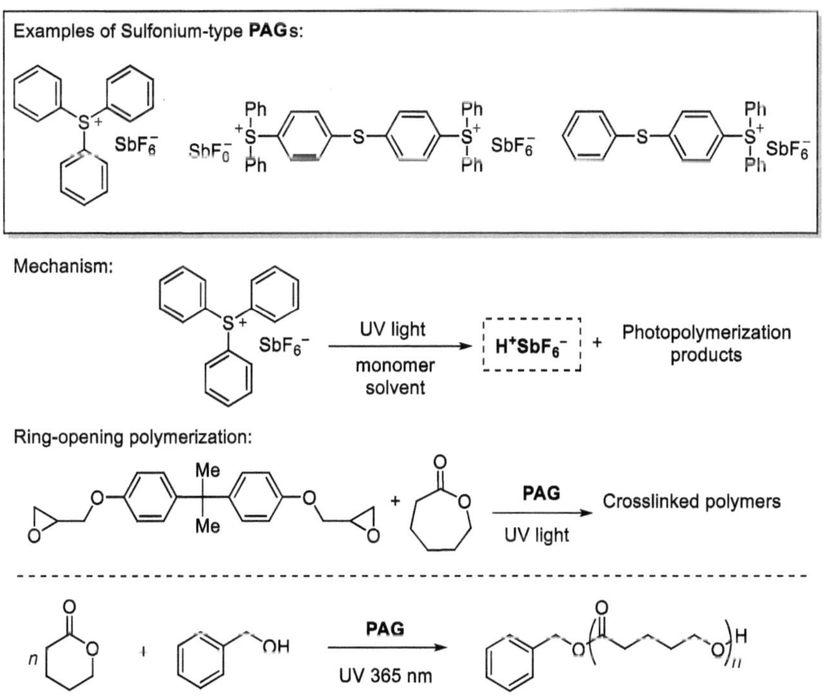

Figure 13.13 Sulfonium-type PAGs and their application in photoinduced ring-opening polymerization.

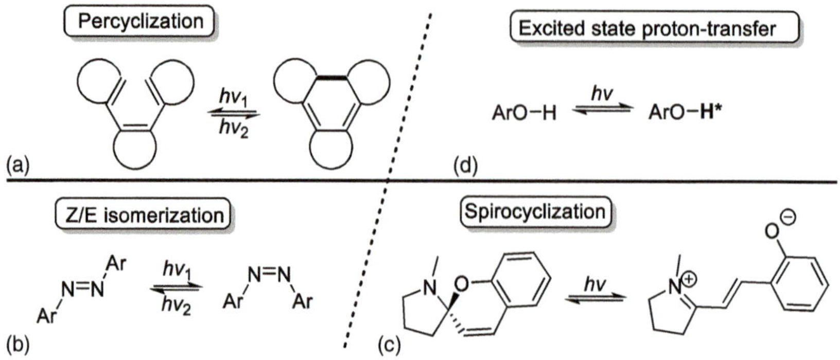

Figure 13.14 Several typical mechanisms employed for the development of photo-switchable catalysts for ring-opening polymerization.

The previously mentioned organic photoacid generators typically involves a photoinduced irreversible decomposition or isomerization, with concurrent extrusion of a strong acid, which could subsequently serve as a catalyst or initiator for the ROP. Whereas, these reversible structure changes make it difficult to establish an activation-deactivation equilibrium that is required for the development of a photocontrolled polymerization, to achieve light regulation on the whole polymerization process and the chain propagation. In the past decade, several organic PCs with photo-switchable catalytic activity have been reported, which demonstrated the great potentials of controlling the ROP process by light. The photo-switchable activity was achieved through different mechanisms (Figure 13.14), but an inherent principle is the switch between active and less/non-active states by light. For example, a) the reversible per-cyclization of triene; b) Z/E isomerization of azo compounds; c) photoinduced spiro-cyclization; and photoinduced excited-state proton transfer (ESPT). The two states often differ not only in catalytic activity but also in their absorption profile, which could thus enable light regulation by wavelength selection [40].

For instance, in 2013, Bielawski and colleagues developed a photo-switchable ROP of lactones by employing a photoresponsive *N*-heterocyclic carbene (NHC) catalyst (Figure 13.15). The NHC catalyst **PC-7** is effective for the ROP of lactones, but it

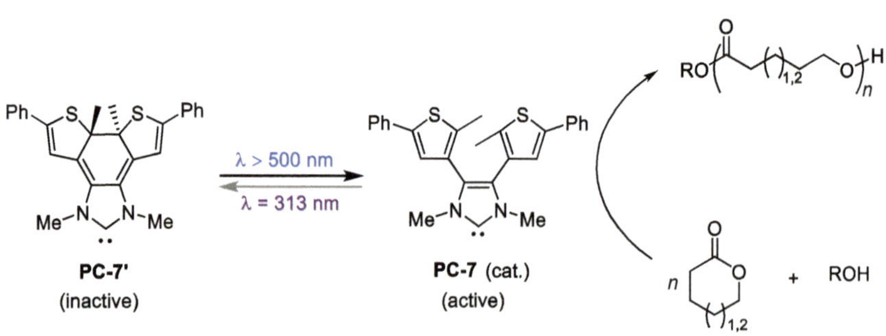

Figure 13.15 Photochromic catalyst for ring-opening polymerization.

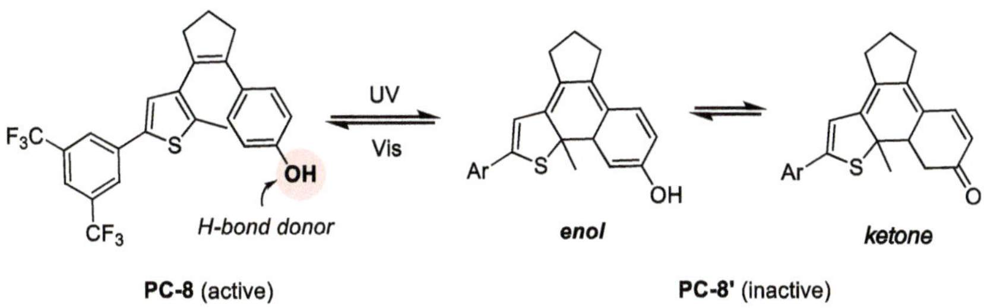

Figure 13.16 Photo-switchable (co)catalyst for photopolymerization of L-lactide. Adapted from [42].

undergoes isomerization/cyclization under UV irradiation. The ring-closure form **PC-7′** was found catalytically inactive for the polymerization, which can be ascribed to the decreased electron density at the corresponding carbene nucleus after cyclization, while exposure to visible light could reverse the cyclization. Thus, the active carbene catalyst can be regenerated in the presence of visible light. As a result, the polymerizations could proceed smoothly under the irradiation of visible light, but UV irradiation will attenuate the polymerization rate. The author also demonstrated the ROP of δ-valerolactone can be reversibly activated and deactivated by switching the irradiation between visible light and UV light [41].

A similar photocyclization mechanism was also utilized for the design of photo-switchable catalysts for the polymerization of lactide monomer. In 2018, Hecht and coworkers developed an effective way to modulate the hydrogen-bonding of **PC-8** with the L-lactide monomer through a light-induced keto-enol tautomerism of the catalyst (Figure 13.16) [42]. In combination with amine cocatalysts, this photo-switchable catalytic system can promote the ROP of lactide. A photocontrol on the polymerization was also demonstrated by switching the light irradiation.

The photoreversible trans-cis isomerization properties of azobenzene compounds have been widely used in tuning the activity of catalysts by excitation with appropriate wavelength light [43], based on the difference in catalytic activity of the corresponding cis and transisomers. In 2016, Wu and coworkers disclosed a thiourea organocatalyst **PC-9** containing a photoresponsive azobenzene-moiety (Figure 13.17) [44], which can be regarded as a thiourea-amine bifunctional catalyst, with the thiourea part to activate the lactide monomer via H-bond activation. The trans configuration with the thiourea is free for H-bonding, while in the cis configuration, this site will be blocked by the meta-nitro group. Accordingly, the active and inactive configuration can inter-convert by switching the light's wavelength (between ambient light and UV light). The authors showed the operation to regulate the polymerization rate by switching on/off the UV irradiation, while a complete shutdown of the polymerization was not achieved, probably due to the slow photoconversion or the presence of a cis/trans equilibrium.

Merocyanines (MEH) are a type of metastable photoacids [45]. Under the irradiation of visible light, MEH can transform to its acidic spiro-form, which can thermally revert back to ring-opened form. Therefore, its acidity under visible light can be harnessed for the development of photoregulated organic transformation

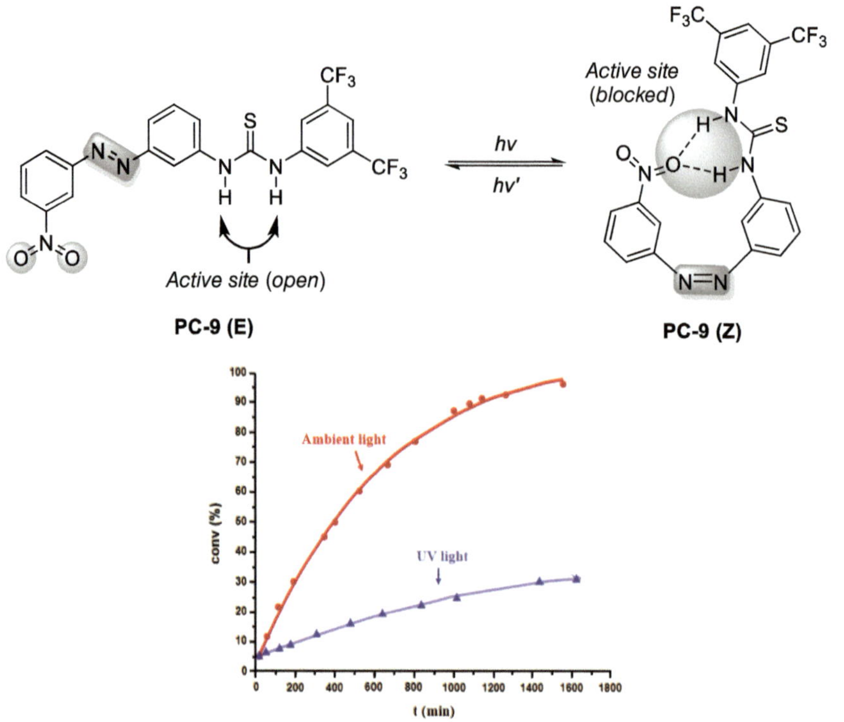

Figure 13.17 Photo-switchable azobenzene-based thiourea catalysts, and the ring-opening polymerization of rac-lactide under UV light (the dark blue line) and ambient light (the red line). Adapted from [44].

and polymerization. In 2016, Boyer and coworkers reported a visible light-regulated ROP of lactones by utilizing the meta-stability of MEH (Figure 13.18) [46]. Under the irradiation of blue light, the photoacid **PC-10a** would undergo an intramolecular cyclization to release proton (H^+), which can be used to catalyze the ROP. As this photoinduced spiro-cyclization is reversible, temporal control on the polymerization can be realized by turning on/off the light. However, significant monomer conversion was also observed when the light was switched off, which is probably due to the slow reversion. In addition, the authors developed an orthogonal polymerization of MEH-mediated photo-ROP and Zn-TPP-catalyzed PET-RAFT polymerization, which was able to alternatively regulate the two different polymerizations by switching the two lights with different wavelengths. It is worth mentioning that Alaniz et al. later examined a series of modified merocyanine photoacids, and found **PC-10b** showing improved solubility in organic solvent as well as an enhanced switching kinetics, which allowed for the realization of a more effective temporal control over the ROP [47]. The influence of the structure on the catalyst solubility, thermal relaxation kinetics, as well as the activation energy were further evaluated by employing ^1H nuclear magnetic resonance (NMR) analysis and pump-probe absorption spectroscopy.

For molecules containing a -OH group like aryl alcohols, excitation by light could cause a perturbation on the population of electron density, and sometimes could

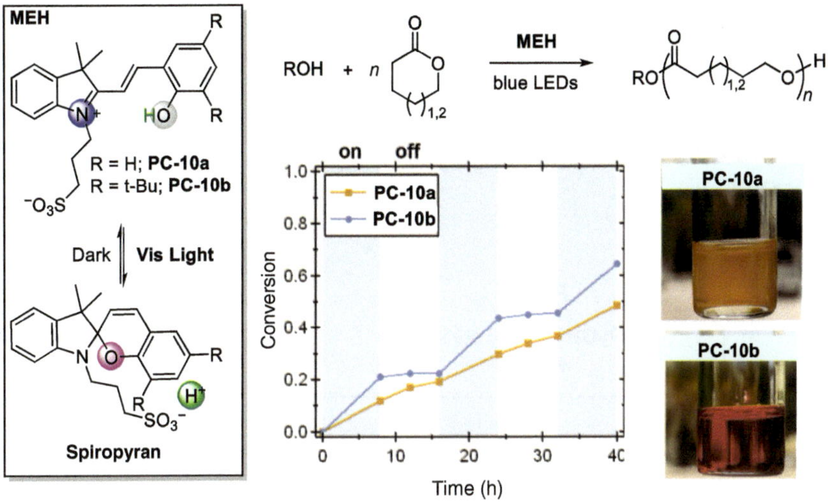

Figure 13.18 Reversible MEH-type photoacids for ring-opening polymerization. Adapted from [46].

result in an elevation of the acidity of the hydroxy protons by several pKa units (even more than 10). This phenomenon is termed as excited-state proton transfer (ESPT) and has been successfully utilized in many fields, including molecule transformation and enzyme activation/inhibition [48, 49]. In 2020, Liao and coworkers first demonstrated that ESPT can be harnessed for the development of photocontrolled polymerization (Figure 13.19) [50]. Initially, they examined a series of aryl alcohols like naphthols as ESPT molecules in combination with photosensitizers (PS) and achieved an effective visible light-regulated ROP of lactones. According to the suggested mechanism, the PS is first excited by light and then undergoes

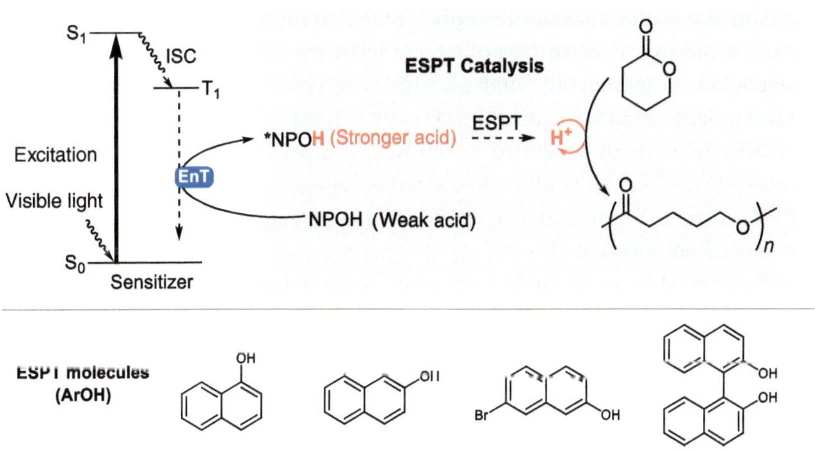

Figure 13.19 ArOH as ESPT catalysts for ring-opening polymerization of lactones in combination with photosensitizers. Adapted from [50].

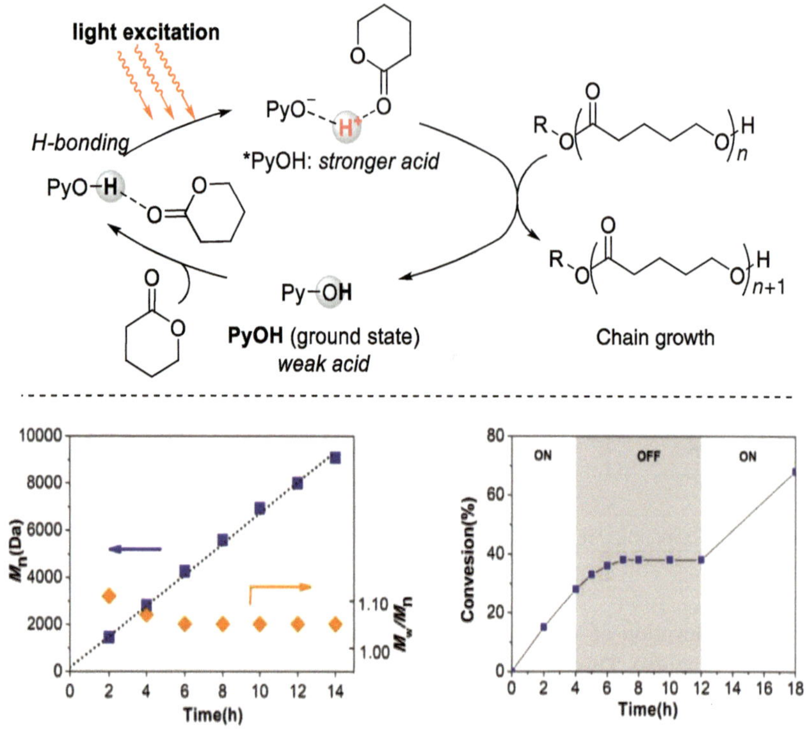

Figure 13.20 1-Hydroxypyrene as a photocatalyst for the visible light-regulated ROP and the light on/off experiments. Zhang et al., [51]/The Royal Society of Chemistry.

intersystem crossing to its triplet excited state. Subsequent triplet energy transfer (EnT) to 1-naphthol (NPOH) could induce the release of protons from NPOH* by the ESPT mechanism, and the protons can serve as an acid catalyst to promote the ROP of lactones. Without PS, NPOH showed low/no activity under the irradiation of blue light, due to the weak/no absorption of the exciting light, while the use of PS could well avoid this disadvantage of simple ESPT molecules like naphthols. This method could well promote the ROP and afford polyester products with controlled molecular weights and low dispersities ($Đ = 1.02$–1.08).

Later on, they found aromatic alcohols like 1-hydroxypyrene (PyOH) with extended conjugation can be directly excited by visible light, and in this case, the use of PS will be not necessary (Figure 13.20) [51]. Compared with NPOH, PyOH with extended conjugation showed an obvious red-shift of absorption maximum and can be excited by visible light (>380 nm). With PyOH as a photoacid catalyst, the polymerization could well proceed under the irradiation of purple light. Kinetic study, chain extension experiments, and block copolymer synthesis further confirmed the living character of this visible light-mediated ROP. The temporal control ability of this system was investigated through light on/off operations by switching on and off the purple light irradiation. When the light was switched off, the polymerization rate decreased and a complete ceasing of reaction can be

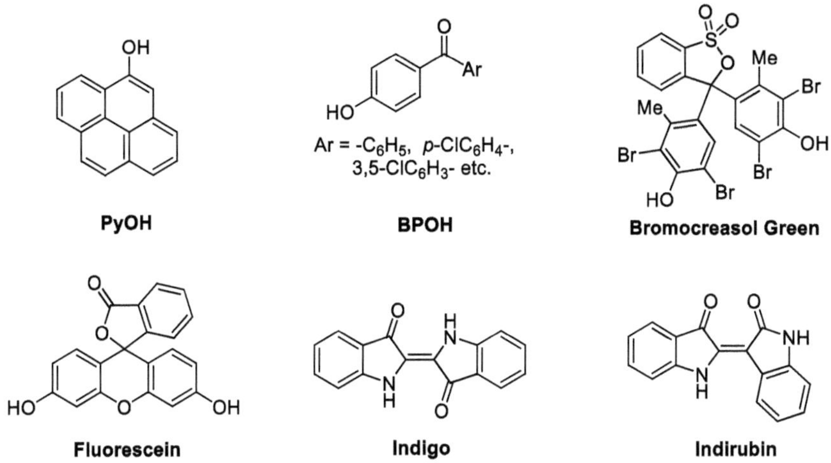

Figure 13.21 Various organic photocatalysts suitable for the development of visible light-regulated ring-opening polymerization of lactones.

realized finally, but with a delay in time. When the reaction was returned to light irradiation, the catalytic activity could recover, proving the possibility of reversibly on–off temporal control over the polymerization by light on/off operations. Control experiments indicated that the photoacid PyOH is a weak acid at ground state, which is not able to promote polymerization. NMR studies suggested H-bonding between PyOH and the carbonyl oxygen of lactones is probably involved in the activation of monomers. The photoinduced increased acidity was further confirmed in the PyOH-catalyzed acetalization reaction between dihydropyran and alcohols, which is a typical acid-catalyzed reaction.

Later on, the same group unveiled a series of molecules, including hydroxybenzophenone (BPOH) and some common organic dyes like Bromocreasol Green and Fluorscein, which could readily promote the ROP of lactones under the irradiation of visible light (Figure 13.21). All these PCs could afford well-defined, biodegradable polyester products with predictable molecular weight and low dispersity (in most cases, $Đ < 1.20$) [52, 53]. In cases of BPOH, modification on the benzophenone structure can further improve their catalytic efficiency, and the Ar group with chloride-substituent at 3,5-positions showed the highest activity, which could afford poly(δ-VL) with controlled molecular mass ($M_n = 4500\,\text{g mol}^{-1}$ at $[\delta\text{-VL}]_0/[\text{Initiator}]_0$ 50 : 1, 88% conversion) and low dispersity ($Đ = 1.05$). Interestingly, an ancient dye, indigo, was also identified as a new and effective photoacid catalyst for the ROP of lactones, which could deliver a well-controlled polymerization under the irradiation of purple light. Excitation by light could enhance the acidity of indigos, which was found crucial to the H-bonding activation of the lactone monomers. Chain extension with successive addition of monomers and block copolymer synthesis were also demonstrated with this method [54]. These studies further expanded the scope of ESPT molecules from aromatic alcohols to organic dyes and amides that lack an –OH group.

13.4 Organocatalysts for Photocontrolled Ring-Opening Metathesis Polymerization (ROMP)

ROMP, which originated from the transition metal catalyst-mediated olefin metathesis, has evolved into one of the most widely employed polymerization technologies that could accommodate a broad range of functionalities in polymer scaffolds and create materials with various topological structures [55, 56]. The system of transition-metal-catalyzed ROMP has been employed for the synthesis of various polymers with applications in many fields, including high-performance plastics, photovoltaics, biomedical engineering, drug delivery, etc. However, the transition-metal-catalyzed processes often suffer from residual metals such as metallic byproducts and metal catalysts, which could result in applications with compromised biological activities, electronic or optical properties, or even unusable. To address this issue, a lot of efforts have been devoted to develop methods to remove these metal components [57]. Whereas, in addition to the direct removal of residual metallic species, the development of a metal-free alternative also represents a good strategy to solve this problem.

In 2006, Chiba reported an electrochemical intermolecular olefin cross-metathesis reaction of vinyl ethers, but an excess of terminal alkene was employed (Figure 13.22, path I) [58]. In this transformation, a single-electron oxidation of the vinyl ether initiator produces the activated radical cation (VE-I), which could subsequently react with the cycloalkene monomer to form a [2 + 2] intermediate (VE-II). VE-II can be reduced to give cyclobutane products (path II) [59], while outcompeting the reduction would lead to the metathesis process. Inspired by these early studies, Boydston and coworkers initially attempted to develop a metal-free ROMP via an electrochemical approach, but the product yields were low. They found the low efficiency can be ascribed to the low solubility of the monomers and also the polymer products in the electrolyte system [60].

They then turned to photoredox catalysis and successfully developed a metal-free photocatalytic system for the ROMP by utilizing organic PCs (Figure 13.23) [60] In the presence of pyrylium **PC-2a**, the ROMP of norbornene could well proceed under the irradiation of blue light LEDs ($\lambda = 450–480$ nm), and high monomer conversions can be obtained, which is in contrast to the low conversion observed under electrochemical conditions. The oxidation potentials of initiators are in the range of 1.43 to

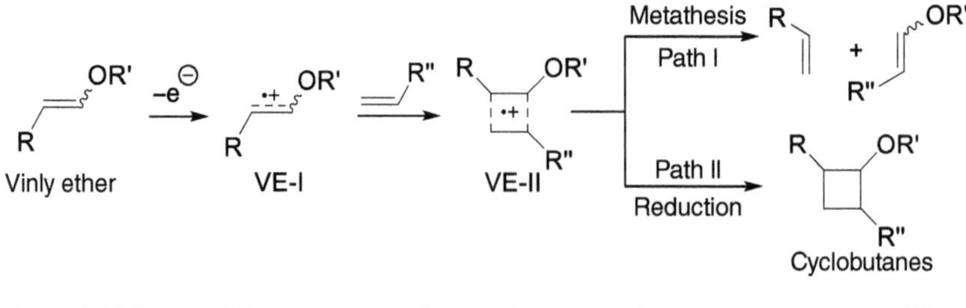

Figure 13.22 Metal-free electrochemical olefin metathesis and the pathways to different products.

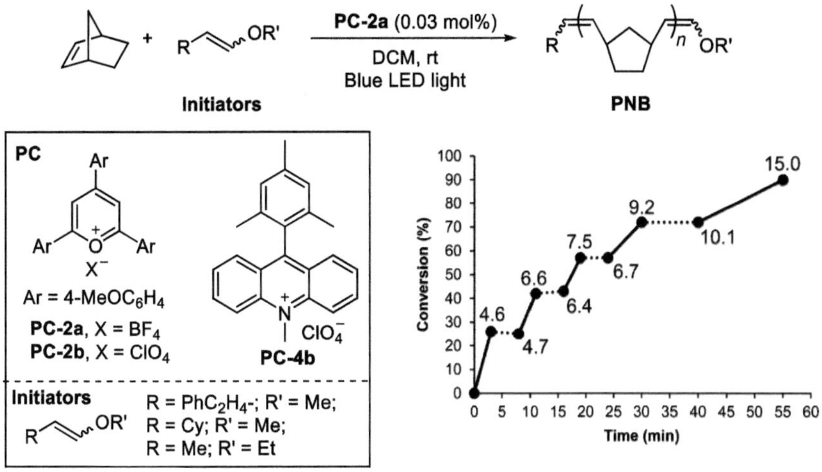

Figure 13.23 Organic photocatalyst-enabled metal-free ROMP and the light on/off experiments.

1.30 V, which are under the oxidizing power of pyrylium and acridinium cations at excited state (**PC-2a** & **PC-4b**, 1.74 and 2.06 V vs SCE, respectively) [61]. Moreover, these PCs could also enable an effective temporal control over the polymerization, by virtue of the establishment of a reversibility between the activation of the enol ethers and deactivation of the active chain ends. As shown in Figure 13.23, this polymerization showed good on/off control upon intermittent exposure to the irradiation of blue LEDs. The polymerization ceased in the dark and can be reinitiated when the light returned.

Although the mechanism of this metal-free ROMP differs from that of traditional ROMP under transition metal catalysis, this photoredox approach produces the same type of polymers. The authors proposed that the vinyl ether undergoes first a single-electron oxidation by the excited pyrylium PC, which affords radical cation **I** (Figure 13.24). Then, a formal [2 + 2] addition of monomer with the radical cation **I** give the key four-membered intermediate **II**. Subsequent metathesis provides/regenerates the propagating radical cation **III**. Notably, the propagating radical cation chain ends and the reduced pyrylium species (PC$^\bullet$) may form a dynamic redox couple, which could regenerate the PC and at the same time afford a dormant species. This termination is reversible, and the reversible reduction and oxidation of the propagating chain ends in **III** is critical for the establishment of an effective activation-deactivation equilibrium of the chain ends, which is a prerequisite for temporal control.

The aforementioned earlier experimental results indicated the formation of a dynamic redox couple in this metal-free ROMP between the propagating radical chain ends and the reduced pyrylium salts. To identify if there is a possible correlation between the catalysts' performance and their excited-state reduction potential (E^*_{red}), the authors later on examined a series of pyrylium and thiopyrylium PCs [57, 62]. Interestingly, these experiments unveiled increasing E^*_{red}

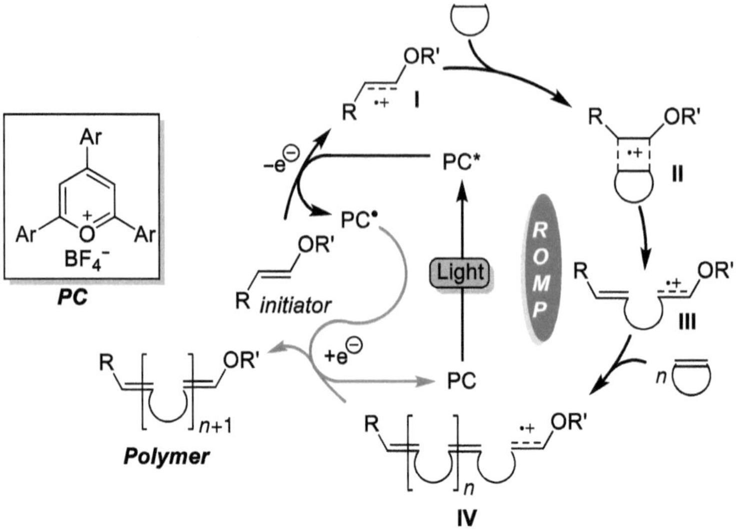

Figure 13.24 The mechanism of metal-free ROMP under photoredox catalysis.

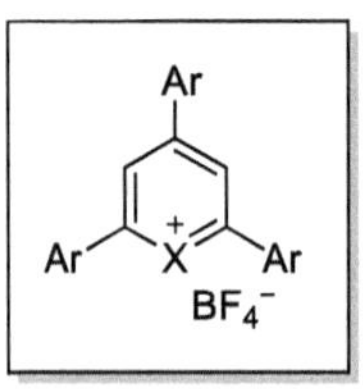

Photocatalyst (PC)	E^* (vs SEC)	Conv. of NB
Ar = 4-MeOC$_6$H$_4$-, X = O	1.89 V	75 ± 4 %
Ar = 4-MeC$_6$H$_4$-, X = O	2.23 V	12 ± 1 %
Ar = C$_6$H$_5$-, X = O	2.46 V	9 ± 1 %
Ar = 4-MeOC$_6$H$_4$-, X = S	1.86 V	84 ± 1 %
Ar = C$_6$H$_5$-, X = S	2.41 V	< 5 %

Figure 13.25 Pyrylium and thiopyrylium photocatalysts examined in the metal-free ROMP.

led to lower monomer conversion (Figure 13.25). When comparing the pyrylium and thiopyrylium salts with the same substituents, higher monomer conversion was observed with the thiopyrylium salt catalysts, suggesting a careful selection of suitable photoredox catalysts is critical to establishing good control over the polymerization. Notably, photooxidants with higher oxidizing capability could also result in over-oxidation of enol ether initiators and even the monomer or main chain olefins.

Compared with traditional transition metal-catalyzed ROMP, this metal-free ROMP system showed a different reactivity and selectivity for certain monomers (Figure 13.26). For example, dicyclopentadiene (DCPD), a common monomer in traditional metal-catalyzed ROMP, could also polymerize under this metal-free ROMP conditions, giving linear, non-cross-linked poly(DCPD) products, albeit with a moderate monomer conversion only. It is worth mentioning that the high activity of metal-based initiators may often result in a cross-linking of poly(DCPD) in traditional ROMP. These results suggest a synthetic potential of the metal-free ROMP, which can be employed for the synthesis of polymer materials previously inaccessible or difficult to synthesize with known methods [63]. The authors further examined a series of functionalized norbornene monomers under the

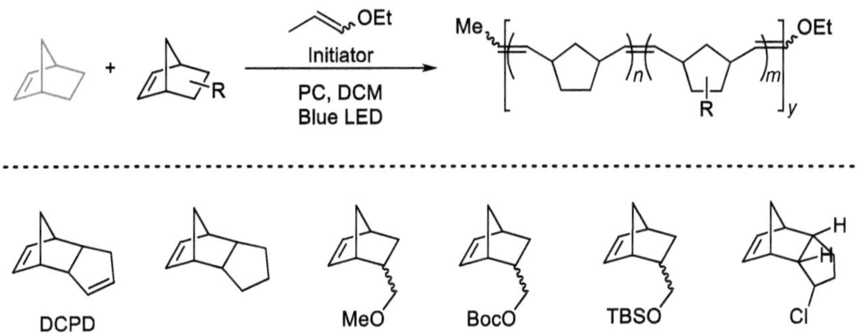

Figure 13.26 Homo- and copolymerizations between norbornene and functionalized monomers.

organocatalytic ROMP conditions, and various homopolymers and copolymers were successfully prepared. The monomers with unprotected alcohol groups were often reluctant to undergo homopolymerization, while some monomers could well proceed in the copolymerizations with norbornene for the synthesis of static or block copolymers. Of note, the *tert*-butyldimethylsilyl (TBS)-protected monomer could also undergo homopolymerization (50% conversion). The chlorinated monomer successfully homopolymerized as well, but this monomer did not react in the copolymerization with norbornene [64]. On the other hand, by virtue of the stability of the reactive radical cation chain ends in a tandem copolymerization, block copolymerization via sequential monomer addition [65] or via a mechanism-switch [66] is also feasible, which could allow access to diverse structures via this metal-free ROMP approach. Recently, the Boydston group examined the continuous flow and droplet flow setups for the metal-free ROMP and compared the results of several common monomers. Of note, by employing the droplet flow reactor, they could achieve comparable conversions to that of batch polymerization [67].

13.5 Organocatalysts for Photocontrolled Step-Growth Polymerization

In contrast to the significant advance in the development of photocontrolled "living" polymerizations, only a few examples of photocontrolled step-growth polymerizations have been reported so far [68]. A typical photocontrolled step-growth polymerization could be the polymerization based on thiol-ene "click" reactions. In 2015, Boyer et al. disclosed an example by employing Ru(bpy)$_3$Cl$_2$ complex as the PC and *p*-toluidine as a redox mediator. This step-growth polymerization method can be used for the synthesis of linear polymers by repeating the thiol-ene addition reaction [69]. In principle, according to the mechanism of the step-wise bond formation in the polymerization, the development of different photocatalytic approaches or activation modes would be required. However, so far few metal-free examples of photocontrolled step-growth polymerizations have been reported based on organic PCs, and this area remains underdeveloped.

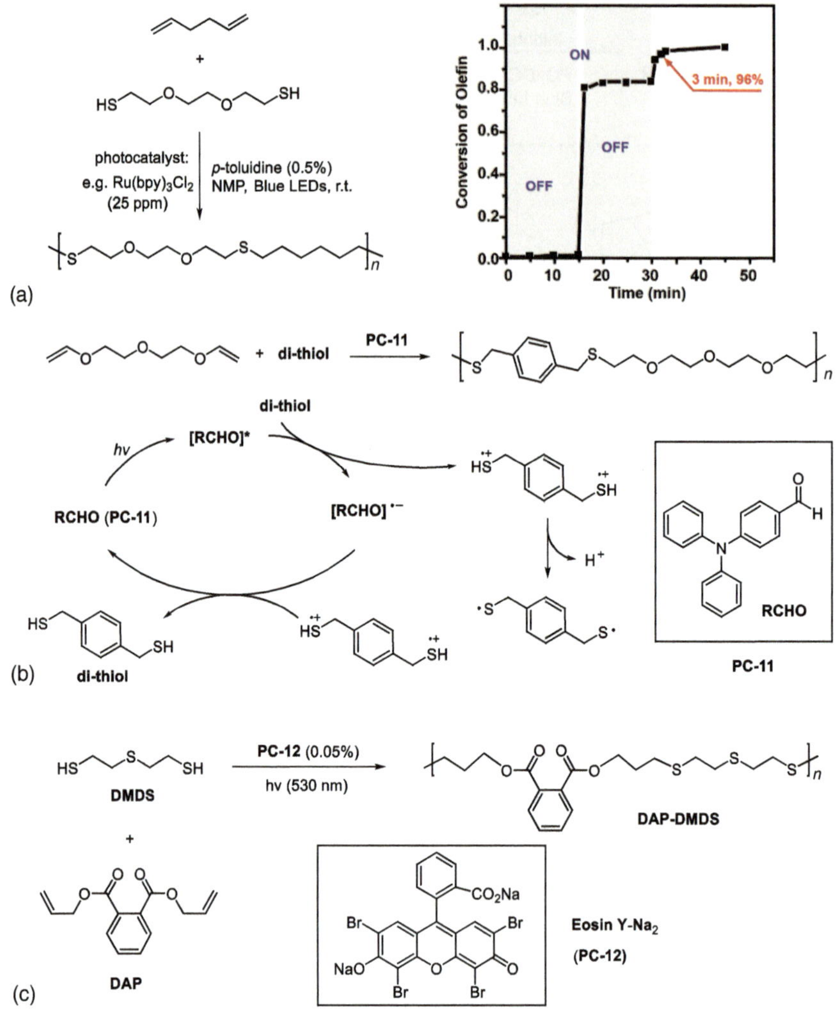

Figure 13.27 Step-growth polymerizations of dithiols and bisolefins under metal-free photoredox catalysis.

In Boyer's study, the thiol-ene click reaction was initially employed for the polymer post-functionalization under visible light and aerobic conditions, and different photoredox catalysts including Ru(bpy)$_3$Cl$_2$ and Eosin Y were examined (Figure 13.27, A). In the case of Eosin Y, a complete modification of the polymer with 1% catalyst was obtained after 24 h of irradiation, by virtue of its high oxidation ability (1.18 V vs SCE), albeit less efficient than that of Ru(bpy)$_3$Cl$_2$. Nevertheless, organic dye PCs provide an advantage in avoiding the involvement of transition metals in the polymer products. Regarding the polymerization mechanism, single-electron oxidation of the thiols was proposed to generate the key sulfur radials, followed by the radical addition to olefins. The use of *p*-toluidine as a redox mediator and *N*-methyl-2-pyrrolidone as the solvent significantly improved the reaction rates and at the same time suppressed the formation of side products. Finally, photocatalytic polymerization was also successfully developed by employing this photocatalytic thiol-ene reaction in the presence of 25 ppm of $Ru(bpy)_3Cl_2$, through a step-growth addition polymerization mechanism [69].

Based on the thiol-ene click reaction, Yang and coworkers also reported the development of an effective step-growth polymerization with 1,4-benzenedimethane thiol (BDMT) and diethylene glycol divinyl ether (DEGVE) as comonomers under the irradiation of visible light by employing 4-(N,N-diphenylamino)-benzaldehyde (DPAB, **PC-11**) as a photoredox catalyst (Figure 13.27, B). The molecular weight of the polythiolether products can be controlled by reaction time and the feeding ratios of monomers. Further, by changing the molar ratios of BDMT to DEGVE, both α,ω-dithiol and α,ω-divinyl telechelic polythiolether oligomers can be synthesized. Of note, polythiolethers with higher molecular weight can be prepared by employing reactive polythiolether oligomers as macro-monomers. For example, α,ω-dithiol oligomer (M_n 2400 g mol^{-1}) could further react with α,ω-divinyl oligomer (M_n 2000 g mol^{-1}) to generate polythiolether with higher molecular weight (M_n 6000 g mol^{-1}) [70]. A photoredox catalytic mechanism involving single-electron oxidation of the di-thiol monomer, followed by the release of protons and subsequent radical addition to the C-C double bonds of DEGVE was proposed. The DPAB PC exhibited a strong UV–vis absorption at 350 nm and a strong fluorescence quenching between BDMT and DPAB was observed.

Recently, Chemtob et al. reported a visible light-mediated thiol-ene aqueous emulsion polymerization, which can be used for the preparation of linear semi-crystalline latexes by employing 2,2′-dimercaptodiethyl sulfide as dithiol monomer in combination with various dienes. The procedure involved LED irradiation, Eosin Y disodium as organic PC (Figure 13.27, C). The polymerization could typically finished in short time scales (<1 h) in the presence of 0.05 mol% catalyst, affording latexes comprising of a linear, thioether repeating structure, with molecular weights of ca. 10 kg mol^{-1} and average diameters of 100 nm. Of note, some polymers obtained from aliphatic dienes, such as di(ethylene glycol) divinyl ether or diallyl adipate, exhibited an apparent crystallization tendency, which was revealed by X-ray diffraction, polarized optical microscopy, and differential scanning calorimetry. Single-electron oxidation of thiols by the excited PC was suggested as a key step in this polymerization [71].

Hydrosilylation is an effective approach for the synthesis of small organosilicon molecules and polymers. Even though transition metal-catalyzed hydrosilylation reactions have been well established, the corresponding metal-free polymerization of dienes and bis(silane)s on the basis of repetitive hydrosilylation remains underdeveloped. In 2021, Pan and coworkers reported a metal-free hydrosilylation polymerization of dienes and bis(silane)s by employing a combination of organic photoredox catalysts and hydrogen atom transfer (HAT) catalysts (Figure 13.28). Well-controlled step-growth hydrosilylation polymerizations can be achieved with both electron-deficient and electron-rich dienes under the irradiation of visible light by choosing appropriate HAT catalysts (**HAT 1** vs **HAT 2**) in combination with an organic photoredox catalyst (**4CzIPN, PC-13**). Based on this approach, a

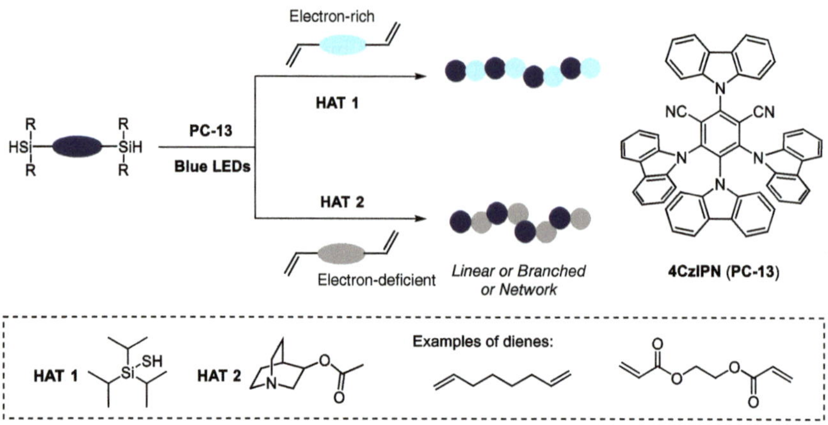

Figure 13.28 Step-growth polymerization of bis(silane)s and bisolefins under organic photoredox and hydrogen atom transfer catalysis.

series of polycarbosilanes from bis(silane)s and dienes were successfully prepared with excellent thermal stability and tunable molecular architectures [72].

For the electron-rich dienes like 1,7-octodiene, the thiol reagent (**HAT 1**) can be first oxidized by the photoredox catalyst 4CzIPN to the sulfur radical, which can then abstract the hydrogen of Si-H in bis(silane) monomer, and give it back to the subsequent radical adducts of silane and alkene (Figure 13.29). Thanks to the selective activation of Si-H bonds and the unfavored essence of radical homopolymerization of electron-rich olefins, linear polymer products can be obtained. For electron-deficient dienes like bisacylates, hydrosilylation polymerization and self-polymerization could occur simultaneously when using the catalyst combination of 4CzIPN and **HAT 2**, which thus provides the opportunity to synthesize linear, hyperbranched, or network polymers. Varying the concentration of electron-deficient dienes as well as the ratio of bis(silane)/diene could alter the proportions of the two polymerizations, thus leading to different products. Later on, the authors further applied this hydrosilylation approach to the post-polymerization modification of various polymer products [73].

The [2 + 2] photopolymerization has been known for more than half a century and widely employed in many fields. The [2 + 2] photopolymerizations were typically carried out in solid states under UV irradiation, while this class of photopolymerizations was rarely achieved in solution under the irradiation of visible light, due to the lack of a monomer preassembly in solution and the monomers having low/no visible light absorption. For example, the study on the photoinduced [2 + 2] polymerization of simple biscinnamate monomers in solid states can be dated back to the 1960s [74], but its solution polymerization remains to be developed. In 2022, Liao and coworkers demonstrated triplet EnT catalysis can be employed to overcome the low efficiency of monomers in the intersystem crossing from singlet to triplet states, and developed the solution [2 + 2] photopolymerization of simple biscinnamate monomers for the first time by employing Ir(ppy)$_3$ as an EnT catalyst [75]. As no monomer preassembly is required in solution polymerization,

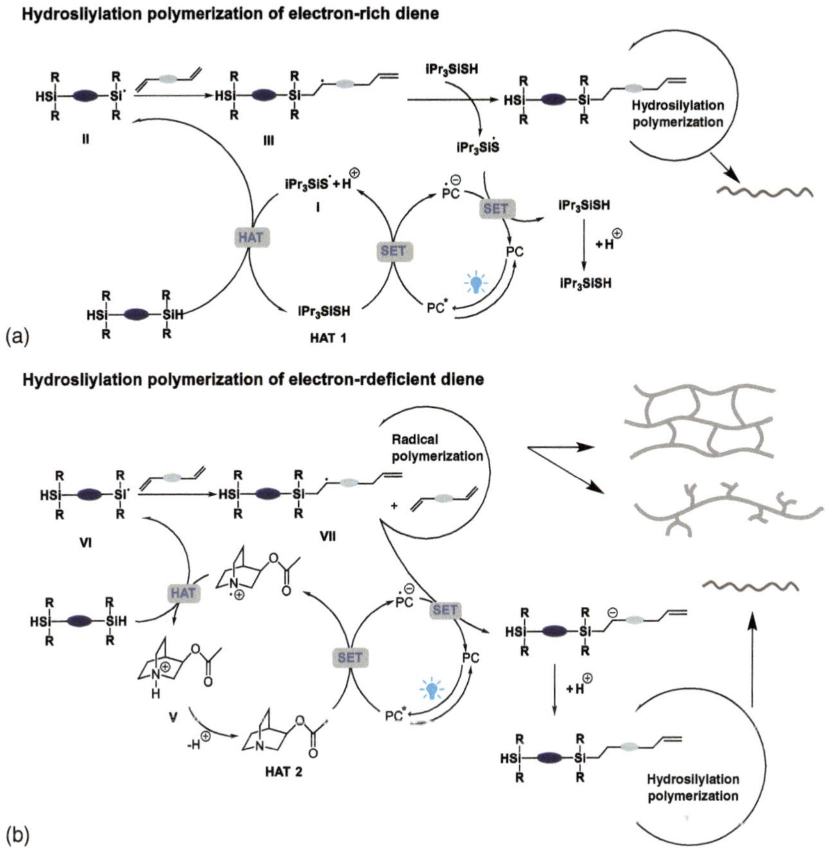

Figure 13.29 Mechanistic proposal for the step-growth polymerization of bis(silane)s and bisolefins.

this new protocol can be applied to biscinnamate monomers with various linker structures. Moreover, the copolymerization of different biscinnamate monomers provides the possibility to further regulate the thermal and mechanical properties of polymers. After this proof-of-concept study, they later developed a metal-free version of the solution [2 + 2] photopolymerization by employing organic PCs (Figure 13.30). A series of common aromatic ketones with higher triplet energy than the monomers, encompassing benzophenone (BP), thioxanthone (TX), and 9-fluorenone (9-FLN), were examined. They further modified the TX catalysts and identified 2,2′-methoxythioxanthone (**PC-14c**) as the most effective triplet EnT PC, which finally enabled the successful development of [2 + 2] solution photopolymerization of biscinnamate monomers under mild conditions and visible light regulation [76]. This organocatalytic system could further expand the monomer scope by replacing the diol linker in biscinnamates with other structures like bisphenol, amino alcohol, diamine, etc. It is worth mentioning that, based on the idea of increasing quantum yield and intersystem crossing by monomer design, they later on developed a visible light-induced [2 + 2] photopolymerization in the

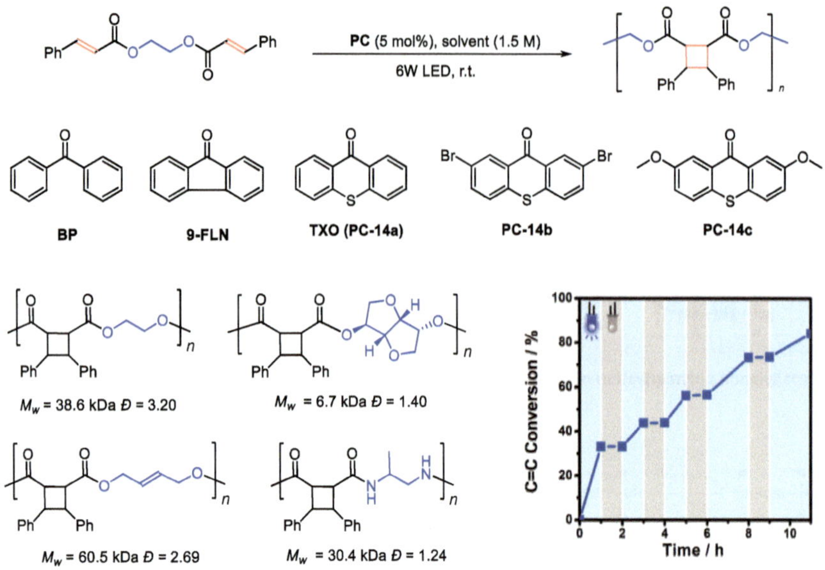

Figure 13.30 Organocatalytic visible light-regulated solution [2 + 2] photopolymerization.

absence of PC [77]. Moreover, very recently, Beeler and Grinstaff found that the continuous flow chemistry could further improve the efficiency of this organocatalytic [2 + 2] photopolymerization in solution in the presence of TX (**PC-14a**) as a PC and successfully synthesize a series of polymers with high molecular weights and low dispersities [78].

References

1 Grahame, J.K. and Erwin, L. (2006). *Thirty Years of Photosynthesis 1974–2004*. Berlin, Heidelberg: Springer.

2 Xiao, P., Zhang, J., Dumur, F. et al. (2015). Visible light sensitive photoinitiating systems: recent progress in cationic and radical photopolymerization reactions under soft conditions. *Prog. Polym. Sci.* 41: 32–66.

3 Garra, P., Fouassier, J.P., Lakhdar, S. et al. (2020). Visible light photoinitiating systems by charge transfer complexes: photochemistry without dyes. *Prog. Polym. Sci.* 107: 101277.

4 Zhou, Y.-N., Li, J.-J., Wu, Y.-Y., and Luo, Z.-H. (2020). Role of external field in polymerization: mechanism and kinetics. *Chem. Rev.* 120 (5): 2950–3048.

5 Fouassier, J.-P. and Lalevée, J. (2021). Reactivity and efficiency of radical photoinitiators. In: *Photoinitiators: Structures, Reactivity and Applications in Polymerization*, 333–337. WILEY-VCH GmbH.

6 Corrigan, N., Shanmugam, S., Xu, J., and Boyer, C. (2016). Photocatalysis in organic and polymer synthesis. *Chem. Soc. Rev.* 45 (22): 6165–6212.

7 Chen, M., Zhong, M., and Johnson, J.A. (2016). Light-controlled radical polymerization: mechanisms, methods, and applications. *Chem. Rev.* 116 (17): 10167–10211.

8 Dadashi-Silab, S., Doran, S., and Yagci, Y. (2016). Photoinduced electron transfer reactions for macromolecular syntheses. *Chem. Rev.* 116 (17): 10212–10275.

9 Corrigan, N., Jung, K., Moad, G. et al. (2020). Reversible-deactivation radical polymerization (controlled/living radical polymerization): from discovery to materials design and applications. *Prog. Polym. Sci.* 111: 101311.

10 Lai, H., Zhang, J., Xing, F., and Xiao, P. (2020). Recent advances in light-regulated non-radical polymerisations. *Chem. Soc. Rev.* 49 (6): 1867–1886.

11 Perkowski, A.J., You, W., and Nicewicz, D.A. (2015). Visible light photoinitiated metal-free living cationic polymerization of 4-methoxystyrene. *J. Am. Chem. Soc.* 137 (24): 7580–7583.

12 Messina, M.S., Axtell, J.C., Wang, Y. et al. (2016). Visible-light-induced olefin activation using 3d aromatic boron-rich cluster photooxidants. *J. Am. Chem. Soc.* 138 (22): 6952–6955.

13 Kottisch, V., Michaudel, Q., and Fors, B.P. (2016). Cationic polymerization of vinyl ethers controlled by visible light. *J. Am. Chem. Soc.* 138 (48): 15535–15538.

14 Uchiyama, M., Satoh, K., and Kamigaito, M. (2015). Cationic raft polymerization using ppm concentrations of organic acid. *Angew. Chem. Int. Ed.* 54 (6): 1924–1928.

15 Michaudel, Q., Chauviré, T., Kottisch, V. et al. (2017). Mechanistic insight into the photocontrolled cationic polymerization of vinyl ethers. *J. Am. Chem. Soc.* 139 (43): 15530–15538.

16 Kottisch, V., Supej, M.J., and Fors, B.P. (2018). Enhancing temporal control and enabling chain-end modification in photoregulated cationic polymerizations by using iridium-based catalysts. *Angew. Chem. Int. Ed.* 57 (27): 8260–8264.

17 Zhang, X., Jiang, Y., Ma, Q. et al. (2021). Metal-free cationic polymerization of vinyl ethers with strict temporal control by employing an organophotocatalyst. *J. Am. Chem. Soc.* 143 (17): 6357–6362.

18 Ma, Q., Song, J., Zhang, X. et al. (2021). Metal-free atom transfer radical polymerization with ppm catalyst loading under sunlight. *Nat. Commun.* 12 (1): 429.

19 Nieto, S., Metola, P., Lynch, V.M., and Anslyn, E.V. (2008). Synthesis of a novel bisphosphonium salt based on 2,2′-bis(diphenylphosphino)-1,1′-binaphthyl (binap). *Organometallics* 27 (14): 3608–3610.

20 Yang, Z., Xiao, W., Zhang, X., and Liao, S. (2022). Organocatalytic cationic degenerate chain transfer polymerization of vinyl ethers with excellent temporal control. *Polym. Chem.* 13 (19): 2776–2781.

21 Yang, Z., Chen, J., and Liao, S. (2022). Monophosphoniums as effective photoredox organocatalysts for visible light-regulated cationic raft polymerization. *ACS Macro Lett.* 11 (9): 1073–1078.

22 Chen, J., Lin, J., Yang, Z. et al. (2023). The development of highly efficient monophosphonium photocatalysts for the visible light-regulated metal-free cationic polymerization of vinyl ethers. *Polym. Chem.* 14 (4): 486–491.

23 Fukuzumi, S., Ohkubo, K., and Suenobu, T. (2014). Long-lived charge separation and applications in artificial photosynthesis. *Acc. Chem. Res.* 47 (5): 1455–1464.

24 Margrey, K.A. and Nicewicz, D.A. (2016). A general approach to catalytic alkene anti-markovnikov hydrofunctionalization reactions via acridinium photoredox catalysis. *Acc. Chem. Res.* 49 (9): 1997–2006.

25 Matsuda, M., Uchiyama, M., Itabashi, Y. et al. (2022). Acridinium salts as photoredox organocatalysts for photomediated cationic raft and dt polymerizations of vinyl ethers. *Polym. Chem.* 13 (8): 1031–1039.

26 Liu, H., Chen, J., Yang, Z. et al. (2024). Light-controlled cationic raft polymerization of vinyl ethers using n-arylacridinium photocatalysts. *Macromolecules* 57 (8): 3751–3756.

27 Wang, L., Xu, Y., Zuo, Q. et al. (2022). Visible light-controlled living cationic polymerization of methoxystyrene. *Nat. Commun.* 13 (1): 3621.

28 Yang, Z., Zhang, X., Jiang, Y. et al. (2022). Organocatalytic stereoselective cationic polymerization of vinyl ethers by employing a confined brønsted acid as the catalyst. *Sci. China: Chem.* 65 (2): 304–308.

29 Zhang, X., Yang, Z., Jiang, Y., and Liao, S. (2022). Organocatalytic, stereoselective, cationic reversible addition–fragmentation chain-transfer polymerization of vinyl ethers. *J. Am. Chem. Soc.* 144 (2): 679–684.

30 Liu, S., Ren, C., Zhao, N. et al. (2018). Phosphazene bases as organocatalysts for ring-opening polymerization of cyclic esters. *Macromol. Rapid Commun.* 39 (24): 1800485.

31 Kamber, N.E., Jeong, W., Waymouth, R.M. et al. (2007). Organocatalytic ring-opening polymerization. *Chem. Rev.* 107 (12): 5813–5840.

32 Dove, A.P. (2012). Organic catalysis for ring-opening polymerization. *ACS Macro Lett.* 1 (12): 1409–1412.

33 Kiesewetter, M.K., Shin, E.J., Hedrick, J.L., and Waymouth, R.M. (2010). Organocatalysis: opportunities and challenges for polymer synthesis. *Macromolecules* 43 (5): 2093–2107.

34 Suyama, K. and Shirai, M. (2009). Photobase generators: recent progress and application trend in polymer systems. *Prog. Polym. Sci.* 34 (2): 194–209.

35 Zivic, N., Kuroishi, P.K., Dumur, F. et al. (2019). Recent advances and challenges in the design of organic photoacid and photobase generators for polymerizations. *Angew. Chem. Int. Ed.* 58 (31): 10410–10422.

36 Barker, I.A. and Dove, A.P. (2013). Triarylsulfonium hexafluorophosphate salts as photoactivated acidic catalysts for ring-opening polymerisation. *Chem. Commun.* 49 (12): 1205–1207.

37 Sangermano, M., Tonin, M., and Yagci, Y. (2010). Degradable epoxy coatings by photoinitiated cationic copolymerization of bisepoxide with ε-caprolactone. *Eur. Polym. J.* 46 (2): 254–259.

38 Li, R., Nakashima, T., and Kawai, T. (2017). A self-contained photoacid generator for super acid based on photochromic terarylene. *Chem. Commun.* 53 (31): 4339–4341.

39 Tian, P., Boomsma, W., Wang, Y. et al. (2015). Structure of a functional amyloid protein subunit computed using sequence variation. *J. Am. Chem. Soc.* 137 (1): 22–25.

40 Neilson, B.M. and Bielawski, C.W. (2013). Illuminating photoswitchable catalysis. *ACS Catal.* 3 (8): 1874–1885.

41 Neilson, B.M. and Bielawski, C.W. (2013). Photoswitchable nhc-promoted ring-opening polymerizations. *Chem. Commun.* 49 (48): 5453–5455.

42 Eisenreich, F., Kathan, M., Dallmann, A. et al. (2018). A photoswitchable catalyst system for remote-controlled (co)polymerization in situ. *Nat. Catal.* 1 (7): 516–522.

43 Léonard, E., Mangin, F., Villette, C. et al. (2016). Azobenzenes and catalysis. *Catal. Sci. Technol.* 6 (2): 379–398.

44 Dai, Z., Cui, Y., Chen, C., and Wu, J. (2016). Photoswitchable ring-opening polymerization of lactide catalyzed by azobenzene-based thiourea. *Chem. Commun.* 52 (57): 8826–8829.

45 Liao, Y. (2017). Design and applications of metastable-state photoacids. *Acc. Chem. Res.* 50 (8): 1956–1964.

46 Fu, C., Xu, J., and Boyer, C. (2016). Photoacid-mediated ring opening polymerization driven by visible light. *Chem. Commun.* 52 (44): 7126–7129.

47 Zayas, M.S., Dolinski, N.D., Self, J.L. et al. (2019). Tuning merocyanine photoacid structure to enhance solubility and temporal control: application in ring opening polymerization. *ChemPhotoChem* 3 (6): 467–472.

48 Tolbert, L.M. and Solntsev, K.M. (2002). Excited-state proton transfer: from constrained systems to "super" photoacids to superfast proton transfer. *Acc. Chem. Res.* 35 (1): 19–27.

49 Demchenko, A.P., Tang, K.-C., and Chou, P.-T. (2013). Excited-state proton coupled charge transfer modulated by molecular structure and media polarization. *Chem. Soc. Rev.* 42 (3): 1379–1408.

50 Zhang, X., Hu, S., Ma, Q., and Liao, S. (2020). Visible light-mediated ring-opening polymerization of lactones based on the excited state acidity of espt molecules. *Polym. Chem.* 11 (22): 3709–3715.

51 Zhang, X., Ma, Q., Jiang, Y. et al. (2021). Visible light-regulated organocatalytic ring-opening polymerization of lactones by harnessing excited state acidity. *Polym. Chem.* 12 (6): 885–892.

52 Hu, S., Chen, J., Zhang, X. et al. (2022). Visible light-regulated organocatalytic ring-opening polymerization of lactones using hydroxybenzophenones as photocatalyst. *ACS Appl. Polym. Mater.* 4 (5): 3361–3368.

53 Liu, Z., Zhang, X., Sun, P. et al. (2023). Harnessing the photoacidity of organic dyes for the development of ring-opening polymerization of lactones under visible light. *Polym. Chem.* 14 (38): 4414–4418.

54 Sun, P., Li, Z., Zhang, X. et al. (2024). Visible light-regulated ring-opening polymerization of lactones by employing indigo as a photoacid catalyst. *Macromol. Rapid Commun.* 45 (11): 2400054.

55 Bielawski, C.W. and Grubbs, R.H. (2007). Living ring-opening metathesis polymerization. *Prog. Polym. Sci.* 32 (1): 1–29.

56 Schrock, R.R. (2014). Synthesis of stereoregular polymers through ring-opening metathesis polymerization. *Acc. Chem. Res.* 47 (8): 2457–2466.

57 Lu, P., Kensy, V.K., Tritt, R.L. et al. (2020). Metal-free ring-opening metathesis polymerization: from concept to creation. *Acc. Chem. Res.* 53 (10): 2325–2335.

58 Miura, T., Kim, S., Kitano, Y. et al. (2006). Electrochemical enol ether/olefin cross-metathesis in a lithium perchlorate/nitromethane electrolyte solution. *Angew. Chem. Int. Ed.* 45 (9): 1461–1463.

59 Chiba, K., Miura, T., Kim, S. et al. (2001). Electrocatalytic intermolecular olefin cross-coupling by anodically induced formal [2+2] cycloaddition between enol ethers and alkenes. *J. Am. Chem. Soc.* 123 (45): 11314–11315.

60 Ogawa, K.A., Goetz, A.E., and Boydston, A.J. (2015). Metal-free ring-opening metathesis polymerization. *J. Am. Chem. Soc.* 137 (4): 1400–1403.

61 Fukuzumi, S., Kotani, H., Ohkubo, K. et al. (2004). Electron-transfer state of 9-mesityl-10-methylacridinium ion with a much longer lifetime and higher energy than that of the natural photosynthetic reaction center. *J. Am. Chem. Soc.* 126 (6): 1600–1601.

62 Pascual, L.M.M., Dunford, D.G., Goetz, A.E. et al. (2016). Comparison of pyrylium and thiopyrylium photooxidants in metal-free ring-opening metathesis polymerization. *Synlett* 27 (05): 759–762.

63 Goetz, A.E. and Boydston, A.J. (2015). Metal-free preparation of linear and cross-linked polydicyclopentadiene. *J. Am. Chem. Soc.* 137 (24): 7572–7575.

64 Goetz, A.E., Pascual, L.M.M., Dunford, D.G. et al. (2016). Expanded functionality of polymers prepared using metal-free ring-opening metathesis polymerization. *ACS Macro Lett.* 5 (5): 579–582.

65 Pascual, L.M.M., Goetz, A.E., Roehrich, A.M., and Boydston, A.J. (2017). Investigation of tacticity and living characteristics of photoredox-mediated metal-free ring-opening metathesis polymerization. *Macromol. Rapid Commun.* 38 (13): 1600766.

66 Lu, P. and Boydston, A.J. (2019). Integration of metal-free ring-opening metathesis polymerization and organocatalyzed ring-opening polymerization through a bifunctional initiator. *Polym. Chem.* 10 (23): 2975–2979.

67 Rigoglioso, V.P. and Boydston, A.J. (2023). Flow optimization of photoredox-mediated metal-free ring-opening metathesis polymerization. *ACS Macro Lett.* 12 (11): 1479–1485.

68 Yilmaz, G. and Yagci, Y. (2020). Light-induced step-growth polymerization. *Prog. Polym. Sci.* 100: 101178.

69 Xu, J. and Boyer, C. (2015). Visible light photocatalytic thiol–ene reaction: an elegant approach for fast polymer postfunctionalization and step-growth polymerization. *Macromolecules* 48 (3): 520–529.

70 Wang, Y., Zhang, X., Ma, Y. et al. (2020). Polythioethers with controlled α,ω-end groups prepared by visible light induced thiol–ene click polymerization of dithiol and divinyl ether with 4-(n,n-diphenylamino)benzaldehyde as organocatalyst. *Macromol. Chem. Phys.* 221 (6): 1900557.

71 Le, C.M.Q., Schrodj, G., Ndao, I. et al. (2022). Semi-crystalline poly(thioether) prepared by visible-light-induced organocatalyzed thiol-ene polymerization in emulsion. *Macromol. Rapid Commun.* 43 (5): 2100740.

72 Huang, Z., Chen, Z., Jiang, Y. et al. (2021). Metal-free hydrosilylation polymerization by merging photoredox and hydrogen atom transfer catalysis. *J. Am. Chem. Soc.* 143 (45): 19167–19177.

73 Huang, Z., Liu, K., Liu, M. et al. (2023). Photocatalytic metal-free radical hydrosilylation for polymer functionalization. *Chin. J. Chem* 41 (18): 2275–2281.

74 Dilling, W.L. (1983). Organic photochemistry. Xvii. Polymerization of unsaturated compounds by photocycloaddition reactions. *Chem. Rev.* 83 (1): 1–47.

75 Jiang, Y., Ma, Q., Zhang, X. et al. (2022). Solution [2 + 2] photopolymerization of biomass-derived nonrigid biscinnamate monomers enabled by energy transfer catalysis. *Polym. Chem.* 13 (17): 2538–2544.

76 Jiang, Y., Zhu, H., Chen, J., and Liao, S. (2023). Organocatalytic [2 + 2] photopolymerization under visible light: accessing sustainable polymers from cinnamic acids. *Macromol. Rapid Commun.* 44 (4): 2200702.

77 Jiang, Y., Zhu, H., Chen, J. et al. (2022). Linear cyclobutane-containing polymer synthesis via [2 + 2] photopolymerization in an unconfined environment under visible light. *ACS Macro Lett.* 11 (12): 1336–1342.

78 El-Arid, S., Lenihan, J.M., Jacobsen, A. et al. (2024). Accessing cyclobutane polymers: overcoming synthetic challenges via efficient continuous flow [2 + 2] photopolymerization. *ACS Macro Lett.* 13 (5): 607–613.

14

Enzyme-Catalyzed Controlled Radical Polymerization

Ruoyu Li[1] *and Zesheng An*[1,2]

[1] *State Key Laboratory of Supramolecular Structure and Materials, College of Chemistry, Jilin University, 2699 Qianjin Street, Changchun 130012, China*
[2] *Key Laboratory for Molecular Enzymology and Engineering of Ministry of Education, School of Life Sciences, Jilin University, 2699 Qianjin Street, Changchun 130012, China*

14.1 Introduction

Free-radical polymerization (FRP) features simple operation, high efficiency and adaptability to a wide range of monomers. Numerous polymer products in our daily life are obtained via FRP. The parameters of polymers such as molecular weight (MW), dispersity ($Đ$), topology, sequence, and end-group functionality play a critical role in determining the properties of polymer materials [1–11]. However, in FRP the propagating radicals have short lifetimes and termination is rapid and irreversible, limiting the precise control of polymer parameters. To solve these problems, reversible deactivation radical polymerization (RDRP), also known as controlled radical polymerization (CRP), has been developed [12–17]. Arguably, the two most popular CRP techniques are reversible addition–fragmentation chain transfer (RAFT) polymerization and atom transfer radical polymerization (ATRP), which are based on the mechanisms of degenerative transfer and reversible deactivation, respectively, to control the equilibrium between active and dormant chains, so that each polymer chain grows almost simultaneously and ultimately the obtained polymers have preset MWs and low dispersities [18–20]. In addition, customization of polymer topology is easily achievable by designing RAFT agents or ATRP initiators with diverse structures [21–23].

With the escalation of global problems such as environmental pollution, climate change, and energy scarcity, increasing demands have been placed on the sustainability of polymer materials from the aspects of both synthesis and application [24–26]. The implementation of green synthetic methods with low energy consumption is an important aspect of the sustainable development of polymer materials. Therefore, CRP is also experiencing an evolutionary trend from traditional thermal initiation and metal catalysis to control by externally applied stimuli (such as light, electricity, and mechanical force), biocatalysis, and others [27–33]. Among these, enzyme-catalyzed CRP has attracted great interest due to its

Organocatalysts in Polymer Chemistry: Synthesis and Applications, First Edition. Edited by Zhibo Li.

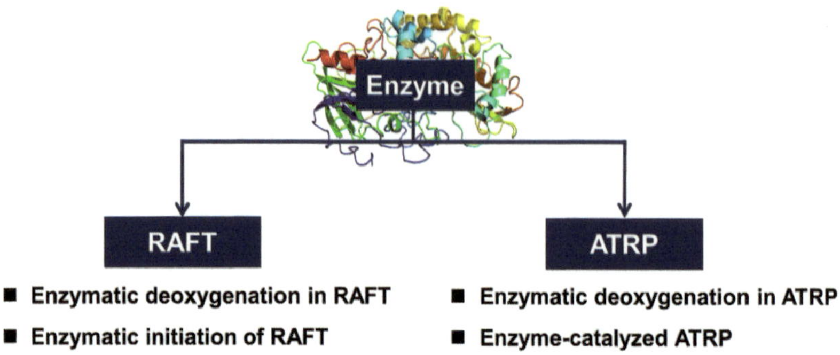

Figure 14.1 Enzyme-catalyzed CRP.

unique advantages of mildness, high efficiency, low toxicity, and oxygen tolerance [34–36]. The use of metalloproteins to catalyze ATRP in 2011 opened a new era of enzyme-catalyzed CRP, and since then, various enzyme catalysis strategies have been developed to enable CRP to be conducted under mild conditions [37–39].

Thanks to the unique advantages of enzyme-catalyzed CRP, some difficult-to-synthesize polymers such as multi-block copolymers [40], polymers based on unconjugated monomers [41], and ultrahigh molecular weight (UHMW) polymers [42] have been successfully obtained. Up to now, enzyme-catalyzed CRP has been used in many fields such as high-throughput polymerization, polymerization-induced self-assembly (PISA), surface brush fabrication, and sensing [43–46]. In this chapter, we summarize the enzyme catalysis strategies developed for RAFT and ATRP and their applications (Figure 14.1).

14.2 Enzymatic Deoxygenation in RAFT Polymerization

Oxygen is a potent radical quencher and inhibits both FRP and CRP, making it challenging to perform radical polymerization without prior deoxygenation [47, 48]. To ensure a successful CRP, oxygen is often removed prior to polymerization by freeze-pump-thaw cycles or purging with inert gas, which are time-consuming steps that are not friendly to inexperienced operators. Biologists discovered long ago that under aerobic conditions, glucose oxidase (GOx) catalyzes the oxidation of glucose by oxygen to generate gluconic acid and hydrogen peroxide (H_2O_2) [49].

Although the first use of GOx deoxygenation in FRP occurred in 1991 [50], it was not used in CRP until a decade ago when Stevens and coworkers used GOx to deoxygenate a RAFT polymerization using VA-044 as a low-temperature initiator [51]. They found that GOx had excellent deoxygenation ability and the polymerization could be carried out smoothly at only 1–4 μM of enzyme. GOx activity assay indicated that this enzyme exhibited excellent tolerance to temperature and organic solvents, where the enzyme activity was almost unchanged after 2.5 h of polymerization at 45 °C in phosphate buffered saline (PBS) with 80% acetonitrile or *tert*-butanol. Polymerizations of various monomers such as 2-hydroxyethyl acrylate

(HEA), *N,N*-dimethylacrylamide (DMA), and *N*-(2-hydroxypropyl) methacrylamide in mixtures of buffer solution and organic solvents achieved high monomer conversions (≥ 90%). Due to its low cost, robustness, and exceptional degassing ability, GOx has been the most popular deoxygenase used in CRP. Rowan's group showed that GOx could even be used for deoxygenation in complex aqueous solvents such as beer, wine, liquor, and fermentation broth [52]. PHEAs synthesized through thermally initiated RAFT polymerization exhibited low dispersities ($Đ \leq 1.2$) and high end-group fidelity in all the complex solvents tested.

While, in the aforementioned RAFT polymerizations, GOx deoxygenation was decoupled from initiation, and it is also possible to utilize the H_2O_2 produced by enzyme deoxygenation to chemically/enzymatically react with other substances to generate radicals. Thus, deoxygenation and initiation can be coupled. Here, we first present examples that utilize H_2O_2 produced by enzymatic deoxygenation to generate radicals through chemical reactions. Enzyme cascade catalysis using H_2O_2 as a bridge will be discussed in the next section. In 2017, our group coupled H_2O_2 produced by enzymatic deoxygenation with vitamin C to create a redox couple, which was used to generate hydroxyl radicals to initiate RAFT polymerization [53]. Well-defined homopolymers and block copolymers were efficiently synthesized in both open and sealed vessels. Besides, the in situ generated H_2O_2 can also be coupled with Fe^{2+} to generate hydroxyl radicals through the Fenton reaction [54]. For example, Qiao's group constructed GOx deoxygenation/Fenton reaction cascade systems using commercially available iron salts [55] and the heme of hemoglobin [56] as the iron source. UHMW poly(*N,N*-dimethylacrylamide) (PDMA) with $M_w \sim 20 \times 10^6\ \text{g mol}^{-1}$ was obtained by this Fenton RAFT polymerization, as this system provided a low concentration of free radicals in a sustained manner [55]. Additionally, they slowly injected H_2O_2 into the solution containing ammonium ferrous sulfate via a syringe pump in order to mimic the deoxygenation process of GOx, which similarly allowed to obtain precisely controlled UHMW polymers. More recently, the same group used this GOx deoxygenation/Fenton reaction cascade system for the automated synthesis of multi-block polymers [57]. The successful synthesis of well-defined polymers ($Đ = 1.06$) with 39 blocks demonstrated the high livingness of this approach.

Strategies to accelerate the rate of radical production and hence the rate of polymerization in these GOx-related redox systems have also been reported. Tang et al. utilized hemin to activate the O-O bond in H_2O_2 generated by GOx deoxygenation [58]. This resulted in an increased rate of hydroxyl radical production from the reaction of H_2O_2 and ascorbic acid, ultimately accelerating the rate of RAFT polymerization. In another report, Rawson and Gurnani used *Cupriavidus metallidurans* CH_{34} to reduce Fe^{3+} produced by the GOx deoxygenation/Fenton reaction cascade system to Fe^{2+}, which accelerated the rate of hydroxyl radical generation [59]. Both water-soluble polymers and block copolymer nanoparticles were synthesized by this method. For these systems with accelerated polymerization rates, attention needs to be paid to maintaining a balance between polymerization rate and livingness.

When enzymes are used for deoxygenation at high temperatures or in more poisonous solvents, immobilization of the enzymes in porous materials or on the surface of solid materials could increase their tolerance to temperature and organic solvents [60–62]. Importantly, the immobilized enzymes could be recycled while simplifying the polymer purification steps. In 2019, Qiao et al. covalently attached GOx to the reaction glassware, which increased the organic solvent resistance and storage stability of GOx [63]. Besides, GOx-functionalized glassware showed excellent recyclability, with deoxygenation activity remaining more than 60% after nine cycles of polymerization. RAFT polymerization using this GOx-functionalized glassware provided the same level of excellent control as that conducted with argon deoxygenation. Tang et al. encapsulated GOx using a metal–organic framework, zeolitic imidazolate framework-8 (ZIF-8), and the enzyme activity assay showed that the relative activity of immobilized GOx remained as high as 73.7% after heating the solution to 80 °C in a mixed solvent of DMF and water (1,1 by vol%) [64]. Given the gradual maturation of enzyme immobilization technology and the growing industrial applications of immobilized enzymes, immobilized enzymes will be used more frequently in CRP to facilitate both synthesis and purification.

In addition to GOx, pyranose oxidase (P2Ox) [40] and formate oxidase (FOx) [65] have also been used for deoxygenation in RAFT polymerization (Figure 14.2). Compared to GOx, P2Ox has a broader substrate range and higher affinity to both glucose and oxygen. Importantly, the 2-dehydro-D-glucose produced by P2Ox-degassing (using glucose as substrate) is more resistant to hydrolysis and does not lower the pH of the solution [66]. FOx utilizes a C_1 substrate formic acid, while GOx uses a C_6 substrate glucose. Thus, the use of FOx instead of GOx is atom–economic. Additionally, CO_2 resulting from FOx deoxygenation can readily escape from the solution and does not accumulate as waste in the polymerization solution [67, 68].

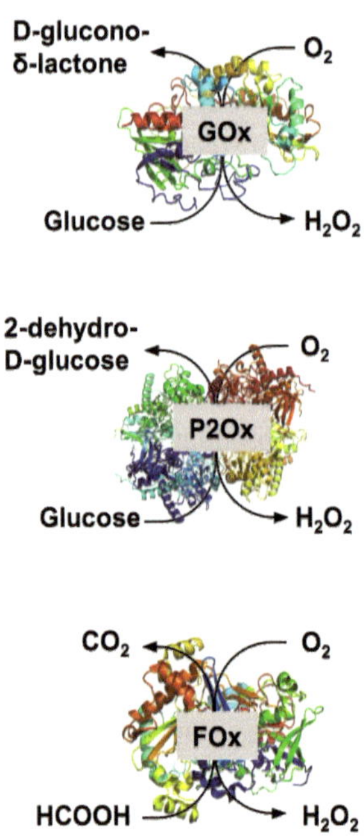

Figure 14.2 Enzymes that have been used for deoxygenation in RAFT polymerization.

The outstanding deoxygenation capability of oxidases makes it possible to conduct polymerization on benchtops with different reactor geometries where deoxygenation by physical processes can be challenging. For example, multi-well plates with reaction volumes ranging from 50 to 350 μL are often chosen as vessels for high-throughput RAFT polymerization because smaller polymerization volumes reduce raw material consumption and costs [69]. However, polymerization in multi-well plates is more difficult than in common reaction vials because oxygen removal from multi-well plates cannot be easily conducted using conventional physical degassing methods. Additionally, the high ratio of surface area to volume of multi-well plate requires a higher oxygen tolerance. For instance, Chapman et al. demonstrated a GOx-degassed high-throughput RAFT polymerization method to generate polymer libraries without complicated robotics or degassing steps [70]. Well-defined homopolymers and block copolymers could be obtained at volumes down to 40 μL. Furthermore, GOx-degassed RAFT polymerization was also utilized for high-throughput PISA to produce polymeric nanoparticles using a photoinitiator sodium phenyl-2,4,6-trimethylbenzoylphosphinate (SPTP) [71]. Specifically, poly(poly(ethylene glycol) methyl ether methacrylate) (PPEGMA) macro-chain transfer agent (CTA) was chain-extended with 2-hydroxypropyl methacrylate (HPMA) via GOx-assisted RAFT-PISA targeting different degrees of polymerization (DPs) of poly(2-hydroxypropyl methacrylate) (PHPMA). $PPEGMA_8$-*b*-PHPMA block copolymer nanoparticles with various morphologies, such as spheres, worms, and vesicles, were obtained. Flow polymerization can be used to generate significant quantities of polymers by continuously passing polymer solution through reaction tubes [72–74]. However, continuous flow RAFT polymerization is more cumbersome as it requires deoxygenation in a closed reaction vessel and then transfer of the polymerization solution to the reaction tube where effective deoxygenation can be challenging, reducing the probability of successful polymerizations. Tan et al. recently extended GOx deoxygenation to flow RAFT-PISA, enabling the production of block copolymer nanoparticles on a large scale [75].

14.3 Enzymatic Initiation of RAFT Polymerization

Electron transfer and change of oxidation state are involved in reactions catalyzed by oxidoreductase and in some cases, these reactions lead to the production of radicals, which can be harnessed to initiate (controlled) radical polymerization. Three enzyme initiation strategies have been developed for RAFT polymerization, including horse radish peroxidase (HRP) initiation [76], photoenzymatic initiation [77], and initiation by tyrosine residue [78].

An et al. reported the first enzymatic initiation of RAFT polymerization using HRP, H_2O_2, and acetylacetone (ACAC) [76]. Later, Konkolewicz's group reported a similar approach [79] and conducted detailed mechanistic and kinetic experiments to study the effect of various reaction parameters on polymerization [80]. It should be noted that the HRP/H_2O_2/ACAC ternary initiation system was previously used to initiate FRP, in which HRP catalyzes the oxidation of β-diketones by H_2O_2,

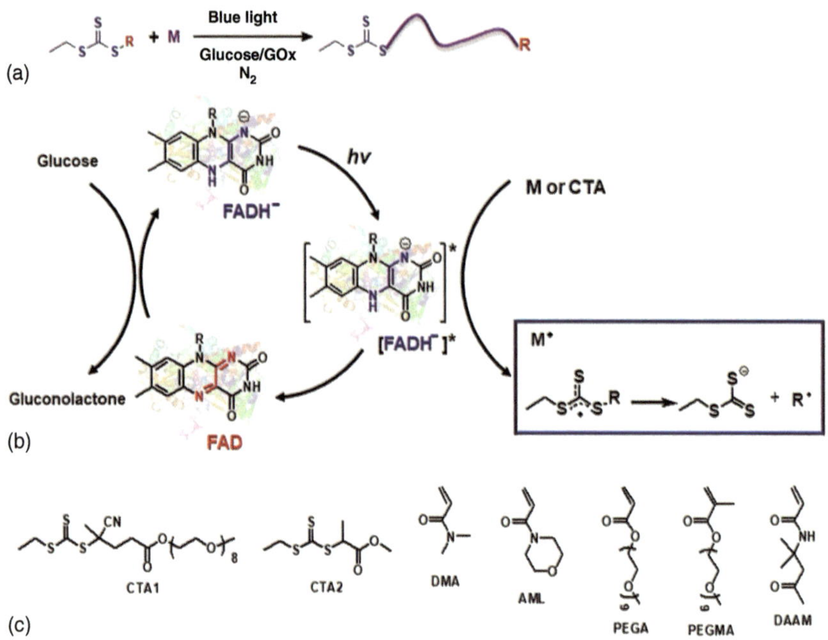

Figure 14.3 (a) Photoenzymatic RAFT polymerization, (b) proposed mechanism, and (c) structures of RAFT agents and monomers. Source: [77]/John Wiley & Sons.

producing free radicals that can be released outside of the enzyme to initiate FRP [81, 82].

Although HRP can be used to initiate RAFT polymerization, this system is oxygen-intolerant. Developing new systems that can simultaneously deoxygenate and initiate polymerization using only one enzyme is much more attractive. In 2019, our group reported a novel photoenzymatic polymerization by combining the natural deoxygenation function of GOx with its non-natural photoenzymatic catalytic function to achieve well-controlled RAFT polymerization with oxygen tolerance using a single enzyme (Figure 14.3) [77]. When the oxygen in the closed reaction vial is completely consumed, the cofactor flavin adenine dinucleotide (FAD) of GOx can still be reduced to $FADH^-$ by glucose in the system, which can be excited by 405 or 460 nm LED light, and the excited state of $FADH^{-*}$ can transfer an electron to the monomer or RAFT agent to initiate RAFT polymerization. This polymerization method is applicable to many types of monomers, including the conjugated monomers DMA, *N*-acryloylmorpholine (NAM), poly(ethylene glycol) methyl ether acrylate (PEGA, $M_n = 480\,\mathrm{g\,mol^{-1}}$), poly(ethylene glycol) methyl ether methacrylate (PEGMA; $M_n = 500\,\mathrm{g\,mol^{-1}}$), methyl acrylate (MA), butyl acrylate (BA), methyl methacrylate (MMA), diacetone acrylamide (DAAM), and styrene (S), as well as the unconjugated monomers *N*-vinyl pyrrolidone (NVP) and *N*-vinyl caprolactam (NVCL) [41, 83]. Significantly, this excellent method has been used to synthesize UHMW (star) polymers, such as $PPEGMA_{2000}$ ($M_n = 973.5\,\mathrm{kg\,mol^{-1}}$ and $Đ = 1.39$), poly(*N*-vinyl pyrrolidone) ($PNVP_{9000}$) ($M_n = 1058.7\,\mathrm{kg\,mol^{-1}}$ and $Đ = 1.27$), 4-arm star-(PNVP$_{2450}$)$_4$ (M_n = 1109.7 kg mol^{-1} and $Đ$ = 1.35), PNVP$_{4650}$-*b*-PNVP$_{4600}$ (M_n = 1058.7 kg mol^{-1} and $Đ$ = 1.30), PDMA$_{5000}$-*b*-PNVP$_{4700}$ (M_n = 998.2 kg mol^{-1} and $Đ$ = 1.35), 4-arm star-(PDMA$_{1250}$-*b*-PNVP$_{1163}$)$_4$ (M_n = 1081.0 kg mol^{-1} and $Đ$ = 1.39), and PNVCL$_{250}$-*b*-PNVP$_{9700}$-*b*-PNVCL$_{250}$ (M_n = 1106.5 kg mol^{-1} and $Đ$ = 1.39). Moreover, this method is also appropriate for heterogeneous polymerization, encompassing dispersion and emulsion polymerizations [83].

The photoenzymatic RAFT polymerization developed by An et al. involves excitation of the FAD cofactor and this system is inherently oxygen-tolerant due to in situ deoxygenation of GOx. Recently, Huang and coworkers found that certain proteins such as bovine serum albumin (BSA), lysozyme, and GOx that contain tyrosine residues could initiate RAFT polymerization by the mechanism of proton-coupled electron transfer (PCET) under the irradiation of 405 nm light [78]. Well-defined polymers were synthesized in biologically related aqueous solution and high biocompatibility was demonstrated by polymerization from yeast cell surface and within yeast cells. This approach of initiating polymerization via tyrosine residues was also applied to dispersion polymerization of 2-methoxyethyl acrylate (MEA) to yield nanoparticles of different diameters ranging from 16 to 80 nm. The advantage of photoinitiation by tyrosine residue is that a variety of proteins can be used as this method is not dependent on the flavin cofactor. However, higher photon energy is required due to the shorter absorption wavelength of tyrosine. Although the authors demonstrated that this polymerization system showed a certain degree of oxygen tolerance, this was based on the waste of active radicals, which is different from the inherent oxygen-scavenging mechanism of photoenzymatic RAFT polymerization using GOx.

In the HRP/H_2O_2/ACAC ternary initiating system, H_2O_2 was externally added and physical deoxygenation was typically required to have a well-controlled polymerization. Since GOx deoxygenation produces H_2O_2, the construction of an enzymatic cascade involving GOx deoxygenation and HRP initiation enables the conversion of a radical inhibitor (oxygen) to a reactant (H_2O_2) and allows polymerization to be conducted in an efficient and atom-economic fashion [76]. In theory, any enzyme capable of deoxygenation and producing H_2O_2 can be cascaded with HRP. Currently, several enzyme cascades have been successfully employed in RAFT polymerization, including GOx-HRP [76], P2Ox-HRP [40], and FOx-HRP [65]. One of the main advantages of enzyme cascade is that the rate of radical production can be controlled by adjusting the concentrations of both enzyme reactions, providing manifold handles to effectively manipulate a polymerization. In particular, by appropriately adjusting the rate of both enzyme reactions, it is possible to sustain the production of ACAC radicals at a sufficiently low concentration throughout the polymerization, which is essential to maintaining a high polymerization livingness. For example, using P2Ox-HRP cascade catalysis, decablock copolymers such as PDMA$_{150}$-(PDMA$_{50}$)$_9$, PNAM$_{150}$-(PNAM$_{50}$)$_9$, PDMA$_{150}$-(PNAM$_{50}$-PDMA$_{50}$)$_4$-PNAM$_{50}$, and PDMA$_{100}$-(PNAM$_{50}$-PNAM$_{50}$-PDMA$_{50}$)$_3$ were obtained with reasonably low dispersities ($Đ$ < 1.5) in open vessels. Well-defined UHMW PDMAs with MWs ranging from 921 to 2676 kg mol^{-1} ($Đ$ < 1.4) were also been obtained. In FOx-HRP cascade catalysis (Figure 14.4), the

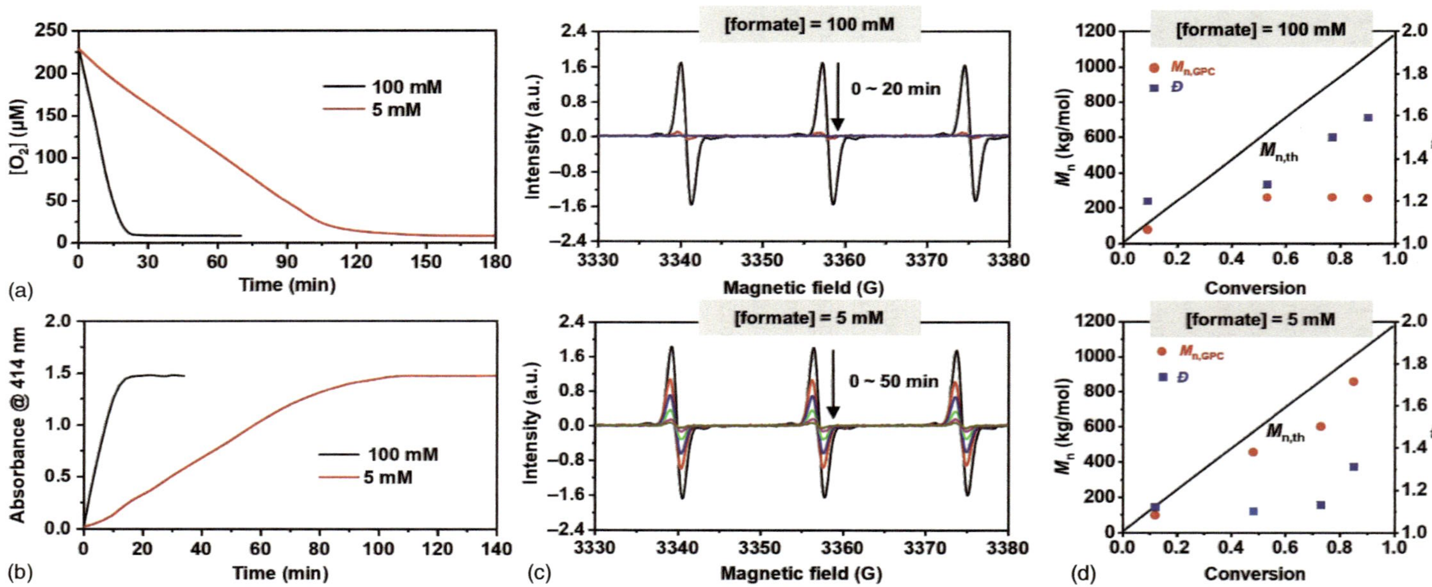

Figure 14.4 The concentration of the formate affects oxygen consumption, H_2O_2 production, and ACAC radical production in FOx-HRP cascade catalysis. (a) Consumption of oxygen with time at different formate concentrations. (b) UV–Vis spectroscopic monitoring of H_2O_2 by 2,2′-azino-bis(3-ethylbenzothiazoline-6-sulfonic acid) diammonium salt (ABTS) chromogenic reaction. (c) Decay of 2,2,6,6-Tetramethyl-1-piperidinyloxyl (TEMPO) electron paramagnetic resonance (EPR) signal due to ACAC radical generation. (d) Plots of MW and dispersity versus conversion. Source: [65]/John Wiley & Sons.

dissolved oxygen in the solution was consumed in 20 minutes at a sodium formate concentration of 100 mM, whereas complete deoxygenation took 150 minutes at a sodium formate concentration of 5 mM. Under otherwise the same conditions, the rate of ACAC radical production was also related to the concentration of sodium formate. The rate of ACAC radical production was 5.69 times higher with 100 mM sodium formate than with 5 mM sodium formate. Polymerization kinetics showed that UHMW polymers with low dispersities could be obtained only at low concentrations of sodium formate (e.g. 5 mM). Although higher concentrations of sodium formate (e.g. 100 mM) expedited the polymerization rate, the MW of the resulting polymers gradually deviated from the theoretical values at the middle and late stages of polymerization. The prominent characteristics of FOx-HRP cascade catalysis such as oxygen tolerance, atom efficiency, and effective manipulation of radical generation enabled the synthesis of UHMW polymers in a high-throughput manner. Among the eleven RAFT agents with different structures selected, eight provided good control for the synthesis of UHMW PDMAs. In addition to UHMW homopolymers, a well-defined UHMW statistical copolymer of P(DMA$_{10800}$-*co*-NAM$_{1200}$) was also obtained (M_n = 939.9 kg mol^{-1} and $Đ$ = 1.39).

In establishing various enzymatic strategies for efficient RAFT polymerizations, our group studied their amenability to both homogeneous and heterogeneous polymerization systems. Both RAFT dispersion and emulsion PISA systems were investigated. In the first enzyme-engaged PISA, HRP-initiated aqueous dispersion RAFT polymerizations produced PDMA$_{90}$-*b*-PMEA$_{980}$ (M_n = 288.2 kg mol^{-1} and $Đ$ = 1.29) and PPEGMA$_{200}$-*b*-PMEA$_{300}$ (M_n = 134.9 kg mol^{-1} and $Đ$ = 1.25) nanoparticles with hydrodynamic diameters (D_h) of 91 nm and 80 nm, respectively [76]. In photoenzymatic RAFT-PISA, PDMA$_{48}$-*b*-PDAAM$_{200}$ (M_n = 39.7 kg mol^{-1} and $Đ$ = 1.28) nanoparticles with a D_h of 53 nm were prepared via aqueous dispersion polymerization [77], while various diblock copolymers of PDMA-*b*-PMA (M_n ranging from 19.8 to 1008.7 kg mol^{-1}, $Đ$ ranging from 1.17 to 1.38), PDMA-*b*-PBA (M_n = 107.1 kg mol^{-1} and $Đ$ = 1.19), PDMA-*b*-PS (M_n = 81.2 kg mol^{-1} and $Đ$ = 1.24), and triblock copolymers of PDMA-*b*-PMA-*b*-PMA (M_n up to 198.2 kg mol^{-1} and $Đ$ = 1.28) were prepared via aqueous emulsion RAFT polymerization [83].

14.4 Enzymatic Deoxygenation in ATRP

In ATRP, oxygen not only reacts with carbon radicals to generate peroxyl radicals but also oxidizes activators in the low valency (such as Cu^{I} and Fe^{II}). Therefore, in most ATRP, oxygen should be strictly removed [84, 85]. In attempting to degas the initiators for continuous activator regeneration (ICAR) ATRP system with GOx, Matyjaszewski et al. observed that the MW of the resulting polymer was only a quarter of the theoretical one [86]. This occurred because the H_2O_2 produced through enzymatic deoxygenation underwent a Fenton-like reaction with the Cu^{I} in the ATRP formulation, resulting in the generation of radicals that initiate new polymer chains. In cell respiration, sodium pyruvate can react with H_2O_2 to produce acetate, CO_2, and water. Inspired by this, polymerizations of oligo(ethylene oxide) methyl

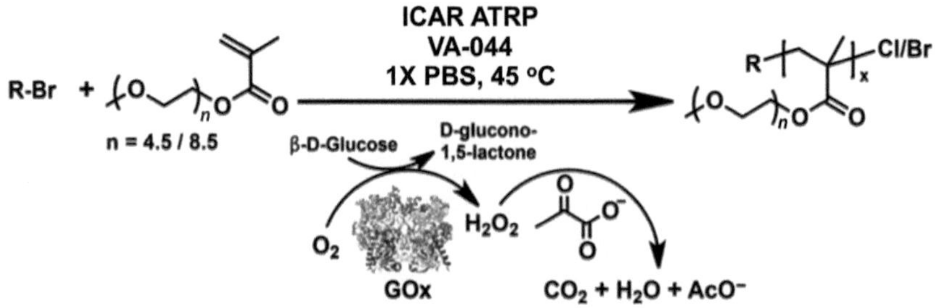

Figure 14.5 ICAR ATRP of OEOMA with GOx deoxygenation. Source: [86]/John Wiley & Sons.

ether methacrylate (OEOMA, M_n = 500 g mol^{-1}), after the addition of sodium pyruvate to the GOx-degassed ATRP system described earlier (Figure 14.5), yielded polymers with MWs close to theoretical values for different DPs (DP = 100–800). Chain-extension experiments using a one-pot or a purification–reinitiation strategy both yielded well-defined p(OEGMA$_{500}$)-*b*-p(OEGMA$_{300}$) block copolymers ($Đ \leq 1.3$), showing high chain-end fidelity. This procedure was also used for the synthesis of protein–polymer conjugates by grafting from BSA functionalized with an ATRP initiator.

After addressing the side reactions associated with H_2O_2, the same group expanded this approach to photoinduced ATRP under irradiation by blue LED light [87] and electrochemically mediated small-volume ATRP [88] for the synthesis of protein–polymer and DNA–polymer conjugates, respectively, demonstrating the excellent biocompatibility of this system. Another application of enzymatic deoxygenation that has been explored is surface-initiated ATRP for the preparation of surface polymer brushes [89, 90]. The results of antifouling performance tests on polymer brush coatings showed that coatings prepared in the presence of GOx exhibited greater resistance to plasma proteins than coatings prepared without GOx, demonstrating the advantages of enzymatic deoxygenation for polymerization from substrate geometries that cannot be effectively degassed via traditional means.

In addition to being consumed by sodium pyruvate, the H_2O_2 produced by GOx deoxygenation can also be used by HRP to generate ACAC radicals, which in turn can activate an ATRP by reducing Cu^{II} to Cu^{I}, thus creating an enzymatic ICAR ATRP (Figure 14.6) [91]. In this case, oxygen is an essential reactant, and only the presence of oxygen in the system can effectively drive the enzymatic ICAR ATRP. On/off polymerization using oxygen as a chemical stimulus was demonstrated. This enzymatic ICAR ATRP provided excellent control over the polymerization of OGEMA$_{500}$ with a target DP of 200 using only 100 ppm of Cu catalyst (M_n = 33.7 kg mol^{-1} and $Đ$ = 1.31).

14.5 Enzyme-Catalyzed ATRP

Enzyme-catalyzed ATRP was introduced independently by Bruns and di Lenas groups [37–39]. Heme enzymes such as HRP, hemoglobin, and catalase, as well

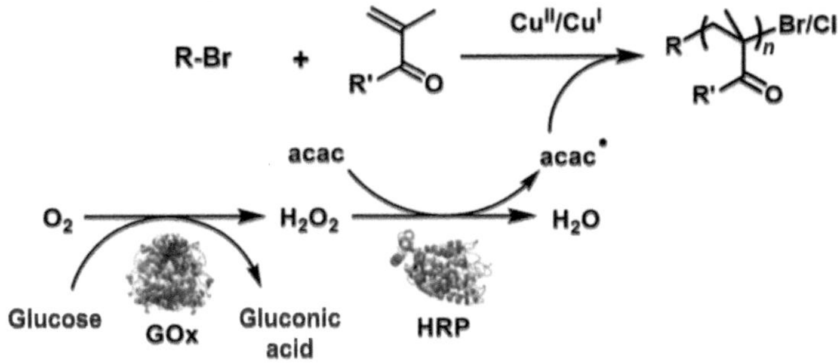

Figure 14.6 ICAR ATRP catalyzed by a GOx/HRP cascade. Source: [91]/John Wiley & Sons.

as the copper-containing laccases, can be used as catalysts to catalyze ATRP, and Bruns et al. referred to the enzymes that possess this novel activity as ATRPases. In Cu-catalyzed ATRP, Cu^{I}/L is used as the activator to abstract bromine from the dormant alkyl bromide, resulting in the formation of an active radical and Cu^{II}/L. The former can propagate with a monomer until deactivation occurs, returning the polymer chain to its dormant state and reforming Cu^{I}/L. Enzyme-catalyzed ATRP was proposed to proceed via a similar mechanism, utilizing the transition metals in these enzymes, except that these transition metals are buried within the proteins rather than being free. Enzyme-catalyzed ATRP generally required the addition of a reducing agent, in a similar fashion to activators regenerated by electron transfer (ARGET) ATRP. Control experiments suggested that the reducing agent, ATRPase, and initiator were all integral components of ARGET ATRP.

In addition to conventional monomers such as PEGA, PEGMA, and *N*-isopropylacrylamide (NIPAM), enzyme-catalyzed ATRP showed unique advantages in polymerizing certain challenging monomers such as *N*-vinylimidazole (NVIm), although, in this case, the polymerization control was inferior [92]. For conventional ATRP, the controlled polymerization of NVIm cannot be achieved, due to the fact that both NVIm and its polymerized product PNVIm are highly susceptible to binding with metal ions, leading to the destruction of ATRP catalyst [93]. As described before, the metal of the ATRPase is firmly bound to the protein. Bruns group achieved polymerization of NVIm with a modest control (M_n ranging from 1660 to 9970 g mol^{-1}, $Đ$ ranging from 1.27 to 1.56) using laccase-catalyzed ATRP [92]. After polymerization, laccase was effectively removed from the polymer by a simple purification step, yielding pure PNVIm.

Enzyme-catalyzed ATRP has been employed in a variety of applications such as surface brush fabrication, sensing, and PISA. Ko et al. investigated the ability of three ATRPases (i.e. catalase, HRP, and laccase) to catalyze the grafting of poly(*N*-isopropylacrylamide) (PNIPAM) polymer brushes from the surface of electrospun lignin nanofibers using surface-initiated ATRP [94]. The thickness of the PNIPAM brushes was affected by reaction conditions and enzyme activity, among other factors. The MW, thickness, and grafting density of polymer brushes could be regulated by controlling the conditions such as the enzyme types, the concentration of the reducing agent, and the pH of the solution. Later, Bruns and Benetti used hemoglobin as ATRPase to catalyze surface-initiated ATRP for the preparation of thermoresponsive PNIPAM [95]. Temperature affected the hydration state of PNIPAM, which in turn affected the affinity of the polymer for the protein. The team of Bruns and Belluati recently reported an innovative approach for yeast cell surface engineering through surface-initiated ATRP using surface-displayed HRP [96]. Decoration of yeast cell surface with synthetic polymers allowed modification of yeast phenotype and aggregation behavior as well as providing means for metal nanoparticle synthesis and functionalization with fluorescent dyes and nonnative enzymes, which expanded the function of yeast cells.

In sensing applications, Bruns et al. used hemoglobin and hemozoin as ATRPases to catalyze ATRP of NIPAM at 37 °C, and the turbidity of the solution was related to the MW of PNIPAM, which could be easily visible by the naked eye and quantitatively measured by UV–vis spectroscopy. The ATRPases could thus be quantified by the turbidity of the formed polymerization solution [97–99]. This method of protein sensing based on the formation of polymer precipitates has the advantage of simple experimental setup, signal amplification, and increased sensitivity.

Compared with enzyme-engaged RAFT-PISA, enzyme-engaged ATRP-PISA has seldom been reported. In a very recent report (Figure 14.7), Bruns et al. used myoglobin as an ATRPase to catalyze the chain extension of a hydrophilic macromolecular ATRP initiator poly(ethylene glycol) methyl ether 2-bromoisobutyrate (mPEG-BiB) with HPMA [100]. The resulting block copolymers self-assembled into nanoparticles of various morphologies, such as micelles, worm-like micelles, vesicles, and giant unilamellar vesicles (GUVs). Interestingly, GUV could encapsulate enzymes, nanoparticles, microparticles, plasmids, and cell lysates during the polymerization.

Compared to enzyme-catalyzed RAFT, enzyme-catalyzed ATRP typically generates polymers with relatively lower MW and higher dispersity. This is probably due to the need for the polymer chains to enter the catalytic center of

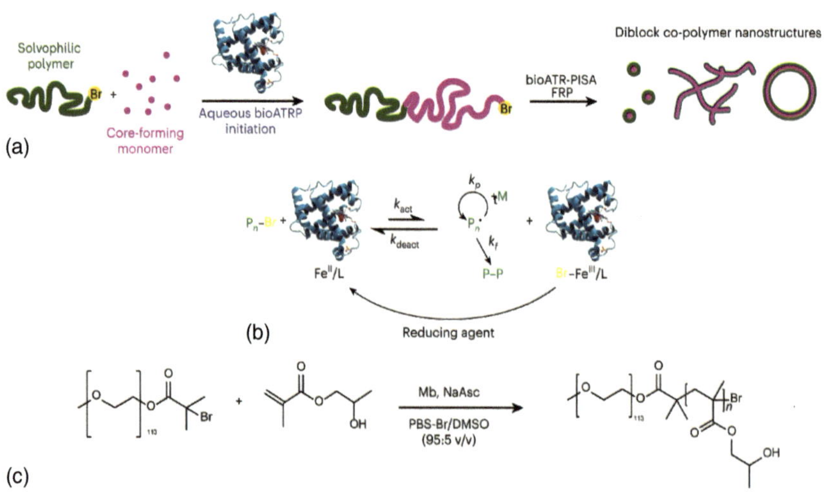

Figure 14.7 Enzyme-catalyzed ATRP-PISA with myoglobin as ATRPase. Source: [100]/Springer Nature/CC BY 4.0.

the enzyme during the polymerization process. As the polymerization progresses, continuous growth of the polymer chains makes it increasingly difficult for them to enter the catalytic center for effective reversible deactivation. We believe that with a deeper understanding of the polymerization mechanism and improved ability to engineer enzymes, enzyme-catalyzed ATRP will yield polymers with higher MWs and lower dispersities.

14.6 Conclusions

Recent achievements in enzyme-catalyzed CRP were highlighted in this chapter. The excellent deoxygenation capability of GOx, P2Ox, and FOx makes successful implementation of CRP easily achievable for the inexperienced and provides the ability to perform high-throughput synthesis. In addition, with the unique advantages of enzyme catalysis, the efficient acquisition of some difficult-to-obtain polymers such as UHMW polymers and PNVIm has become possible. This permits us to further investigate the unique properties of these polymer materials. Moreover, enzymes are ubiquitously present in living organisms, which provides interesting new ways to initiate polymerization directly in living organisms [7, 101]. With a deeper understanding of the mechanism of enzyme catalysis and advancements in the field of enzyme engineering, more cost-effective, robust, and efficient enzymes will become readily available, which will boost further progress of enzyme-catalyzed CRP [102–104].

Acknowledgments

We thank the financial support of the Key R&D Program of China (2021YFA1501600) and the National Natural Science Foundation of China (22193020, 22193024, and 22371089).

References

1 Zhao, Y., Chen, Y., Zhou, H. et al. (2023). Controlled radical copolymerization of fluoroalkenes by using light-driven redox-relay catalysis. *Nat. Synth.* 2 (7): 653–662.

2 Chen, K., Guo, X., and Chen, M. (2023). Controlled radical copolymerization toward well-defined fluoropolymers. *Angew. Chem. Int. Ed.* 62 (48): e202310636.

3 Gentekos, D.T., Sifri, R.J., and Fors, B.P. (2019). Controlling polymer properties through the shape of the molecular-weight distribution. *Nat. Rev. Mater.* 4 (12): 761–774.

4 Ma, Q., Qiao, G.G., and An, Z. (2023). Visible light photoiniferter polymerization for Dispersity control in high molecular weight polymers. *Angew. Chem. Int. Ed.* 62 (48): e202314729.

5 Sumerlin, B.S. (2018). Next-generation self-healing materials. *Science* 362 (6411): 150–151.

6 Cornel Erik, J., Jiang, J., Chen, S., and Du, J. (2020). Principles and characteristics of polymerization-induced self-assembly with various polymerization techniques. *CCS Chemistry* 3 (4): 2104–2125.

7 Geng, J., Li, W., Zhang, Y. et al. (2019). Radical polymerization inside living cells. *Nat. Chem.* 11 (6): 578–586.

8 Ni, C., Chen, D., Yin, Y. et al. (2023). Shape memory polymer with programmable recovery onset. *Nature* 622 (7984): 748–753.

9 Urban, M.W., Davydovich, D., Yang, Y. et al. (2018). Key-and-lock commodity self-healing copolymers. *Science* 362 (6411): 220–225.

10 Shi, Q., Yin, H., Song, R. et al. (2023). Digital micelles of encoded polymeric amphiphiles for direct sequence reading and ex vivo label-free quantification. *Nat. Chem.* 15 (2): 257–270.

11 Mapas, J.K.D., Thomay, T., Cartwright, A.N. et al. (2016). Ultrahigh molecular weight linear block copolymers: rapid access by reversible-deactivation radical polymerization and self-assembly into large domain nanostructures. *Macromolecules* 49 (10): 3733–3738.

12 Parkatzidis, K., Wang, H.S., Truong, N.P., and Anastasaki, A. (2020). Recent developments and future challenges in controlled radical polymerization: a 2020 update. *Chem* 6 (7): 1575–1588.

13 Perrier, S. (2017). 50th anniversary perspective: RAFT polymerization—a user guide. *Macromolecules* 50 (19): 7433–7447.

14 Lodge, T.P. (2017). Celebrating 50 years of macromolecules. *Macromolecules* 50 (24): 9525–9527.

15 Corrigan, N., Jung, K., Moad, G. et al. (2020). Reversible-deactivation radical polymerization (controlled/living radical polymerization): from discovery to materials design and applications. *Prog. Polym. Sci.* 111: 101311.

16 Wang, J.-S. and Matyjaszewski, K. (1995). Controlled/“living” radical polymerization. Atom transfer radical polymerization in the presence of transition-metal complexes. *J. Am. Chem. Soc.* 117 (20): 5614–5615.

17 Yamago, S. (2009). Precision polymer synthesis by degenerative transfer controlled/living radical polymerization using Organotellurium, Organostibine, and Organobismuthine chain-transfer agents. *Chem. Rev.* 109 (11): 5051–5068.

18 Chiefari, J., Chong, Y.K., Ercole, F. et al. (1998). Living free-radical polymerization by reversible addition–fragmentation chain transfer: the RAFT process. *Macromolecules* 31 (16): 5559–5562.

19 Truong, N.P., Jones, G.R., Bradford, K.G.E. et al. (2021). A comparison of RAFT and ATRP methods for controlled radical polymerization. *Nat. Rev. Chem.* 5 (12): 859–869.

20 Treat, N.J., Sprafke, H., Kramer, J.W. et al. (2014). Metal-free atom transfer radical polymerization. *J. Am. Chem. Soc.* 136 (45): 16096–16101.

21 Ren, J.M., McKenzie, T.G., Fu, Q. et al. (2016). Star polymers. *Chem. Rev.* 116 (12): 6743–6836.

22 Cao, M., Liu, Y., Zhang, X. et al. (2022). Expanding the toolbox of controlled/living branching radical polymerization through simulation-informed reaction design. *Chem* 8 (5): 1460–1475.

23 Pal, D., Miao, Z., Garrison, J.B. et al. (2020). Ultra-high-molecular-weight macrocyclic bottlebrushes via post-polymerization modification of a cyclic polymer. *Macromolecules* 53 (22): 9717–9724.

24 O'Dea, R.M., Willie, J.A., and Epps III, T.H. (2020). 100th anniversary of macromolecular science viewpoint: polymers from Lignocellulosic biomass. Current challenges and future opportunities. *ACS Macro Lett.* 9 (4): 476–493.

25 Meier, M.A.R., Metzger, J.O., and Schubert, U.S. (2007). Plant oil renewable resources as green alternatives in polymer science. *Chem. Soc. Rev.* 36 (11): 1788–1802.

26 Gandini, A. (2008). Polymers from renewable resources: a challenge for the future of macromolecular materials. *Macromolecules* 41 (24): 9491–9504.

27 Piogé, S., Tran, T.N., McKenzie, T.G. et al. (2018). Sono-RAFT polymerization-induced self-assembly in aqueous dispersion: synthesis of LCST-type Thermosensitive Nanogels. *Macromolecules* 51 (21): 8862–8869.

28 Allegrezza, M.L. and Konkolewicz, D. (2021). PET-RAFT polymerization: mechanistic perspectives for future materials. *ACS Macro Lett.* 10 (4): 433–446.

29 Lee, Y., Boyer, C., and Kwon, M.S. (2023). Photocontrolled RAFT polymerization: past, present, and future. *Chem. Soc. Rev.* 52 (9): 3035–3097.

30 Magenau, A.J.D., Strandwitz, N.C., Gennaro, A., and Matyjaszewski, K. (2011). Electrochemically mediated atom transfer radical polymerization. *Science* 332 (6025): 81–84.

31 Wu, C., Corrigan, N., Lim, C.-H. et al. (2022). Rational Design of Photocatalysts for controlled polymerization: effect of structures on photocatalytic activities. *Chem. Rev.* 122 (6): 5476–5518.

32 Xie, W., Zhao, L., Wei, Y., and Yuan, J. (2021). Advances in enzyme-catalysis-mediated RAFT polymerization. *Cell Rep. Phys. Sci.* 2 (7): 100487.

33 Theriot, J.C., Lim, C.-H., Yang, H. et al. (2016). Organocatalyzed atom transfer radical polymerization driven by visible light. *Science* 352 (6289): 1082–1086.

34 Li, R., Kong, W., and An, Z. (2023). Controlling radical polymerization with biocatalysts. *Macromolecules* 56 (3): 751–761.

35 Rodriguez, K.J., Gajewska, B., Pollard, J. et al. (2018). Repurposing biocatalysts to control radical polymerizations. *ACS Macro Lett.* 7 (9): 1111–1119.

36 Li, R., Kong, W., and An, Z. (2022). Enzyme catalysis for reversible deactivation radical polymerization. *Angew. Chem. Int. Ed.* 61 (26): e202202033.

37 Ng, Y.-H., di Lena, F., and Chai, C.L.L. (2011). Metalloenzymatic radical polymerization using alkyl halides as initiators. *Polym. Chem.* 2 (3): 589–594.

38 Ng, Y.-H., di Lena, F., and Chai, C.L.L. (2011). PolyPEGA with predetermined molecular weights from enzyme-mediated radical polymerization in water. *Chem. Commun.* 47 (22): 6464–6466.

39 Sigg, S.J., Seidi, F., Renggli, K. et al. (2011). Horseradish peroxidase as a catalyst for atom transfer radical polymerization. *Macromol. Rapid Commun.* 32 (21): 1710–1715.

40 Liu, Z., Lv, Y., and An, Z. (2017). Enzymatic Cascade catalysis for the synthesis of multiblock and ultrahigh-molecular-weight polymers with oxygen tolerance. *Angew. Chem. Int. Ed.* 56 (44): 13852–13856.

41 Li, R. and An, Z. (2020). Achieving ultrahigh molecular weights with diverse architectures for unconjugated monomers through oxygen-tolerant Photoenzymatic RAFT polymerization. *Angew. Chem. Int. Ed.* 59 (49): 22258–22264.

42 An, Z. (2020). 100th anniversary of macromolecular science viewpoint: achieving ultrahigh molecular weights with reversible deactivation radical polymerization. *ACS Macro Lett.* 9 (3): 350–357.

43 Lv, F., An, Z., and Wu, P. (2019). Scalable preparation of alternating block copolymer particles with inverse bicontinuous mesophases. *Nat. Commun.* 10 (1): 1397.

44 Meier, M.A.R., Hoogenboom, R., and Schubert, U.S. (2004). Combinatorial methods, automated synthesis and high-throughput screening in polymer research: the evolution continues. *Macromol. Rapid Commun.* 25 (1): 21–33.

45 Kim, S. and Sikes, H.D. (2020). Radical polymerization reactions for amplified biodetection signals. *Polym. Chem.* 11 (8): 1424–1444.

46 Liu, X., Liu, B., Ding, J. et al. (2022). Building a library for catalysts research using high-throughput approaches. *Adv. Funct. Mater.* 32 (1): 2107862.

47 Wu, C., Jung, K., Ma, Y. et al. (2021). Unravelling an oxygen-mediated reductive quenching pathway for photopolymerisation under long wavelengths. *Nat. Commun.* 12 (1): 478.

48 Li, N. and Pan, X.-C. (2021). Controlled radical polymerization: from oxygen inhibition and tolerance to oxygen initiation. *Chin. J. Polym. Sci.* 39 (9): 1084–1092.

49 Richter, M., Schulenburg, C., Jankowska, D. et al. (2015). Novel materials through Nature's catalysts. *Mater. Today* 18 (8): 459–467.

50 Iwata, H., Hata, Y., Matsuda, T., and Ikada, Y. (1991). Initiation of radical polymerization by glucose oxidase utilizing dissolved oxygen. *J. Polym. Sci. A Polym. Chem.* 29 (8): 1217–1218.

51 Chapman, R., Gormley, A.J., Herpoldt, K.-L., and Stevens, M.M. (2014). Highly controlled open vessel RAFT polymerizations by enzyme degassing. *Macromolecules* 47 (24): 8541–8547.

52 Schneiderman, D.K., Ting, J.M., Purchel, A.A. et al. (2018). Open-to-air RAFT polymerization in complex solvents: from whisky to fermentation broth. *ACS Macro Lett.* 7 (4): 406–411.

53 Lv, Y., Liu, Z., Zhu, A., and An, Z. (2017). Glucose oxidase deoxygenation–redox initiation for RAFT polymerization in air. *J. Polym. Sci. A Polym. Chem.* 55 (1): 164–174.

54 Reyhani, A., McKenzie, T.G., Fu, Q., and Qiao, G.G. (2019). Fenton-chemistry-mediated radical polymerization. *Macromol. Rapid Commun.* 40 (18): 1900220.

55 Reyhani, A., Allison-Logan, S., Ranji-Burachaloo, H. et al. (2019). Synthesis of ultra-high molecular weight polymers by controlled production of initiating radicals. *J. Polym. Sci. A Polym. Chem.* 57 (18): 1922–1930.

56 Reyhani, A., Nothling, M.D., Ranji-Burachaloo, H. et al. (2018). Blood-catalyzed RAFT polymerization. *Angew. Chem. Int. Ed.* 57 (32): 10288–10292.

57 Jafari, V.F., Mossayebi, Z., Allison-Logan, S. et al. (2023). The power of automation in polymer chemistry: precision synthesis of multiblock copolymers with block sequence control. *Chem. Eur. J.* 29 (53): e202301767.

58 Yuan, B., Huang, T., Lv, X. et al. (2022). Bioenhanced rapid redox initiation for RAFT polymerization in the air. *Macromol. Rapid Commun.* 43 (19): 2200218.

59 Bennett, M.R., Moloney, C., Catrambone, F. et al. (2022). Oxygen-tolerant RAFT polymerization initiated by living bacteria. *ACS Macro Lett.* 11 (8): 954–960.

60 Klibanov, A.M. (2001). Improving enzymes by using them in organic solvents. *Nature* 409 (6817): 241–246.

61 Chen, W.-H., Vázquez-González, M., Zoabi, A. et al. (2018). Biocatalytic cascades driven by enzymes encapsulated in metal–organic framework nanoparticles. *Nat. Catal.* 1 (9): 689–695.

62 Sheldon, R.A. and van Pelt, S. (2013). Enzyme immobilisation in biocatalysis: why, what and how. *Chem. Soc. Rev.* 42 (15): 6223–6235.

63 Nothling, M.D., McKenzie, T.G., Eastland, I.A. et al. (2019). Self-deoxygenating glassware. *Chem. Commun.* 55 (59): 8544–8547.

64 Yuan, B., Huang, T., Wang, X. et al. (2022). Oxygen-tolerant RAFT polymerization catalyzed by a recyclable biomimetic mineralization enhanced biological Cascade system. *Macromol. Rapid Commun.* 43 (1): 2100559.

65 Li, R., Zhang, S., Li, Q. et al. (2022). An atom-economic enzymatic Cascade catalysis for high-throughput RAFT synthesis of ultrahigh molecular weight polymers. *Angew. Chem. Int. Ed.* 61 (46): e202213396.

66 Prongjit, M., Sucharitakul, J., Wongnate, T. et al. (2009). Kinetic mechanism of pyranose 2-oxidase from Trametes multicolor. *Biochemistry* 48 (19): 4170–4180.

67 Robbins, J.M., Souffrant, M.G., Hamelberg, D. et al. (2017). Enzyme-mediated conversion of Flavin adenine dinucleotide (FAD) to 8-formyl FAD in Formate oxidase results in a modified cofactor with enhanced catalytic properties. *Biochemistry* 56 (29): 3800–3807.

68 Tieves, F., Willot, S.J.-P., van Schie, M.M.C.H. et al. (2019). Formate oxidase (FOx) from aspergillus oryzae: one catalyst enables diverse H_2O_2-dependent biocatalytic oxidation reactions. *Angew. Chem. Int. Ed.* 58 (23): 7873–7877.

69 Wang, M., Zhang, J., Guerrero-Sanchez, C. et al. (2019). Enzyme degassing for oxygen-sensitive reactions in open vessels of an automated parallel synthesizer: RAFT polymerizations. *ACS Comb. Sci.* 21 (10): 643–649.

70 Chapman, R., Gormley, A.J., Stenzel, M.H., and Stevens, M.M. (2016). Combinatorial low-volume synthesis of well-defined polymers by enzyme degassing. *Angew. Chem. Int. Ed.* 55 (14): 4500–4503.

71 Tan, J., Liu, D., Bai, Y. et al. (2017). Enzyme-assisted Photoinitiated polymerization-induced self-assembly: An oxygen-tolerant method for preparing block copolymer Nano-objects in open vessels and multiwell plates. *Macromolecules* 50 (15): 5798–5806.

72 Corrigan, N., Rosli, D., Jones, J.W.J. et al. (2016). Oxygen tolerance in living radical polymerization: investigation of mechanism and implementation in continuous flow polymerization. *Macromolecules* 49 (18): 6779–6789.

73 Zhong, F., Zhou, Y., and Chen, M. (2019). The influence of mixing on chain extension by photo-controlled/living radical polymerization under continuous-flow conditions. *Polym. Chem.* 10 (35): 4879–4886.

74 Zhou, Y., Gu, Y., Jiang, K., and Chen, M. (2019). Droplet-flow photopolymerization aided by computer: overcoming the challenges of viscosity and facilitating the generation of copolymer libraries. *Macromolecules* 52 (15): 5611–5617.

75 Cai, W.-B., Liu, D.-D., Chen, Y. et al. (2021). Enzyme-assisted Photoinitiated polymerization-induced self-assembly in continuous flow reactors with oxygen tolerance. *Chin. J. Polym. Sci.* 39 (9): 1127–1137.

76 Zhang, B., Wang, X., Zhu, A. et al. (2015). Enzyme-initiated reversible addition–fragmentation chain transfer polymerization. *Macromolecules* 48 (21): 7792–7802.

77 Zhou, F., Li, R., Wang, X. et al. (2019). Non-natural photoenzymatic controlled radical polymerization inspired by DNA photolyase. *Angew. Chem. Int. Ed.* 58 (28): 9479–9484.

78 Zhu, M., Wang, S., Li, Z. et al. (2023). Tyrosine residues initiated photopolymerization in living organisms. *Nat. Commun.* 14 (1): 3598.

79 Danielson, A.P., Van Kuren, D.B., Lucius, M.E. et al. (2016). Well-defined macromolecules using horseradish peroxidase as a RAFT Initiase. *Macromol. Rapid Commun.* 37 (4): 362–367.

80 Danielson, A.P., Van-Kuren, D.B., Bornstein, J.P. et al. (2018). Investigating the mechanism of horseradish peroxidase as a RAFT-Initiase. *Polymers* 10 (7): 741.

81 Qi, G., Jones, C.W., and Schork, F.J. (2006). Enzyme-initiated miniemulsion polymerization. *Biomacromolecules* 7 (11): 2927–2930.

82 Teixeira, D., Lalot, T., Brigodiot, M., and Maréchal, E. (1999). β-Diketones as key compounds in free-radical polymerization by enzyme-mediated initiation. *Macromolecules* 32 (1): 70–72.

83 Li, R.-Y. and An, Z.-S. (2021). Photoenzymatic RAFT emulsion polymerization with oxygen tolerance. *Chin. J. Polym. Sci.* 39 (9): 1138–1145.

84 Szczepaniak, G., Fu, L., Jafari, H. et al. (2021). Making ATRP more practical: oxygen tolerance. *Acc. Chem. Res.* 54 (7): 1779–1790.

85 Lorandi, F., Fantin, M., and Matyjaszewski, K. (2022). Atom transfer radical polymerization: a mechanistic perspective. *J. Am. Chem. Soc.* 144 (34): 15413–15430.

86 Enciso, A.E., Fu, L., Russell, A.J., and Matyjaszewski, K. (2018). A breathing atom-transfer radical polymerization: fully oxygen-tolerant polymerization inspired by aerobic respiration of cells. *Angew. Chem. Int. Ed.* 57 (4): 933–936.

87 Fu, L., Wang, Z., Lathwal, S. et al. (2018). Synthesis of polymer bioconjugates via photoinduced atom transfer radical polymerization under blue light irradiation. *ACS Macro Lett.* 7 (10): 1248–1253.

88 Sun, Y., Lathwal, S., Wang, Y. et al. (2019). Preparation of well-defined polymers and DNA–polymer bioconjugates via small-volume eATRP in the presence of air. *ACS Macro Lett.* 8 (5): 603–609.

89 Chen, W.-L., Cordero, R., Tran, H., and Ober, C.K. (2017). 50th anniversary perspective: polymer brushes: novel surfaces for future materials. *Macromolecules* 50 (11): 4089–4113.

90 Navarro, L.A., Enciso, A.E., Matyjaszewski, K., and Zauscher, S. (2019). Enzymatically degassed surface-initiated atom transfer radical polymerization with real-time monitoring. *J. Am. Chem. Soc.* 141 (7): 3100–3109.

91 Enciso, A.E., Fu, L., Lathwal, S. et al. (2018). Biocatalytic "oxygen-fueled" atom transfer radical polymerization. *Angew. Chem. Int. Ed.* 57 (49): 16157–16161.

92 Fodor, C., Gajewska, B., Rifaie-Graham, O. et al. (2016). Laccase-catalyzed controlled radical polymerization of N-vinylimidazole. *Polym. Chem.* 7 (43): 6617–6625.

93 Allen, M.H. Jr., Hemp, S.T., Smith, A.E., and Long, T.E. (2012). Controlled radical polymerization of 4-Vinylimidazole. *Macromolecules* 45 (9): 3669–3676.

94 Gao, G., Karaaslan, M.A., Kadla, J.F., and Ko, F. (2014). Enzymatic synthesis of ionic responsive lignin nanofibres through surface poly(N-isopropylacrylamide) immobilization. *Green Chem.* 16 (8): 3890–3898.

95 Divandari, M., Pollard, J., Dehghani, E. et al. (2017). Controlling enzymatic polymerization from surfaces with switchable bioaffinity. *Biomacromolecules* 18 (12): 4261–4270.

96 Belluati, A., Happel, D., Erbe, M. et al. (2023). Self-decorating cells via surface-initiated enzymatic controlled radical polymerization. *Nanoscale* 15 (48): 19486–19492.

97 Rifaie-Graham, O., Pollard, J., Raccio, S. et al. (2019). Hemozoin-catalyzed precipitation polymerization as an assay for malaria diagnosis. *Nat. Commun.* 10 (1): 1369.

98 Pollard, J., Rifaie-Graham, O., Raccio, S. et al. (2020). Biocatalytically initiated precipitation atom transfer radical polymerization (ATRP) as a quantitative method for hemoglobin detection in biological fluids. *Anal. Chem.* 92 (1): 1162–1170.

99 Raccio, S., Pollard, J., Djuhadi, A. et al. (2020). Rapid quantification of the malaria biomarker hemozoin by improved biocatalytically initiated precipitation atom transfer radical polymerizations. *Analyst* 145 (23): 7741–7751.

100 Belluati, A.; Jimaja, S.; Chadwick, R. J.; Glynn, C.; Chami, M.; Happel, D.; Guo, C.; Kolmar, H.; Bruns, N., Artificial cell synthesis using biocatalytic polymerization-induced self-assembly. *Nat. Chem.* 2023, https://doi.org/10.1038/s41557-023-01391-y.

101 Zhang, Y., Gao, Q., Li, W. et al. (2022). Controlled intracellular polymerization for cancer treatment. *JACS Au* 2 (3): 579–589.

102 Scrutton, N.S. (2017). Enzymes make light work of hydrocarbon production. *Science* 357 (6354): 872–873.

103 Zhou, Q., Chin, M., Fu, Y. et al. (2021). Stereodivergent atom-transfer radical cyclization by engineered cytochromes P450. *Science* 374 (6575): 1612–1616.

104 Kong, D., Moon, P.J., Lui, E.K.J. et al. (2020). Direct reversible decarboxylation from stable organic acids in dimethylformamide solution. *Science* 369 (6503): 557–561.

15

Organocatalyst-Catalyzed Degradation of Polymers

Ge Yang[1] *and Zhibo Li*[1,2]

[1]*College of Chemical Engineering, Qingdao University of Science and Technology, No.53 Zhengzhou Rd, Qingdao, 266042, China*
[2]*Key Laboratory of Biobased Polymer Materials, Shandong Provincial Education Department; College of Polymer Science and Engineering, Qingdao University of Science and Technology, No.53 Zhengzhou Rd, Qingdao, 260042, China*

15.1 Introduction

Polymers have become an integral part of our society, providing immense convenience in our daily lives. Nevertheless, a crisis concerning this has already emerged due to the excessive production and improper disposal of plastics. By 2022, global plastic production had soared to 367 million tons. Shockingly, around 50% of all plastic products are discarded after a single use or a short period of circulation [1]. Despite global efforts to curb plastic waste, including initiatives like the Basel Convention, plastic emissions are projected to rise [2, 3]. The mismanagement of these disposed polymers leads to their accumulation in landfills or their entry into the natural environment. Of greater concern is that plastics exhibit characteristics of "poorly reversible pollutant"remaining in the environment for extended periods without significant degradation [4]. The accumulation of plastic pollution outpaces natural removal processes and clean-up efforts, leading to irreversible harm when the environmental threshold is breached. Consequently, it is urgent to develop effective plastic degradation technologies.

Chemical recycling is emerging as a crucial solution to address the issue of waste plastic. This process involves transforming polymers into high-purity monomers or valued-added fuel oil. However, due to the high structural stability of polymers, the use of catalysts is necessary for the degradation process. Metal-based catalysts are widely acknowledged for their effectiveness in plastic degradation, offering numerous active sites that enhance electron transport and facilitate the formation of electron–hole pairs [5, 6]. Extensive discussions on the efficacy of metal-based catalysts have been extensively covered in various reviews [7, 8], and will not be reiterated within this chapter.

As a promising alternative, organocatalyst have emerged as environmentally friendly substitutes for traditional organometallic complexes. While they have

Organocatalysts in Polymer Chemistry: Synthesis and Applications, First Edition. Edited by Zhibo Li.

been well-established in polymer polymerization as introduced in the previous chapters, their applications in polymer degradation are still nascent. In 2019, Dove published a review titled "Organocatalysis for depolymerisation" [9] and a book chapter titled "Organic catalysis for polymerization" [10], which comprehensively reviewed the application of organic catalysts in polymer degradation. Building upon these influential works, this chapter will primarily focus on highlighting the latest research progress published since then.

15.2 Chemical Degradation of Functional Polymers

Polymers with functional groups, mainly polycarbonates, polyesters, and polyurethanes (PUs), are extensively utilized in the market; however, their recycling remains inadequate. The presence of highly reactive functional groups within the polymer chain offers favorable reaction sites for catalytic degradation.

15.2.1 Organic Bases and Organic Acids

Organic bases have emerged as a potent tool extensively employed in various transesterification reactions. Since Fukushima et al. reported the use of commercially available guanidine catalyst 1,5,7-triazecyclic[4.4.0]dec-5-ene (TBD) for organic catalytic depolymerization of polyethylene terephthalate (PET), organic bases have gradually been applied in the field of plastic degradation [11]. This efficiency of TBD was found to be comparable to that of commonly used metal acetate/alkoxide catalysts for PET depolymerization. Computational and experimental investigations revealed that both the TBD catalyst and ethylene glycol (EG) depolymerizing agent, facilitated the depolymerization reaction through hydrogen bond.

This group further focused on the organic catalytic degradation of aliphatic polycarbonates, with a particular emphasis on polytrimethyl carbonate (PTMC) as a standard model for hydrolysis and methanolysis [12]. Organocatalysts play a crucial role in PTMC depolymerization by activating carbonyl groups with organoacids or enhancing nucleophilicity through hydrogen bonding with organobases [13]. The catalytic activity of 1-*tert*-butyl-2,2,4,4,4-pentakis(dimethylamino)-2λ^5,4λ^5 catenadi(phosphazene) (P_2-*t*-Bu), 1,8-diazabicyclo[5.4.0]undec-7-ene (DBU), and dodecylbenzenesulfonic acid (DBSA) for the inhomogeneous hydrolysis of PTMC is higher than that of TBD, while methanesulfonic acid (MSA) exhibits the lowest activity. This suggests that their acidity and basicity do not directly correlate with catalytic activity. However, DBU shows limited potency in homogeneous hydrolysis. In contrast to hydrolysis reactions, TBD exhibited the highest catalytic activity in methanolysis reactions, followed by DBU and P_2-*t*-Bu. However, organoacids such as MSA, diphenyl phosphate (DPP), and DBSA only yielded oligomers. Considering the higher basicity of P_2-*t*-Bu compared to TBD, it is likely that the dual hydrogen bonding activation mechanism accounts for the superior activity of TBD, which could explain the significant disparity in TBD's catalytic activity between hydrolysis and methanolysis. The present study further demonstrates the greater research

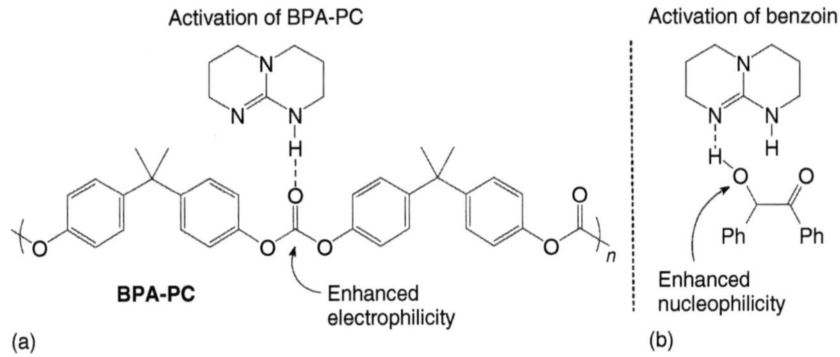

Figure 15.1 (a) Hydrogen bond activation mechanism of TBD, and (b) The impact of the chemical and configurational properties of α-hydroxyketones depolymerizer on the BPA-PC degradation reaction. Source: [14]/Royal Society of Chemistry.

potential of organic base catalysts compared to organic acid catalysts in the realm of polymer degradation.

In subsequent studies, TBD has been successfully employed for the degradation of various polymers. In 2023, Duguet and coworkers further elucidated the efficacy of TBD catalyst in the degradation of poly(bisphenol A carbonate) (BPA-PC) by highlighting its dual functionality – activation of both the carbonyl group of BPA-PC and the hydroxyl group of benzoin (used as a depolymerizing agent) through hydrogen bonding (Figure 15.1(a)) [14]. This leads to an augmentation in electrophilicity and nucleophilicity, respectively. The impact of the chemical and configurational properties of α-hydroxyketones depolymerizer on the reaction was also studied (Figure 15.1(b)). Results indicate that an increase in halogen electronegativity leads to a decrease in the nucleophilicity of the α-hydroxyl group of depolymerizer, resulting in reduced yield. Additionally, depolymerizers with substituents exhibiting excessive steric hindrance are unfavorable for the reaction. This finding further supports the hydrogen bond activation mechanism of TBD. Furthermore, in the study of biobased poly(limonene carbonate), TBD has demonstrated its ability to facilitate controlled polymer degradation through base-mediated activation of end-groups or main-chain breaks facilitated by hydrogen bonding [15]. Fehér et al. catalyzed the degradation of BPA-PC using silica-supported TBD, further enhancing the recyclability of this organocatalyst [16].

The dual functional mechanism of TBD enables the activation of various polymers and depolymerizers, leading to the production of value-added products. As in the previous work, the resulting degradation products encompass not only BPA-PC monomer but also valuable vinylene carbonates that can be directly utilized or serve as precursors in industries such as lithium batteries or 3D printing [17, 18]. Tomović and coworkers also utilized the exceptional depolymerization catalytic activity of TBD to prepare a closed-cycle poly(imine-carbonate) (P(ImC)s) [19]. In this study, commercial BPA-PC was effectively depolymerized using TBD as the catalyst and hydroxyethylated derivative of vanillin (HEV) as the depolymerizing agent. This approach not only yielded repolymerizable BPA monomer but also bis(2-[4-formyl-2-methoxyphenoxy]-ethyl) carbonate (DVEC). By condensing DVEC with various primary amines, both linear and cross-linked P(ImC)s were successfully prepared. The resulting polymeric network showed excellent malleability at high temperature and demonstrated remarkable reprocessability. Moreover, the polymeric network demonstrated exceptional selectivity for degradability in mixed waste plastic streams. After treatment with 0.1 M HCl and tetrahydrofuran (THF) (v : v = 9 : 1) for two hours, the imine bonds were cleaved, allowing depolymerization of the P(ImC)s into their constituent DVEC and primary amine parts, which could be separated from other stable polymers by filtration.

The concept of selectively degrading a specific plastic in a mixed plastic stream was also successfully realized by Liptrot's group. They utilized *N*-heterocyclic phosphine (NHP) as an organic catalyst and methanol as a depolymerizer to achieve precise degradation of poly(lactic acid) (PLA) within PLA/poly(ε-caprolactone) (PCL) and PLA/PET mixtures, resulting in the production of methyl lactate (MeLA) [20]. Furthermore, it was discovered that the enediamine, tBuN(H)CH=CHN(H)tBu, was generated through methanolysis of the NHP and proved to be an effective catalyst for selective PLA degradation. However, it should be noted that this active catalyst is sensitive to the atmosphere, necessitating the use of an air-stabilizer (Figure 15.2). The specificity of this system toward PLA depolymerization can be attributed to the rigid structure of the catalyst, which offers a specific spatial arrangement of hydrogen bond donor groups. Although the activity of the catalyst was sacrificed, its manipulability was enhanced.

In addition to homopolymers, organic catalysts have also been employed for the depolymerization of copolymers consisting of degradable blocks. These copolymers exhibit exceptional physicochemical and mechanical properties, along with adjustable biodegradability in comparison to those formed from their parent components. However, due to the intricate nature of copolymer monomers, selectively shearing each block, preventing side reactions, and minimizing product distribution pose significant challenges in copolymer degradation studies. Liu's group has successfully demonstrated a potentially replicable method for upcycling *cis-(exo,exo)*-poly(cyclohexene carbonate/norbornene anhydride)-*b*-poly(cyclohexene carbonate) (poly[CHO/NA]-*block*-PCHC) copolymer. This approach involves selectively breaking down the PCHC and poly(CHO/NA)

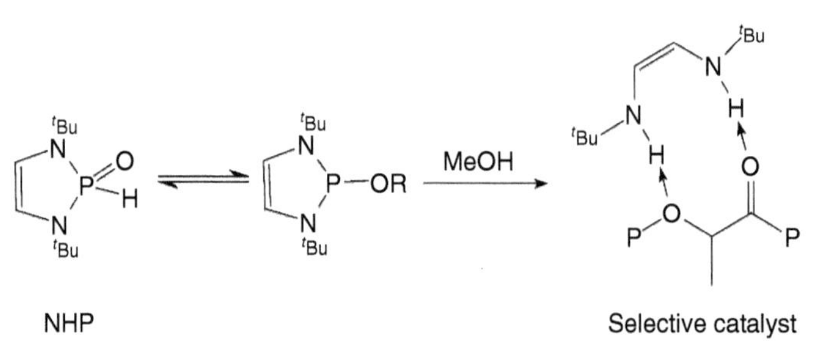

Figure 15.2 Selective organocatalyst for specific PLA depolymerization. Source: [20]/John Wiley & Sons/CC BY 4.0.

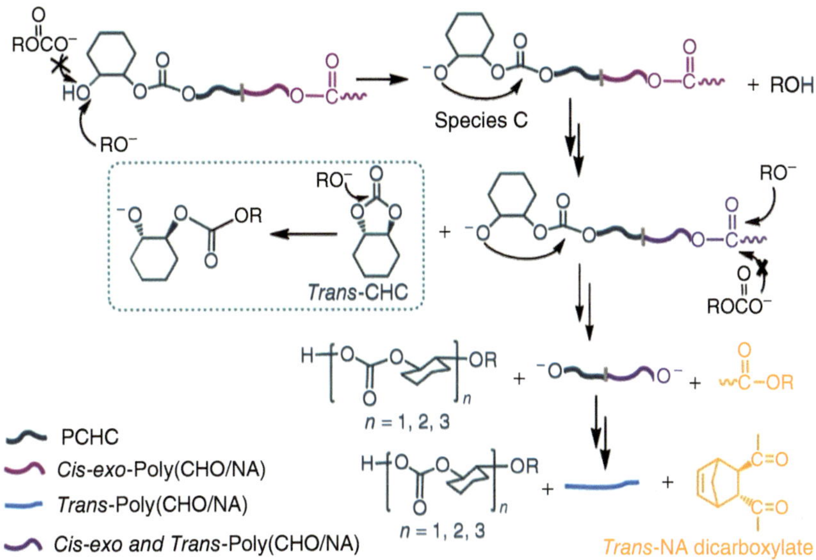

Figure 15.3 The degradation process of poly(CHO/NA)-*block*-PCHC copolymer under DMAP/CHO. Source: [21]/with permission of ELSEVIER.

blocks with the assistance of 4-dimethylaminopyridine (DMAP)/cyclohexene oxide (CHO) organocatalytic system [21]. The upcycling process involved three stages (Figure 15.3): first, the PCHC block depolymerized to produce trans-CHC through an unzipping mechanism; second, both the PCHC and poly(CHO/NA) segments decomposed; finally, any remaining poly(CHO/NA) blocks broken down into monomers. During the degradation process, DMAP functioned as an organic base and nucleophilically attacked the epoxide group of CHO, resulting in the formation of basic cyclohexyloxy anions. These anions then deprotonated the terminal -OH group of copolymers to generate polymeric alkoxide anions, which interacted with the carbonyl group of PCHC to generate trans-CHC until the consumption of PCHC block [22].

Dong et al.'s study presents an alternative organocatalyst-based copolymer upcycling approach, wherein TBD was employed for controlled degradation and sequential transesterification of PLA and PCL, resulting in the synthesis of a high-performance PLA-PCL-PLA triblock polyester [23].

15.2.2 Lonic Liquids and Acid–Base Salts

Organic catalysts, particularly organic base catalysts, show excellent polymer degradation properties; however, their limited thermal stability hinders their application. In recent years, significant attention has been devoted to the investigation of organic acid–base salts or ionic liquids in the field of polymer depolymerization due to their superior thermal stability compared to individual acids or bases.

Dove's group has successfully developed a series of acid–base organic salts for the closed-loop chemical recycling of PET [24]. The thermal stability of these salts,

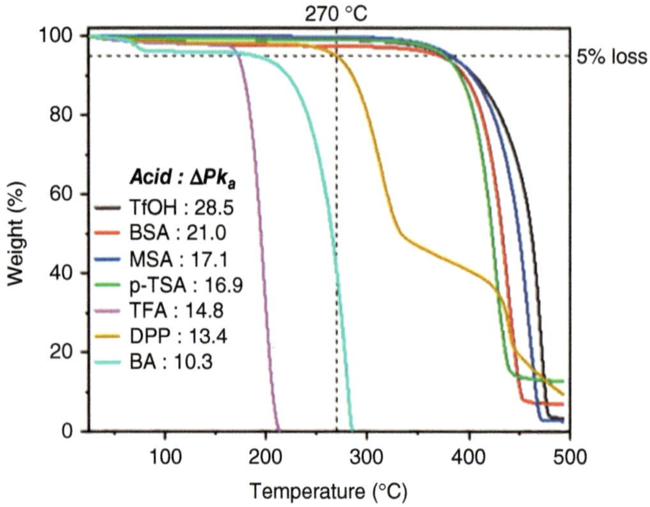

Figure 15.4 TGA thermograms for various TBD-based salts: dynamic heating at a rate of $10\,°C\,min^{-1}$. Source: [24]/American Chemical Society/CC BY 4.0.

as indicated by their ΔpK_a values, plays a crucial role in determining their performance. Notably, the salts with higher ΔpK_a values exhibit enhanced stability due to stronger acid–base interactions [25, 26] (Figure 15.4). Among them, TBD:TSA (*p*-toluenesulfonic acid) demonstrates exceptional thermal stability with only 2.5% weight loss observed at 270 °C, enabling its utilization within a structurally stable temperature range that encompasses both PET polymerization and depolymerization processes. By employing TBD:TSA as a catalyst, bis(2-hydroxyethyl) terephthalate (BHET) can be effectively polymerized into PET under vacuum conditions at 270 °C while remaining the catalyst within the polymer. Furthermore, at 180 °C, PET can be efficiently depolymerized back into BHET solely through the addition of EG. These findings demonstrate the feasibility and efficacy of implementing this organocatalysis system for simple closed-loop recycling of PET.

Jehanno, Sardon, and coworkers investigated the application of an acid–base organocatalyst system in the aminolysis of PU [27]. TBD:MSA has already exhibited its ability to enhance both polymerization and depolymerization reactions conducted at elevated temperature, showing remarkable thermal stability up to 400 °C, thus rendering it suitable for the temperature range employed in this study (130 ~ 190 °C). Based on this premise, the influence of the ammonia depolymerizer on PU depolymerization was examined. The results demonstrated that secondary amines selectively cleave the C–O bond of the urethane group during depolymerization reactions, thereby preventing the release of toxic amines and yielding valuable monomers. In contrast, primary amines non-selectively break both C–O and C–N bonds. Furthermore, similar catalytic products can be obtained with TBD, DBU, and DBU:BA catalysts; however, their catalytic efficiency is lower than that of TBD:MSA. These findings indicated that the selective cleavage of the PU chain is governed by the nature of the depolymerizer employed rather than by the catalyst.

A wide range of value-added products can be obtained using unconventional depolymerizers, as discussed in section 15.2.1. However, the utilization of acid–base organocatalysts offers a more stable reaction system for this design concept. Cyclic carbonates featuring six-membered rings hold great potential for the synthesis of novel polymers via ring-opening polymerization. Meanwhile, their current synthetic approach involves the utilization of highly hazardous reagents. Dove and Sardon et al. successfully depolymerized BPA-PC based on the TBD:MSA system to obtain functionalized six-membered cyclic carbonates [28]. Various organocatalysts have been effectively employed for the depolymerization of BPA-PC, including TBD [14], DBU [29], and DMAP [30]. Nevertheless, their limited thermal stability is disadvantageous when high temperatures are required. By employing the TBD:MSA complex as the catalyst, the group was able to conduct BPA degradation reactions at 130 °C. Despite the increased ring strain observed in the six-membered carbonates compared to five-membered cyclic carbonates, depolymerizers with longer alkyl chains exhibited a faster reaction rate and thus displayed a higher propensity for closed-loop reactions. The substitution of functional groups in the β position of depolymerizers' hydroxyl groups enabled the generation of corresponding cyclic carbonate compounds, thereby indicating that the degradation reaction effectively preserves product reactivity and yields high-value monomers suitable for synthesizing high-performance polymeric materials. Moreover, the configurational distortion of depolymerizers with functional groups may promote the ring closure (Figure 15.5).

Imidazolium-based ionenes are also known to be high-performance materials for a great variety of applications [31]. The preparation of these polymers requires the use of bis-imidazole starting monomers, which are commonly prepared by using toxic chloride reagents [32]. Sardon and coworkers took advantage of the structural diversity in plastic waste, obtaining bis-imidazoles featuring terephthalamide or urea moieties from PET and BPA-PC, respectively, via aminolysis [33] (Figure 15.6). TBD:MSA was selected as a catalyst due to its thermal stability and recyclability. The resulting terephthalamide- and urea-based bis-imidazole products were then used as starting materials for step-growth polymerizations, proceeding by the Menshutkin reaction with p-xylene. Following anion exchange to the hydrophobic NTf_2^- form, these terephthalamide- and urea-based ionenes were achieved, showing excellent elastomeric and self-healing behavior.

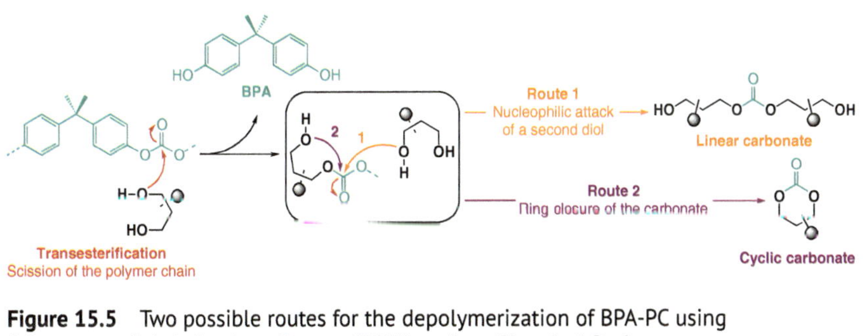

Figure 15.5 Two possible routes for the depolymerization of BPA-PC using 1,3-propanediol derivatives. Source: [28]/American Chemical Society.

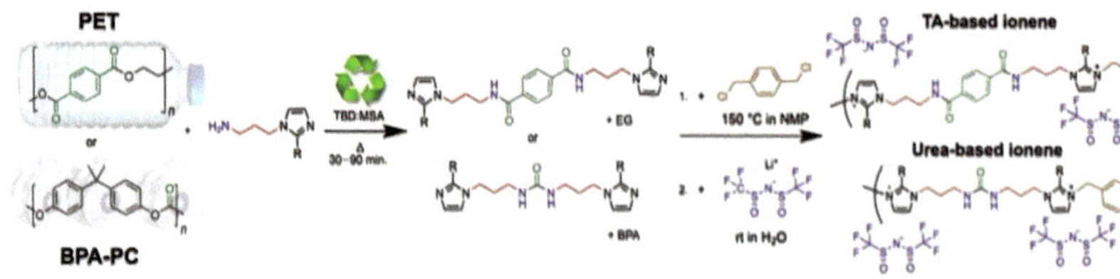

Figure 15.6 Synthesis of imidazolium-based self-supported elastomeric ionenes from plastic wastes. Source: Reproduced with permission from [33]/John Wiley & Sons.

15.3 Photocatalytic Degradation of Non-Functional Polymers

Polystyrene (PS) belongs to polyolefins, whose backbone consists only of C-C bonds. The degradation of PS requires more stringent conditions due to the absence of C-O, C-N, or other functional groups. Generally, temperatures exceeding 350 °C are necessary, which surpasses the survival range for typical organic catalysts [34, 35]. Therefore, the early research in the field of organocatalyzed PS degradation was limited. However, with the advancement of photocatalysis science, researchers have discovered the viability of utilizing $FeCl_3$ or acid as drivers for photocatalytic decomposition of PS [36, 37]. The reaction can be conducted at ambient temperatures, thus establishing a foundation for employing organic catalysts.

Reisner and coworkers have developed a visible-driven reaction using Fluorenone as a photocatalyst and H_2SO_4 as an assistant to degrade PS at 50 °C under ambient pressure [38]. This reaction employs blue LED irradiation ($\lambda_{max} \sim 450$ nm) and requires an O_2 atmosphere (Figure 15.7(a)). The reaction proceeds through the HAT process, which can break two C-H bonds of a secondary sp^3 carbon (-CH_2-) and form a carbonyl group. Fluorenone is active by the blue LED, initiating the HAT process between the excited state of Fluorenone and PS, resulting in the generation of PS radicals and ketyl radicals. The PS radicals undergo a series of oxidation, HAT, and β-scission processes to yield monomeric aromatic products consisting of 40%

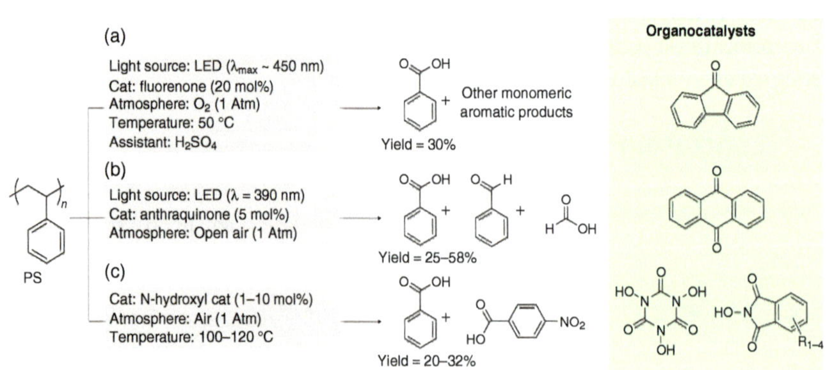

Figure 15.7 Degradation of PS catalyzed by organocatalysts. (a) Reisner et al.'s work (b) Kokotos et al.'s work, and (c) Lim et al.'s work. Source: Adapted from [39].

benzoic acid (BA) and 20% other by-products. The density functional theory (DFT) calculation shows that H_2SO_4 is essential in this system. The kinetic energy barrier of C-C bond cleavage is 3.9 kcal mol^{-1} and 66.1 kcal mol^{-1} in the presence and absence of H^+, respectively. Furthermore, this work has demonstrated the potential of photocatalytic oxidation reaction for PS depolymerization at the gram-scale. However, this work is subject to two significant limitations, including the utilization of H_2SO_4 and the reliance on pure O_2 atmospheres.

Kokotos and coworkers used anthraquinone as the photocatalyst to upgrade the reaction conditions to LED irradiation ($\lambda = 390$ nm) and open air, thus solving the drawback of photocatalytic oxidation of PS (Figure 15.7(b)) [39]. The optimized process is not only applicable to a wide range of commercial PS plastic products, but it has also successfully overcome the challenge of gram-scale depolymerization. In the investigation of the reaction mechanism, it was discovered that PS undergoes degradation through both HAT cycle and single oxygen cycle simultaneously under the catalysis of anthraquinone (Figure 15.8).

It is evident that the primary factor enabling photocatalysis to accomplish polymer degradation under mild conditions lies in its ability to induce the formation of carbon-center radicals within the polymer's main chain. If alternative methods can be employed to generate carbon-center radicals under mild conditions, it would further obviate the reliance on LED radiation for the oxidation reactions of PS. Lim and coworkers applied *N*-hydroxyl catalysts to generate C-centered radicals without the need for LED irradiation, such as *N*, *N′*, *N″*-trihydroxyisocyanuric acid (THICA) and *N*-hydroxyphthalimides (NHPIs) (Figure 15.7(c)) [40]. These catalysts serve as precursors for N-O radicals, which readily engage in HAT reactions with PS. The participation of O_2 and N_xO_y species in the radical propagation reactions explains the requirement for both an aerobic atmosphere and nitrate decomposition to achieve significant BA yields. This highlights that the potential release of toxic N_xO_y compounds represents a major concern associated with this process.

In conclusion, the successful implementation of the HAT process has effectively linked organic catalysis and PS degradation. This strategy holds great potential for application in other polymers, such as polythene (PE) and polypropylene (PP). However, it is important to note for oxidizing polymers with higher inertness than PS, a HAT organocatalyst possessing stronger oxidation ability would be required.

15.4 Conclusion

The application of organocatalysts in plastic degradation has been highly effective, showcasing significant advancements. Notably, organocatalysts with high bond-breaking efficiency have been designed for specific chemical bonds, especially C-O bond in esters. Consequently, recent research efforts have gradually shifted from catalyst development to the exploration and appreciation of degradation products. It has been shown that the selective chain cleavage of polymer skeleton is primarily affected by the choice of depolymerizers. In addition to the traditional depolymerizers capable of obtaining simple monomers like EG and methanol,

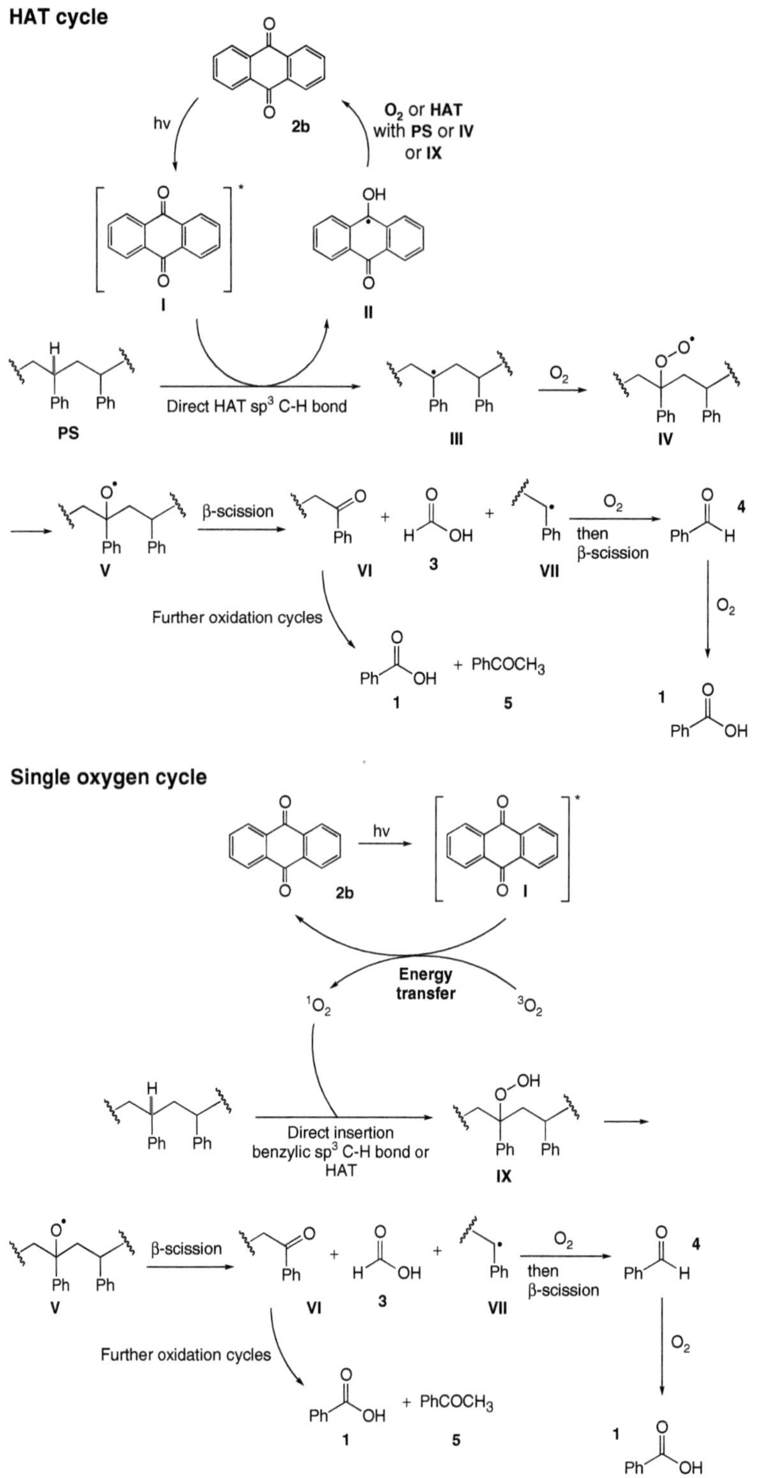

Figure 15.8 HAT cycle and single oxygen cycle mechanism. Source: [39]/Royal Society of Chemistry.

researchers have successfully devised a range of functional depolymerizers with the aid of organocatalysts to produce valuable products. This direction holds great promise for future research endeavors.

Furthermore, with the advancement of photocatalysis science, organocatalysts have demonstrated their potential in photodegrading non-functional polyolefins. Although the current study has solely achieved photocatalytic oxidation degradation of PS, it is anticipated that this approach will eventually be employed for treating more stable polyolefins such as PP and PE by developing higher effective oxidation catalysts.

References

1 Zou, L., Xu, R., Wang, H. et al. (2023). Chemical recycling of polyolefins: a closed-loop cycle of waste to olefins. *Natl. Sci. Rev.* 10 (9): nwad207.

2 Borrelle, S.B., Ringma, J., Law, K.L. et al. (2020). Predicted growth in plastic waste exceeds efforts to mitigate plastic pollution. *Science* 369 (6510): 1515–1518.

3 Lau, W., Shiran, Y., Bailey, R.M. et al. (2020). Evaluating scenarios toward zero plastic pollution. *Science* 369 (6510): 1455–1461.

4 Macleod, M., Arp, H.P.H., Tekman, M.B., and Jahnke, A. (2021). The global threat from plastic pollution. *Science* 373 (6550): 61–65.

5 Yang, R., Xu, G., Dong, B. et al. (2022). Selective, sequential, and "one-pot" depolymerization strategies for chemical recycling of commercial plastics and mixed plastics. *ACS Sustain. Chem. Eng.* 10 (30): 9860–9871.

6 Burkart, L., Eith, A., Hoffmann, A., and Herres, P.S. (2023). Open loop recycling – guanidine iron(ii) polymerization catalyst for the depolymerization of polylactide. *Chem. Asian J.* 18 (5): e202201195.

7 Deng, L., Guo, W., Ngo, H.H. et al. (2023). Novel catalysts in catalytic upcycling of common polymer wastes. *J. Chem. Eng.* (Lausanne, Switzerland : 1996*)* 471: 144350.

8 Barnard, E., Rubio, A.J., and J., Thielemans W. (2021). Chemolytic depolymerisation of PET: a review. *Green Chem.* 23 (11): 3765–3789.

9 Jehanno, C., Pérez-Madrigal, M.M., Demarteau, J. et al. (2019). Organocatalysis for depolymerisation. *Polym. Chem.* 1 (2): 172–186.

10 Jehanno, C., Demarteau, J., and Dove, A.P. (2019). Organocatalysis for depolymerisation. *Polym. Chem.* 10 (2): 172–186.

11 Fukushima, K., Coulembier, O., Lecuyer, J.M. et al. (2011). Organocatalytic depolymerization of poly(ethylene terephthalate). *J. Polym. Sci.* 5 (49): 1273–1281.

12 Fukushima, K., Watanabe, Y., Ueda, T. et al. (2022). Organocatalytic depolymerization of poly(trimethylene carbonate). *J. Polym. Sci.* 60 (24): 3489–3500.

13 Fukushima, K. and Nozaki, K. (2020). Organocatalysis: a paradigm shift in the synthesis of aliphatic polyesters and polycarbonates. *Macromolecules* 53 (13): 5018–5022.

14 Onida, K., Fayad, M., Norsic, S. et al. (2023). Chemical upcycling of poly(bisphenol a carbonate) to vinylene carbonates through organocatalysis. *Green Chem.* 25 (11): 4282–4291.

15 Lamparelli, D.H., Villar, Y.A., Dittrich, L. et al. (2023). Bicyclic guanidine promoted mechanistically divergent depolymerization and recycling of a biobased polycarbonate. *Angew. Chem. Int. Ed. Engl.* 62 (51): e202314659.

16 Fehér, Z., Németh, R., Kiss, J. et al. (2024). A silica-supported organocatalyst for polycarbonate methanolysis under mild and economic conditions. *Chem. Eng. J.* 485: 149832.

17 Stolz, B., Mönkemeyer, F., Mader, M. et al. (2020). Polyhydroxymethylenes as multifunctional high molecular weight sugar alcohols tailored for 3d printing and medical applications. *Macromol. Chem. Phys.* 221 (15): 2000132.

18 Aurbach, D., Gamolsky, K., Markovsky, B. et al. (2002). On the use of vinylene carbonate (VC) as an additive to electrolyte solutions for Li-ion batteries. *Electrochim. Acta* 47 (9): 1423–1439.

19 Saito, K., Eisenreich, F., Turel, T., and Tomovic, Z. (2022). Closed-loop recycling of poly(imine-carbonate) derived from plastic waste and bio-based resources. *Angew. Chem. Int. Ed. Engl.* 61 (43): e202211806.

20 English, L.E., Jones, M.D., and Liptrot, D.J. (2022). N-heterocyclic phosphines as precatalysts for the highly selective degradation of poly(lactic acid). *ChemCatChem* 14 (5): e202101904.

21 Ding, H., Wu, X., Han, B. et al. (2022). Organocatalyzed chemo-selective one-pot upcycling of polyester-*block*-polycarbonate. *Polym. Degrad. Stab.* 203: 110053.

22 Darensbourg, D.J. and Yeung, A.D. (2013). Thermodynamics of the carbon dioxide–epoxide copolymerization and kinetics of the metal-free degradation: a computational study. *Macromolecules* 46 (1): 83–95.

23 Dong, B., Xu, G., Yang, R. et al. (2023). Preparation of block copolymers by a sequential transesterification strategy: a feasible route for upcycling end-of-life polyester plastics to elastomers. *Macromolecules* 56 (24): 10143–10152.

24 Kaiho, S., Hmayed, A.A.R., Delle, C.K., and R., Worch J. C., Dove A. P. (2022). Designing thermally stable organocatalysts for poly(ethylene terephthalate) synthesis: toward a one-pot, closed-loop chemical recycling system for PET. *Macromolecules* 55 (23): 10628–10639.

25 Xue, Z., Qin, L., Jiang, J. et al. (2018). Thermal, electrochemical and radiolytic stabilities of ionic liquids. *Phys. Chem. Chem. Phys.* 20 (13): 8382–8402.

26 Miran, M.S., Hoque, M., Yasuda, T. et al. (2019). Key factor governing the physicochemical properties and extent of proton transfer in protic ionic liquids: Δpka or chemical structure? *Phys. Chem. Chem. Phys.* 21 (1): 418–426.

27 Olazabal, I., González, A., Vallejos, S. et al. (2023). Upgrading polyurethanes into functional ureas through the asymmetric chemical deconstruction of carbamates. *ACS Sustain. Chem. Eng.* 11 (1): 332–342.

28 Jehanno, C., Demarteau, J., Mantione, D. et al. (2020). Synthesis of functionalized cyclic carbonates through commodity polymer upcycling. *ACS Macro Lett.* 9 (4): 443–447.

29 Quaranta, E., Sgherza, D., and Tartaro, G. (2017). Depolymerization of poly(bisphenol a carbonate) under mild conditions by solvent-free alcoholysis catalyzed by 1,8-diazabicyclo[5.4.0]undec-7-ene as a recyclable organocatalyst: a route to chemical recycling of waste polycarbonate. *Green Chem.* 19 (22): 5422–5434.

30 Alberti, C. and Enthaler, S. (2020). Depolymerization of end-of-life poly (bisphenol a carbonate) via 4-dimethylaminopyridine-catalyzed methanolysis. *Waste Biomass Valori.* 11 (9): 4621–4629.

31 Kammakakam, I., O'Harra, K.E., Dennis, G.P. et al. (2019). Self-healing imidazolium-based ionene-polyamide membranes: an experimental study on physical and gas transport properties. *Polym. Int.* 68 (6): 1123–1129.

32 Bara, J.E., Harra, O., and K. E., Durbin M. M., Dennis G. P., Jackson E. M., Thomas B., Odutola J. A. (2018). Synthesis and characterization of ionene-polyamide materials as candidates for new gas separation membranes. *MRS Adv.* 3 (52): 3091–3102.

33 Demarteau, J., O'Harra, K.E., Bara, J.E., and Sardon, H. (2020). Valorization of plastic wastes for the synthesis of imidazolium-based self-supported elastomeric ionenes. *ChemSusChem* 13 (12): 3122–3126.

34 Marquez, C., Martin, C., Linares, N., and De Vos, D. (2023). Catalytic routes towards polystyrene recycling. *Mater. Horiz.* 10 (5): 1625–1640.

35 Maafa, I.M. (2021). Pyrolysis of polystyrene waste: a review. *Polymers* 13 (2): 225.

36 Huang, Z., Shanmugam, M., Liu, Z. et al. (2022). Chemical recycling of polystyrene to valuable chemicals via selective acid-catalyzed aerobic oxidation under visible light. *J. Am. Chem. Soc.* 144 (14): 6532–6542.

37 Oh, S. and Stache, E.E. (2022). Chemical upcycling of commercial polystyrene via catalyst-controlled photooxidation. *J. Am. Chem. Soc.* 144 (13): 5745–5749.

38 Li, T., Vijeta, A., Casadevall, C. et al. (2022). Bridging plastic recycling and organic catalysis: photocatalytic deconstruction of polystyrene via a C-H oxidation pathway. *ACS Catal.* 12 (14): 8155–8163.

39 Nikitas, N.F., Skolia, E., Gkizis, P.L. et al. (2023). Photochemical aerobic upcycling of polystyrene plastics to commodity chemicals using anthraquinone as the photocatalyst. *Green Chem.* 25 (12): 4750–4759.

40 Ong, A., Teo, J.Y.Q., Feng, Z. et al. (2023). Organocatalytic aerobic oxidative degradation of polystyrene to aromatic acids. *ACS Sustain. Chem. Eng.* 11 (34): 12514–12522.

Index

d

e

t